Statistical Physics

Statistical Physics

Volume 1 of *Modern Classical Physics*

KIP S. THORNE *and* **ROGER D. BLANDFORD**

PRINCETON UNIVERSITY PRESS

Princeton and Oxford

Published by Princeton University Press
41 William Street, Princeton, New Jersey 08540
6 Oxford Street, Woodstock, Oxfordshire OX20 1TR

press.princeton.edu

All Rights Reserved
ISBN (pbk.) 978-0-691-20612-7
ISBN (e-book) 978-0-691-21555-6

British Library Cataloging-in-Publication Data is available

Editorial: Ingrid Gnerlich and Arthur Werneck
Production Editorial: Mark Bellis
Text and Cover Design: Wanda España
Production: Jacqueline Poirier
Publicity: Matthew Taylor and Amy Stewart
Copyeditor: Cyd Westmoreland

This book has been composed in MinionPro, Whitney, and Ratio Modern by Windfall Software, Carlisle, Massachusetts, using ZzTEX

Printed on acid-free paper.

Printed in China

10 9 8 7 6 5 4 3 2 1

A NOTE TO READERS
This book is the first in a series of volumes that together comprise a unified work titled *Modern Classical Physics*. Each volume is designed to be read independently of the others and can be used as a textbook in an advanced undergraduate- to graduate-level course on the subject of the title or for self-study. However, as the five volumes are highly complementary to one another, we hope that reading one volume may inspire the reader to investigate others in the series—or the full, unified work—and thereby explore the rich scope of modern classical physics.

To Carolee and Liz

CONTENTS

T2 Track Two; see page xvii

N Nonrelativistic (Newtonian) kinetic theory; see page 96

R Relativistic theory; see page 96

BOXES

PREFACE

The tools of statistical physics have enormous power and a remarkably wide variety of applications, ranging over condensed matter, geophysics, biology and biophysics, chemistry, engineering, and more. For this reason, we (the authors, Roger Blandford and Kip Thorne) believe that every student in the natural sciences and engineering should be exposed to the basic concepts and a few applications of statistical physics. We have written this book to facilitate this exposure for students at the advanced undergraduate, master's, or PhD level, and to serve as a resource for working scientists and engineers who want to improve their understanding of statistical physics.

Many other textbooks have been written on statistical physics, so why this one? Because of our unique approach to the subject. Our approach

- is geometric;
- embraces and elucidates the intimate links between the *quantum* domain and the *classical* domain;
- embraces and treats the *relativistic* domain as well as the *Newtonian* domain;
- exhibits the power of statistical techniques (particularly statistical mechanics) by presenting applications not only to the usual kinds of phenomena (e.g., gases, liquids, solids, magnetic materials) but also to a much wider range of phenomena (e.g., black holes, the entire universe, information and its communication, noise, and signal processing amid noise); and
- is brief enough to be covered easily in one quarter of a year.

Of necessity, so brief a treatment will not go into as much depth as desired and needed by readers with a sharp focus on some special area of natural science, such as condensed matter physics, or readers who want a thorough mastery of statistical physics. Such readers would best do most of their study using some other textbook, but they could find ours useful and fun for a first introduction or for broadening their horizons.

THE CLASSICAL-QUANTUM CONNECTION

Although this book's principal focus is classical statistical physics, inevitably we make frequent reference to quantum mechanical concepts and phenomena, and we often use quantum concepts and techniques in the classical domain. This is because classical physics arises from quantum physics as an approximation. The roots of classical physics are in the quantum domain, and perhaps nowhere in physics do quantum mechanical tendrils extend so far into the classical realm as they do in statistical physics. Moreover, quantum statistical physics is so important that, without apology and with enthusiasm, we devote some effort to explaining its most important concepts and presenting a few powerful examples—though we do not develop its beautiful mathematical formalism.

OUR GEOMETRIC APPROACH TO PHYSICS

We (the authors) embrace a coordinate-independent, geometric approach to physics in our roles as teachers, mentors, and research scientists. The essence of this approach is that all the laws of physics must be expressible in coordinate-independent, geometric language (in terms of coordinate-independent scalars, vectors, tensors, and their associated differential operators). This geometric language has great power. It usually elucidates the physics of a situation more clearly than coordinate-based language, and it often circumvents lengthy, coordinate-based calculations. The geometric language is most powerful, perhaps, in relativity and in continuum mechanics, but it also permeates our approach to two major branches of statistical physics: kinetic theory and statistical mechanics.

For those students and other readers who are not familiar with this geometric approach, we have provided a detailed introduction to it at the beginning of this book—in Part I: Foundations. This geometric introduction has two chapters: Chap. 1 for Newtonian (nonrelativistic) physics and Chap. 2 for relativistic physics. Chapter 2 is a rather thorough introduction to special relativity.

It is by no means necessary for readers to study these geometric chapters in detail before plunging into the core of this book, Part II: Statistical Physics. Many or most readers who ignore Part I will get along just fine without it, at least most of the time. When readers do run into difficulty due to not having read Part I, they can easily dip back into the relevant sections of that part for clarification. However, familiarity with Part I will give readers a deeper understanding of and appreciation for our approach to a number of important ideas and results in statistical physics.

RELATIVISTIC AND NEWTONIAN DOMAINS

Much of modern physics deals with particles (usually photons, gravitons, neutrinos, electrons, protons, and heavier atomic nuclei) that have relativistic speeds—that is, speeds that are a nonnegligible fraction of the speed of light. In this book, we embrace treating the statistical physics of such relativistic particles via relativistic kinetic theory and relativistic statistical mechanics, though we devote far less space to this than to nonrelativistic (Newtonian) statistical physics. Relativistic sections are marked by **R** and can easily be skipped, though we love them so much that we may shed a tear or two if most readers skip them. Newtonian sections are marked by **N** or are unmarked.

GUIDANCE FOR READERS

The amount and variety of material covered in this book may seem overwhelming. If so, keep in mind that

- *the primary goals of this book* are to teach the fundamental concepts and principles of statistical physics (which are not so extensive that they should overwhelm); to illustrate those concepts and principles in action; and through our illustrations, to give the reader some physical and intuitive understanding of statistical physics.

We do not intend the reader to master all of our many illustrative applications.

We have aimed this book at advanced undergraduates and first-year graduate students, of whom we expect only (1) a typical physics or engineering student's facility with applied mathematics, and (2) a typical undergraduate-level understanding of classical mechanics, electromagnetism, elementary quantum mechanics, and elementary thermodynamics. (In Chap. 5, we deduce the principles and some details of thermodynamics from statistical mechanics, but a prior familiarity with thermodynamics will be helpful.)

This book contains four chapters devoted to statistical physics and can be taught in a one-quarter course or even a half semester, though a more leisurely full semester course will leave the student with a broader and deeper understanding of the wide scope of statistical physics. For those readers who seek a briefer introduction to statistical physics, we have labeled as "Track Two" sections that can be skipped on a first reading, or skipped entirely, but are sufficiently interesting that most readers may choose to browse or study them. Track-Two sections are identified by the symbol **T2**. For readers who want more detailed and comprehensive introductions than ours, we offer some recommendations at the end of each chapter.

This book is the first of five volumes that together constitute a single treatise, *Modern Classical Physics* (or "MCP," as we shall call it). The full treatise was published in 2017 as an embarrassingly thick single book (the electronic edition is a good deal lighter). For readers' convenience, we have placed, at the end of this volume, the Table of Contents, Preface, and Acknowledgments of MCP. The five separate textbooks of this decomposition are

- Volume 1: *Statistical Physics,*
- Volume 2: *Optics,*
- Volume 3: *Elasticity and Fluid Dynamics,*
- Volume 4: *Plasma Physics,* and
- Volume 5: *Relativity and Cosmology.*

These individual volumes are much more suitable for human transport and for use in individual courses than their one-volume parent treatise, MCP.

The present volume is enriched by extensive cross-references to the other four volumes—cross-references that elucidate the rich interconnections of various areas of physics.

In this and the other four volumes, we have retained the chapter numbers from MCP, and, for the body of each volume, MCP's pagination. In fact, the body of this volume is identical to the corresponding MCP chapters, aside from corrections of errata (which are tabulated at the MCP website http://press.princeton.edu/titles/MCP .html), and a small amount of updating that has not changed pagination. For readers' cross-referencing convenience, a list of the chapters in each of the five volumes appears immediately after this Preface.

EXERCISES

Exercises are a major component of this volume, as well as of the other four volumes of MCP. The exercises come in five types:

1. *Practice*. Exercises that provide practice at mathematical manipulations (e.g., of tensors).

2. *Derivation*. Exercises that fill in details of arguments skipped over in the text.

3. *Example*. Exercises that lead the reader step by step through the details of some important extension or application of the material in the text.

4. *Problem*. Exercises with few, if any, hints, in which the task of figuring out how to set up the calculation and get started on it often is as difficult as doing the calculation itself.

5. *Challenge*. Especially difficult exercises whose solution may require reading other books or articles as a foundation for getting started.

We urge readers to try working many of the exercises—especially the examples, which should be regarded as continuations of the text and which contain many of the most illuminating applications. Exercises that we regard as especially important are designated by **.

BRIEF OUTLINE OF THIS BOOK

After presenting our geometric viewpoint on physics in Part I (Newtonian in Chap. 1, relativistic in Chap. 2), we turn to statistical physics in the book's core, Part II.

We begin, in Chap. 3, with *kinetic theory*, both Newtonian and relativistic. Here our central focus is a probability distribution for large numbers of particles: their distribution both in space and in momentum, and the distribution's evolution in time. We meet the power of kinetic theory in a variety of applications, for example:

1. the macroscopic properties of solids, liquids, gases, plasmas, and radiation—their equations of state and specific heats, and the transport coefficients that quantify the diffusion of heat, electric charge, and momentum through them: thermal conductivity, electrical conductivity, and viscosity;

2. neutron diffusion in a nuclear reactor, and the use of a moderator to prevent runaway chain reactions;

3. for a large collection of particles, the dividing line between regimes of classical wave behavior and individual particle behavior;

4. in a thermalized medium, reciprocity relations among the rates for spontaneous emission of radiation, stimulated emission, and absorption;

5. dipole anisotropy of the cosmic microwave radiation observed from earth, due to the earth's motion through the universe; and

6. solar heating of the earth and the greenhouse effect that underlies global warming.

In Chap. 4, we turn to a remarkable generalization of kinetic theory, called *statistical mechanics*. Here, instead of dealing with a probability distribution for particles, we focus on a probability distribution for *systems*. "System" is an amazingly general concept. Examples include a balloon full of gas, a sapphire crystal, a normal mode of vibration of a sapphire crystal, a star, a galaxy, the universe, a black hole, and the quantum mechanical atmosphere of a black hole (whose leakage produces Hawking radiation). In statistical mechanics, having chosen a system to study, one imagines an ensemble of such systems and a probability distribution for the physical states of the systems in that ensemble—a distribution often chosen to represent the stochastic (ergodic) wandering, over time, of a single system of the ensemble (the only one that exists in reality).

In Chap. 4, we explore the power of statistical mechanics by applying it to elucidate, for example:

1. the nature of entropy, its inexorable increase, and its power in studying the evolution of physical systems;

2. the maximal-entropy distribution of particles in a single quantum state (or classical normal mode)—that is, the Fermi-Dirac distribution for fermions and the Bose-Einstein distribution for bosons;

3. Bose-Einstein condensation in a dilute gas;

4. the relationship between quantum decoherence and entropy increase;

5. the relationship between entropy and information, and the quantification of information and of information transfer through communication channels;

6. the behavior and role of entropy in galaxies and in our expanding universe; and

7. the quantum atmospheres of spinning black holes, black-hole thermodynamics, and Hawking radiation.

In Chap. 5, we specialize statistical mechanics to ensembles of systems that are in or near *statistical equilibrium* (thermal equilibrium; maximal entropy), and we use such ensembles to derive the laws of thermodynamics and to study probabilities for fluctuations away from statistical equilibrium. We illustrate the power of this *statistical thermodynamics* (as it is called) by applying it, for example, to

1. the equations of state of a perfect gas, and of more realistic gases: one with excitable internal degrees of freedom (an ideal gas) and one, like water, with attractive and repulsive forces between its molecules (a van der Waals gas);

2. in gases and other statistical-equilibrium systems, probability distributions for spontaneous fluctuations of temperature, density, and the like;

3. chemical reactions, nuclear reactions, and fluctuations in them;

4. phase transitions (e.g., the freezing of water and melting of ice) and their fluctuations, nucleation of water droplets from steam, and surface tension in water;

5. spontaneous symmetry breaking in a crystal's phase transitions; and

6. a ferromagnet, whose phase transition we describe by the *Ising model* and use to illustrate *Monte Carlo methods* and the *renormalization group*.

Finally, in Chap. 6, we develop the theory of *random processes*, that is, of the statistical time evolution of physical entities that are influenced by factors over which we have little control and little knowledge, except for their statistical properties. We demonstrate the power of the random-process formalism by using it to explore and elucidate, for example,

1. Brownian motion of dust particles buffeted by air molecules;

2. density fluctuations in our cosmologically expanding universe;

3. noise in the ticking rates of atomic clocks, in photon arrival times at photodetectors (shot noise), in the flow rates of rivers, and in many other systems;

4. the extraction of weak signals from noisy data (signal processing);

5. the intimate connection between friction and fluctuations: the *fluctuation-dissipation theorem*, and its applications to thermal noise in electronics and in a gravitational-wave interferometer, such as LIGO; and

6. the Fokker-Planck equation and its applications to thermal noise in an oscillator and to Doppler cooling of atomic motions in a very cold gas (optical molasses).

Volume 4: Plasma Physics

Volume 5: Relativity and Cosmology

I

FOUNDATIONS

In this book, a central theme will be a *Geometric Principle: The laws of physics must all be expressible as geometric (coordinate-independent and reference-frame-independent) relationships between geometric objects (scalars, vectors, tensors, . . .) that represent physical entities.*

There are three different conceptual frameworks for the classical laws of physics, and correspondingly, three different geometric arenas for the laws; see Fig. 1. General relativity is the most accurate classical framework; it formulates the laws as geometric relationships among geometric objects in the arena of curved 4-dimensional space-time. Special relativity is the limit of general relativity in the complete absence of gravity; its arena is flat, 4-dimensional Minkowski spacetime.[1] Newtonian physics is the limit of general relativity when

- gravity is weak but not necessarily absent,
- relative speeds of particles and materials are small compared to the speed of light c, and
- all stresses (pressures) are small compared to the total density of mass-energy.

Its arena is flat, 3-dimensional Euclidean space with time separated off and made universal (by contrast with relativity's reference-frame-dependent time).

In Parts II–VI of this book (covering statistical physics, optics, elasticity, fluid mechanics, and plasma physics), we confine ourselves to the Newtonian formulations of the laws (plus special relativistic formulations in portions of Track Two), and accordingly, our arena will be flat Euclidean space (plus flat Minkowski spacetime in portions of Track Two). In Part VII, we extend many of the laws we have studied into the domain of strong gravity (general relativity)—the arena of curved spacetime.

---

1. This is so-called because Hermann Minkowski (1908) identified the special relativistic invariant interval as defining a metric in spacetime and elucidated the resulting geometry of flat spacetime.

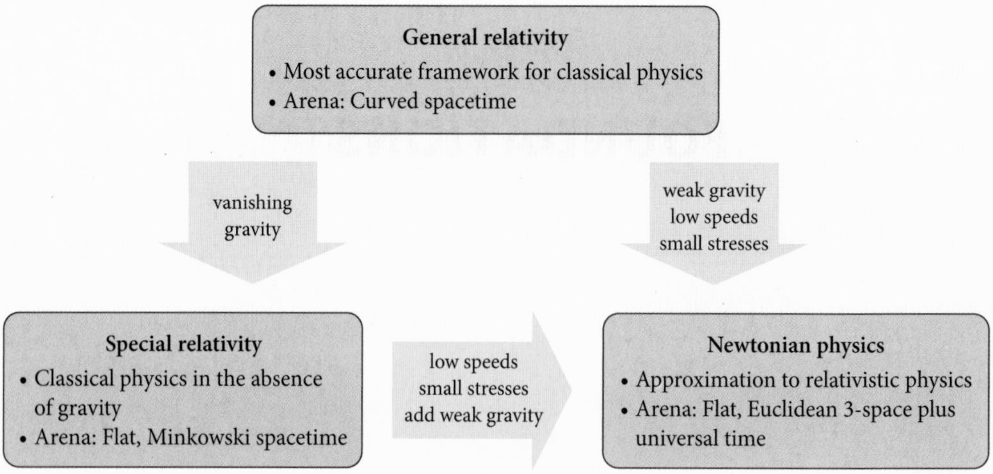

FIGURE 1 The three frameworks and arenas for the classical laws of physics and their relationship to one another.

In Parts II and III (on statistical physics and optics), in addition to confining ourselves to flat space (plus flat spacetime in Track Two), we avoid any sophisticated use of curvilinear coordinates. Correspondingly, when using coordinates in nontrivial ways, we confine ourselves to Cartesian coordinates in Euclidean space (and Lorentz coordinates in Minkowski spacetime).

Part I of this book contains just two chapters. Chapter 1 is an introduction to our geometric viewpoint on Newtonian physics and to all the geometric mathematical tools that we shall need in Parts II and III for Newtonian physics in its arena, 3-dimensional Euclidean space. Chapter 2 introduces our geometric viewpoint on special relativistic physics and extends our geometric tools into special relativity's arena, flat Minkowski spacetime. Readers whose focus is Newtonian physics will have no need for Chap. 2; and if they are already familiar with the material in Chap. 1 but not from our geometric viewpoint, they can successfully study Parts II–VI without reading Chap. 1. However, in doing so, they will miss some deep insights; so we recommend they at least browse Chap. 1 to get some sense of our viewpoint, then return to the chapter occasionally, as needed, when encountering an unfamiliar geometric argument.

In Parts IV, V, and VI, when studying elasticity, fluid dynamics, and plasma physics, we use curvilinear coordinates in nontrivial ways. As a foundation for this, at the beginning of Part IV, we extend our flat-space geometric tools to curvilinear coordinate systems (e.g., cylindrical and spherical coordinates). Finally, at the beginning of Part VII, we extend our geometric tools to the arena of curved spacetime.

Throughout this book, we pay close attention to the relationship between classical physics and quantum physics. Indeed, we often find it powerful to use quantum mechanical language or formalism when discussing and analyzing classical phenomena.

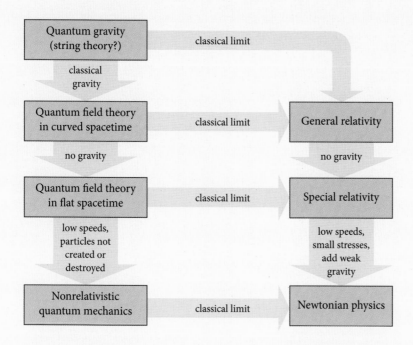

FIGURE 2 The relationship of the three frameworks for classical physics (on right) to four frameworks for quantum physics (on left). Each arrow indicates an approximation. All other frameworks are approximations to the ultimate laws of quantum gravity (whatever they may be—perhaps a variant of string theory).

This quantum power in classical domains arises because quantum physics is primary and classical physics is secondary. Today we see classical physics as arising from quantum physics, though historically the linkage was inverted. The relationship between quantum frameworks and arenas for the laws of physics, and classical frameworks, is sketched in Fig. 2.

Newtonian Physics: Geometric Viewpoint

Geometry postulates the solution of these problems from mechanics and teaches the use of the problems thus solved. And geometry can boast that with so few principles obtained from other fields, it can do so much.

ISAAC NEWTON, 1687

1.1 Introduction

1.1.1 The Geometric Viewpoint on the Laws of Physics

In this book, we adopt a different viewpoint on the laws of physics than that in many elementary and intermediate texts. In most textbooks, physical laws are expressed in terms of quantities (locations in space, momenta of particles, etc.) that are measured in some coordinate system. For example, Newtonian vectorial quantities are expressed as triplets of numbers [e.g., $\mathbf{p} = (p_x, p_y, p_z) = (1, 9, -4)$], representing the components of a particle's momentum on the axes of a Cartesian coordinate system; and tensors are expressed as arrays of numbers (e.g.,

$$\mathbf{I} = \begin{bmatrix} I_{xx} & I_{xy} & I_{xz} \\ I_{yx} & I_{yy} & I_{yz} \\ I_{zx} & I_{zy} & I_{zz} \end{bmatrix} \tag{1.1}$$

for the moment of inertia tensor).

By contrast, in this book we express all physical quantities and laws in *geometric forms,* i.e., in forms that are *independent of any coordinate system or basis vectors.* For example, a particle's velocity $\mathbf{v}$ and the electric and magnetic fields $\mathbf{E}$ and $\mathbf{B}$ that it encounters will be vectors described as arrows that live in the 3-dimensional, flat Euclidean space of everyday experience.[1] They require no coordinate system or basis vectors for their existence or description—though often coordinates will be useful. In other words, $\mathbf{v}$ represents the vector itself and is not just shorthand for an ordered list of numbers.

---

1. This interpretation of a vector is close to the ideas of Newton and Faraday. Lagrange, Hamilton, Maxwell, and many others saw vectors in terms of Cartesian components. The vector notation was streamlined by Gibbs, Heaviside, and others, but the underlying coordinate system was still implicit, and $\mathbf{v}$ was usually regarded as shorthand for (v_x, v_y, v_z).

We insist that the Newtonian laws of physics all obey a *Geometric Principle:* they are all geometric relationships among geometric objects (primarily scalars, vectors, and tensors), expressible without the aid of any coordinates or bases. An example is the Lorentz force law $md\mathbf{v}/dt = q(\mathbf{E} + \mathbf{v} \times \mathbf{B})$—a (coordinate-free) relationship between the geometric (coordinate-independent) vectors $\mathbf{v}$, $\mathbf{E}$, and $\mathbf{B}$ and the particle's scalar mass m and charge q. As another example, a body's moment of inertia tensor $\mathbf{I}$ can be viewed as a vector-valued linear function of vectors (a coordinate-independent, basis-independent geometric object). Insert into the tensor $\mathbf{I}$ the body's angular velocity vector $\mathbf{\Omega}$, and you get out the body's angular momentum vector: $\mathbf{J} = \mathbf{I}(\mathbf{\Omega})$. No coordinates or basis vectors are needed for this law of physics, nor is any description of $\mathbf{I}$ as a matrix-like entity with components I_{ij} required. Components are secondary; they only exist after one has chosen a set of basis vectors. Components (we claim) are an impediment to a clear and deep understanding of the laws of classical physics. The coordinate-free, component-free description is deeper, and—once one becomes accustomed to it—much more clear and understandable.[2]

---

2. This philosophy is also appropriate for quantum mechanics (see Box 1.2) and, especially, quantum field theory, where it is the invariance of the description under gauge and other symmetry operations that is the powerful principle. However, its implementation there is less direct, simply because the spaces in which these symmetries lie are more abstract and harder to conceptualize.

By adopting this geometric viewpoint, we gain great conceptual power and often also computational power. For example, when we ignore experiment and simply ask what forms the laws of physics can possibly take (what forms are allowed by the requirement that the laws be geometric), we shall find that there is remarkably little freedom. Coordinate independence and basis independence strongly constrain the laws of physics.[3]

This power, together with the elegance of the geometric formulation, suggests that in some deep sense, Nature's physical laws are geometric and have nothing whatsoever to do with coordinates or components or vector bases.

1.1.2 Purposes of This Chapter

The principal purpose of this foundational chapter is to teach the reader this geometric viewpoint.

The mathematical foundation for our geometric viewpoint is *differential geometry* (also called "tensor analysis" by physicists). Differential geometry can be thought of as an extension of the vector analysis with which all readers should be familiar. *A second purpose of this chapter is to develop key parts of differential geometry in a simple form well adapted to Newtonian physics.*

1.1.3 Overview of This Chapter

In this chapter, we lay the geometric foundations for the Newtonian laws of physics in flat Euclidean space. We begin in Sec. 1.2 by introducing some foundational geometric concepts: points, scalars, vectors, inner products of vectors, and the distance between points. Then in Sec. 1.3, we introduce the concept of a tensor as a linear function of vectors, and we develop a number of geometric tools: the tools of coordinate-free tensor algebra. In Sec. 1.4, we illustrate our tensor-algebra tools by using them to describe—without any coordinate system—the kinematics of a charged point particle that moves through Euclidean space, driven by electric and magnetic forces.

In Sec. 1.5, we introduce, for the first time, Cartesian coordinate systems and their basis vectors, and also the components of vectors and tensors on those basis vectors; and we explore how to express geometric relationships in the language of components. In Sec. 1.6, we deduce how the components of vectors and tensors transform when one rotates the chosen Cartesian coordinate axes. (These are the transformation laws that most physics textbooks use to define vectors and tensors.)

In Sec. 1.7, we introduce directional derivatives and gradients of vectors and tensors, thereby moving from tensor algebra to true differential geometry (in Euclidean space). We also introduce the Levi-Civita tensor and use it to define curls and cross

---

3. Examples are the equation of elastodynamics (12.4b) and the Navier-Stokes equation of fluid mechanics (13.69), which are both dictated by momentum conservation plus the form of the stress tensor [Eqs. (11.18), (13.43), and (13.68)]—forms that are dictated by the irreducible tensorial parts (Box 11.2) of the strain and rate of strain.

products, and we learn how to use *index gymnastics* to derive, quickly, formulas for multiple cross products. In Sec. 1.8, we use the Levi-Civita tensor to define vectorial areas, scalar volumes, and integration over surfaces. These concepts then enable us to formulate, in geometric, coordinate-free ways, integral and differential conservation laws. In Sec. 1.9, we discuss, in particular, the law of momentum conservation, formulating it in a geometric way with the aid of a geometric object called the *stress tensor*. As important examples, we use this geometric conservation law to derive and discuss the equations of Newtonian fluid dynamics, and the interaction between a charged medium and an electromagnetic field. We conclude in Sec. 1.10 with some concepts from special relativity that we shall need in our discussions of Newtonian physics.

1.2

1.2 Foundational Concepts

In this section, we sketch the foundational concepts of Newtonian physics without using any coordinate system or basis vectors. This is the geometric viewpoint that we advocate.

space and time

The arena for the Newtonian laws of physics is a spacetime composed of the familiar 3-dimensional Euclidean space of everyday experience (which we call *3-space*) and a universal time t. We denote points (locations) in 3-space by capital script letters, such as $\mathcal{P}$ and $\mathcal{Q}$. These points and the 3-space in which they live require no coordinates for their definition.

scalar

A *scalar* is a single number. We are most interested in scalars that directly represent physical quantities (e.g., temperature T). As such, they are real numbers, and when they are functions of location $\mathcal{P}$ in space [e.g., $T(\mathcal{P})$], we call them *scalar fields*. However, sometimes we will work with complex numbers—most importantly in quantum mechanics, but also in various Fourier representations of classical physics.

vector

A *vector* in Euclidean 3-space can be thought of as a straight arrow (or more formally a directed line segment) that reaches from one point, $\mathcal{P}$, to another, $\mathcal{Q}$ (e.g., the arrow $\Delta\mathbf{x}$ in Fig. 1.1a). Equivalently, $\Delta\mathbf{x}$ can be thought of as a direction at $\mathcal{P}$ and a number, the vector's length. Sometimes we shall select one point $\mathcal{O}$ in 3-space as an "origin" and identify all other points, say, $\mathcal{Q}$ and $\mathcal{P}$, by their vectorial separations $\mathbf{x}_{\mathcal{Q}}$ and $\mathbf{x}_{\mathcal{P}}$ from that origin.

distance and length

The Euclidean distance $\Delta\sigma$ between two points $\mathcal{P}$ and $\mathcal{Q}$ in 3-space can be measured with a ruler and so, of course, requires no coordinate system for its definition. (If one does have a Cartesian coordinate system, then $\Delta\sigma$ can be computed by the Pythagorean formula, a precursor to the invariant interval of flat spacetime; Sec. 2.2.3.) This distance $\Delta\sigma$ is also the *length* $|\Delta\mathbf{x}|$ of the vector $\Delta\mathbf{x}$ that reaches from $\mathcal{P}$ to $\mathcal{Q}$, and the square of that length is denoted

$$|\Delta\mathbf{x}|^2 \equiv (\Delta\mathbf{x})^2 \equiv (\Delta\sigma)^2. \tag{1.2}$$

Of particular importance is the case when $\mathcal{P}$ and $\mathcal{Q}$ are neighboring points and $\Delta\mathbf{x}$ is a differential (infinitesimal) quantity $d\mathbf{x}$. This *infinitesimal displacement* is a more fundamental physical quantity than the finite $\Delta\mathbf{x}$. To create a finite vector out

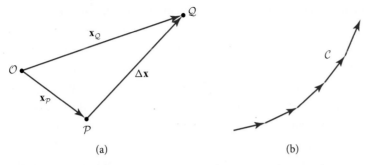

FIGURE 1.1 (a) A Euclidean 3-space diagram depicting two points $\mathcal{P}$ and $\mathcal{Q}$, their respective vectorial separations $\mathbf{x}_\mathcal{P}$ and $\mathbf{x}_\mathcal{Q}$ from the (arbitrarily chosen) origin $\mathcal{O}$, and the vector $\Delta\mathbf{x} = \mathbf{x}_\mathcal{Q} - \mathbf{x}_\mathcal{P}$ connecting them. (b) A curve $\mathcal{P}(\lambda)$ generated by laying out a sequence of infinitesimal vectors, tail-to-tip.

of infinitesimal vectors, one has to add several infinitesimal vectors head to tail, head to tail, and so on, and then take a limit. This involves *translating* a vector from one point to the next. There is no ambiguity about doing this in flat Euclidean space using the geometric notion of parallelism.[4] This simple property of Euclidean space enables us to add (and subtract) vectors at a point. We attach the tail of a second vector to the head of the first vector and then construct the sum as the vector from the tail of the first to the head of the second, or vice versa, as should be quite familiar. The point is that we do not need to add the Cartesian components to sum vectors.

We can also rotate vectors about their tails by pointing them along a different direction in space. Such a rotation can be specified by two angles. The space that is defined by all possible changes of length and direction at a point is called that point's *tangent space*. Again, we generally view the rotation as being that of a physical vector in space, and not, as it is often useful to imagine, the rotation of some coordinate system's basis vectors, with the chosen vector itself kept fixed.

tangent space

We can also construct a path through space by laying down a sequence of infinitesimal $d\mathbf{x}$s, tail to head, one after another. The resulting path is a *curve* to which these $d\mathbf{x}$s are tangent (Fig. 1.1b). The curve can be denoted $\mathcal{P}(\lambda)$, with λ a parameter along the curve and $\mathcal{P}(\lambda)$ the point on the curve whose parameter value is λ, or $\mathbf{x}(\lambda)$ where $\mathbf{x}$ is the vector separation of $\mathcal{P}$ from the arbitrary origin $\mathcal{O}$. The infinitesimal vectors that map the curve out are $d\mathbf{x} = (d\mathcal{P}/d\lambda)\, d\lambda = (d\mathbf{x}/d\lambda)\, d\lambda$, and $d\mathcal{P}/d\lambda = d\mathbf{x}/d\lambda$ is the tangent vector to the curve.

curve

tangent vector

If the curve followed is that of a particle, and the parameter λ is time t, then we have defined the *velocity* $\mathbf{v} \equiv d\mathbf{x}/dt$. In effect we are multiplying the vector $d\mathbf{x}$ by the scalar $1/dt$ and taking the limit. Performing this operation at every point $\mathcal{P}$ in the space occupied by a fluid defines the fluid's *velocity field* $\mathbf{v}(\mathbf{x})$. Multiplying a particle's velocity $\mathbf{v}$ by its scalar mass gives its *momentum* $\mathbf{p} = m\mathbf{v}$. Similarly, the difference $d\mathbf{v}$

---

4. The statement that there is just one choice of line parallel to a given line, through a point not lying on the line, is the famous fifth axiom of Euclid.

of two velocity measurements during a time interval dt, multiplied by $1/dt$, generates the particle's *acceleration* $\mathbf{a} = d\mathbf{v}/dt$. Multiplying by the particle's mass gives the force $\mathbf{F} = m\mathbf{a}$ that produced the acceleration; dividing an electrically produced force by the particle's charge q gives the electric field $\mathbf{E} = \mathbf{F}/q$. And so on.

We can define inner products [see Eq. (1.4a) below] and cross products [Eq. (1.22a)] of pairs of vectors at the same point geometrically; then using those vectors we can define, for example, the rate that work is done by a force and a particle's angular momentum about a point.

These two products can be expressed geometrically as follows. If we allow the two vectors to define a parallelogram, then their cross product is the vector orthogonal to the parallelogram with length equal to the parallelogram's area. If we first rotate one vector through a right angle in a plane containing the other, and then define the parallelogram, its area is the vectors' inner product.

We can also define spatial derivatives. We associate the difference of a scalar between two points separated by $d\mathbf{x}$ at the same time with a *gradient* and, likewise, go on to define the scalar *divergence* and the vector *curl*. The freedom to translate vectors from one point to the next also underlies the association of a single vector (e.g., momentum) with a group of particles or an extended body. One simply adds all the individual momenta, taking a limit when necessary.

In this fashion (which should be familiar to the reader and will be elucidated, formalized, and generalized below), we can construct all the standard scalars and vectors of Newtonian physics. What is important is that *these physical quantities require no coordinate system for their definition.* They are geometric (coordinate-independent) objects residing in Euclidean 3-space at a particular time.

It is a fundamental (though often ignored) principle of physics that *the Newtonian physical laws are all expressible as geometric relationships among these types of geometric objects, and these relationships do not depend on any coordinate system or orientation of axes, nor on any reference frame* (i.e., on any purported velocity of the Euclidean space in which the measurements are made).[5] We call this the *Geometric Principle* for the laws of physics, and we use it throughout this book. It is the Newtonian analog of Einstein's Principle of Relativity (Sec. 2.2.2).

1.3 Tensor Algebra without a Coordinate System

In preparation for developing our geometric view of physical laws, we now introduce, in a coordinate-free way, some fundamental concepts of differential geometry: tensors, the inner product, the metric tensor, the tensor product, and contraction of tensors.

We have already defined a vector $\mathbf{A}$ as a straight arrow from one point, say $\mathcal{P}$, in our space to another, say $\mathcal{Q}$. Because our space is flat, there is a unique and obvious way to

---

5. By changing the velocity of Euclidean space, one adds a constant velocity to all particles, but this leaves the laws (e.g., Newton's $\mathbf{F} = m\mathbf{a}$) unchanged.

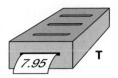

FIGURE 1.2 A rank-3 tensor **T**.

transport such an arrow from one location to another, keeping its length and direction unchanged.[6] Accordingly, we shall regard vectors as unchanged by such transport. This enables us to ignore the issue of where in space a vector actually resides; it is completely determined by its direction and its length.

A *rank-n tensor* **T** is, by definition, a real-valued linear function of n vectors.[7] **tensor** Pictorially we regard **T** as a box (Fig. 1.2) with n slots in its top, into which are inserted n vectors, and one slot in its end, which prints out a single real number: the value that the tensor **T** has when evaluated as a function of the n inserted vectors. Notationally we denote the tensor by a boldfaced sans-serif character **T**:

$$\mathsf{T}(\underbrace{\_\_, \_\_, \_\_, \_\_}) \tag{1.3a}$$

$$\nwarrow \; n \text{ slots in which to put the vectors.}$$

This definition of a tensor is very different (and far simpler) than the one found in most standard physics textbooks (e.g., Marion and Thornton, 1995; Jackson, 1999; Griffiths, 1999). There, a tensor is an array of numbers that transform in a particular way under rotations. We shall learn the connection between these definitions in Sec. 1.6 below.

To illustrate this approach, if **T** is a rank-3 tensor (has 3 slots) as in Fig. 1.2, then its value on the vectors **A**, **B**, **C** is denoted **T**(**A**, **B**, **C**). Linearity of this function can be expressed as

$$\mathsf{T}(e\mathbf{E} + f\mathbf{F}, \mathbf{B}, \mathbf{C}) = e\mathsf{T}(\mathbf{E}, \mathbf{B}, \mathbf{C}) + f\mathsf{T}(\mathbf{F}, \mathbf{B}, \mathbf{C}), \tag{1.3b}$$

where e and f are real numbers, and similarly for the second and third slots.

We have already defined the squared length $(\mathbf{A})^2 \equiv \mathbf{A}^2$ of a vector **A** as the squared **inner product** distance between the points at its tail and its tip. The *inner product* (also called the dot product) $\mathbf{A} \cdot \mathbf{B}$ of two vectors is defined in terms of this squared length by

$$\boxed{\mathbf{A} \cdot \mathbf{B} \equiv \frac{1}{4}\left[(\mathbf{A} + \mathbf{B})^2 - (\mathbf{A} - \mathbf{B})^2\right].} \tag{1.4a}$$

In Euclidean space, this is the standard inner product, familiar from elementary geometry and discussed above in terms of the area of a parallelogram.

---

6. This is not so in curved spaces, as we shall see in Sec. 24.3.4.
7. This is a different use of the word *rank* than for a matrix, whose rank is its number of linearly independent rows or columns.

One can show that the inner product (1.4a) is a real-valued linear function of each of its vectors. Therefore, we can regard it as a tensor of rank 2. When so regarded, the

inner product is denoted $\mathbf{g}(\_\_, \_\_)$ and is called the *metric tensor*. In other words, the metric tensor $\mathbf{g}$ is that linear function of two vectors whose value is given by

$$\boxed{\mathbf{g}(\mathbf{A}, \mathbf{B}) \equiv \mathbf{A} \cdot \mathbf{B}.} \tag{1.4b}$$

Notice that, because $\mathbf{A} \cdot \mathbf{B} = \mathbf{B} \cdot \mathbf{A}$, the metric tensor is *symmetric* in its two slots—one gets the same real number independently of the order in which one inserts the two vectors into the slots:

$$\mathbf{g}(\mathbf{A}, \mathbf{B}) = \mathbf{g}(\mathbf{B}, \mathbf{A}). \tag{1.4c}$$

With the aid of the inner product, we can regard any vector $\mathbf{A}$ as a tensor of rank one: the real number that is produced when an arbitrary vector $\mathbf{C}$ is inserted into $\mathbf{A}$'s single slot is

$$\boxed{\mathbf{A}(\mathbf{C}) \equiv \mathbf{A} \cdot \mathbf{C}.} \tag{1.4d}$$

In Newtonian physics, we rarely meet tensors of rank higher than two. However, second-rank tensors appear frequently—often in roles where one sticks a single vector into the second slot and leaves the first slot empty, thereby producing a single-slotted entity, a vector. An example that we met in Sec. 1.1.1 is a rigid body's moment-of-inertia tensor $\mathsf{I}(\_\_, \_\_)$, which gives us the body's angular momentum $\mathbf{J}(\_\_) = \mathsf{I}(\_\_, \boldsymbol{\Omega})$ when its angular velocity $\boldsymbol{\Omega}$ is inserted into its second slot.[8] Another example is the stress tensor of a solid, a fluid, a plasma, or a field (Sec. 1.9 below).

From three vectors $\mathbf{A}$, $\mathbf{B}$, $\mathbf{C}$, we can construct a tensor, their *tensor product* (also called *outer product* in contradistinction to the inner product $\mathbf{A} \cdot \mathbf{B}$), defined as follows:

$$\boxed{\mathbf{A} \otimes \mathbf{B} \otimes \mathbf{C}(\mathbf{E}, \mathbf{F}, \mathbf{G}) \equiv \mathbf{A}(\mathbf{E})\mathbf{B}(\mathbf{F})\mathbf{C}(\mathbf{G}) = (\mathbf{A} \cdot \mathbf{E})(\mathbf{B} \cdot \mathbf{F})(\mathbf{C} \cdot \mathbf{G}).} \tag{1.5a}$$

Here the first expression is the notation for the value of the new tensor, $\mathbf{A} \otimes \mathbf{B} \otimes \mathbf{C}$ evaluated on the three vectors $\mathbf{E}$, $\mathbf{F}$, $\mathbf{G}$; the middle expression is the ordinary product of three real numbers, the value of $\mathbf{A}$ on $\mathbf{E}$, the value of $\mathbf{B}$ on $\mathbf{F}$, and the value of $\mathbf{C}$ on $\mathbf{G}$; and the third expression is that same product with the three numbers rewritten as scalar products. Similar definitions can be given (and should be obvious) for the tensor product of any number of vectors, and of any two or more tensors of any rank; for example, if T has rank 2 and S has rank 3, then

$$\mathsf{T} \otimes \mathsf{S}(\mathbf{E}, \mathbf{F}, \mathbf{G}, \mathbf{H}, \mathbf{J}) \equiv \mathsf{T}(\mathbf{E}, \mathbf{F})\mathsf{S}(\mathbf{G}, \mathbf{H}, \mathbf{J}). \tag{1.5b}$$

One last geometric (i.e., frame-independent) concept we shall need is *contraction*. We illustrate this concept first by a simple example, then give the general definition.

---

8. Actually, it doesn't matter which slot, since I is symmetric.

From two vectors $\mathbf{A}$ and $\mathbf{B}$ we can construct the tensor product $\mathbf{A} \otimes \mathbf{B}$ (a second-rank tensor), and we can also construct the scalar product $\mathbf{A} \cdot \mathbf{B}$ (a real number, i.e., a *scalar*, also known as a *rank-0 tensor*). The process of contraction is the construction of $\mathbf{A} \cdot \mathbf{B}$ from $\mathbf{A} \otimes \mathbf{B}$:

$$\boxed{\text{contraction}(\mathbf{A} \otimes \mathbf{B}) \equiv \mathbf{A} \cdot \mathbf{B}.} \tag{1.6a}$$

One can show fairly easily using component techniques (Sec. 1.5 below) that any second-rank tensor $\mathbf{T}$ can be expressed as a sum of tensor products of vectors, $\mathbf{T} = \mathbf{A} \otimes \mathbf{B} + \mathbf{C} \otimes \mathbf{D} + \ldots$. Correspondingly, it is natural to define the contraction of $\mathbf{T}$ to be contraction$(\mathbf{T}) = \mathbf{A} \cdot \mathbf{B} + \mathbf{C} \cdot \mathbf{D} + \ldots$. Note that this contraction process lowers the rank of the tensor by two, from 2 to 0. Similarly, for a tensor of rank n one can construct a tensor of rank $n - 2$ by contraction, but in this case one must specify which slots are to be contracted. For example, if $\mathbf{T}$ is a third-rank tensor, expressible as $\mathbf{T} = \mathbf{A} \otimes \mathbf{B} \otimes \mathbf{C} + \mathbf{E} \otimes \mathbf{F} \otimes \mathbf{G} + \ldots$, then the contraction of $\mathbf{T}$ on its first and third slots is the rank-1 tensor (vector)

$$1\&3\text{contraction}(\mathbf{A} \otimes \mathbf{B} \otimes \mathbf{C} + \mathbf{E} \otimes \mathbf{F} \otimes \mathbf{G} + \ldots) \equiv (\mathbf{A} \cdot \mathbf{C})\mathbf{B} + (\mathbf{E} \cdot \mathbf{G})\mathbf{F} + \ldots.$$

$$\tag{1.6b}$$

Unfortunately, there is no simple index-free notation for contraction in common use.

All the concepts developed in this section (vector, tensor, metric tensor, inner product, tensor product, and contraction of a tensor) can be carried over, with no change whatsoever, into any vector space[9] that is endowed with a concept of squared length—for example, to the 4-dimensional spacetime of special relativity (next chapter).

1.4 Particle Kinetics and Lorentz Force in Geometric Language

In this section, we illustrate our geometric viewpoint by formulating Newton's laws of motion for particles.

In Newtonian physics, a classical particle moves through Euclidean 3-space as universal time t passes. At time t it is located at some point $\mathbf{x}(t)$ (its *position*). The function $\mathbf{x}(t)$ represents a curve in 3-space, the particle's *trajectory*. The particle's *velocity* $\mathbf{v}(t)$ is the time derivative of its position, its *momentum* $\mathbf{p}(t)$ is the product of its mass m and velocity, its *acceleration* $\mathbf{a}(t)$ is the time derivative of its velocity, and its *kinetic energy* $E(t)$ is half its mass times velocity squared:

trajectory, velocity, momentum, acceleration, and energy

$$\mathbf{v}(t) = \frac{d\mathbf{x}}{dt}, \quad \mathbf{p}(t) = m\mathbf{v}(t), \quad \mathbf{a}(t) = \frac{d\mathbf{v}}{dt} = \frac{d^2\mathbf{x}}{dt^2}, \quad E(t) = \frac{1}{2}m\mathbf{v}^2. \tag{1.7a}$$

---

9. Or, more precisely, any vector space over the real numbers. If the vector space's scalars are complex numbers, as in quantum mechanics, then slight changes are needed.

Since points in 3-space are geometric objects (defined independently of any coordinate system), so also are the trajectory $\mathbf{x}(t)$, the velocity, the momentum, the acceleration, and the energy. (Physically, of course, the velocity has an ambiguity; it depends on one's standard of rest.)

Newton's second law of motion states that the particle's momentum can change only if a force $\mathbf{F}$ acts on it, and that its change is given by

$$d\mathbf{p}/dt = m\mathbf{a} = \mathbf{F}. \tag{1.7b}$$

If the force is produced by an electric field $\mathbf{E}$ and magnetic field $\mathbf{B}$, then this law of motion in SI units takes the familiar Lorentz-force form

$$d\mathbf{p}/dt = q(\mathbf{E} + \mathbf{v} \times \mathbf{B}). \tag{1.7c}$$

(Here we have used the vector cross product, with which the reader should be familiar, and which will be discussed formally in Sec. 1.7.)

laws of motion

The laws of motion (1.7) are geometric relationships among geometric objects. Let us illustrate this using something very familiar, planetary motion. Consider a light planet orbiting a heavy star. If there were no gravitational force, the planet would continue in a straight line with constant velocity $\mathbf{v}$ and speed $v = |\mathbf{v}|$, sweeping out area A at a rate $dA/dt = rv_t/2$, where r is the radius, and v_t is the tangential speed. Elementary geometry equates this to the constant $vb/2$, where b is the impact parameter—the smallest separation from the star. Now add a gravitational force $\mathbf{F}$ and let it cause a small radial impulse. A second application of geometry showed Newton that the product $rv_t/2$ is unchanged to first order in the impulse, and he recovered Kepler's second law ($dA/dt = $ const) without introducing coordinates.[10]

Contrast this approach with one relying on coordinates. For example, one introduces an (r, ϕ) coordinate system, constructs a lagrangian and observes that the coordinate ϕ is ignorable; then the Euler-Lagrange equations immediately imply the conservation of angular momentum, which is equivalent to Kepler's second law. So, which of these two approaches is preferable? The answer is surely "both!" Newton wrote the *Principia* in the language of geometry at least partly for a reason that remains valid today: it brought him a quick understanding of fundamental laws of physics. Lagrange followed his coordinate-based path to the function that bears his name, because he wanted to solve problems in celestial mechanics that would not yield to

10. Continuing in this vein, when the force is inverse square, as it is for gravity and electrostatics, we can use Kepler's second law to argue that when the orbit turns through a succession of equal angles $d\theta$, its successive changes in velocity $d\mathbf{v} = \mathbf{a}dt$ (with $\mathbf{a}$ the gravitational acceleration) all have the same magnitude $|d\mathbf{v}|$ and have the same angles $d\theta$ from one to another. So, if we trace the head of the velocity vector in velocity space, it follows a circle. The circle is not centered on zero velocity when the eccentricity is nonzero but there exists a reference frame in which the speed of the planet is constant. This graphical representation is known as a *hodograph*, and similar geometrical approaches are used in fluid mechanics. For Richard Feynman's masterful presentation of these ideas to first-year undergraduates, see Goodstein and Goodstein (1996).

Newton's approach. So it is today. Geometry and analysis are both indispensible. In the domain of classical physics, the geometry is of greater importance in deriving and understanding fundamental laws and has arguably been underappreciated; coordinates hold sway when we apply these laws to solve real problems. Today, both old and new laws of physics are commonly expressed geometrically, using lagrangians, hamiltonians, and actions, for example Hamilton's action principle $\delta \int L dt = 0$ where L is the coordinate-independent lagrangian. Indeed, being able to do this without introducing coordinates is a powerful guide to deriving these laws and a tool for comprehending their implications.

symmetry and
conservation laws

A comment is needed on the famous connection between *symmetry* and *conservation laws*. In our example above, angular momentum conservation followed from axial symmetry which was embodied in the lagrangian's independence of the angle ϕ; but we also deduced it geometrically. This is usually the case in classical physics; typically, we do not need to introduce a specific coordinate system to understand symmetry and to express the associated conservation laws. However, symmetries are sometimes well hidden, for example with a nutating top, and coordinate transformations are then usually the best approach to uncover them.

Often in classical physics, real-world factors invalidate or complicate Lagrange's and Hamilton's coordinate-based analytical dynamics, and so one is driven to geometric considerations. As an example, consider a spherical marble rolling on a flat horizontal table. The analytical dynamics approach is to express the height of the marble's center of mass and the angle of its rotation as constraints and align the basis vectors so there is a single horizontal coordinate defined by the initial condition. It is then deduced that linear and angular momenta are conserved. Of course that result is trivial and just as easily gotten without this formalism. However, this model is also used for many idealized problems where the outcome is far from obvious and the approach is brilliantly effective. But consider the real world in which tables are warped and bumpy, marbles are ellipsoidal and scratched, air imposes a resistance, and wood and glass comprise polymers that attract one another. And so on. When one includes these factors, it is to geometry that one quickly turns to understand the real marble's actual dynamics. Even ignoring these effects and just asking what happens when the marble rolls off the edge of a table introduces a *nonholonomic* constraint, and figuring out where it lands and how fast it is spinning are best addressed not by the methods of Lagrange and Hamilton, but instead by considering the geometry of the gravitational and reaction forces. In the following chapters, we shall encounter many examples where we have to deal with messy complications like these.

Exercise 1.1 *Practice: Energy Change for Charged Particle*

EXERCISES

Without introducing any coordinates or basis vectors, show that when a particle with charge q interacts with electric and magnetic fields, its kinetic energy changes at a rate

$$dE/dt = q \, \mathbf{v} \cdot \mathbf{E}. \qquad (1.8)$$

Exercise 1.2 *Practice: Particle Moving in a Circular Orbit*

Consider a particle moving in a circle with uniform speed $v = |\mathbf{v}|$ and uniform magnitude $a = |\mathbf{a}|$ of acceleration. Without introducing any coordinates or basis vectors, do the following.

(a) At any moment of time, let $\mathbf{n} = \mathbf{v}/v$ be the unit vector pointing along the velocity, and let s denote distance that the particle travels in its orbit. By drawing a picture, show that $d\mathbf{n}/ds$ is a unit vector that points to the center of the particle's circular orbit, divided by the radius of the orbit.

(b) Show that the vector (not unit vector) pointing from the particle's location to the center of its orbit is $(v/a)^2\mathbf{a}$.

1.5 Component Representation of Tensor Algebra

In the Euclidean 3-space of Newtonian physics, there is a unique set of *orthonormal basis vectors* $\{\mathbf{e}_x, \mathbf{e}_y, \mathbf{e}_z\} \equiv \{\mathbf{e}_1, \mathbf{e}_2, \mathbf{e}_3\}$ associated with any *Cartesian coordinate system* $\{x, y, z\} \equiv \{x^1, x^2, x^3\} \equiv \{x_1, x_2, x_3\}$. (In Cartesian coordinates in Euclidean space, we usually place indices down, but occasionally we place them up. It doesn't matter. By definition, in Cartesian coordinates a quantity is the same whether its index is down or up.) The basis vector $\mathbf{e}_j$ points along the x_j coordinate direction, which is orthogonal to all the other coordinate directions, and it has unit length (Fig. 1.3), so

$$\boxed{\mathbf{e}_j \cdot \mathbf{e}_k = \delta_{jk},} \tag{1.9a}$$

where δ_{jk} is the Kronecker delta.

Any vector $\mathbf{A}$ in 3-space can be expanded in terms of this basis:

$$\mathbf{A} = A_j \mathbf{e}_j. \tag{1.9b}$$

Here and throughout this book, we adopt the *Einstein summation convention*: repeated indices (in this case j) are to be summed (in this 3-space case over $j = 1, 2, 3$), unless otherwise instructed. By virtue of the orthonormality of the basis, the components A_j of $\mathbf{A}$ can be computed as the scalar product

$$A_j = \mathbf{A} \cdot \mathbf{e}_j. \tag{1.9c}$$

[The proof of this is straightforward: $\mathbf{A} \cdot \mathbf{e}_j = (A_k \mathbf{e}_k) \cdot \mathbf{e}_j = A_k(\mathbf{e}_k \cdot \mathbf{e}_j) = A_k \delta_{kj} = A_j$.]

Any tensor, say, the third-rank tensor $\mathbf{T}(\_, \_, \_)$, can be expanded in terms of tensor products of the basis vectors:

$$\boxed{\mathbf{T} = T_{ijk}\mathbf{e}_i \otimes \mathbf{e}_j \otimes \mathbf{e}_k.} \tag{1.9d}$$

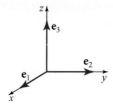

FIGURE 1.3 The orthonormal basis vectors $\mathbf{e}_j$ associated with a Euclidean coordinate system in Euclidean 3-space.

The components T_{ijk} of **T** can be computed from **T** and the basis vectors by the generalization of Eq. (1.9c):

Cartesian components of a tensor

$$\boxed{T_{ijk} = \mathbf{T}(\mathbf{e}_i, \mathbf{e}_j, \mathbf{e}_k).} \tag{1.9e}$$

[This equation can be derived using the orthonormality of the basis in the same way as Eq. (1.9c) was derived.] As an important example, the components of the metric tensor are $g_{jk} = \mathbf{g}(\mathbf{e}_j, \mathbf{e}_k) = \mathbf{e}_j \cdot \mathbf{e}_k = \delta_{jk}$ [where the first equality is the method (1.9e) of computing tensor components, the second is the definition (1.4b) of the metric, and the third is the orthonormality relation (1.9a)]:

$$\boxed{g_{jk} = \delta_{jk}.} \tag{1.9f}$$

The components of a tensor product [e.g., $\mathbf{T}(\_\_, \_\_, \_\_) \otimes \mathbf{S}(\_\_, \_\_)$] are easily deduced by inserting the basis vectors into the slots [Eq. (1.9e)]; they are $\mathbf{T}(\mathbf{e}_i, \mathbf{e}_j, \mathbf{e}_k) \otimes \mathbf{S}(\mathbf{e}_l, \mathbf{e}_m) = T_{ijk}S_{lm}$ [cf. Eq. (1.5a)]. In words, the components of a tensor product are equal to the ordinary arithmetic product of the components of the individual tensors.

In component notation, the inner product of two vectors and the value of a tensor when vectors are inserted into its slots are given by

$$\boxed{\mathbf{A} \cdot \mathbf{B} = A_j B_j, \qquad \mathbf{T}(\mathbf{A}, \mathbf{B}, \mathbf{C}) = T_{ijk} A_i B_j C_k,} \tag{1.9g}$$

as one can easily show using previous equations. Finally, the contraction of a tensor [say, the fourth-rank tensor $\mathbf{R}(\_\_, \_\_, \_\_, \_\_)$] on two of its slots (say, the first and third) has components that are easily computed from the tensor's own components:

$$\text{components of } [1\&3\text{contraction of } \mathbf{R}] = R_{ijik}. \tag{1.9h}$$

Note that R_{ijik} is summed on the i index, so it has only two free indices, j and k, and thus is the component of a second-rank tensor, as it must be if it is to represent the contraction of a fourth-rank tensor.

1.5.1 Slot-Naming Index Notation

We now pause in our development of the component version of tensor algebra to introduce a very important new viewpoint.

BOX 1.2. VECTORS AND TENSORS IN QUANTUM THEORY T2

The laws of quantum theory, like all other laws of Nature, can be expressed as geometric relationships among geometric objects. Most of quantum theory's geometric objects, like those of classical theory, are vectors and tensors: the quantum state $|\psi\rangle$ of a physical system (e.g., a particle in a harmonic-oscillator potential) is a Hilbert-space vector—a generalization of a Euclidean-space vector **A**. There is an inner product, denoted $\langle\phi|\psi\rangle$, between any two states $|\phi\rangle$ and $|\psi\rangle$, analogous to **B** · **A**; but **B** · **A** is a real number, whereas $\langle\phi|\psi\rangle$ is a complex number (and we add and subtract quantum states with complex-number coefficients). The Hermitian operators that represent observables (e.g., the hamiltonian $\hat{H}$ for the particle in the potential) are two-slotted (second-rank), complex-valued functions of vectors; $\langle\phi|\hat{H}|\psi\rangle$ is the complex number that one gets when one inserts ϕ and ψ into the first and second slots of $\hat{H}$. Just as, in Euclidean space, we get a new vector (first-rank tensor) **T**(__, **A**) when we insert the vector **A** into the second slot of **T**, so in quantum theory we get a new vector (physical state) $\hat{H}|\psi\rangle$ (the result of letting $\hat{H}$ "act on" $|\psi\rangle$) when we insert $|\psi\rangle$ into the second slot of $\hat{H}$. In these senses, we can regard **T** as a linear map of Euclidean vectors into Euclidean vectors and $\hat{H}$ as a linear map of states (Hilbert-space vectors) into states.

For the electron in the hydrogen atom, we can introduce a set of orthonormal basis vectors $\{|1\rangle, |2\rangle, |3\rangle, \ldots\}$, that is, the atom's energy eigenstates, with $\langle m|n\rangle = \delta_{mn}$. But by contrast with Newtonian physics, where we only need three basis vectors (because our Euclidean space is 3-dimensional), for the particle in a harmonic-oscillator potential, we need an infinite number of basis vectors (since the Hilbert space of all states is infinite-dimensional). In the particle's quantum-state basis, any observable (e.g., the particle's position $\hat{x}$ or momentum $\hat{p}$) has components computed by inserting the basis vectors into its two slots: $x_{mn} = \langle m|\hat{x}|n\rangle$, and $p_{mn} = \langle m|\hat{p}|n\rangle$. In this basis, the operator $\hat{x}\hat{p}$ (which maps states into states) has components $x_{jk}p_{km}$ (a matrix product), and the noncommutation of position and momentum $[\hat{x}, \hat{p}] = i\hbar$ (an important physical law) is expressible in terms of components as $x_{jk}p_{km} - p_{jk}x_{km} = i\hbar\delta_{jm}$.

Consider the rank-2 tensor **F**(__, __). We can define a new tensor **G**(__, __) to be the same as **F**, but with the slots interchanged: i.e., for any two vectors **A** and **B**, it is true that **G**(**A**, **B**) = **F**(**B**, **A**). We need a simple, compact way to indicate that **F** and **G** are equal except for an interchange of slots. The best way is to give the slots names, say a and b—i.e., to rewrite **F**(__, __) as **F**(__$_a$, __$_b$) or more conveniently as F_{ab}, and then to write the relationship between **G** and **F** as $G_{ab} = F_{ba}$. "NO!" some readers might object. This notation is indistinguishable from our notation for components on a particular basis. "GOOD!" a more astute reader will exclaim. The relation $G_{ab} = F_{ba}$ in a particular basis is a true statement if and only if "$\mathbf{G} = \mathbf{F}$ with slots interchanged" is true, so why not use the same notation to symbolize both? In fact, we shall do this. We ask our readers to look at any "index equation," such as $G_{ab} = F_{ba}$, like they would look at an Escher drawing: momentarily think of it as a relationship between components of tensors in a specific basis; then do a quick mind-flip and regard it quite differently, as a relationship between geometric, basis-independent tensors with the indices playing the roles of slot names. This mind-flip approach to tensor algebra will pay substantial dividends.

As an example of the power of this *slot-naming index notation*, consider the contraction of the first and third slots of a third-rank tensor $\mathbf{T}$. In any basis the components of 1&3contraction($\mathbf{T}$) are T_{aba}; cf. Eq. (1.9h). Correspondingly, in slot-naming index notation we denote 1&3contraction($\mathbf{T}$) by the simple expression T_{aba}. We can think of the first and third slots as annihilating each other by the contraction, leaving free only the second slot (named b) and therefore producing a rank-1 tensor (a vector).

slot-naming index notation

We should caution that the phrase "slot-naming index notation" is unconventional. You are unlikely to find it in any other textbooks. However, we like it. It says precisely what we want it to say.

1.5.2 Particle Kinetics in Index Notation

1.5.2

As an example of slot-naming index notation, we can rewrite the equations of particle kinetics (1.7) as follows:

$$v_i = \frac{dx_i}{dt}, \quad p_i = mv_i, \quad a_i = \frac{dv_i}{dt} = \frac{d^2x_i}{dt^2},$$

$$E = \frac{1}{2}mv_jv_j, \quad \frac{dp_i}{dt} = q(E_i + \epsilon_{ijk}v_jB_k). \tag{1.10}$$

(In the last equation ϵ_{ijk} is the so-called Levi-Civita tensor, which is used to produce the cross product; we shall learn about it in Sec. 1.7. And note that the scalar energy E must not be confused with the electric field vector E_i.)

Equations (1.10) can be viewed in either of two ways: (i) as the basis-independent geometric laws $\mathbf{v} = d\mathbf{x}/dt$, $\mathbf{p} = m\mathbf{v}$, $\mathbf{a} = d\mathbf{v}/dt = d^2\mathbf{x}/dt^2$, $E = \frac{1}{2}mv^2$, and $d\mathbf{p}/dt = q(\mathbf{E} + \mathbf{v} \times \mathbf{B})$ written in slot-naming index notation; or (ii) as equations for the components of $\mathbf{v}$, $\mathbf{p}$, $\mathbf{a}$, $\mathbf{E}$, and $\mathbf{B}$ in some particular Cartesian coordinate system.

EXERCISES

Exercise 1.3 *Derivation: Component Manipulation Rules*
Derive the component manipulation rules (1.9g) and (1.9h).

Exercise 1.4 *Example and Practice: Numerics of Component Manipulations*
The third-rank tensor $\mathbf{S}(\_\_, \_\_, \_\_)$ and vectors $\mathbf{A}$ and $\mathbf{B}$ have as their only nonzero components $S_{123} = S_{231} = S_{312} = +1$, $A_1 = 3$, $B_1 = 4$, $B_2 = 5$. What are the

components of the vector $\mathbf{C} = \mathbf{S}(\mathbf{A}, \mathbf{B}, \_\_)$, the vector $\mathbf{D} = \mathbf{S}(\mathbf{A}, \_\_, \mathbf{B})$, and the tensor $\mathbf{W} = \mathbf{A} \otimes \mathbf{B}$?

[Partial solution: In component notation, $C_k = S_{ijk} A_i B_j$, where (of course) we sum over the repeated indices i and j. This tells us that $C_1 = S_{231} A_2 B_3$, because S_{231} is the only component of $\mathbf{S}$ whose last index is a 1; this in turn implies that $C_1 = 0$, since $A_2 = 0$. Similarly, $C_2 = S_{312} A_3 B_1 = 0$ (because $A_3 = 0$). Finally, $C_3 = S_{123} A_1 B_2 = +1 \times 3 \times 5 = 15$. Also, in component notation $W_{ij} = A_i B_j$, so $W_{11} = A_1 \times B_1 = 3 \times 4 = 12$, and $W_{12} = A_1 \times B_2 = 3 \times 5 = 15$. Here the $\times$ stands for numerical multiplication, not the vector cross product.]

Exercise 1.5 *Practice: Meaning of Slot-Naming Index Notation*

(a) The following expressions and equations are written in slot-naming index notation. Convert them to geometric, index-free notation: $A_i B_{jk}$, $A_i B_{ji}$, $S_{ijk} = S_{kji}$, $A_i B_i = A_i B_j g_{ij}$.

(b) The following expressions are written in geometric, index-free notation. Convert them to slot-naming index notation: $\mathbf{T}(\_\_, \_\_, \mathbf{A})$, $\mathbf{T}(\_\_, \mathbf{S}(\mathbf{B}, \_\_), \_\_)$.

1.6 Orthogonal Transformations of Bases

Consider two different Cartesian coordinate systems $\{x, y, z\} \equiv \{x_1, x_2, x_3\}$, and $\{\bar{x}, \bar{y}, \bar{z}\} \equiv \{x_{\bar{1}}, x_{\bar{2}}, x_{\bar{3}}\}$. Denote by $\{\mathbf{e}_i\}$ and $\{\mathbf{e}_{\bar{p}}\}$ the corresponding bases. It is possible to expand the basis vectors of one basis in terms of those of the other. We denote the expansion coefficients by the letter R and write

$$\mathbf{e}_i = \mathbf{e}_{\bar{p}} R_{\bar{p}i}, \qquad \mathbf{e}_{\bar{p}} = \mathbf{e}_i R_{i\bar{p}}. \qquad (1.11)$$

The quantities $R_{\bar{p}i}$ and $R_{i\bar{p}}$ are not the components of a tensor; rather, they are the elements of transformation matrices

$$[R_{\bar{p}i}] = \begin{bmatrix} R_{\bar{1}1} & R_{\bar{1}2} & R_{\bar{1}3} \\ R_{\bar{2}1} & R_{\bar{2}2} & R_{\bar{2}3} \\ R_{\bar{3}1} & R_{\bar{3}2} & R_{\bar{3}3} \end{bmatrix}, \qquad [R_{i\bar{p}}] = \begin{bmatrix} R_{1\bar{1}} & R_{1\bar{2}} & R_{1\bar{3}} \\ R_{2\bar{1}} & R_{2\bar{2}} & R_{2\bar{3}} \\ R_{3\bar{1}} & R_{3\bar{2}} & R_{3\bar{3}} \end{bmatrix}. \qquad (1.12a)$$

(Here and throughout this book we use square brackets to denote matrices.) These two matrices must be the inverse of each other, since one takes us from the barred basis to the unbarred, and the other in the reverse direction, from unbarred to barred:

$$R_{\bar{p}i} R_{i\bar{q}} = \delta_{\bar{p}\bar{q}}, \qquad R_{i\bar{p}} R_{\bar{p}j} = \delta_{ij}. \qquad (1.12b)$$

The orthonormality requirement for the two bases implies that $\delta_{ij} = \mathbf{e}_i \cdot \mathbf{e}_j = (\mathbf{e}_{\bar{p}} R_{\bar{p}i}) \cdot (\mathbf{e}_{\bar{q}} R_{\bar{q}j}) = R_{\bar{p}i} R_{\bar{q}j} (\mathbf{e}_{\bar{p}} \cdot \mathbf{e}_{\bar{q}}) = R_{\bar{p}i} R_{\bar{q}j} \delta_{\bar{p}\bar{q}} = R_{\bar{p}i} R_{\bar{p}j}$. This says that the transpose of $[R_{\bar{p}i}]$ is its inverse—which we have already denoted by $[R_{i\bar{p}}]$:

$$[R_{i\bar{p}}] \equiv \text{inverse}\left([R_{\bar{p}i}]\right) = \text{transpose}\left([R_{\bar{p}i}]\right). \qquad (1.12c)$$

This property implies that the transformation matrix is orthogonal, so the transformation is a reflection or a rotation (see, e.g., Goldstein, Poole, and Safko, 2002). Thus (as should be obvious and familiar), the bases associated with any two Euclidean coordinate systems are related by a reflection or rotation, and the matrices (1.12a) are called *rotation matrices*. Note that Eq. (1.12c) does not say that $[R_{i\bar{p}}]$ is a symmetric matrix. In fact, most rotation matrices are not symmetric [see, e.g., Eq. (1.14)].

orthogonal transformation and rotation

The fact that a vector **A** is a geometric, basis-independent object implies that $\mathbf{A} = A_i \mathbf{e}_i = A_i (\mathbf{e}_{\bar{p}} R_{\bar{p}i}) = (R_{\bar{p}i} A_i) \mathbf{e}_{\bar{p}} = A_{\bar{p}} \mathbf{e}_{\bar{p}}$:

$$A_{\bar{p}} = R_{\bar{p}i} A_i, \quad \text{and similarly,} \quad A_i = R_{i\bar{p}} A_{\bar{p}}; \tag{1.13a}$$

and correspondingly for the components of a tensor:

$$T_{\bar{p}\bar{q}\bar{r}} = R_{\bar{p}i} R_{\bar{q}j} R_{\bar{r}k} T_{ijk}, \quad T_{ijk} = R_{i\bar{p}} R_{j\bar{q}} R_{k\bar{r}} T_{\bar{p}\bar{q}\bar{r}}. \tag{1.13b}$$

It is instructive to compare the transformation law (1.13a) for the components of a vector with Eqs. (1.11) for the bases. To make these laws look natural, we have placed the transformation matrix on the left in the former and on the right in the latter. In Minkowski spacetime (Chap. 2), the placement of indices, up or down, will automatically tell us the order.

If we choose the origins of our two coordinate systems to coincide, then the vector **x** reaching from the common origin to some point $\mathcal{P}$, whose coordinates are x_j and $x_{\bar{p}}$, has components equal to those coordinates; and as a result, the coordinates themselves obey the same transformation law as any other vector:

$$x_{\bar{p}} = R_{\bar{p}i} x_i, \quad x_i = R_{i\bar{p}} x_{\bar{p}}. \tag{1.13c}$$

The product of two rotation matrices $[R_{i\bar{p}} R_{\bar{p}\bar{s}}]$ is another rotation matrix $[R_{i\bar{s}}]$, which transforms the Cartesian bases $\mathbf{e}_{\bar{s}}$ to $\mathbf{e}_i$. Under this product rule, the rotation matrices form a mathematical *group*: the *rotation group*, whose *group representations* play an important role in quantum theory.

rotation group

Exercise 1.6 ***Example and Practice: Rotation in x-y Plane***

EXERCISES

Consider two Cartesian coordinate systems rotated with respect to each other in the x-y plane as shown in Fig. 1.4.

(a) Show that the rotation matrix that takes the barred basis vectors to the unbarred basis vectors is

$$[R_{\bar{p}i}] = \begin{bmatrix} \cos\phi & \sin\phi & 0 \\ -\sin\phi & \cos\phi & 0 \\ 0 & 0 & 1 \end{bmatrix}, \tag{1.14}$$

and show that the inverse of this rotation matrix is, indeed, its transpose, as it must be if this is to represent a rotation.

(b) Verify that the two coordinate systems are related by Eq. (1.13c).

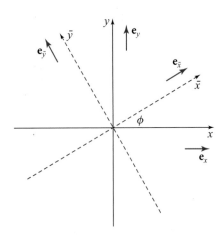

FIGURE 1.4 Two Cartesian coordinate systems $\{x, y, z\}$ and $\{\bar{x}, \bar{y}, \bar{z}\}$ and their basis vectors in Euclidean space, rotated by an angle ϕ relative to each other in the x-y plane. The z- and $\bar{z}$-axes point out of the paper or screen and are not shown.

(c) Let A_j be the components of the electromagnetic vector potential that lies in the x-y plane, so that $A_z = 0$. The two nonzero components A_x and A_y can be regarded as describing the two polarizations of an electromagnetic wave propagating in the z direction. Show that $A_{\bar{x}} + iA_{\bar{y}} = (A_x + iA_y)e^{-i\phi}$. One can show (cf. Sec. 27.3.3) that the factor $e^{-i\phi}$ implies that the quantum particle associated with the wave—the photon—has spin one [i.e., spin angular momentum $\hbar = (\text{Planck's constant})/2\pi$].

(d) Let h_{jk} be the components of a symmetric tensor that is *trace-free* (its contraction h_{jj} vanishes) and is confined to the x-y plane (so $h_{zk} = h_{kz} = 0$ for all k). Then the only nonzero components of this tensor are $h_{xx} = -h_{yy}$ and $h_{xy} = h_{yx}$. As we shall see in Sec. 27.3.1, this tensor can be regarded as describing the two polarizations of a gravitational wave propagating in the z direction. Show that $h_{\bar{x}\bar{x}} + ih_{\bar{x}\bar{y}} = (h_{xx} + ih_{xy})e^{-2i\phi}$. The factor $e^{-2i\phi}$ implies that the quantum particle associated with the gravitational wave (the graviton) has spin two (spin angular momentum $2\hbar$); cf. Eq. (27.31) and Sec. 27.3.3.

1.7 Differentiation of Scalars, Vectors, and Tensors; Cross Product and Curl

Consider a tensor field $\mathbf{T}(\mathcal{P})$ in Euclidean 3-space and a vector $\mathbf{A}$. We define the *directional derivative* of $\mathbf{T}$ along $\mathbf{A}$ by the obvious limiting procedure

$$\nabla_{\mathbf{A}}\mathbf{T} \equiv \lim_{\epsilon \to 0} \frac{1}{\epsilon}[\mathbf{T}(\mathbf{x}_\mathcal{P} + \epsilon\mathbf{A}) - \mathbf{T}(\mathbf{x}_\mathcal{P})] \tag{1.15a}$$

and similarly for the directional derivative of a vector field $\mathbf{B}(\mathcal{P})$ and a scalar field $\psi(\mathcal{P})$. [Here we have denoted points, e.g., $\mathcal{P}$, by the vector $\mathbf{x}_\mathcal{P}$ that reaches from some arbitrary origin to the point, and $\mathbf{T}(\mathbf{x}_\mathcal{P})$ denotes the field's dependence on location in space; $\mathbf{T}$'s slots and dependence on what goes into the slots are suppressed; and the units of ϵ are chosen to ensure that $\epsilon \mathbf{A}$ has the same units as $\mathbf{x}_\mathcal{P}$. There is no other appearance of vectors in this chapter.] In definition (1.15a), the quantity in square brackets is simply the difference between two linear functions of vectors (two tensors), so the quantity on the left-hand side is also a tensor with the same rank as $\mathbf{T}$.

It should not be hard to convince oneself that this directional derivative $\nabla_{\mathbf{A}}\mathbf{T}$ of any tensor field $\mathbf{T}$ is linear in the vector $\mathbf{A}$ along which one differentiates. Correspondingly, if $\mathbf{T}$ has rank n (n slots), then there is another tensor field, denoted $\nabla\mathbf{T}$, with rank $n+1$, such that

$$\boxed{\nabla_{\mathbf{A}}\mathbf{T} = \nabla\mathbf{T}(\_\,, \_\,, \_\,, \mathbf{A}).}$$ (1.15b)

Here on the right-hand side the first n slots (3 in the case shown) are left empty, and $\mathbf{A}$ is put into the last slot (the "differentiation slot"). The quantity $\nabla\mathbf{T}$ is called the *gradient* of $\mathbf{T}$. In slot-naming index notation, it is conventional to denote this gradient by $T_{abc;d}$, where in general the number of indices preceding the semicolon is the rank of $\mathbf{T}$. Using this notation, the directional derivative of $\mathbf{T}$ along $\mathbf{A}$ reads [cf. Eq. (1.15b)] $T_{abc;j}A_j$.

It is not hard to show that in any Cartesian coordinate system, the components of the gradient are nothing but the partial derivatives of the components of the original tensor, which we denote by a comma:

$$\boxed{T_{abc;j} = \frac{\partial T_{abc}}{\partial x_j} \equiv T_{abc,j}.}$$ (1.15c)

In a non-Cartesian basis (e.g., the spherical and cylindrical bases often used in electromagnetic theory), the components of the gradient typically are not obtained by simple partial differentiation [Eq. (1.15c) fails] because of turning and/or length changes of the basis vectors as we go from one location to another. In Sec. 11.8, we shall learn how to deal with this by using objects called *connection coefficients*. Until then, we confine ourselves to Cartesian bases, so subscript semicolons and subscript commas (partial derivatives) can be used interchangeably.

Because the gradient and the directional derivative are defined by the same standard limiting process as one uses when defining elementary derivatives, they obey the standard (Leibniz) rule for differentiating products:

$$\nabla_{\mathbf{A}}(\mathbf{S} \otimes \mathbf{T}) = (\nabla_{\mathbf{A}}\mathbf{S}) \otimes \mathbf{T} + \mathbf{S} \otimes \nabla_{\mathbf{A}}\mathbf{T},$$

$$\text{or} \quad (S_{ab}T_{cde})_{;j}A_j = (S_{ab;j}A_j)T_{cde} + S_{ab}(T_{cde;j}A_j);$$ (1.16a)

and

$$\nabla_{\mathbf{A}}(f\mathbf{T}) = (\nabla_{\mathbf{A}}f)\mathbf{T} + f\nabla_{\mathbf{A}}\mathbf{T}, \quad \text{or} \quad (fT_{abc})_{;j}A_j = (f_{;j}A_j)T_{abc} + fT_{abc;j}A_j.$$ (1.16b)

In an orthonormal basis these relations should be obvious: they follow from the Leibniz rule for partial derivatives.

Because the components g_{ab} of the metric tensor are constant in any Cartesian coordinate system, Eq. (1.15c) (which is valid in such coordinates) guarantees that $g_{ab;j} = 0$; i.e., the metric has vanishing gradient:

$$\nabla \mathbf{g} = 0, \quad \text{or} \quad g_{ab;j} = 0. \tag{1.17}$$

From the gradient of any vector or tensor we can construct several other important derivatives by contracting on slots:

1. Since the gradient $\nabla \mathbf{A}$ of a vector field $\mathbf{A}$ has two slots, $\nabla \mathbf{A}(\_\_, \_\_)$, we can contract its slots on each other to obtain a scalar field. That scalar field is the *divergence* of $\mathbf{A}$ and is denoted

$$\nabla \cdot \mathbf{A} \equiv (\text{contraction of } \nabla \mathbf{A}) = A_{a;a}. \tag{1.18}$$

2. Similarly, if $\mathbf{T}$ is a tensor field of rank 3, then $T_{abc;c}$ is its divergence on its third slot, and $T_{abc;b}$ is its divergence on its second slot.

3. By taking the double gradient and then contracting on the two gradient slots we obtain, from any tensor field $\mathbf{T}$, a new tensor field with the same rank,

$$\nabla^2 \mathbf{T} \equiv (\nabla \cdot \nabla)\mathbf{T}, \quad \text{or} \quad T_{abc;jj}. \tag{1.19}$$

Here and henceforth, all indices following a semicolon (or comma) represent gradients (or partial derivatives): $T_{abc;jj} \equiv T_{abc;j;j}$, $T_{abc,jk} \equiv \partial^2 T_{abc}/\partial x_j \partial x_k$. The operator ∇^2 is called the *laplacian*.

The metric tensor is a fundamental property of the space in which it lives; it embodies the inner product and hence the space's notion of distance. In addition to the metric, there is one (and only one) other fundamental tensor that describes a piece of Euclidean space's geometry: the *Levi-Civita tensor* $\boldsymbol{\epsilon}$, which embodies the space's notion of volume.

In a Euclidean space with dimension n, the Levi-Civita tensor $\boldsymbol{\epsilon}$ is a completely antisymmetric tensor with rank n (with n slots). A parallelepiped whose edges are the n vectors $\mathbf{A}, \mathbf{B}, \dots, \mathbf{F}$ is said to have the *volume*

$$\boxed{\text{volume} = \boldsymbol{\epsilon}(\mathbf{A}, \mathbf{B}, \dots, \mathbf{F}).} \tag{1.20}$$

(We justify this definition in Sec. 1.8.) Notice that this volume can be positive or negative, and if we exchange the order of the parallelepiped's legs, the volume's sign changes: $\boldsymbol{\epsilon}(\mathbf{B}, \mathbf{A}, \dots, \mathbf{F}) = -\boldsymbol{\epsilon}(\mathbf{A}, \mathbf{B}, \dots, \mathbf{F})$ by antisymmetry of $\boldsymbol{\epsilon}$.

It is easy to see (Ex. 1.7) that (i) the volume vanishes unless the legs are all linearly independent, (ii) once the volume has been specified for one parallelepiped (one set of linearly independent legs), it is thereby determined for all parallelepipeds, and therefore, (iii) we require only one number plus antisymmetry to determine $\boldsymbol{\epsilon}$

fully. If the chosen parallelepiped has legs that are orthonormal (all are orthogonal to one another and all have unit length—properties determined by the metric **g**), then it must have unit volume, or more precisely volume ±1. This is a compatibility relation between **g** and ϵ. It is easy to see (Ex. 1.7) that (iv) ϵ is fully determined by its antisymmetry, compatibility with the metric, and a single sign: the choice of which parallelepipeds have positive volume and which have negative. It is conventional in Euclidean 3-space to give right-handed parallelepipeds positive volume and left-handed ones negative volume: $\epsilon(\mathbf{A}, \mathbf{B}, \mathbf{C})$ is positive if, when we place our right thumb along **C** and the fingers of our right hand along **A**, then bend our fingers, they sweep toward **B** and not −**B**.

These considerations dictate that in a right-handed orthonormal basis of Euclidean 3-space, the only nonzero components of ϵ are

$$\epsilon_{123} = +1,$$

$$\epsilon_{abc} = \begin{cases} +1 & \text{if } a, b, c \text{ is an even permutation of } 1, 2, 3 \\ -1 & \text{if } a, b, c \text{ is an odd permutation of } 1, 2, 3 \\ 0 & \text{if } a, b, c \text{ are not all different;} \end{cases} \tag{1.21}$$

and in a left-handed orthonormal basis, the signs of these components are reversed.

The Levi-Civita tensor is used to define the cross product and the curl:

cross product and curl

$$\mathbf{A} \times \mathbf{B} \equiv \epsilon(\_\_, \mathbf{A}, \mathbf{B}); \quad \text{in slot-naming index notation, } \epsilon_{ijk} A_j B_k; \tag{1.22a}$$

$$\nabla \times \mathbf{A} \equiv \text{(the vector field whose slot-naming index form is } \epsilon_{ijk} A_{k;j}). \tag{1.22b}$$

[Equation (1.22b) is an example of an expression that is complicated if stated in index-free notation; it says that $\nabla \times \mathbf{A}$ is the double contraction of the rank-5 tensor $\epsilon \otimes \nabla \mathbf{A}$ on its second and fifth slots, and on its third and fourth slots.]

Although Eqs. (1.22a) and (1.22b) look like complicated ways to deal with concepts that most readers regard as familiar and elementary, they have great power. The power comes from the following property of the Levi-Civita tensor in Euclidean 3-space [readily derivable from its components (1.21)]:

$$\epsilon_{ijm}\epsilon_{klm} = \delta^{ij}_{kl} \equiv \delta^i_k \delta^j_l - \delta^i_l \delta^j_k. \tag{1.23}$$

Here δ^i_k is the Kronecker delta. Examine the 4-index delta function δ^{ij}_{kl} carefully; it says that either the indices above and below each other must be the same ($i = k$ and $j = l$) with a + sign, or the diagonally related indices must be the same ($i = l$ and $j = k$) with a − sign. [We have put the indices ij of δ^{ij}_{kl} up solely to facilitate remembering this rule. Recall (first paragraph of Sec. 1.5) that in Euclidean space and Cartesian coordinates, it does not matter whether indices are up or down.] With the aid of Eq. (1.23) and the index-notation expressions for the cross product and curl, one can quickly and easily derive a wide variety of useful vector identities; see the very important Ex. 1.8.

Exercise 1.7 *Derivation: Properties of the Levi-Civita Tensor*

From its complete antisymmetry, derive the four properties of the Levi-Civita tensor, in n-dimensional Euclidean space, that are claimed in the text following Eq. (1.20).

Exercise 1.8 **Example and Practice: Vectorial Identities for the Cross Product and Curl*

Here is an example of how to use index notation to derive a vector identity for the double cross product $\mathbf{A} \times (\mathbf{B} \times \mathbf{C})$: in index notation this quantity is $\epsilon_{ijk} A_j (\epsilon_{klm} B_l C_m)$. By permuting the indices on the second ϵ and then invoking Eq. (1.23), we can write this as $\epsilon_{ijk} \epsilon_{lmk} A_j B_l C_m = \delta_{ij}^{lm} A_j B_l C_m$. By then invoking the meaning of the 4-index delta function [Eq. (1.23)], we bring this into the form $A_j B_i C_j - A_j B_j C_i$, which is the slot-naming index-notation form of $(\mathbf{A} \cdot \mathbf{C})\mathbf{B} - (\mathbf{A} \cdot \mathbf{B})\mathbf{C}$. Thus, it must be that $\mathbf{A} \times (\mathbf{B} \times \mathbf{C}) = (\mathbf{A} \cdot \mathbf{C})\mathbf{B} - (\mathbf{A} \cdot \mathbf{B})\mathbf{C}$. Use similar techniques to evaluate the following quantities.

(a) $\nabla \times (\nabla \times \mathbf{A})$.

(b) $(\mathbf{A} \times \mathbf{B}) \cdot (\mathbf{C} \times \mathbf{D})$.

(c) $(\mathbf{A} \times \mathbf{B}) \times (\mathbf{C} \times \mathbf{D})$.

Exercise 1.9 **Example and Practice: Levi-Civita Tensor in 2-Dimensional Euclidean Space*

In Euclidean 2-space, let $\{\mathbf{e}_1, \mathbf{e}_2\}$ be an orthonormal basis with positive volume.

(a) Show that the components of ϵ in this basis are

$$\epsilon_{12} = +1, \qquad \epsilon_{21} = -1, \qquad \epsilon_{11} = \epsilon_{22} = 0. \tag{1.24a}$$

(b) Show that

$$\epsilon_{ik} \epsilon_{jk} = \delta_{ij}. \tag{1.24b}$$

1.8

1.8 Volumes, Integration, and Integral Conservation Laws

In Cartesian coordinates of 2-dimensional Euclidean space, the basis vectors are orthonormal, so (with a conventional choice of sign) the components of the Levi-Civita tensor are given by Eqs. (1.24a). Correspondingly, the area (i.e., 2-dimensional volume) of a parallelogram whose sides are $\mathbf{A}$ and $\mathbf{B}$ is

$$2\text{-volume} = \epsilon(\mathbf{A}, \mathbf{B}) = \epsilon_{ab} A_a B_b = A_1 B_2 - A_2 B_1 = \det \begin{bmatrix} A_1 & B_1 \\ A_2 & B_2 \end{bmatrix}, \tag{1.25}$$

a relation that should be familiar from elementary geometry. Equally familiar should be the following expression for the 3-dimensional volume of a parallelepiped with legs

Chapter 1. Newtonian Physics: Geometric Viewpoint

A, **B**, and **C** [which follows from the components (1.21) of the Levi-Civita tensor]:

$$3\text{-volume} = \epsilon(\mathbf{A}, \mathbf{B}, \mathbf{C}) = \epsilon_{ijk} A_i B_j C_k = \mathbf{A} \cdot (\mathbf{B} \times \mathbf{C}) = \det \begin{bmatrix} A_1 & B_1 & C_1 \\ A_2 & B_2 & C_2 \\ A_3 & B_3 & C_3 \end{bmatrix}. \quad (1.26)$$

Our formal definition (1.20) of volume is justified because it gives rise to these familiar equations.

Equations (1.25) and (1.26) are foundations from which one can derive the usual formulas $dA = dx\,dy$ and $dV = dx\,dy\,dz$ for the area and volume of elementary surface and volume elements with Cartesian side lengths dx, dy, and dz (Ex. 1.10).

In Euclidean 3-space, we define the vectorial surface area of a 2-dimensional parallelogram with legs **A** and **B** to be

$$\boxed{\mathbf{\Sigma} = \mathbf{A} \times \mathbf{B} = \epsilon(\_\_, \mathbf{A}, \mathbf{B}).} \quad (1.27)$$

This vectorial surface area has a magnitude equal to the area of the parallelogram and a direction perpendicular to it. Notice that this surface area $\epsilon(\_\_, \mathbf{A}, \mathbf{B})$ can be thought of as an object that is waiting for us to insert a third leg, **C**, so as to compute a 3-volume $\epsilon(\mathbf{C}, \mathbf{A}, \mathbf{B})$—the volume of the parallelepiped with legs **C**, **A**, and **B**.

A parallelogram's surface has two faces (two sides), called the *positive face* and the *negative face*. If the vector **C** sticks out of the positive face, then $\mathbf{\Sigma}(\mathbf{C}) = \epsilon(\mathbf{C}, \mathbf{A}, \mathbf{B})$ is positive; if **C** sticks out of the negative face, then $\mathbf{\Sigma}(\mathbf{C})$ is negative.

1.8.1 Gauss's and Stokes' Theorems

Such vectorial surface areas are the foundation for surface integrals in 3-dimensional space and for the familiar *Gauss's theorem*,

$$\boxed{\int_{\mathcal{V}_3} (\mathbf{\nabla} \cdot \mathbf{A}) dV = \int_{\partial \mathcal{V}_3} \mathbf{A} \cdot d\mathbf{\Sigma}} \quad (1.28a)$$

(where $\mathcal{V}_3$ is a compact 3-dimensional region, and $\partial \mathcal{V}_3$ is its closed 2-dimensional boundary) and *Stokes' theorem*,

$$\boxed{\int_{\mathcal{V}_2} \mathbf{\nabla} \times \mathbf{A} \cdot d\mathbf{\Sigma} = \int_{\partial \mathcal{V}_2} \mathbf{A} \cdot d\mathbf{l}} \quad (1.28b)$$

(where $\mathcal{V}_2$ is a compact 2-dimensional region, $\partial \mathcal{V}_2$ is the 1-dimensional closed curve that bounds it, and the last integral is a line integral around that curve); see, e.g., Arfken, Weber, and Harris (2013).

This mathematics is illustrated by the integral and differential conservation laws for electric charge and for particles: The total charge and the total number of particles inside a 3-dimensional region of space $\mathcal{V}_3$ are $\int_{\mathcal{V}_3} \rho_e\,dV$ and $\int_{\mathcal{V}_3} n\,dV$, where ρ_e is the charge density and n the number density of particles. The rates that charge and particles flow out of $\mathcal{V}_3$ are the integrals of the current density **j** and the particle flux vector **S** over its boundary ∂V_3. Therefore, the *integral laws of charge conservation and particle conservation* are

integral conservation laws

$$\boxed{\frac{d}{dt} \int_{V_3} \rho_e \, dV + \int_{\partial V_3} \mathbf{j} \cdot d\mathbf{\Sigma} = 0,} \qquad \boxed{\frac{d}{dt} \int_{V_3} n \, dV + \int_{\partial V_3} \mathbf{S} \cdot d\mathbf{\Sigma} = 0.}$$ (1.29)

Pull the time derivative inside each volume integral (where it becomes a partial derivative), and apply Gauss's law to each surface integral; the results are $\int_{V_3} (\partial \rho_e / \partial t + \nabla \cdot \mathbf{j}) dV = 0$ and similarly for particles. The only way these equations can be true for all choices of V_3 is for the integrands to vanish:

differential conservation laws

$$\boxed{\partial \rho_e / \partial t + \nabla \cdot \mathbf{j} = 0,} \qquad \boxed{\partial n / \partial t + \nabla \cdot \mathbf{S} = 0.}$$ (1.30)

These are the *differential conservation laws for charge and for particles*. They have a standard, universal form: the time derivative of the density of a quantity plus the divergence of its flux vanishes.

Note that the integral conservation laws (1.29) and the differential conservation laws (1.30) require no coordinate system or basis for their description, and no coordinate system or basis was used in deriving the differential laws from the integral laws. This is an example of the fundamental principle that *the Newtonian physical laws are all expressible as geometric relationships among geometric objects*.

EXERCISES

Exercise 1.10 *Derivation and Practice: Volume Elements in Cartesian Coordinates*
Use Eqs. (1.25) and (1.26) to derive the usual formulas $dA = dx\,dy$ and $dV = dx\,dy\,dz$ for the 2-dimensional and 3-dimensional integration elements, respectively, in right-handed Cartesian coordinates. [Hint: Use as the edges of the integration volumes $dx\,\mathbf{e}_x$, $dy\,\mathbf{e}_y$, and $dz\,\mathbf{e}_z$.]

Exercise 1.11 *Example and Practice: Integral of a Vector Field over a Sphere*
Integrate the vector field $\mathbf{A} = z\mathbf{e}_z$ over a sphere with radius a, centered at the origin of the Cartesian coordinate system (i.e., compute $\int \mathbf{A} \cdot d\mathbf{\Sigma}$). Hints:

(a) Introduce spherical polar coordinates on the sphere, and construct the vectorial integration element $d\mathbf{\Sigma}$ from the two legs $a\,d\theta\,\mathbf{e}_{\hat{\theta}}$ and $a \sin\theta\,d\phi\,\mathbf{e}_{\hat{\phi}}$. Here $\mathbf{e}_{\hat{\theta}}$ and $\mathbf{e}_{\hat{\phi}}$ are unit-length vectors along the θ and ϕ directions. (Here as in Sec. 1.6 and throughout this book, we use accents on indices to indicate which basis the index is associated with: hats here for the spherical orthonormal basis, bars in Sec. 1.6 for the barred Cartesian basis.) Explain the factors $a\,d\theta$ and $a \sin\theta\,d\phi$ in the definitions of the legs. Show that

$$d\mathbf{\Sigma} = \boldsymbol{\epsilon}(\underline{}, \mathbf{e}_{\hat{\theta}}, \mathbf{e}_{\hat{\phi}}) a^2 \sin\theta\,d\theta\,d\phi.$$ (1.31)

(b) Using $z = a \cos\theta$ and $\mathbf{e}_z = \cos\theta\,\mathbf{e}_{\hat{r}} - \sin\theta\,\mathbf{e}_{\hat{\theta}}$ on the sphere (where $\mathbf{e}_{\hat{r}}$ is the unit vector pointing in the radial direction), show that

$$\mathbf{A} \cdot d\mathbf{\Sigma} = a \cos^2\theta\,\boldsymbol{\epsilon}(\mathbf{e}_{\hat{r}}, \mathbf{e}_{\hat{\theta}}, \mathbf{e}_{\hat{\phi}})\,a^2 \sin\theta\,d\theta\,d\phi.$$

(c) Explain why $\epsilon(\mathbf{e}_{\hat{r}}, \mathbf{e}_{\hat{\theta}}, \mathbf{e}_{\hat{\phi}}) = 1$.

(d) Perform the integral $\int \mathbf{A} \cdot d\mathbf{\Sigma}$ over the sphere's surface to obtain your final answer $(4\pi/3)a^3$. This, of course, is the volume of the sphere. Explain pictorially why this had to be the answer.

Exercise 1.12 *Example: Faraday's Law of Induction*
One of Maxwell's equations says that $\mathbf{\nabla} \times \mathbf{E} = -\partial \mathbf{B}/\partial t$ (in SI units), where $\mathbf{E}$ and $\mathbf{B}$ are the electric and magnetic fields. This is a geometric relationship between geometric objects; it requires no coordinates or basis for its statement. By integrating this equation over a 2-dimensional surface $\mathcal{V}_2$ with boundary curve $\partial \mathcal{V}_2$ and applying Stokes' theorem, derive Faraday's law of induction—again, a geometric relationship between geometric objects.

1.9 The Stress Tensor and Momentum Conservation

Press your hands together in the y-z plane and feel the force that one hand exerts on the other across a tiny area A—say, one square millimeter of your hands' palms (Fig. 1.5). That force, of course, is a vector $\mathbf{F}$. It has a normal component (along the x direction). It also has a tangential component: if you try to slide your hands past each other, you feel a component of force along their surface, a "shear" force in the y and z directions. Not only is the force $\mathbf{F}$ vectorial; so is the 2-surface across which it acts, $\mathbf{\Sigma} = A\,\mathbf{e}_x$. (Here $\mathbf{e}_x$ is the unit vector orthogonal to the tiny area A, and we have chosen the negative side of the surface to be the $-x$ side and the positive side to be $+x$. With this choice, the force $\mathbf{F}$ is that which the negative hand, on the $-x$ side, exerts on the positive hand.)

force vector

Now, it should be obvious that the force $\mathbf{F}$ is a linear function of our chosen surface $\mathbf{\Sigma}$. Therefore, there must be a tensor, the *stress tensor,* that reports the force to us when we insert the surface into its second slot:

stress tensor

$$\mathbf{F}(\underline{}) = \mathbf{T}(\underline{}, \mathbf{\Sigma}), \quad \text{or} \quad F_i = T_{ij}\Sigma_j. \tag{1.32}$$

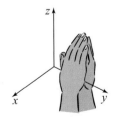

FIGURE 1.5 Hands, pressed together, exert a force on each other.

Newton's law of action and reaction tells us that the force that the positive hand exerts on the negative hand must be equal and opposite to that which the negative hand exerts on the positive. This shows up trivially in Eq. (1.32): by changing the sign of $\mathbf{\Sigma}$, one reverses which hand is regarded as negative and which positive, and since $\mathbf{T}$ is linear in $\mathbf{\Sigma}$, one also reverses the sign of the force.

The definition (1.32) of the stress tensor gives rise to the following physical meaning of its components:

$$T_{jk} = \left(\begin{array}{c} j \text{ component of force per unit area} \\ \text{across a surface perpendicular to } \mathbf{e}_k \end{array} \right)$$

$$= \left(\begin{array}{c} j \text{ component of momentum that crosses a unit} \\ \text{area that is perpendicular to } \mathbf{e}_k, \text{ per unit time,} \\ \text{with the crossing being from } -x_k \text{ to } +x_k \end{array} \right). \qquad (1.33)$$

meaning of components of stress tensor

The stresses inside a table with a heavy weight on it are described by the stress tensor $\mathbf{T}$, as are the stresses in a flowing fluid or plasma, in the electromagnetic field, and in any other physical medium. Accordingly, we shall use the stress tensor as an important mathematical tool in our study of force balance in kinetic theory (Chap. 3), elasticity (Part IV), fluid dynamics (Part V), and plasma physics (Part VI).

symmetry of stress tensor

It is not obvious from its definition, but the stress tensor $\mathbf{T}$ is always symmetric in its two slots. To see this, consider a small cube with side L in any medium (or field) (Fig. 1.6). The medium outside the cube exerts forces, and hence also torques, on the cube's faces. The z-component of the torque is produced by the shear forces on the front and back faces and on the left and right. As shown in the figure, the shear forces on the front and back faces have magnitudes $T_{xy}L^2$ and point in opposite directions, so they exert identical torques on the cube, $N_z = T_{xy}L^2(L/2)$ (where $L/2$ is the distance of each face from the cube's center). Similarly, the shear forces on the left and right faces have magnitudes $T_{yx}L^2$ and point in opposite directions, thereby exerting identical torques on the cube, $N_z = -T_{yx}L^2(L/2)$. Adding the torques from all four faces and equating them to the rate of change of angular momentum, $\frac{1}{6}\rho L^5 d\Omega_z/dt$ (where ρ is the mass density, $\frac{1}{6}\rho L^5$ is the cube's moment of inertia, and Ω_z is the z component of its angular velocity), we obtain $(T_{xy} - T_{yx})L^3 = \frac{1}{6}\rho L^5 d\Omega_z/dt$. Now, let the cube's edge length become arbitrarily small, $L \to 0$. If $T_{xy} - T_{yx}$ does not vanish, then the cube will be set into rotation with an infinitely large angular acceleration, $d\Omega_z/dt \propto 1/L^2 \to \infty$— an obviously unphysical behavior. Therefore, $T_{yx} = T_{xy}$, and similarly for all other components: *the stress tensor is always symmetric under interchange of its two slots.*

1.9.1 Examples: Electromagnetic Field and Perfect Fluid

Two examples will make the concept of the stress tensor more concrete.

- **Electromagnetic field:** See Ex. 1.14.

perfect fluid

- **Perfect fluid:** A *perfect fluid* is a medium that can exert an isotropic pressure P but no shear stresses, so the only nonzero components of its stress tensor

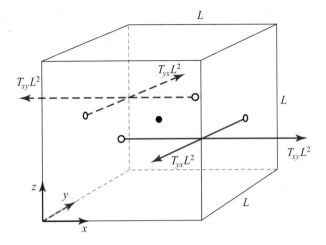

FIGURE 1.6 The shear forces exerted on the left, right, front, and back faces of a vanishingly small cube of side length L. The resulting torque about the z direction will set the cube into rotation with an arbitrarily large angular acceleration unless the stress tensor is symmetric.

in a Cartesian basis are $T_{xx} = T_{yy} = T_{zz} = P$. (Examples of nearly perfect fluids are air and water, but not molasses.) We can summarize this property by $T_{ij} = P\delta_{ij}$ or equivalently, since δ_{ij} are the components of the Euclidean metric, $T_{ij} = Pg_{ij}$. The frame-independent version of this is

$$\mathbf{T} = P\mathbf{g} \quad \text{or, in slot-naming index notation,} \quad T_{ij} = Pg_{ij}. \quad (1.34)$$

Note that, as always, the formula in slot-naming index notation looks identical to the formula $T_{ij} = Pg_{ij}$ for the components in our chosen Cartesian coordinate system. To check Eq. (1.34), consider a 2-surface $\mathbf{\Sigma} = A\mathbf{n}$ with area A oriented perpendicular to some arbitrary unit vector $\mathbf{n}$. The vectorial force that the fluid exerts across $\mathbf{\Sigma}$ is, in index notation, $F_j = T_{jk}\Sigma_k = Pg_{jk}An_k = PAn_j$ (i.e., it is a normal force with magnitude equal to the fluid pressure P times the surface area A). This is what it should be.

1.9.2 Conservation of Momentum

The stress tensor plays a central role in the Newtonian law of momentum conservation because (by definition) the force acting across a surface is the same as the rate of flow of momentum, per unit area, across the surface: *the stress tensor is the flux of momentum.*

Consider the 3-dimensional region of space $\mathcal{V}_3$ used above in formulating the integral laws of charge and particle conservation (1.29). The total momentum in $\mathcal{V}_3$ is $\int_{\mathcal{V}_3} \mathbf{G}\,dV$, where $\mathbf{G}$ is the momentum density. This quantity changes as a result of momentum flowing into and out of $\mathcal{V}_3$. The net rate at which momentum flows outward is the integral of the stress tensor over the surface $\partial\mathcal{V}_3$ of $\mathcal{V}_3$. Therefore, by analogy with charge and particle conservation (1.29), *the integral law of momentum conservation* says

$$\frac{d}{dt}\int_{\mathcal{V}_3}\mathbf{G}\,dV + \int_{\partial\mathcal{V}_3}\mathbf{T}\cdot d\mathbf{\Sigma} = 0. \tag{1.35}$$

By pulling the time derivative inside the volume integral (where it becomes a partial derivative) and applying the vectorial version of Gauss's law to the surface integral, we obtain $\int_{\mathcal{V}_3}(\partial\mathbf{G}/\partial t + \mathbf{\nabla}\cdot\mathbf{T})\,dV = 0$. This can be true for all choices of $\mathcal{V}_3$ only if the integrand vanishes:

$$\frac{\partial\mathbf{G}}{\partial t} + \mathbf{\nabla}\cdot\mathbf{T} = 0, \quad \text{or} \quad \frac{\partial G_j}{\partial t} + T_{jk;k} = 0. \tag{1.36}$$

(Because $\mathbf{T}$ is symmetric, it does not matter which of its slots the divergence acts on.) This is *the differential law of momentum conservation*. It has the standard form for any local conservation law: the time derivative of the density of some quantity (here momentum), plus the divergence of the flux of that quantity (here the momentum flux is the stress tensor), is zero. We shall make extensive use of this Newtonian law of momentum conservation in Part IV (elasticity), Part V (fluid dynamics), and Part VI (plasma physics).

EXERCISES

Exercise 1.13 **Example: Equations of Motion for a Perfect Fluid*
(a) Consider a perfect fluid with density ρ, pressure P, and velocity $\mathbf{v}$ that vary in time and space. Explain why the fluid's momentum density is $\mathbf{G} = \rho\mathbf{v}$, and explain why its momentum flux (stress tensor) is

$$\mathbf{T} = P\mathbf{g} + \rho\mathbf{v}\otimes\mathbf{v}, \quad \text{or, in slot-naming index notation,} \quad T_{ij} = Pg_{ij} + \rho v_i v_j. \tag{1.37a}$$

(b) Explain why the law of mass conservation for this fluid is

$$\frac{\partial\rho}{\partial t} + \mathbf{\nabla}\cdot(\rho\mathbf{v}) = 0. \tag{1.37b}$$

(c) Explain why the derivative operator

$$\frac{d}{dt} \equiv \frac{\partial}{\partial t} + \mathbf{v}\cdot\mathbf{\nabla} \tag{1.37c}$$

describes the rate of change as measured by somebody who moves locally with the fluid (i.e., with velocity $\mathbf{v}$). This is sometimes called the fluid's *advective time derivative* or *convective time derivative* or *material derivative*.

(d) Show that the fluid's law of mass conservation (1.37b) can be rewritten as

$$\frac{1}{\rho}\frac{d\rho}{dt} = -\nabla \cdot \mathbf{v}, \qquad\qquad (1.37d)$$

which says that the divergence of the fluid's velocity field is minus the fractional rate of change of its density, as measured in the fluid's local rest frame.

(e) Show that the differential law of momentum conservation (1.36) for the fluid can be written as

$$\frac{d\mathbf{v}}{dt} = -\frac{\nabla P}{\rho}. \qquad\qquad (1.37e)$$

This is called the fluid's *Euler equation*. Explain why this Euler equation is Newton's second law of motion, $\mathbf{F} = m\mathbf{a}$, written on a per unit mass basis.

In Part V of this book, we use Eqs. (1.37) to study the dynamical behaviors of fluids. For many applications, the Euler equation will need to be augmented by the force per unit mass exerted by the fluid's internal viscosity.

Exercise 1.14 **Problem: Electromagnetic Stress Tensor*
(a) An electric field $\mathbf{E}$ exerts (in SI units) a pressure $\epsilon_o \mathbf{E}^2/2$ orthogonal to itself and a tension of this same magnitude along itself. Similarly, a magnetic field $\mathbf{B}$ exerts a pressure $\mathbf{B}^2/2\mu_o = \epsilon_o c^2 \mathbf{B}^2/2$ orthogonal to itself and a tension of this same magnitude along itself. Verify that the following stress tensor embodies these stresses:

$$\boxed{\mathbf{T} = \frac{\epsilon_o}{2}\left[(\mathbf{E}^2 + c^2\mathbf{B}^2)\mathbf{g} - 2(\mathbf{E}\otimes\mathbf{E} + c^2\mathbf{B}\otimes\mathbf{B})\right].} \qquad (1.38)$$

(b) Consider an electromagnetic field interacting with a material that has a charge density ρ_e and a current density $\mathbf{j}$. Compute the divergence of the electromagnetic stress tensor (1.38) and evaluate the derivatives using Maxwell's equations. Show that the result is the negative of the force density that the electromagnetic field exerts on the material. Use momentum conservation to explain why this has to be so.

1.10 Geometrized Units and Relativistic Particles for Newtonian Readers

Readers who are skipping the relativistic parts of this book will need to know two important pieces of relativity: (i) geometrized units and (ii) the relativistic energy and momentum of a moving particle.

1.10.1 Geometrized Units

The speed of light is independent of one's reference frame (i.e., independent of how fast one moves). This is a fundamental tenet of special relativity, and in the era before 1983, when the meter and the second were defined independently, it was tested and confirmed experimentally with very high precision. By 1983, this constancy had become so universally accepted that it was used to redefine the meter (which is hard to measure precisely) in terms of the second (which is much easier to measure with modern technology).[11] The meter is now related to the second in such a way that the speed of light is precisely $c = 299{,}792{,}458$ m s^{-1} (i.e., 1 meter is the distance traveled by light in $1/299{,}792{,}458$ seconds). Because of this constancy of the light speed, it is permissible when studying special relativity to set c to unity. Doing so is equivalent to the relationship

$$c = 2.99792458 \times 10^8 \text{ m s}^{-1} = 1 \tag{1.39a}$$

between seconds and centimeters; i.e., equivalent to

$$1 \text{ s} = 2.99792458 \times 10^8 \text{ m}. \tag{1.39b}$$

geometrized units

We refer to units in which $c = 1$ as *geometrized units*, and we adopt them throughout this book when dealing with relativistic physics, since they make equations look much simpler. Occasionally it will be useful to restore the factors of c to an equation, thereby converting it to ordinary (SI or cgs) units. This restoration is achieved easily using dimensional considerations. For example, the equivalence of mass m and relativistic energy $\mathcal{E}$ is written in geometrized units as $\mathcal{E} = m$. In SI units $\mathcal{E}$ has dimensions of joule = kg m^2 s^{-2}, while m has dimensions of kg, so to make $\mathcal{E} = m$ dimensionally correct we must multiply the right side by a power of c that has dimensions m^2 s^{-2} (i.e., by c^2); thereby we obtain $\mathcal{E} = mc^2$.

1.10.2

1.10.2 Energy and Momentum of a Moving Particle

A particle with rest mass m, moving with velocity $\mathbf{v} = d\mathbf{x}/dt$ and speed $v = |\mathbf{v}|$, has a relativistic energy $\mathcal{E}$ (including its rest mass), relativistic kinetic energy E (excluding its rest mass), and relativistic momentum $\mathbf{p}$ given by

relativistic energy and momentum

$$\boxed{\mathcal{E} = \frac{m}{\sqrt{1 - v^2}} \equiv \frac{m}{\sqrt{1 - v^2/c^2}} \equiv E + m,} \qquad \boxed{\mathbf{p} = \mathcal{E}\mathbf{v} = \frac{m\mathbf{v}}{\sqrt{1 - v^2}};}$$

$$\text{(1.40)}$$

$$\text{so } \boxed{\mathcal{E} = \sqrt{m^2 + \mathbf{p}^2}.}$$

In the low-velocity (Newtonian) limit, the energy E with rest mass removed (kinetic energy) and the momentum $\mathbf{p}$ take their familiar Newtonian forms:

$$\text{When } v \ll c \equiv 1, \quad E \rightarrow \frac{1}{2}mv^2 \quad \text{and } \mathbf{p} \rightarrow m\mathbf{v}. \tag{1.41}$$

---

11. The second is defined as the duration of 9,192,631,770 periods of the radiation produced by a certain hyperfine transition in the ground state of a ^{133}Cs atom that is at rest in empty space. Today (2016) all fundamental physical units except mass units (e.g., the kilogram) are defined similarly in terms of fundamental constants of Nature.

A particle with zero rest mass (a photon or a graviton)[12] always moves with the speed of light $v = c = 1$, and like other particles it has momentum $\mathbf{p} = \mathcal{E}\mathbf{v}$, so the magnitude of its momentum is equal to its energy: $|\mathbf{p}| = \mathcal{E}v = \mathcal{E}c = \mathcal{E}$.

When particles interact (e.g., in chemical reactions, nuclear reactions, and elementary-particle collisions) the sum of the particle energies $\mathcal{E}$ is conserved, as is the sum of the particle momenta $\mathbf{p}$.

For further details and explanations, see Chap. 2.

EXERCISES

Exercise 1.15 *Practice: Geometrized Units*

Convert the following equations from the geometrized units in which they are written to SI units.

(a) The "Planck time" t_P expressed in terms of Newton's gravitation constant G and Planck's reduced constant $\hbar$, $t_P = \sqrt{G\hbar}$. What is the numerical value of t_P in seconds? in meters?

(b) The energy $\mathcal{E} = 2m$ obtained from the annihilation of an electron and a positron, each with rest mass m.

(c) The Lorentz force law $m d\mathbf{v}/dt = e(\mathbf{E} + \mathbf{v} \times \mathbf{B})$.

(d) The expression $\mathbf{p} = \hbar\omega\mathbf{n}$ for the momentum $\mathbf{p}$ of a photon in terms of its angular frequency ω and direction $\mathbf{n}$ of propagation.

How tall are you, in seconds? How old are you, in meters?

Bibliographic Note

Most of the concepts developed in this chapter are treated, though from rather different viewpoints, in intermediate and advanced textbooks on classical mechanics or electrodynamics, such as Marion and Thornton (1995); Jackson (1999); Griffiths (1999); Goldstein, Poole, and Safko (2002).

Landau and Lifshitz's (1976) advanced text *Mechanics* is famous for its concise and precise formulations; it lays heavy emphasis on symmetry principles and their implications. A similar approach is followed in the next volume in their Course of Theoretical Physics series, *The Classical Theory of Fields* (Landau and Lifshitz, 1975), which is rooted in special relativity and goes on to cover general relativity. We refer to other volumes in this remarkable series in subsequent chapters.

The three-volume *Feynman Lectures on Physics* (Feynman, Leighton, and Sands, 2013) had a big influence on several generations of physicists, and even more so on their teachers. Both of us (Blandford and Thorne) are immensely indebted to Richard Feynman for shaping our own approaches to physics. His insights on the foundations

---

12. We do not know for sure that photons and gravitons are massless, but the laws of physics as currently understood require them to be massless, and there are tight experimental limits on their rest masses.

of classical physics and its relationship to quantum mechanics, and on calculational techniques, are as relevant today as in 1963, when his course was first delivered.

The geometric viewpoint on the laws of physics, which we present and advocate in this chapter, is not common (but it should be because of its great power). For example, the vast majority of mechanics and electrodynamics textbooks, including all those listed above, define a tensor as a matrix-like entity whose components transform under rotations in the manner described by Eq. (1.13b). This is a complicated definition that hides the great simplicity of a tensor as nothing more than a linear function of vectors; it obscures thinking about tensors geometrically, without the aid of any coordinate system or basis.

The geometric viewpoint comes to the physics community from mathematicians, largely by way of relativity theory. By now, most relativity textbooks espouse it. See the Bibliographic Note to Chap. 2. Fortunately, this viewpoint is gradually seeping into the nonrelativistic physics curriculum (e.g., Kleppner and Kolenkow, 2013). We hope this chapter will accelerate that seepage.

2

Special Relativity: Geometric Viewpoint T2

Henceforth space by itself, and time by itself, are doomed to fade away into mere shadows,
and only a kind of union of the two will preserve an independent reality.

HERMANN MINKOWSKI, 1908

2.1 Overview

This chapter is a fairly complete introduction to special relativity at an intermediate
level. We extend the geometric viewpoint, developed in Chap. 1 for Newtonian physics, to the domain of special relativity; and we extend the tools of differential geometry,
developed in Chap. 1 for the arena of Newtonian physics (3-dimensional Euclidean
space) to that of special relativity (4-dimensional Minkowski spacetime).

We begin in Sec. 2.2 by defining inertial (Lorentz) reference frames and then introducing fundamental, geometric, reference-frame-independent concepts: events,
4-vectors, and the invariant interval between events. Then in Sec. 2.3, we develop
the basic concepts of tensor algebra in Minkowski spacetime (tensors, the metric
tensor, the inner product and tensor product, and contraction), patterning our development on the corresponding concepts in Euclidean space. In Sec. 2.4, we illustrate
our tensor-algebra tools by using them to describe—without any coordinate system or
reference frame—the kinematics (world lines, 4-velocities, 4-momenta) of point particles that move through Minkowski spacetime. The particles are allowed to collide
with one another and be accelerated by an electromagnetic field. In Sec. 2.5, we introduce components of vectors and tensors in an inertial reference frame and rewrite
our frame-independent equations in slot-naming index notation; then in Sec. 2.6,
we use these extended tensorial tools to restudy the motions, collisions, and electromagnetic accelerations of particles. In Sec. 2.7, we discuss Lorentz transformations in
Minkowski spacetime, and in Sec. 2.8, we develop spacetime diagrams and use them
to study length contraction, time dilation, and simultaneity breakdown. In Sec. 2.9,
we illustrate the tools we have developed by asking whether the laws of physics permit
a highly advanced civilization to build time machines for traveling backward in time
as well as forward. In Sec. 2.10, we introduce directional derivatives, gradients, and
the Levi-Civita tensor in Minkowski spacetime, and in Sec. 2.11, we use these tools to
discuss Maxwell's equations and the geometric nature of electric and magnetic fields.

BOX 2.1. READERS' GUIDE

- Parts II (Statistical Physics), III (Optics), IV (Elasticity), V (Fluid Dynamics), and VI (Plasma Physics) of this book deal almost entirely with Newtonian physics; only a few sections and exercises are relativistic. Readers who are inclined to skip those relativistic items (which are all labeled Track Two) can skip this chapter and then return to it just before embarking on Part VII (General Relativity). Accordingly, this chapter is Track Two for readers of Parts II–VI and Track One for readers of Part VII—and in this spirit we label it Track Two.

- More specifically, this chapter is a prerequisite for the following: sections on relativistic kinetic theory in Chap. 3, Sec. 13.8 on relativistic fluid dynamics, Ex. 17.9 on relativistic shocks in fluids, many comments in Parts II–VI about relativistic effects and connections between Newtonian physics and relativistic physics, and all of Part VII (General Relativity).

- We recommend that those readers for whom relativity is relevant— and who already have a strong understanding of special relativity— not skip this chapter entirely. Instead, we suggest they browse it, especially Secs. 2.2–2.4, 2.8, and 2.11–2.13, to make sure they understand this book's geometric viewpoint and to ensure their familiarity with such concepts as the stress-energy tensor that they might not have met previously.

In Sec. 2.12, we develop our final set of geometric tools: volume elements and the integration of tensors over spacetime; finally, in Sec. 2.13, we use these tools to define the stress-energy tensor and to formulate very general versions of the conservation of 4-momentum.

2.2 Foundational Concepts

2.2.1 Inertial Frames, Inertial Coordinates, Events, Vectors, and Spacetime Diagrams

Because the nature and geometry of Minkowski spacetime are far less obvious intuitively than those of Euclidean 3-space, we need a crutch in our development of the geometric viewpoint for physics in spacetime. That crutch will be inertial reference frames.

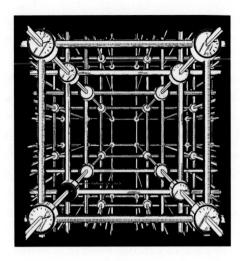

FIGURE 2.1 An inertial reference frame. From Taylor and Wheeler (1966). Used with permission of E. F. Taylor and the estate of J. A. Wheeler.

An inertial reference frame is a 3-dimensional latticework of measuring rods and clocks (Fig. 2.1) with the following properties:

inertial reference frame

- The latticework is purely conceptual and has arbitrarily small mass, so it does not gravitate.

- The latticework moves freely through spacetime (i.e., no forces act on it) and is attached to gyroscopes, so it is inertially nonrotating.

- The measuring rods form an orthogonal lattice, and the length intervals marked on them are uniform when compared to, for example, the wavelength of light emitted by some standard type of atom or molecule. Therefore, the rods form an orthonormal Cartesian coordinate system with the coordinate x measured along one axis, y along another, and z along the third.

- The clocks are densely packed throughout the latticework so that, ideally, there is a separate clock at every lattice point.

- The clocks tick uniformly when compared to the period of the light emitted by some standard type of atom or molecule (i.e., they are *ideal clocks*).

ideal clocks and their synchronization

- The clocks are synchronized by the Einstein synchronization process: if a pulse of light, emitted by one of the clocks, bounces off a mirror attached to another and then returns, the time of bounce t_b, as measured by the clock that does the bouncing, is the average of the times of emission and reception, as measured by the emitting and receiving clock: $t_b = \frac{1}{2}(t_e + t_r)$.[1]

---

1. For a deeper discussion of the nature of ideal clocks and ideal measuring rods see, for example, Misner, Thorne, and Wheeler (1973, pp. 23–29 and 395–399).

(That inertial frames with these properties can exist, when gravity is unimportant, is an empirical fact; it tells us that, in the absence of gravity, spacetime is truly Minkowski.)

Our first fundamental, frame-independent relativistic concept is the *event*. An event is a precise location in space at a precise moment of time—a precise location (or *point*) in 4-dimensional spacetime. We sometimes denote events by capital script letters, such as $\mathcal{P}$ and $\mathcal{Q}$—the same notation used for points in Euclidean 3-space.

A *4-vector* (also often referred to as a *vector in spacetime* or just a *vector*) is a straight[2] arrow $\Delta\vec{x}$ reaching from one event $\mathcal{P}$ to another, $\mathcal{Q}$. We often deal with 4-vectors and ordinary (3-space) vectors simultaneously, so we shall use different notations for them: boldface Roman font for 3-vectors, $\Delta\mathbf{x}$, and arrowed italic font for 4-vectors, $\Delta\vec{x}$. Sometimes we identify an event $\mathcal{P}$ in spacetime by its vectorial separation $\vec{x}_{\mathcal{P}}$ from some arbitrarily chosen event in spacetime, the origin $\mathcal{O}$.

An inertial reference frame provides us with a coordinate system for spacetime. The coordinates $(x^0, x^1, x^2, x^3) = (t, x, y, z)$ that it associates with an event $\mathcal{P}$ are $\mathcal{P}$'s location (x, y, z) in the frame's latticework of measuring rods and the time t of $\mathcal{P}$ *as measured by the clock that sits in the lattice at the event's location.* (Many apparent paradoxes in special relativity result from failing to remember that the time t of an event is always measured by a clock that resides at the event—never by clocks that reside elsewhere in spacetime.)

It is useful to depict events on *spacetime diagrams,* in which the time coordinate $t = x^0$ of some inertial frame is plotted upward; two of the frame's three spatial coordinates, $x = x^1$ and $y = x^2$, are plotted horizontally; and the third coordinate $z = x^3$ is omitted. Figure 2.2 is an example. Two events $\mathcal{P}$ and $\mathcal{Q}$ are shown there, along with their vectorial separations $\vec{x}_{\mathcal{P}}$ and $\vec{x}_{\mathcal{Q}}$ from the origin and the vector $\Delta\vec{x} = \vec{x}_{\mathcal{Q}} - \vec{x}_{\mathcal{P}}$ that separates them from each other. The coordinates of $\mathcal{P}$ and $\mathcal{Q}$, which are the same as the components of $\vec{x}_{\mathcal{P}}$ and $\vec{x}_{\mathcal{Q}}$ in this coordinate system, are $(t_{\mathcal{P}}, x_{\mathcal{P}}, y_{\mathcal{P}}, z_{\mathcal{P}})$ and $(t_{\mathcal{Q}}, x_{\mathcal{Q}}, y_{\mathcal{Q}}, z_{\mathcal{Q}})$. Correspondingly, the components of $\Delta\vec{x}$ are

$$\Delta x^0 = \Delta t = t_{\mathcal{Q}} - t_{\mathcal{P}}, \qquad \Delta x^1 = \Delta x = x_{\mathcal{Q}} - x_{\mathcal{P}},$$

$$\Delta x^2 = \Delta y = y_{\mathcal{Q}} - y_{\mathcal{P}}, \qquad \Delta x^3 = \Delta z = z_{\mathcal{Q}} - z_{\mathcal{P}}. \tag{2.1}$$

We denote these components of $\Delta\vec{x}$ more compactly by Δx^α, where the index α and all other lowercased Greek indices range from 0 (for t) to 3 (for z).

When the physics or geometry of a situation being studied suggests some preferred inertial frame (e.g., the frame in which some piece of experimental apparatus is at rest), then we typically use as axes for our spacetime diagrams the coordinates of that preferred frame. By contrast, when our situation provides no preferred inertial frame, or when we wish to emphasize a frame-independent viewpoint, we use as axes

---

2. By "straight" we mean that in any inertial reference frame, the coordinates along $\Delta\vec{x}$ are linear functions of one another.

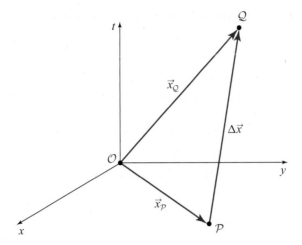

FIGURE 2.2 A spacetime diagram depicting two events $\mathcal{P}$ and $\mathcal{Q}$, their vectorial separations $\vec{x}_\mathcal{P}$ and $\vec{x}_\mathcal{Q}$ from an (arbitrarily chosen) origin $\mathcal{O}$, and the vector $\Delta\vec{x} = \vec{x}_\mathcal{Q} - \vec{x}_\mathcal{P}$ connecting them. The laws of physics cannot involve the arbitrary origin; we introduce it only as a conceptual aid.

the coordinates of a completely arbitrary inertial frame and think of the diagram as depicting spacetime in a coordinate-independent, frame-independent way.

We use the terms *inertial coordinate system* and *Lorentz coordinate system* inter-changeably[3] to mean the coordinate system (t, x, y, z) provided by an inertial frame; we also use the term *Lorentz frame* interchangeably with *inertial frame*. A physicist or other intelligent being who resides in a Lorentz frame and makes measurements using its latticework of rods and clocks will be called an *observer*.

inertial coordinates
(Lorentz coordinates)

observer

Although events are often described by their coordinates in a Lorentz reference frame, and 4-vectors by their components (coordinate differences), it should be obvious that the concepts of an event and a 4-vector need not rely on any coordinate system whatsoever for their definitions. For example, the event $\mathcal{P}$ of the birth of Isaac Newton and the event $\mathcal{Q}$ of the birth of Albert Einstein are readily identified without coordinates. They can be regarded as points in spacetime, and their separation vector is the straight arrow reaching through spacetime from $\mathcal{P}$ to $\mathcal{Q}$. Different observers in different inertial frames will attribute different coordinates to each birth and different components to the births' vectorial separation, but all observers can agree that they are talking about the same events $\mathcal{P}$ and $\mathcal{Q}$ in spacetime and the same separation vector $\Delta\vec{x}$. In this sense, $\mathcal{P}$, $\mathcal{Q}$, and $\Delta\vec{x}$ are *frame-independent, geometric objects* (points and arrows) that reside in spacetime.

---

3. It was Lorentz (1904) who first wrote down the relationship of one such coordinate system to another: the Lorentz transformation.

2.2.2 The Principle of Relativity and Constancy of Light Speed

Einstein's Principle of Relativity, stated in modern form, says that *Every (special relativistic) law of physics must be expressible as a geometric, frame-independent relationship among geometric, frame-independent objects* (i.e., such objects as points in spacetime and 4-vectors and tensors, which represent physical quantities, such as events, particle momenta, and the electromagnetic field). This is nothing but our Geometric Principle for physical laws (Chap. 1), lifted from the Euclidean-space arena of Newtonian physics to the Minkowski-spacetime arena of special relativity.

Since the laws are all geometric (i.e., unrelated to any reference frame or coordinate system), they can't distinguish one inertial reference frame from any other. This leads to an alternative form of the Principle of Relativity (one commonly used in elementary textbooks and equivalent to the above): *All the (special relativistic) laws of physics are the same in every inertial reference frame everywhere in spacetime.* This, in fact, is Einstein's own version of his Principle of Relativity; only in the sixty years since his death have we physicists reexpressed it in geometric language.

Because inertial reference frames are related to one another by Lorentz transformations (Sec. 2.7), we can restate Einstein's version of this Principle as *All the (special relativistic) laws of physics are Lorentz invariant.*

A more operational version of this Principle is: Give identical instructions for a specific physics experiment to two different observers in two different inertial reference frames at the same or different locations in Minkowski (i.e., gravity-free) spacetime. The experiment must be self-contained; that is, it must not involve observations of the external universe's properties (the "environment"). For example, an unacceptable experiment would be a measurement of the anisotropy of the universe's cosmic microwave radiation and a computation therefrom of the observer's velocity relative to the radiation's mean rest frame; such an experiment studies the universal environment, not the fundamental laws of physics. An acceptable experiment would be a measurement of the speed of light using the rods and clocks of the observer's own frame, or a measurement of cross sections for elementary particle reactions using particles moving in the reference frame's laboratory. The Principle of Relativity says that in these or any other similarly self-contained experiments, the two observers in their two different inertial frames must obtain identical experimental results—to within the accuracy of their experimental techniques. Since the experimental results are governed by the (nongravitational) laws of physics, this is equivalent to the statement that all physical laws are the same in the two inertial frames.

constancy of light speed

Perhaps the most central of special relativistic laws is the one stating that *The speed of light c in vacuum is frame independent;* that is, it is a constant, independent of the inertial reference frame in which it is measured. In other words, there is no "aether" that supports light's vibrations and in the process influences its speed—a remarkable fact that came as a great experimental surprise to physicists at the end of the nineteenth century.

The constancy of the speed of light, in fact, is built into Maxwell's equations. For these equations to be frame independent, the speed of light, which appears in them, must be frame independent. In this sense, the constancy of the speed of light follows from the Principle of Relativity; it is not an independent postulate. This is illustrated in Box 2.2.

T2

BOX 2.2. MEASURING THE SPEED OF LIGHT WITHOUT LIGHT

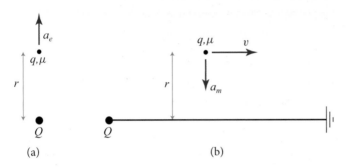

(a)   (b)

In some inertial reference frame, we perform two thought experiments using two particles, one with a large charge Q; the other, a test particle, with a much smaller charge q and mass μ. In the first experiment, we place the two particles at rest, separated by a distance $|\Delta x| \equiv r$, and measure the electrical repulsive acceleration a_e of q (panel a in the diagram). In Gaussian units (where the speed of light shows up explicitly instead of via $\epsilon_o \mu_o = 1/c^2$), the acceleration is $a_e = q Q/r^2 \mu$. In the second experiment, we connect Q to ground by a long wire, and we place q at the distance $|\Delta x| = r$ from the wire and set it moving at speed v parallel to the wire. The charge Q flows down the wire with an e-folding time τ, so the current is $I = dQ/d\tau = (Q/\tau)e^{-t/\tau}$. At early times $0 < t \ll \tau$, this current $I = Q/\tau$ produces a solenoidal magnetic field at q with field strength $B = (2/cr)(Q/\tau)$, and this field exerts a magnetic force on q, giving it an acceleration $a_m = q(v/c)B/\mu = 2vq Q/c^2\tau r \mu$. The ratio of the electric acceleration in the first experiment to the magnetic acceleration in the second experiment is $a_e/a_m = c^2\tau/2rv$. Therefore, we can measure the speed of light c in our chosen inertial frame by performing this pair of experiments; carefully measuring the separation r, speed v, current Q/τ, and accelerations; and then simply computing $c = \sqrt{(2rv/\tau)(a_e/a_m)}$. The Principle of Relativity insists that the result of this pair of experiments should be independent of the inertial frame in which they are performed. Therefore, the speed of light c that appears in Maxwell's equations must be frame independent. In this sense, the constancy of the speed of light follows from the Principle of Relativity as applied to Maxwell's equations.

What makes light so special? What about the propagation speeds of other types of waves? Are they or should they be the same as light's speed? For a digression on this topic, see Box 2.3.

The constancy of the speed of light underlies our ability to use the geometrized units introduced in Sec. 1.10. Any reader who has not studied that section should do so now. We use geometrized units throughout this chapter (and also throughout this book) when working with relativistic physics.

BOX 2.3. PROPAGATION SPEEDS OF OTHER WAVES

Electromagnetic radiation is not the only type of wave in Nature. In this book, we encounter dispersive media, such as optical fibers and plasmas, where electromagnetic signals travel slower than c. We also analyze sound waves and seismic waves, whose governing laws do not involve electromagnetism at all. How do these fit into our special relativistic framework? The answer is simple. Each of these waves involves an underlying medium that is at rest in one particular frame (not necessarily inertial), and the velocity at which the wave's information propagates (the group velocity) is most simply calculated in this frame *from the wave's and medium's fundamental laws*. We can then use the kinematic rules of Lorentz transformations to compute the velocity in another frame. However, if we had chosen to compute the wave speed in the second frame directly, using the same fundamental laws, we would have gotten the same answer, albeit perhaps with greater effort. All waves are in full compliance with the Principle of Relativity. What is special about vacuum electromagnetic waves and, by extension, photons, is that no medium (or "aether," as it used to be called) is needed for them to propagate. Their speed is therefore the same in all frames. (Although some physicists regard the cosmological constant, discussed in Chap. 28, as a modern aether, we must emphasize that, unlike its nineteenth-century antecedent, its presence does not alter the propagation of photons through Lorentz frames.)

This raises an interesting question. What about other waves that do not require an underlying medium? What about electron de Broglie waves? Here the fundamental wave equation, Schrödinger's or Dirac's, is mathematically different from Maxwell's and contains an important parameter, the electron rest mass. This rest mass allows the fundamental laws of relativistic quantum mechanics to be written in a form that is the same in all inertial reference frames and at the same time allows an electron, considered as either a wave or a particle, to travel at a different speed when measured in a different frame.

(continued)

BOX 2.3. (continued)

Some particles that have been postulated (such as gravitons, the quanta of gravitational waves; Chap. 27) are believed to exist without a rest mass (or an aether!), just like photons. Must these travel at the same speed as photons? The answer, according to the Principle of Relativity, is "yes." Why? Suppose there were two such waves or particles whose governing laws led to different speeds, c and $c' < c$, with each speed claimed to be the same in all reference frames. Such a claim produces insurmountable conundrums. For example, if we move with speed c' in the direction of propagation of the second wave, we will bring it to rest, in conflict with our hypothesis that its speed is frame independent. Therefore, all signals whose governing laws require them to travel with a speed that has no governing parameters (no rest mass and no underlying physical medium) must travel with a unique speed, which we call c. The speed of light is more fundamental to relativity than light itself!

2.2.3 The Interval and Its Invariance

Next we turn to another fundamental concept, the *interval* $(\Delta s)^2$ between the two events $\mathcal{P}$ and $\mathcal{Q}$ whose separation vector is $\Delta \vec{x}$. In a specific but arbitrary inertial reference frame and in geometrized units, $(\Delta s)^2$ is given by

$$(\Delta s)^2 \equiv -(\Delta t)^2 + (\Delta x)^2 + (\Delta y)^2 + (\Delta z)^2 = -(\Delta t)^2 + \sum_{i,j} \delta_{ij} \Delta x^i \Delta x^j;$$

(2.2a)

cf. Eq. (2.1). If $(\Delta s)^2 > 0$, the events $\mathcal{P}$ and $\mathcal{Q}$ are said to have a *spacelike* separation; if $(\Delta s)^2 = 0$, their separation is *null* or *lightlike;* and if $(\Delta s)^2 < 0$, their separation is *timelike*. For timelike separations, $(\Delta s)^2 < 0$ implies that Δs is imaginary; to avoid dealing with imaginary numbers, we describe timelike intervals by

$$(\Delta \tau)^2 \equiv -(\Delta s)^2,$$

(2.2b)

whose square root $\Delta \tau$ is real.

The coordinate separation between $\mathcal{P}$ and $\mathcal{Q}$ depends on one's reference frame: if $\Delta x^{\alpha'}$ and Δx^α are the coordinate separations in two different frames, then $\Delta x^{\alpha'} \neq$

Δx^α. Despite this frame dependence, the Principle of Relativity forces the interval $(\Delta s)^2$ to be the same in all frames:

$$(\Delta s)^2 = -(\Delta t)^2 + (\Delta x)^2 + (\Delta y)^2 + (\Delta z)^2$$
$$= -(\Delta t')^2 + (\Delta x')^2 + (\Delta y')^2 + (\Delta z')^2. \tag{2.3}$$

In Box 2.4, we sketch a proof for the case of two events $\mathcal{P}$ and $\mathcal{Q}$ whose separation is timelike.

Because of its frame invariance, the interval $(\Delta s)^2$ can be regarded as a geometric property of the vector $\Delta \vec{x}$ that reaches from $\mathcal{P}$ to $\mathcal{Q}$; we call it the *squared length* $(\Delta \vec{x})^2$ of $\Delta \vec{x}$:

$$(\Delta \vec{x})^2 \equiv (\Delta s)^2. \tag{2.4}$$

BOX 2.4. PROOF OF INVARIANCE OF THE INTERVAL FOR A TIMELIKE SEPARATION

A simple demonstration that the interval is invariant is provided by a thought experiment in which a photon is emitted at event $\mathcal{P}$, reflects off a mirror, and is then detected at event $\mathcal{Q}$. We consider the interval between these events in two reference frames, primed and unprimed, that move with respect to each other. Choose the spatial coordinate systems of the two frames in such a way that (i) their relative motion (with speed β, which will not enter into our analysis) is along the x and x' directions, (ii) event $\mathcal{P}$ lies on the x and x' axes, and (iii) event $\mathcal{Q}$ lies in the x-y and x'-y' planes, as depicted below. Then evaluate the interval between $\mathcal{P}$ and $\mathcal{Q}$ in the unprimed frame by the following construction: Place the mirror parallel to the x-z plane at precisely the height h that permits a photon, emitted from $\mathcal{P}$, to travel along the dashed line to the mirror, then reflect off the mirror and continue along the dashed path, arriving at event $\mathcal{Q}$. If the mirror were placed lower, the photon would arrive at the spatial location of $\mathcal{Q}$ sooner than the time of $\mathcal{Q}$; if placed higher, it would arrive later. Then the distance the photon travels (the length of the two-segment dashed line) is equal to $c\Delta t = \Delta t$, where Δt is the time between events $\mathcal{P}$ and $\mathcal{Q}$ as measured in the unprimed frame. If the mirror had not been present, the photon would have arrived at event $\mathcal{R}$ after time Δt, so $c\Delta t$ is the distance between $\mathcal{P}$ and $\mathcal{R}$. From the diagram, it is easy to see that the height of $\mathcal{R}$ above the x-axis is $2h - \Delta y$, and the Pythagorean theorem then implies that

$$(\Delta s)^2 = -(\Delta t)^2 + (\Delta x)^2 + (\Delta y)^2 = -(2h - \Delta y)^2 + (\Delta y)^2. \tag{1a}$$

The same construction in the primed frame must give the same formula, but with primes:

$$(\Delta s')^2 = -(\Delta t')^2 + (\Delta x')^2 + (\Delta y')^2 = -(2h' - \Delta y')^2 + (\Delta y')^2. \tag{1b}$$

(continued)

BOX 2.4. (continued)

The proof that $(\Delta s')^2 = (\Delta s)^2$ then reduces to showing that the Principle of Relativity requires that distances perpendicular to the direction of relative motion of two frames be the same as measured in the two frames: $h' = h$, $\Delta y' = \Delta y$. We leave it to the reader to develop a careful argument for this (Ex. 2.2).

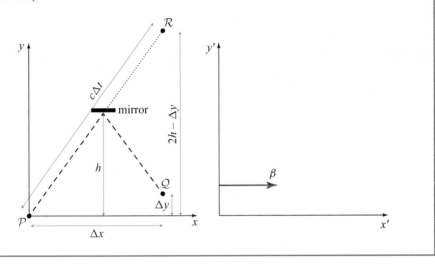

Note that this squared length, despite its name, can be negative (for timelike $\Delta \vec{x}$) or zero (for null $\Delta \vec{x}$) as well as positive (for spacelike $\Delta \vec{x}$).

The invariant interval $(\Delta s)^2$ between two events is as fundamental to Minkowski spacetime as the Euclidean distance between two points is to flat 3-space. Just as the Euclidean distance gives rise to the geometry of 3-space (as embodied, e.g., in Euclid's axioms), so the interval gives rise to the geometry of spacetime, which we shall be exploring. If this spacetime geometry were as intuitively obvious to humans as is Euclidean geometry, we would not need the crutch of inertial reference frames to arrive at it. Nature (presumably) has no need for such a crutch. To Nature (it seems evident), the geometry of Minkowski spacetime, as embodied in the invariant interval, is among the most fundamental aspects of physical law.

EXERCISES

Exercise 2.1 *Practice: Geometrized Units*
Do Ex. 1.15 in Chap. 1.

Exercise 2.2 *Derivation and Example: Invariance of the Interval*
Complete the derivation of the invariance of the interval given in Box 2.4, using the

Principle of Relativity in the form that the laws of physics must be the same in the primed and unprimed frames. Hints (if you need them):

(a) Having carried out the construction in the unprimed frame, depicted at the bottom left of Box 2.4, use the same mirror and photons for the analogous construction in the primed frame. Argue that, independently of the frame in which the mirror is at rest (unprimed or primed), the fact that the reflected photon has (angle of reflection) = (angle of incidence) in its rest frame implies that this is also true for the same photon in the other frame. Thereby conclude that the construction leads to Eq. (1b) in Box 2.4, as well as to Eq. (1a).

(b) Then argue that the perpendicular distance of an event from the common x- and x'-axes must be the same in the two reference frames, so $h' = h$ and $\Delta y' = \Delta y$; whence Eqs. (1b) and (1a) in Box 2.4 imply the invariance of the interval. [Note: For a leisurely version of this argument, see Taylor and Wheeler (1992, Secs. 3.6 and 3.7).]

2.3

2.3 Tensor Algebra without a Coordinate System

Having introduced points in spacetime (interpreted physically as events), the invariant interval $(\Delta s)^2$ between two events, 4-vectors (as arrows between two events), and the squared length of a vector (as the invariant interval between the vector's tail and tip), we can now introduce the remaining tools of tensor algebra for Minkowski spacetime in precisely the same way as we did for the Euclidean 3-space of Newtonian physics (Sec. 1.3), with the invariant interval between events playing the same role as the squared length between Euclidean points.

tensor

In particular: a *tensor* $\boldsymbol{T}(\_\_, \_\_, \_\_)$ is a real-valued linear function of vectors in Minkowski spacetime. (We use slanted letters $\boldsymbol{T}$ for tensors in spacetime and unslanted letters $\mathbf{T}$ in Euclidean space.) A tensor's *rank* is equal to its number of slots. The *inner product* (also called the dot product) of two 4-vectors is

inner product

$$\vec{A} \cdot \vec{B} \equiv \frac{1}{4}\left[(\vec{A} + \vec{B})^2 - (\vec{A} - \vec{B})^2\right], \tag{2.5}$$

where $(\vec{A} + \vec{B})^2$ is the squared length of this vector (i.e., the invariant interval between its tail and its tip). The *metric tensor* of spacetime is that linear function of 4-vectors whose value is the inner product of the vectors:

metric tensor

$$\boldsymbol{g}(\vec{A}, \vec{B}) \equiv \vec{A} \cdot \vec{B}. \tag{2.6}$$

Using the inner product, we can regard any vector $\vec{A}$ as a rank-1 tensor: $\vec{A}(\vec{C}) \equiv \vec{A} \cdot \vec{C}$.

tensor product

Similarly, the *tensor product* $\otimes$ is defined precisely as in the Euclidean domain, Eqs. (1.5), as is the *contraction* of two slots of a tensor against each other, Eqs. (1.6),

contraction

which lowers the tensor's rank by two.

2.4.1 Relativistic Particle Kinetics: World Lines, 4-Velocity, 4-Momentum and Its Conservation, 4-Force

In this section, we illustrate our geometric viewpoint by formulating the special relativistic laws of motion for particles.

An accelerated particle moving through spacetime carries an *ideal clock*. By "ideal" we mean that the clock is unaffected by accelerations: it ticks at a uniform rate when compared to unaccelerated atomic oscillators that are momentarily at rest beside the clock and are well protected from their environments. The builders of inertial guidance systems for airplanes and missiles try to make their clocks as ideal as possible in just this sense. We denote by τ the time ticked by the particle's ideal clock, and we call it the particle's *proper time*.

ideal clock

proper time

The particle moves through spacetime along a curve, called its *world line,* which we can denote equally well by $\mathcal{P}(\tau)$ (the particle's spacetime location $\mathcal{P}$ at proper time τ), or by $\vec{x}(\tau)$ (the particle's vector separation from some arbitrarily chosen origin at proper time τ).[4]

world line

We refer to the inertial frame in which the particle is momentarily at rest as its *momentarily comoving inertial frame* or *momentary rest frame*. Now, the particle's clock (which measures τ) is ideal, and so are the inertial frame's clocks (which measure coordinate time t). Therefore, a tiny interval $\Delta\tau$ of the particle's proper time is equal to the lapse of coordinate time in the particle's momentary rest frame $\Delta\tau = \Delta t$. Moreover, since the two events $\vec{x}(\tau)$ and $\vec{x}(\tau + \Delta\tau)$ on the clock's world line occur at the same spatial location in its momentary rest frame ($\Delta x^i = 0$, where $i = 1, 2, 3$) to first order in $\Delta\tau$, the invariant interval between those events is $(\Delta s)^2 = -(\Delta t)^2 + \sum_{i,j} \Delta x^i \Delta x^j \delta_{ij} = -(\Delta t)^2 = -(\Delta\tau)^2$. Thus, *the particle's proper time τ is equal to the square root of the negative of the invariant interval, $\tau = \sqrt{-s^2}$, along its world line.*

momentary rest frame

Figure 2.3 shows the world line of the accelerated particle in a spacetime diagram where the axes are coordinates of an arbitrary Lorentz frame. This diagram is intended to emphasize the world line as a frame-independent, geometric object. Also shown in the figure is the particle's *4-velocity* $\vec{u}$, which (by analogy with velocity in 3-space) is the time derivative of its position

4-velocity

$$\boxed{\vec{u} \equiv d\mathcal{P}/d\tau = d\vec{x}/d\tau} \tag{2.7}$$

and is the tangent vector to the world line. The derivative is defined by the usual limiting process

$$\frac{d\mathcal{P}}{d\tau} = \frac{d\vec{x}}{d\tau} \equiv \lim_{\Delta\tau \to 0} \frac{\mathcal{P}(\tau + \Delta\tau) - \mathcal{P}(\tau)}{\Delta\tau} = \lim_{\Delta\tau \to 0} \frac{\vec{x}(\tau + \Delta\tau) - \vec{x}(\tau)}{\Delta\tau}. \tag{2.8}$$

---

4. One of the basic ideas in string theory is that an elementary particle is described as a 1-dimensional loop in space rather than a 0-dimensional point. This means that it becomes a cylinder-like surface in spacetime—a world tube.

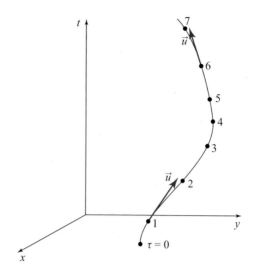

FIGURE 2.3 Spacetime diagram showing the world line $\vec{x}(\tau)$ and 4-velocity $\vec{u}$ of an accelerated particle. Note that the 4-velocity is tangent to the world line.

Here $\mathcal{P}(\tau + \Delta\tau) - \mathcal{P}(\tau)$ and $\vec{x}(\tau + \Delta\tau) - \vec{x}(\tau)$ are just two different ways to denote the same vector that reaches from one point on the world line to another.

The squared length of the particle's 4-velocity is easily seen to be -1:

$$\vec{u}^2 \equiv \boldsymbol{g}(\vec{u}, \vec{u}) = \frac{d\vec{x}}{d\tau} \cdot \frac{d\vec{x}}{d\tau} = \frac{d\vec{x} \cdot d\vec{x}}{(d\tau)^2} = -1. \tag{2.9}$$

The last equality follows from the fact that $d\vec{x} \cdot d\vec{x}$ is the squared length of $d\vec{x}$, which equals the invariant interval $(\Delta s)^2$ along it, and $(d\tau)^2$ is the negative of that invariant interval.

4-momentum

The particle's *4-momentum* is the product of its 4-velocity and rest mass:

$$\boxed{\vec{p} \equiv m\vec{u} = md\vec{x}/d\tau \equiv d\vec{x}/d\zeta.} \tag{2.10}$$

Here the parameter ζ is a renormalized version of proper time,

$$\zeta \equiv \tau/m. \tag{2.11}$$

affine parameter

This ζ and any other renormalized version of proper time with a position-independent renormalization factor are called *affine parameters* for the particle's world line. Expression (2.10), together with $\vec{u}^2 = -1$, implies that the squared length of the 4-momentum is

$$\boxed{\vec{p}^2 = -m^2.} \tag{2.12}$$

In quantum theory, a particle is described by a relativistic wave function, which, in the geometric optics limit (Chap. 7), has a wave vector $\vec{k}$ that is related to the classical particle's 4-momentum by

$$\vec{k} = \vec{p}/\hbar. \tag{2.13}$$

The above formalism is valid only for particles with nonzero rest mass, $m \neq 0$. The corresponding formalism for a *particle with zero rest mass* (e.g., a photon or a graviton) can be obtained from the above by taking the limit as $m \to 0$ and $d\tau \to 0$ with the quotient $d\zeta = d\tau/m$ held finite. More specifically, the 4-momentum of a zero-rest-mass particle is well defined (and participates in the conservation law to be discussed below), and it is expressible in terms of the particle's affine parameter ζ by Eq. (2.10):

$$\vec{p} = d\vec{x}/d\zeta. \tag{2.14}$$

By contrast, the particle's 4-velocity $\vec{u} = \vec{p}/m$ is infinite and thus undefined, and proper time $\tau = m\zeta$ ticks vanishingly slowly along its world line and thus is undefined. Because proper time is the square root of the invariant interval along the world line, the interval between two neighboring points on the world line vanishes. Therefore, *the world line of a zero-rest-mass particle is null*. (By contrast, since $d\tau^2 > 0$ and $ds^2 < 0$ along the world line of a particle with finite rest mass, *the world line of a finite-rest-mass particle is timelike*.)

The 4-momenta of particles are important because of the *law of conservation of 4-momentum* (which, as we shall see in Sec. 2.6, is equivalent to the conservation laws for energy and ordinary momentum): If a number of "initial" particles, named $A = 1, 2, 3, \ldots$, enter a restricted region of spacetime $\mathcal{V}$ and there interact strongly to produce a new set of "final" particles, named $\bar{A} = \bar{1}, \bar{2}, \bar{3}, \ldots$ (Fig. 2.4), then the total 4-momentum of the final particles must be the same as the total 4-momentum of the initial ones:

$$\boxed{\sum_{\bar{A}} \vec{p}_{\bar{A}} = \sum_{A} \vec{p}_{A}.} \tag{2.15}$$

Note that this law of 4-momentum conservation is expressed in frame-independent, geometric language—in accord with Einstein's insistence that all the laws of physics should be so expressible. As we shall see in Part VII, 4-momentum conservation is a consequence of the translation symmetry of flat, 4-dimensional spacetime. In general relativity's curved spacetime, where that translation symmetry is lost, we lose 4-momentum conservation except under special circumstances; see Eq. (25.56) and associated discussion.

If a particle moves freely (no external forces and no collisions with other particles), then its 4-momentum $\vec{p}$ will be conserved along its world line, $d\vec{p}/d\zeta = 0$. Since $\vec{p}$ is tangent to the world line, this conservation means that the direction of the world line in spacetime never changes: the free particle moves along a straight line through spacetime. To change the particle's 4-momentum, one must act on it with a *4-force* $\vec{F}$,

$$d\vec{p}/d\tau = \vec{F}. \tag{2.16}$$

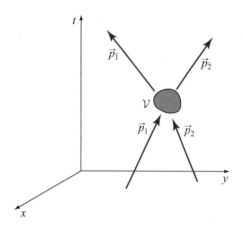

FIGURE 2.4 Spacetime diagram depicting the law of 4-momentum conservation for a situation where two particles, numbered 1 and 2, enter an interaction region $\mathcal{V}$ in spacetime, and there interact strongly and produce two new particles, numbered $\bar{1}$ and $\bar{2}$. The sum of the final 4-momenta, $\vec{p}_{\bar{1}} + \vec{p}_{\bar{2}}$, must be equal to the sum of the initial 4-momenta, $\vec{p}_1 + \vec{p}_2$.

If the particle is a fundamental one (e.g., photon, electron, proton), then the 4-force must leave its rest mass unchanged,

$$0 = dm^2/d\tau = -d\vec{p}^{\,2}/d\tau = -2\vec{p}\cdot d\vec{p}/d\tau = -2\vec{p}\cdot\vec{F}; \tag{2.17}$$

that is, the 4-force must be orthogonalxpage4-force|orthogonal to 4–velocity to the 4-momentum in the 4-dimensional sense that their inner product vanishes.

2.4.2 Geometric Derivation of the Lorentz Force Law

As an illustration of these physical concepts and mathematical tools, we use them to deduce the relativistic version of the Lorentz force law. From the outset, in accord with the Principle of Relativity, we insist that the law we seek be expressible in geometric, frame-independent language, that is, in terms of vectors and tensors.

electromagnetic field tensor

Consider a particle with charge q and rest mass $m \neq 0$ interacting with an electromagnetic field. It experiences an electromagnetic 4-force whose mathematical form we seek. The Newtonian version of the electromagnetic force $\mathbf{F} = q(\mathbf{E} + \mathbf{v} \times \mathbf{B})$ is proportional to q and contains one piece (electric) that is independent of velocity $\mathbf{v}$ and a second piece (magnetic) that is linear in $\mathbf{v}$. It is reasonable to expect that, to produce this Newtonian limit, the relativistic 4-force $\vec{F}$ will be proportional to q and will be linear in the 4-velocity $\vec{u}$. Linearity means there must exist some second-rank tensor $\mathbf{F}(\_\,, \_)$, the *electromagnetic field tensor*, such that

$$d\vec{p}/d\tau = \vec{F}(\_) = q\mathbf{F}(\_\,, \vec{u}). \tag{2.18}$$

Because the 4-force $\vec{F}$ must be orthogonal to the particle's 4-momentum and thence also to its 4-velocity, $\vec{F} \cdot \vec{u} \equiv \mathbf{F}(\vec{u}) = 0$, expression (2.18) must vanish when $\vec{u}$ is inserted into its empty slot. In other words, for all timelike unit-length vectors $\vec{u}$,

$$\mathbf{F}(\vec{u}, \vec{u}) = 0. \tag{2.19}$$

It is an instructive exercise (Ex. 2.3) to show that this is possible only if $\mathbf{F}$ is *antisymmetric*, so the electromagnetic 4-force is

$$\boxed{d\vec{p}/d\tau = q\mathbf{F}(\underline{\quad}, \vec{u}), \quad \text{where} \quad \mathbf{F}(\vec{A}, \vec{B}) = -\mathbf{F}(\vec{B}, \vec{A}) \quad \text{for all } \vec{A} \text{ and } \vec{B}.} \tag{2.20}$$

electromagnetic 4-force

Equation (2.20) must be the relativistic form of the Lorentz force law. In Sec. 2.11 we deduce the relationship of the electromagnetic field tensor $\mathbf{F}$ to the more familiar electric and magnetic fields, and the relationship of this relativistic Lorentz force to its Newtonian form (1.7c).

The discussion of particle kinematics and the electromagnetic force in this Sec. 2.4 is elegant but perhaps unfamiliar. In Secs. 2.6 and 2.11, we shall see that it is equivalent to the more elementary (but more complex) formalism based on components of vectors in Euclidean 3-space.

EXERCISES

Exercise 2.3 *Derivation and Example: Antisymmetry of Electromagnetic Field Tensor*
Show that Eq. (2.19) can be true for all timelike, unit-length vectors $\vec{u}$ if and only if $\mathbf{F}$ is antisymmetric. [Hints: (i) Show that the most general second-rank tensor $\mathbf{F}$ can be written as the sum of a symmetric tensor $\mathbf{S}$ and an antisymmetric tensor $\mathbf{A}$, and that the antisymmetric piece contributes nothing to Eq. (2.19), so $\mathbf{S}(\vec{u}, \vec{u})$ must vanish for every timelike $\vec{u}$. (ii) Let $\vec{a}$ be a timelike vector, let $\vec{b}$ be an artibrary vector (timelike, null, or spacelike), and let ϵ be a number small enough that $\vec{A}_{\pm} \equiv \vec{a} \pm \epsilon \vec{b}$ are both timelike. From the fact that $\mathbf{S}(\vec{A}_{+}, \vec{A}_{+})$, $\mathbf{S}(\vec{A}_{-}, \vec{A}_{-})$, and $\mathbf{S}(\vec{a}, \vec{a})$ all vanish, deduce that $\mathbf{S}(\vec{b}, \vec{b}) = 0$ for the arbitrary vector $\vec{b}$. (iii) From this, deduce that $\mathbf{S}$ vanishes (i.e., it gives zero when any two vectors are inserted into its slots).]

Exercise 2.4 *Problem: Relativistic Gravitational Force Law*
In Newtonian theory, the gravitational potential Φ exerts a force $\mathbf{F} = d\mathbf{p}/dt = -m\nabla\Phi$ on a particle with mass m and momentum $\mathbf{p}$. Before Einstein formulated general relativity, some physicists constructed relativistic theories of gravity in which a Newtonian-like scalar gravitational field Φ exerted a 4-force $\vec{F} = d\vec{p}/d\tau$ on any particle with rest mass m, 4-velocity $\vec{u}$, and 4-momentum $\vec{p} = m\vec{u}$. What must that force law have been for it to (i) obey the Principle of Relativity, (ii) reduce to Newton's law in the nonrelativistic limit, and (iii) preserve the particle's rest mass as time passes?

2.5.1   ### 2.5.1 Lorentz Coordinates

In Minkowski spacetime, associated with any inertial reference frame (Fig. 2.1 and Sec. 2.2.1), there is a Lorentz coordinate system $\{t, x, y, z\} = \{x^0, x^1, x^2, x^3\}$ generated by the frame's rods and clocks. (Note the use of superscripts.) And associated

Lorentz (orthonormal) basis

with these coordinates is a set of *Lorentz basis vectors* $\{\vec{e}_t, \vec{e}_x, \vec{e}_y, \vec{e}_z\} = \{\vec{e}_0, \vec{e}_1, \vec{e}_2, \vec{e}_3\}$. (Note the use of subscripts. The reason for this convention will become clear below.) The basis vector $\vec{e}_\alpha$ points along the x^α coordinate direction, which is orthogonal to all the other coordinate directions, and it has squared length -1 for $\alpha = 0$ (vector pointing in a timelike direction) and $+1$ for $\alpha = 1, 2, 3$ (spacelike):

$$\vec{e}_\alpha \cdot \vec{e}_\beta = \eta_{\alpha\beta}. \tag{2.21}$$

Here $\eta_{\alpha\beta}$ (a spacetime analog of the Kronecker delta) are defined by

$$\eta_{00} \equiv -1, \qquad \eta_{11} \equiv \eta_{22} \equiv \eta_{33} \equiv 1, \qquad \eta_{\alpha\beta} \equiv 0 \text{ if } \alpha \neq \beta. \tag{2.22}$$

Any basis in which $\vec{e}_\alpha \cdot \vec{e}_\beta = \eta_{\alpha\beta}$ is said to be *orthonormal* (by analogy with the Euclidean notion of orthonormality, $\mathbf{e}_j \cdot \mathbf{e}_k = \delta_{jk}$).

Because $\vec{e}_\alpha \cdot \vec{e}_\beta \neq \delta_{\alpha\beta}$, many of the Euclidean-space component-manipulation formulas (1.9b)–(1.9h) do not hold in Minkowski spacetime. There are two approaches to recovering these formulas. One approach, used in many older textbooks (including the first and second editions of Goldstein's *Classical Mechanics* and Jackson's *Classical Electrodynamics*), is to set $x^0 = it$, where $i = \sqrt{-1}$ and correspondingly make the time basis vector be imaginary, so that $\vec{e}_\alpha \cdot \vec{e}_\beta = \delta_{\alpha\beta}$. When this approach is adopted, the resulting formalism does not depend on whether indices are placed up or down; one can place them wherever one's stomach or liver dictates without asking one's brain. However, this $x^0 = it$ approach has severe disadvantages: (i) it hides the true physical geometry of Minkowski spacetime, (ii) it cannot be extended in any reasonable manner to nonorthonormal bases in flat spacetime, and (iii) it cannot be extended in any reasonable manner to the curvilinear coordinates that must be used in general relativity. For these reasons, most modern texts (including the third editions of Goldstein and Jackson) take an alternative approach, one always used in general relativity. This alternative, which we shall adopt, requires introducing two different types of components for vectors (and analogously for tensors): *contravariant components* denoted by superscripts (e.g., $T^{\alpha\beta\gamma}$) and *covariant components* denoted by subscripts (e.g., $T_{\alpha\beta\gamma}$). In Parts I–VI of this book, we introduce these components only for orthonormal bases; in Part VII, we develop a more sophisticated version of them, valid for nonorthonormal bases.

2.5.2   ### 2.5.2 Index Gymnastics

contravariant components

A vector's or tensor's contravariant components are defined as its expansion coefficients in the chosen basis [analogs of Eqs. (1.9b) and (1.9d) in Euclidean 3-space]:

$$\vec{A} \equiv A^\alpha \vec{e}_\alpha, \qquad \mathbf{T} \equiv T^{\alpha\beta\gamma} \vec{e}_\alpha \otimes \vec{e}_\beta \otimes \vec{e}_\gamma. \qquad (2.23a)$$

Here and throughout this book, *Greek (spacetime) indices are to be summed when they are repeated with one up and the other down*; this is called the Einstein summation convention.

The covariant components are defined as the numbers produced by evaluating the vector or tensor on its basis vectors [analog of Eq. (1.9e) in Euclidean 3-space]:

$$A_\alpha \equiv \vec{A}(\vec{e}_\alpha) = \vec{A} \cdot \vec{e}_\alpha, \qquad T_{\alpha\beta\gamma} \equiv \mathbf{T}(\vec{e}_\alpha, \vec{e}_\beta, \vec{e}_\gamma). \qquad (2.23b)$$

(Just as there are contravariant and covariant components A^α and A_α, so also there is a second set of basis vectors $\vec{e}^\alpha$ dual to the set $\vec{e}_\alpha$. However, for economy of notation we delay introducing them until Part VII.)

These definitions have a number of important consequences. We derive them one after another and then summarize them succinctly with equation numbers:

(i) The covariant components of the metric tensor are $g_{\alpha\beta} = \mathbf{g}(\vec{e}_\alpha, \vec{e}_\beta) = \vec{e}_\alpha \cdot \vec{e}_\beta = \eta_{\alpha\beta}$. Here the first equality is the definition (2.23b) of the covariant components, the second equality is the definition (2.6) of the metric tensor, and the third equality is the orthonormality relation (2.21) for the basis vectors.

(ii) The covariant components of any tensor can be computed from the contravariant components by

$$T_{\lambda\mu\nu} = \mathbf{T}(\vec{e}_\lambda, \vec{e}_\mu, \vec{e}_\nu) = T^{\alpha\beta\gamma} \vec{e}_\alpha \otimes \vec{e}_\beta \otimes \vec{e}_\gamma (\vec{e}_\lambda, \vec{e}_\mu, \vec{e}_\nu)$$

$$= T^{\alpha\beta\gamma} (\vec{e}_\alpha \cdot \vec{e}_\lambda)(\vec{e}_\beta \cdot \vec{e}_\mu)(\vec{e}_\gamma \cdot \vec{e}_\nu) = T^{\alpha\beta\gamma} g_{\alpha\lambda} g_{\beta\mu} g_{\gamma\nu}.$$

The first equality is the definition (2.23b) of the covariant components, the second is the expansion (2.23a) of $\mathbf{T}$ on the chosen basis, the third is the definition (1.5a) of the tensor product, and the fourth is one version of our result (i) for the covariant components of the metric.

(iii) This result, $T_{\lambda\mu\nu} = T^{\alpha\beta\gamma} g_{\alpha\lambda} g_{\beta\mu} g_{\gamma\nu}$, together with the numerical values (i) of $g_{\alpha\beta}$, implies that when one lowers a spatial index there is no change in the numerical value of a component, and when one lowers a temporal index, the sign changes: $T_{ijk} = T^{ijk}, T_{0jk} = -T^{0jk}, T_{0j0} = +T^{0j0}, T_{000} = -T^{000}$. We call this the "sign-flip-if-temporal" rule. As a special case, $-1 = g_{00} = g^{00}, 0 = g_{0j} = -g^{0j}, \delta_{jk} = g_{jk} = g^{jk}$—that is, the metric's covariant and contravariant components are numerically identical; they are both equal to the orthonormality values $\eta_{\alpha\beta}$.

(iv) It is easy to see that this sign-flip-if-temporal rule for lowering indices implies the same sign-flip-if-temporal rule for raising them, which in turn can be written in terms of metric components as $T^{\alpha\beta\gamma} = T_{\lambda\mu\nu} g^{\lambda\alpha} g^{\mu\beta} g^{\nu\gamma}$.

(v) It is convenient to define *mixed components* of a tensor, components with some indices up and others down, as having numerical values obtained by raising or lowering some but not all of its indices using the metric, for example, $T^{\alpha}{}_{\mu\nu} = T^{\alpha\beta\gamma} g_{\beta\mu} g_{\gamma\nu} = T_{\lambda\mu\nu} g^{\lambda\alpha}$. Numerically, this continues to follow the sign-flip-if-temporal rule: $T^{0}{}_{0k} = -T^{00k}$, $T^{0}{}_{jk} = T^{0jk}$, and it implies, in particular, that the mixed components of the metric are $g^{\alpha}{}_{\beta} = \delta_{\alpha\beta}$ (the Kronecker-delta values; $+1$ if $\alpha = \beta$ and 0 otherwise).

These important results can be summarized as follows. *The numerical values of the components of the metric in Minkowski spacetime are expressed in terms of the matrices* $[\delta_{\alpha\beta}]$ *and* $[\eta_{\alpha\beta}]$ *as*

$$g_{\alpha\beta} = \eta_{\alpha\beta}, \quad g^{\alpha}{}_{\beta} = \delta_{\alpha\beta}, \quad g_{\alpha}{}^{\beta} = \delta_{\alpha\beta}, \quad g^{\alpha\beta} = \eta_{\alpha\beta}; \tag{2.23c}$$

indices on all vectors and tensors can be raised and lowered using these components of the metric:

$$A_{\alpha} = g_{\alpha\beta} A^{\beta}, \quad A^{\alpha} = g^{\alpha\beta} A_{\beta}, \quad T^{\alpha}{}_{\mu\nu} \equiv g_{\mu\beta} g_{\nu\gamma} T^{\alpha\beta\gamma}, \quad T^{\alpha\beta\gamma} \equiv g^{\beta\mu} g^{\gamma\nu} T^{\alpha}{}_{\mu\nu}, \tag{2.23d}$$

which is equivalent to the sign-flip-if-temporal rule.

This index notation gives rise to formulas for tensor products, inner products, values of tensors on vectors, and tensor contractions that are obvious analogs of those in Euclidean space:

$$[\text{Contravariant components of } \mathbf{T}(\_,\_,\_) \otimes \mathbf{S}(\_,\_)] = T^{\alpha\beta\gamma} S^{\delta\epsilon}, \tag{2.23e}$$

$$\vec{A} \cdot \vec{B} = A^{\alpha} B_{\alpha} = A_{\alpha} B^{\alpha}, \quad \mathbf{T}(\mathbf{A}, \mathbf{B}, \mathbf{C}) = T_{\alpha\beta\gamma} A^{\alpha} B^{\beta} C^{\gamma} = T^{\alpha\beta\gamma} A_{\alpha} B_{\beta} C_{\gamma}, \tag{2.23f}$$

$$\text{Covariant components of } [1\&3\text{contraction of } \mathbf{R}] = R^{\mu}{}_{\alpha\mu\beta},$$

$$\text{Contravariant components of } [1\&3\text{contraction of } \mathbf{R}] = R^{\mu\alpha}{}_{\mu}{}^{\beta}. \tag{2.23g}$$

Notice the very simple pattern in Eqs. (2.23b) and (2.23d), which universally permeates the rules of index gymnastics, a pattern that permits one to reconstruct the rules without any memorization: *Free indices (indices not summed over) must agree in position (up versus down) on the two sides of each equation.* In keeping with this pattern, one can regard the two indices in a pair that is summed as "annihilating each other by contraction," and one speaks of "lining up the indices" on the two sides of an equation to get them to agree. These rules provide helpful checks when performing calculations.

In Part VII, when we use nonorthonormal bases, all these index-notation equations (2.23) will remain valid except for the numerical values [Eq. (2.23c)] of the metric components and the sign-flip-if-temporal rule.

2.5.3 Slot-Naming Notation

In Minkowski spacetime, as in Euclidean space, we can (and often do) use slot-naming index notation to represent frame-independent geometric objects and equations and physical laws. (Readers who have not studied Sec. 1.5.1 on slot-naming index notation should do so now.)

For example, we often write the frame-independent Lorentz force law $d\vec{p}/d\tau = q\mathbf{F}(\_\_, \vec{u})$ [Eq. (2.20)] as $dp_\mu/d\tau = qF_{\mu\nu}u^\nu$.

Notice that, because the components of the metric in any Lorentz basis are $g_{\alpha\beta} = \eta_{\alpha\beta}$, we can write the invariant interval between two events x^α and $x^\alpha + dx^\alpha$ as

$$ds^2 = g_{\alpha\beta}dx^\alpha dx^\beta = -dt^2 + dx^2 + dy^2 + dz^2. \tag{2.24}$$

This is called the special relativistic *line element*.

line element

EXERCISES

Exercise 2.5 *Derivation: Component Manipulation Rules*
Derive the relativistic component manipulation rules (2.23e)–(2.23g).

Exercise 2.6 *Practice: Numerics of Component Manipulations*
In some inertial reference frame, the vector $\vec{A}$ and second-rank tensor $\mathbf{T}$ have as their only nonzero components $A^0 = 1$, $A^1 = 2$; $T^{00} = 3$, $T^{01} = T^{10} = 2$, $T^{11} = -1$. Evaluate $\mathbf{T}(\vec{A}, \vec{A})$ and the components of $\mathbf{T}(\vec{A}, \_\_)$ and $\vec{A} \otimes \mathbf{T}$.

Exercise 2.7 *Practice: Meaning of Slot-Naming Index Notation*
(a) Convert the following expressions and equations into geometric, index-free notation: $A^\alpha B_{\gamma\delta}$; $A_\alpha B_\gamma{}^\delta$; $S_\alpha{}^{\beta\gamma} = S^{\gamma\beta}{}_\alpha$; $A^\alpha B_\alpha = A_\alpha B^\beta g^\alpha{}_\beta$.
(b) Convert $\mathbf{T}(\_\_, \mathbf{S}(\mathbf{R}(\vec{C}, \_\_), \_\_), \_\_)$ into slot-naming index notation.

Exercise 2.8 *Practice: Index Gymnastics*
(a) Simplify the following expression so the metric does not appear in it:
$$A^{\alpha\beta\gamma}g_{\beta\rho}S_{\gamma\lambda}g^{\rho\delta}g^\lambda{}_\alpha.$$
(b) The quantity $g_{\alpha\beta}g^{\alpha\beta}$ is a scalar since it has no free indices. What is its numerical value?
(c) What is wrong with the following expression and equation?
$$A_\alpha{}^{\beta\gamma}S_{\alpha\gamma}; \qquad A_\alpha{}^{\beta\gamma}S_\beta T_\gamma = R_{\alpha\beta\delta}S^\beta.$$

2.6 Particle Kinetics in Index Notation and in a Lorentz Frame

2.6

As an illustration of the component representation of tensor algebra, let us return to the relativistic, accelerated particle of Fig. 2.3 and, from the frame-independent equations for the particle's 4-velocity $\vec{u}$ and 4-momentum $\vec{p}$ (Sec. 2.4), derive the component description given in elementary textbooks.

We introduce a specific inertial reference frame and associated Lorentz coordinates x^α and basis vectors $\{\vec{e}_\alpha\}$. In this Lorentz frame, the particle's world line

$\vec{x}(\tau)$ is represented by its coordinate location $x^\alpha(\tau)$ as a function of its proper time τ. The contravariant components of the separation vector $d\vec{x}$ between two neighboring events along the particle's world line are the events' coordinate separations dx^α [Eq. (2.1)]; and correspondingly, the components of the particle's 4-velocity $\vec{u} = d\vec{x}/d\tau$ are

$$u^\alpha = dx^\alpha/d\tau \qquad (2.25\text{a})$$

(the time derivatives of the particle's spacetime coordinates). Note that Eq. (2.25a) implies

$$v^j \equiv \frac{dx^j}{dt} = \frac{dx^j/d\tau}{dt/d\tau} = \frac{u^j}{u^0}. \qquad (2.25\text{b})$$

This relation, together with $-1 = \vec{u}^2 = g_{\alpha\beta}u^\alpha u^\beta = -(u^0)^2 + \delta_{ij}u^i u^j = -(u^0)^2(1 - \delta_{ij}v^i v^j)$, implies that the components of the 4-velocity have the forms familiar from elementary textbooks:

$$u^0 = \gamma, \quad u^j = \gamma v^j, \quad \text{where} \quad \gamma = \frac{1}{(1 - \delta_{ij}v^i v^j)^{\frac{1}{2}}}. \qquad (2.25\text{c})$$

It is useful to think of v^j as the components of a 3-dimensional vector $\mathbf{v}$, the *ordinary velocity*, that lives in the 3-dimensional Euclidean space $t = \text{const}$ of the chosen Lorentz frame (the green plane in Fig. 2.5). This 3-space is sometimes called the frame's *slice of simultaneity* or *3-space of simultaneity*, because all events lying in it are simultaneous, as measured by the frame's observers. This 3-space is not well defined until a Lorentz frame has been chosen, and correspondingly, $\mathbf{v}$ relies for its existence on a specific choice of frame. However, once the frame has been chosen, $\mathbf{v}$ can be regarded as a coordinate-independent, basis-independent 3-vector lying in the frame's slice of simultaneity. Similarly, the spatial part of the 4-velocity $\vec{u}$ (the part with components u^j in our chosen frame) can be regarded as a 3-vector $\mathbf{u}$ lying in the frame's 3-space; and Eqs. (2.25c) become the component versions of the coordinate-independent, basis-independent 3-space relations

$$\mathbf{u} = \gamma\mathbf{v}, \quad \gamma = \frac{1}{\sqrt{1 - \mathbf{v}^2}}, \qquad (2.25\text{d})$$

where $\mathbf{v}^2 = \mathbf{v} \cdot \mathbf{v}$. This γ is called the "Lorentz factor."

The components of the particle's 4-momentum $\vec{p}$ in our chosen Lorentz frame have special names and special physical significances: The time component of the 4-momentum is the particle's (relativistic) *energy* $\mathcal{E}$ as measured in that frame:

$$\mathcal{E} \equiv p^0 = mu^0 = m\gamma = \frac{m}{\sqrt{1 - \mathbf{v}^2}} = \text{(the particle's energy)}$$

$$\simeq m + \frac{1}{2}m\mathbf{v}^2 \quad \text{for } |\mathbf{v}| \ll 1. \qquad (2.26\text{a})$$

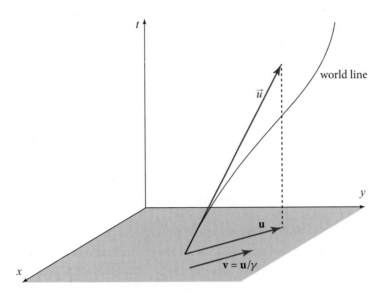

FIGURE 2.5 Spacetime diagram in a specific Lorentz frame, showing the frame's 3-space $t = 0$ (green region), the world line of a particle, the 4-velocity $\vec{u}$ of the particle as it passes through the 3-space, and two 3-dimensional vectors that lie in the 3-space: the spatial part $\mathbf{u}$ of the particle's 4-velocity and the particle's ordinary velocity $\mathbf{v}$.

Note that this energy is the sum of the particle's *rest mass-energy* $m = mc^2$ and its *kinetic energy*

$$E \equiv \mathcal{E} - m = m \left(\frac{1}{\sqrt{1 - \mathbf{v}^2}} - 1 \right)$$

$$\simeq \frac{1}{2} m \mathbf{v}^2 \quad \text{for } |\mathbf{v}| \ll 1. \tag{2.26b}$$

The spatial components of the 4-momentum, when regarded from the viewpoint of 3-dimensional physics, are the same as the components of the *momentum*, a 3-vector residing in the chosen Lorentz frame's 3-space:

$$p^j = mu^j = m\gamma v^j = \frac{mv^j}{\sqrt{1 - \mathbf{v}^2}} = \mathcal{E} v_j$$

$$= (j \text{ component of particle's momentum}); \tag{2.26c}$$

or, in basis-independent, 3-dimensional vector notation,

$$\mathbf{p} = m\mathbf{u} = m\gamma \mathbf{v} = \frac{m\mathbf{v}}{\sqrt{1 - \mathbf{v}^2}} = \mathcal{E}\mathbf{v} = (\text{particle's momentum}). \tag{2.26d}$$

For a zero-rest-mass particle, as for one with finite rest mass, we identify the time component of the 4-momentum, in a chosen Lorentz frame, as the particle's energy, and the spatial part as its momentum. Moreover, if—appealing to quantum theory—we regard a zero-rest-mass particle as a quantum associated with a monochromatic wave, then quantum theory tells us that the wave's angular frequency ω as measured in a chosen Lorentz frame is related to its energy by

$$\mathcal{E} \equiv p^0 = \hbar\omega = \text{(particle's energy)}; \tag{2.27a}$$

and, since the particle has $\vec{p}^2 = -(p^0)^2 + \mathbf{p}^2 = -m^2 = 0$ (in accord with the lightlike nature of its world line), its momentum as measured in the chosen Lorentz frame is

$$\mathbf{p} = \mathcal{E}\mathbf{n} = \hbar\omega\mathbf{n}. \tag{2.27b}$$

Here $\mathbf{n}$ is the unit 3-vector that points in the direction of the particle's travel, as measured in the chosen frame; that is (since the particle moves at the speed of light $v = 1$), $\mathbf{n}$ is the particle's ordinary velocity. Eqs. (2.27a) and (2.27b) are respectively the temporal and spatial components of the geometric, frame-independent relation $\vec{p} = \hbar\vec{k}$ [Eq. (2.13), which is valid for zero-rest-mass particles as well as finite-mass ones].

3+1 split of spacetime into space plus time

The introduction of a specific Lorentz frame into spacetime can be said to produce a 3+1 split of every 4-vector into a 3-dimensional vector plus a scalar (a real number). The 3+1 split of a particle's 4-momentum $\vec{p}$ produces its momentum $\mathbf{p}$ plus its energy $\mathcal{E} = p^0$. Correspondingly, the 3+1 split of the law of 4-momentum conservation (2.15) produces a law of conservation of momentum plus a law of conservation of energy:

$$\sum_{\bar{A}} \mathbf{p}_{\bar{A}} = \sum_{A} \mathbf{p}_A, \qquad \sum_{\bar{A}} \mathcal{E}_{\bar{A}} = \sum_{A} \mathcal{E}_A. \tag{2.28}$$

The unbarred quantities in Eqs. (2.28) are momenta and energies of the particles entering the interaction region, and the barred quantities are those of the particles leaving (see Fig. 2.4).

Because the concept of energy does not even exist until one has chosen a Lorentz frame—and neither does that of momentum—the laws of energy conservation and momentum conservation separately are frame-dependent laws. In this sense, they are far less fundamental than their combination, the frame-independent law of 4-momentum conservation.

By learning to think about the 3+1 split in a geometric, frame-independent way, one can gain conceptual and computational power. As an example, consider a particle with 4-momentum $\vec{p}$, being studied by an observer with 4-velocity $\vec{U}$. In the observer's own Lorentz reference frame, her 4-velocity has components $U^0 = 1$ and $U^j = 0$, and therefore, her 4-velocity is $\vec{U} = U^\alpha \vec{e}_\alpha = \vec{e}_0$; that is, it is identically equal to the time basis vector of her Lorentz frame. Thus the particle energy that she measures is $\mathcal{E} = p^0 = -p_0 = -\vec{p} \cdot \vec{e}_0 = -\vec{p} \cdot \vec{U}$. This equation, derived in the observer's Lorentz frame, is actually a geometric, frame-independent relation: the inner product of two 4-vectors. It says that *when an observer with 4-velocity $\vec{U}$ measures the energy of a* *particle with 4-momentum $\vec{p}$, the result she gets (the time part of the 3+1 split of $\vec{p}$ as seen by her) is*

$$\boxed{\mathcal{E} = -\vec{p} \cdot \vec{U}.}$$ (2.29)

We shall use this equation in later chapters. In Exs. 2.9 and 2.10, the reader can gain experience deriving and interpreting other frame-independent equations for 3+1 splits. Exercise 2.11 exhibits the power of this geometric way of thinking by using it to derive the Doppler shift of a photon.

Exercise 2.9 **Practice: Frame-Independent Expressions for Energy, Momentum, and Velocity*

An observer with 4-velocity $\vec{U}$ measures the properties of a particle with 4-momentum $\vec{p}$. The energy she measures is $\mathcal{E} = -\vec{p} \cdot \vec{U}$ [Eq. (2.29)].

(a) Show that the particle's rest mass can be expressed in terms of $\vec{p}$ as

$$m^2 = -\vec{p}^2.$$ (2.30a)

(b) Show that the momentum the observer measures has the magnitude

$$|\mathbf{p}| = [(\vec{p} \cdot \vec{U})^2 + \vec{p} \cdot \vec{p}]^{\frac{1}{2}}.$$ (2.30b)

(c) Show that the ordinary velocity the observer measures has the magnitude

$$|\mathbf{v}| = \frac{|\mathbf{p}|}{\mathcal{E}},$$ (2.30c)

where $|\mathbf{p}|$ and $\mathcal{E}$ are given by the above frame-independent expressions.

(d) Show that the ordinary velocity $\mathbf{v}$, thought of as a 4-vector that happens to lie in the observer's slice of simultaneity, is given by

$$\vec{v} = \frac{\vec{p} + (\vec{p} \cdot \vec{U})\vec{U}}{-\vec{p} \cdot \vec{U}}.$$ (2.30d)

Exercise 2.10 **Example: 3-Metric as a Projection Tensor*

Consider, as in Ex. 2.9, an observer with 4-velocity $\vec{U}$ who measures the properties of a particle with 4-momentum $\vec{p}$.

(a) Show that the Euclidean metric of the observer's 3-space, when thought of as a tensor in 4-dimensional spacetime, has the form

$$\boxed{\mathbf{P} \equiv \mathbf{g} + \vec{U} \otimes \vec{U}.}$$ (2.31a)

Show, further, that if $\vec{A}$ is an arbitrary vector in spacetime, then $-\vec{A} \cdot \vec{U}$ is the component of $\vec{A}$ along the observer's 4-velocity $\vec{U}$, and

$$\mathbf{P}(\_, \vec{A}) = \vec{A} + (\vec{A} \cdot \vec{U})\vec{U}$$ (2.31b)

is the projection of $\vec{A}$ into the observer's 3-space (i.e., it is the spatial part of $\vec{A}$ as seen by the observer). For this reason, $\mathbf{P}$ is called a *projection tensor*. In quantum mechanics, the concept of a *projection operator* $\hat{P}$ is introduced as one that satisfies the equation $\hat{P}^2 = \hat{P}$. Show that the projection tensor $\mathbf{P}$ is a projection operator in the same sense:

$$P_{\alpha\mu} P^{\mu}{}_{\beta} = P_{\alpha\beta}. \tag{2.31c}$$

(b) Show that Eq. (2.30d) for the particle's ordinary velocity, thought of as a 4-vector, can be rewritten as

$$\vec{v} = \frac{\mathbf{P}(\underline{\ }, \vec{p})}{-\vec{p} \cdot \vec{U}}. \tag{2.32}$$

Exercise 2.11 **Example: Doppler Shift Derived without Lorentz Transformations*

(a) An observer at rest in some inertial frame receives a photon that was emitted in direction $\mathbf{n}$ by an atom moving with ordinary velocity $\mathbf{v}$ (Fig. 2.6). The photon frequency and energy as measured by the emitting atom are ν_{em} and $\mathcal{E}_{\mathrm{em}}$; those measured by the receiving observer are ν_{rec} and $\mathcal{E}_{\mathrm{rec}}$. By a calculation carried out solely in the receiver's inertial frame (the frame of Fig. 2.6), and without the aid of any Lorentz transformation, derive the standard formula for the photon's Doppler shift:

$$\frac{\nu_{\mathrm{rec}}}{\nu_{\mathrm{em}}} = \frac{\sqrt{1 - v^2}}{1 - \mathbf{v} \cdot \mathbf{n}}. \tag{2.33}$$

[Hint: Use Eq. (2.29) to evaluate $\mathcal{E}_{\mathrm{em}}$ using receiver-frame expressions for the emitting atom's 4-velocity $\vec{U}$ and the photon's 4-momentum $\vec{p}$.]

(b) Suppose that instead of emitting a photon, the emitter produces a particle with finite rest mass m. Using the same method as in part (a), derive an expression for the ratio of received energy to emitted energy, $\mathcal{E}_{\mathrm{rec}}/\mathcal{E}_{\mathrm{em}}$, expressed in terms of the emitter's ordinary velocity $\mathbf{v}$ and the particle's ordinary velocity $\mathbf{V}$ (both as measured in the receiver's frame).

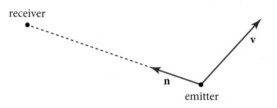

FIGURE 2.6 Geometry for Doppler shift, drawn in a slice of simultaneity of the receiver's inertial frame.

2.7 Lorentz Transformations

Consider two different inertial reference frames in Minkowski spacetime. Denote their Lorentz coordinates by $\{x^\alpha\}$ and $\{x^{\bar\mu}\}$ and their bases by $\{\mathbf{e}_\alpha\}$ and $\{\mathbf{e}_{\bar\mu}\}$, respectively, and write the transformation from one basis to the other as

$$\vec{e}_\alpha = \vec{e}_{\bar\mu} L^{\bar\mu}{}_\alpha, \qquad \vec{e}_{\bar\mu} = \vec{e}_\alpha L^\alpha{}_{\bar\mu}. \tag{2.34}$$

As in Euclidean 3-space (the rotation matrices of Sec 1.6), $L^{\bar\mu}{}_\alpha$ and $L^\alpha{}_{\bar\mu}$ are elements of two different transformation matrices, and since these matrices operate in opposite directions, they must be the inverse of each other:

transformation matrix

$$L^{\bar\mu}{}_\alpha L^\alpha{}_{\bar\nu} = \delta^{\bar\mu}{}_{\bar\nu}, \qquad L^\alpha{}_{\bar\mu} L^{\bar\mu}{}_\beta = \delta^\alpha{}_\beta. \tag{2.35a}$$

Notice the up/down placement of indices on the elements of the transformation matrices: the first index is always up, and the second is always down. This is just a convenient convention, which helps systematize the index shuffling rules in a way that can easily be remembered. Our rules about summing on the same index when up and down, and matching unsummed indices on the two sides of an equation automatically dictate the matrix to use in each of the transformations (2.34); and similarly for all other equations in this section.

In Euclidean 3-space the orthonormality of the two bases dictates that the transformations must be orthogonal (i.e., must be reflections or rotations). In Minkowski spacetime, orthonormality implies $g_{\alpha\beta} = \vec{e}_\alpha \cdot \vec{e}_\beta = (\vec{e}_{\bar\mu} L^{\bar\mu}{}_\alpha) \cdot (\vec{e}_{\bar\nu} L^{\bar\nu}{}_\beta) = L^{\bar\mu}{}_\alpha L^{\bar\nu}{}_\beta g_{\bar\mu\bar\nu}$; that is,

$$g_{\bar\mu\bar\nu} L^{\bar\mu}{}_\alpha L^{\bar\nu}{}_\beta = g_{\alpha\beta}, \quad \text{and similarly,} \quad g_{\alpha\beta} L^\alpha{}_{\bar\mu} L^\beta{}_{\bar\nu} = g_{\bar\mu\bar\nu}. \tag{2.35b}$$

Any matrix whose elements satisfy these equations is a *Lorentz transformation*.

Lorentz transformation

From the fact that vectors and tensors are geometric, frame-independent objects, one can derive the Minkowski-space analogs of the Euclidean transformation laws for components (1.13a) and (1.13b):

$$A^{\bar\mu} = L^{\bar\mu}{}_\alpha A^\alpha, \qquad T^{\bar\mu\bar\nu\bar\rho} = L^{\bar\mu}{}_\alpha L^{\bar\nu}{}_\beta L^{\bar\rho}{}_\gamma T^{\alpha\beta\gamma}, \tag{2.36a}$$

and similarly in the opposite direction. Notice that here, as elsewhere, these equations can be constructed by lining up indices in accord with our standard rules.

If (as is conventional) we choose the spacetime origins of the two Lorentz coordinate systems to coincide, then the vector $\vec{x}$ extending from the origin to some event $\mathcal{P}$, whose coordinates are x^α and $x^{\bar\alpha}$, has components equal to those coordinates. As a result, the transformation law for the coordinates takes the same form as Eq. (2.36a) for the components of a vector:

$$x^\alpha = L^\alpha{}_{\bar\mu} x^{\bar\mu}, \qquad x^{\bar\mu} = L^{\bar\mu}{}_\alpha x^\alpha. \tag{2.36b}$$

The product $L^\alpha{}_{\bar\mu} L^{\bar\mu}{}_{\bar\rho}$ of two Lorentz transformation matrices is a Lorentz transformation matrix. Under this product rule, the Lorentz transformations form a mathematical group, the *Lorentz group*, whose representations play an important role in quantum field theory (cf. the rotation group in Sec. 1.6).

An important specific example of a Lorentz transformation is:

$$\left[L^\alpha{}_{\bar\mu}\right] = \begin{bmatrix} \gamma & \beta\gamma & 0 & 0 \\ \beta\gamma & \gamma & 0 & 0 \\ 0 & 0 & 1 & 0 \\ 0 & 0 & 0 & 1 \end{bmatrix}, \quad \left[L^{\bar\mu}{}_\alpha\right] = \begin{bmatrix} \gamma & -\beta\gamma & 0 & 0 \\ -\beta\gamma & \gamma & 0 & 0 \\ 0 & 0 & 1 & 0 \\ 0 & 0 & 0 & 1 \end{bmatrix}, \quad (2.37a)$$

where β and γ are related by

$$|\beta| < 1, \quad \gamma \equiv (1 - \beta^2)^{-\frac{1}{2}}. \tag{2.37b}$$

[Notice that γ is the Lorentz factor associated with β; cf. Eq. (2.25d).]

One can readily verify (Ex. 2.12) that these matrices are the inverses of each other and that they satisfy the Lorentz-transformation relation (2.35b). These transformation matrices produce the following change of coordinates [Eq. (2.36b)]:

$$t = \gamma(\bar t + \beta\bar x), \quad x = \gamma(\bar x + \beta\bar t), \quad y = \bar y, \quad z = \bar z,$$
$$\bar t = \gamma(t - \beta x), \quad \bar x = \gamma(x - \beta t), \quad \bar y = y, \quad \bar z = z. \tag{2.37c}$$

These expressions reveal that any particle at rest in the unbarred frame (a particle with fixed, time-independent x, y, z) is seen in the barred frame to move along the world line $\bar x = \text{const} - \beta\bar t, \bar y = \text{const}, \bar z = \text{const}$. In other words, the unbarred frame is seen by observers at rest in the barred frame to move with uniform velocity $\mathbf{v} = -\beta\mathbf{e}_{\bar x}$, and correspondingly the barred frame is seen by observers at rest in the unbarred frame to move with the opposite uniform velocity $\mathbf{v} = +\beta\mathbf{e}_x$. This special Lorentz transformation is called a *pure boost* along the x direction.

EXERCISES

Exercise 2.12 *Derivation: Lorentz Boosts*
Show that the matrices (2.37a), with β and γ satisfying Eq. (2.37b), are the inverses of each other, and that they obey the condition (2.35b) for a Lorentz transformation.

Exercise 2.13 *Example: General Boosts and Rotations*
(a) Show that, if n^j is a 3-dimensional unit vector and β and γ are defined as in Eq. (2.37b), then the following is a Lorentz transformation [i.e., it satisfies Eq. (2.35b)]:

$$L^0{}_{\bar 0} = \gamma, \quad L^0{}_{\bar j} = L^j{}_{\bar 0} = \beta\gamma n^j, \quad L^j{}_{\bar k} = L^k{}_{\bar j} = (\gamma - 1)n^j n^k + \delta^{jk}. \tag{2.38}$$

Show, further, that this transformation is a pure boost along the direction $\mathbf{n}$ with speed β, and show that the inverse matrix $L^{\bar\mu}{}_\alpha$ for this boost is the same as $L^\alpha{}_{\bar\mu}$, but with β changed to $-\beta$.

(b) Show that the following is also a Lorentz transformation:

$$[L^{\alpha}{}_{\bar{\mu}}] = \begin{bmatrix} 1 & 0 & 0 & 0 \\ 0 & & & \\ 0 & & [R_{i\bar{j}}] & \\ 0 & & & \end{bmatrix},$$

(2.39)

where $[R_{i\bar{j}}]$ is a 3-dimensional rotation matrix for Euclidean 3-space. Show, further, that this Lorentz transformation rotates the inertial frame's spatial axes (its latticework of measuring rods) while leaving the frame's velocity unchanged (i.e., the new frame is at rest with respect to the old).

One can show (not surprisingly) that the general Lorentz transformation [i.e., the general solution of Eqs. (2.35b)] can be expressed as a sequence of pure boosts, pure rotations, and pure inversions (in which one or more of the coordinate axes are reflected through the origin, so $x^{\alpha} = -x^{\bar{\alpha}}$).

2.8 Spacetime Diagrams for Boosts

Figure 2.7 illustrates the pure boost (2.37c). Panel a in that figure is a 2-dimensional spacetime diagram, with the y- and z-coordinates suppressed, showing the $\bar{t}$- and $\bar{x}$-axes of the boosted Lorentz frame $\bar{\mathcal{F}}$ in the t, x Lorentz coordinate system of the unboosted frame $\mathcal{F}$. That the barred axes make angles $\tan^{-1}\beta$ with the unbarred axes, as shown, can be inferred from the Lorentz transformation (2.37c). Note that the orthogonality of the $\bar{t}$- and $\bar{x}$-axes to each other ($\vec{e}_{\bar{t}} \cdot \vec{e}_{\bar{x}} = 0$) shows up as the two axes making the same angle $\pi/2 - \beta$ with the null line $x = t$. The invariance of the interval guarantees that, as shown for $a = 1$ or 2 in Fig. 2.7a, the event $\bar{x} = a$ on the $\bar{x}$-axis lies at the intersection of that axis with the dashed hyperbola $x^2 - t^2 = a^2$; and similarly, the event $\bar{t} = a$ on the $\bar{t}$-axis lies at the intersection of that axis with the dashed hyperbola $t^2 - x^2 = a^2$.

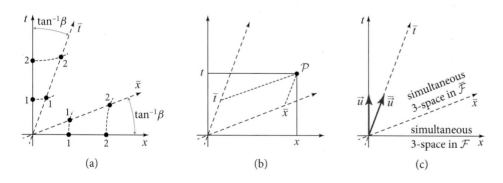

FIGURE 2.7 Spacetime diagrams illustrating the pure boost (2.37c) from one Lorentz reference frame to another.

As shown in Fig. 2.7b, the barred coordinates $\{\bar{t}, \bar{x}\}$ of an event $\mathcal{P}$ can be inferred by projecting from $\mathcal{P}$ onto the $\bar{t}$- and $\bar{x}$-axes, with the projection parallel to the $\bar{x}$- and $\bar{t}$-axes, respectively. Figure 2.7c shows the 4-velocity $\vec{u}$ of an observer at rest in frame $\mathcal{F}$ and that, $\vec{\bar{u}}$, of an observer at rest in frame $\overline{\mathcal{F}}$. The events that observer $\mathcal{F}$ regards as all simultaneous, with time $t = 0$, lie in a 3-space that is orthogonal to $\vec{u}$ and includes the x-axis. This is a slice of simultaneity of reference frame $\mathcal{F}$. Similarly, the events that observer $\overline{\mathcal{F}}$ regards as all simultaneous, with $\bar{t} = 0$, live in the 3-space that is orthogonal to $\vec{\bar{u}}$ and includes the $\bar{x}$-axis (i.e., in a slice of simultaneity of frame $\overline{\mathcal{F}}$).

length contraction, time dilation, and breakdown of simultaneity

Exercise 2.14 uses spacetime diagrams similar to those in Fig. 2.7 to deduce a number of important relativistic phenomena, including the contraction of the length of a moving object (length contraction), the breakdown of simultaneity as a universally agreed-on concept, and the dilation of the ticking rate of a moving clock (time dilation). This exercise is extremely important; every reader who is not already familiar with it should study it.

EXERCISES

Exercise 2.14 **Example: Spacetime Diagrams*
Use spacetime diagrams to prove the following:

(a) Two events that are simultaneous in one inertial frame are not necessarily simultaneous in another. More specifically, if frame $\overline{\mathcal{F}}$ moves with velocity $\vec{v} = \beta \vec{e}_x$ as seen in frame $\mathcal{F}$, where $\beta > 0$, then of two events that are simultaneous in $\overline{\mathcal{F}}$ the one farther "back" (with the more negative value of $\bar{x}$) will occur in $\mathcal{F}$ before the one farther "forward."

(b) Two events that occur at the same spatial location in one inertial frame do not necessarily occur at the same spatial location in another.

(c) If $\mathcal{P}_1$ and $\mathcal{P}_2$ are two events with a timelike separation, then there exists an inertial reference frame in which they occur at the same spatial location, and in that frame the time lapse between them is equal to the square root of the negative of their invariant interval, $\Delta t = \Delta \tau \equiv \sqrt{-(\Delta s)^2}$.

(d) If $\mathcal{P}_1$ and $\mathcal{P}_2$ are two events with a spacelike separation, then there exists an inertial reference frame in which they are simultaneous, and in that frame the spatial distance between them is equal to the square root of their invariant interval, $\sqrt{g_{ij} \Delta x^i \Delta x^j} = \Delta s \equiv \sqrt{(\Delta s)^2}$.

(e) If the inertial frame $\overline{\mathcal{F}}$ moves with speed β relative to the frame $\mathcal{F}$, then a clock at rest in $\overline{\mathcal{F}}$ ticks more slowly as viewed from $\mathcal{F}$ than as viewed from $\overline{\mathcal{F}}$—more slowly by a factor $\gamma^{-1} = (1 - \beta^2)^{\frac{1}{2}}$. This is called *relativistic time dilation*. As one consequence, the lifetimes of unstable particles moving with speed β are increased by the Lorentz factor γ.

(f) If the inertial frame $\overline{\mathcal{F}}$ moves with velocity $\vec{v} = \beta \vec{e}_x$ relative to the frame $\mathcal{F}$, then an object at rest in $\overline{\mathcal{F}}$ as studied in $\mathcal{F}$ appears shortened by a factor $\gamma^{-1} = (1 - \beta^2)^{\frac{1}{2}}$ along the x direction, but its length along the y and z directions is unchanged. This is called *Lorentz contraction*. As one consequence, heavy ions moving at high speeds in a particle accelerator appear to act like pancakes, squashed along their directions of motion.

Exercise 2.15 *Problem: Allowed and Forbidden Electron-Photon Reactions*
Show, using spacetime diagrams and also using frame-independent calculations, that the law of conservation of 4-momentum forbids a photon to be absorbed by an electron, $e + \gamma \rightarrow e$, and also forbids an electron and a positron to annihilate and produce a single photon, $e^+ + e^- \rightarrow \gamma$ (in the absence of any other particles to take up some of the 4-momentum); but the annihilation to form two photons, $e^+ + e^- \rightarrow 2\gamma$, is permitted.

2.9 Time Travel

2.9.1 Measurement of Time; Twins Paradox

Time dilation is one facet of a more general phenomenon: time, as measured by ideal clocks, is a personal thing, different for different observers who move through spacetime on different world lines. This is well illustrated by the infamous "twins paradox," in which one twin, Methuselah, remains forever at rest in an inertial frame and the other, Florence, makes a spacecraft journey at high speed and then returns to rest beside Methuselah.

twins paradox

The twins' world lines are depicted in Fig. 2.8a, a spacetime diagram whose axes are those of Methuselah's inertial frame. The time measured by an ideal clock that Methuselah carries is the coordinate time t of his inertial frame; and its total time lapse, from Florence's departure to her return, is $t_{\text{return}} - t_{\text{departure}} \equiv T_{\text{Methuselah}}$. By contrast, the time measured by an ideal clock that Florence carries is her proper time τ (i.e., the square root of the invariant interval (2.4) along her world line). Thus her total time lapse from departure to return is

$$T_{\text{Florence}} = \int d\tau = \int \sqrt{dt^2 - \delta_{ij}dx^i dx^j} = \int_0^{T_{\text{Methuselah}}} \sqrt{1 - v^2} \, dt. \quad (2.40)$$

Here (t, x^i) are the time and space coordinates of Methuselah's inertial frame, and v is Florence's ordinary speed, $v = \sqrt{\delta_{ij}(dx^i/dt)(dx^j/dt)}$, as measured in Methuselah's frame. Obviously, Eq. (2.40) predicts that T_{Florence} is less than $T_{\text{Methuselah}}$. In fact (Ex. 2.16), even if Florence's acceleration is kept no larger than one Earth gravity throughout her trip, and her trip lasts only $T_{\text{Florence}} = $ (a few tens of years), $T_{\text{Methuselah}}$ can be hundreds or thousands or millions or billions of years.

Does this mean that Methuselah actually "experiences" a far longer time lapse, and actually ages far more than Florence? Yes! The time experienced by humans and the aging of the human body are governed by chemical processes, which in turn are governed by the natural oscillation rates of molecules, rates that are constant to high accuracy when measured in terms of ideal time (or, equivalently, proper time τ).

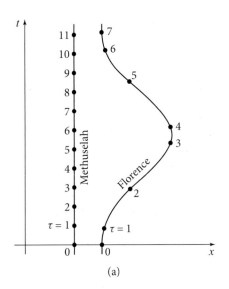
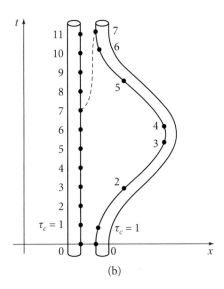

(a)   (b)

FIGURE 2.8 (a) Spacetime diagram depicting the so-called "twins paradox." Marked along the two world lines are intervals of proper time as measured by the two twins. (b) Spacetime diagram depicting the motions of the two mouths of a wormhole. Marked along the mouths' world tubes are intervals of proper time τ_c as measured by the single clock that sits in the common mouths.

Therefore, a human's experiential time and aging time are the same as the human's proper time—so long as the human is not subjected to such high accelerations as to damage her body.

In effect, then, Florence's spacecraft has functioned as a time machine to carry her far into Methuselah's future, with only a modest lapse of her own proper time (i.e., ideal, experiential, or aging time). This may be a "paradox" in the sense that it is surprising. However, it is in no sense a contradiction. This type of time dilation is routinely measured in high-energy physics storage rings.

2.9.2 Wormholes

Is it also possible, at least in principle, for Florence to construct a time machine that carries her into Methuselah's past—and also her own past? At first sight, the answer would seem to be "yes." Figure 2.8b shows one possible method, using a *wormhole*. [See Frolov and Novikov (1990), Friedman and Higuchi (2006), Everett and Roman (2011) for other approaches.]

wormhole   Wormholes are hypothetical handles in the topology of space. A simple model of a wormhole can be obtained by taking a flat 3-dimensional space, removing from it the interiors of two identical spheres, and identifying the spheres' surfaces so that if one enters the surface of one of the spheres, one immediately finds oneself exiting through the surface of the other. When this is done, there is a bit of strongly localized spatial curvature at the spheres' common surface, so to analyze such a wormhole properly, one must use general relativity rather than special relativity. In particular, it is the laws of general relativity, combined with the laws of quantum field theory, that tell one how to construct such a wormhole and what kinds of materials are required to hold it open, so things can pass through it. Unfortunately, despite considerable effort, theoretical physicists have not yet deduced definitively whether those laws permit such wormholes to exist and stay open, though indications are pessimistic (Everett and Roman, 2011; Friedman and Higuchi, 2006). However, assuming such wormholes *can* exist, the following special relativistic analysis (Morris et al., 1988) shows how one might be used to construct a machine for backward time travel.

2.9.3 Wormhole as Time Machine

The two identified spherical surfaces are called the wormhole's mouths. Ask Methuselah to keep one mouth with him, forever at rest in his inertial frame, and ask Florence to take the other mouth with her on her high-speed journey. The two mouths' *world tubes* (analogs of world lines for a 3-dimensional object) then have the forms shown in Fig. 2.8b. Suppose that a single ideal clock sits in the wormhole's identified mouths, so that from the external universe one sees it both in Methuselah's wormhole mouth and in Florence's. As seen in Methuselah's mouth, the clock measures his proper time, which is equal to the coordinate time t (see tick marks along the left world tube in Fig. 2.8b). As seen in Florence's mouth, the clock measures her proper time, Eq. (2.40) (see tick marks along the right world tube in Fig. 2.8b). The result should be obvious, if surprising: When Florence returns to rest beside Methuselah, the wormhole has become a time machine. If she travels through the wormhole when the clock reads $\tau_c = 7$, she goes backward in time as seen in Methuselah's (or anyone else's) inertial frame; and then, in fact, traveling along the everywhere timelike world line (dashed in Fig. 2.8b), she is able to meet her younger self before she entered the wormhole.

world tube

This scenario is profoundly disturbing to most physicists because of the dangers of science-fiction-type paradoxes (e.g., the older Florence might kill her younger self, thereby preventing herself from making the trip through the wormhole and killing herself). Fortunately perhaps, it seems likely (though far from certain) that vacuum fluctuations of quantum fields will destroy the wormhole at the moment its mouths' motion first makes backward time travel possible. It may be that this mechanism will always prevent the construction of backward-travel time machines, no matter what tools one uses for their construction (Kay et al., 1997; Kim and Thorne, 1991); but see also contrary indications in research reviewed by Everett and Roman (2011) and Friedman and Higuchi (2006). Whether this is so we likely will not know until the laws of quantum gravity have been mastered.

chronology protection

Exercise 2.16 *Example: Twins Paradox*

EXERCISES

(a) The 4-acceleration of a particle or other object is defined by $\vec{a} \equiv d\vec{u}/d\tau$, where $\vec{u}$ is its 4-velocity and τ is proper time along its world line. Show that, if an observer carries an accelerometer, the magnitude $|\mathbf{a}|$ of the 3-dimensional acceleration $\mathbf{a}$ measured by the accelerometer will always be equal to the magnitude of the observer's 4-acceleration, $|\mathbf{a}| = |\vec{a}| \equiv \sqrt{\vec{a} \cdot \vec{a}}$.

(b) In the twins paradox of Fig. 2.8a, suppose that Florence begins at rest beside Methuselah, then accelerates in Methuselah's x-direction with an acceleration a equal to one Earth gravity, g, for a time $T_{\text{Florence}}/4$ as measured by her, then accelerates in the $-x$-direction at g for a time $T_{\text{Florence}}/2$, thereby reversing her motion; then she accelerates in the $+x$-direction at g for a time $T_{\text{Florence}}/4$, thereby returning to rest beside Methuselah. (This is the type of motion shown in the figure.) Show that the total time lapse as measured by Methuselah is

$$T_{\text{Methuselah}} = \frac{4}{g} \sinh\left(\frac{g T_{\text{Florence}}}{4}\right). \tag{2.41}$$

(c) Show that in the geometrized units used here, Florence's acceleration (equal to acceleration of gravity at the surface of Earth) is $g = 1.033/\text{yr}$. Plot $T_{\text{Methuselah}}$ as a function of T_{Florence}, and from your plot estimate T_{Florence} if $T_{\text{Methuselah}}$ is the age of the Universe, 14 billion years.

Exercise 2.17 *Challenge: Around the World on TWA*
In a long-ago era when an airline named Trans World Airlines (TWA) flew around the world, Josef Hafele and Richard Keating (1972a) carried out a real live twins paradox experiment: They synchronized two atomic clocks and then flew one around the world eastward on TWA, and on a separate trip, around the world westward, while the other clock remained at home at the Naval Research Laboratory near Washington, D.C. When the clocks were compared after each trip, they were found to have aged differently. Making reasonable estimates for the airplane routing and speeds, compute the difference in aging, and compare your result with the experimental data in Hafele and Keating (1972b). [Note: The rotation of Earth is important, as is the general relativistic gravitational redshift associated with the clocks' altitudes; but the gravitational redshift drops out of the difference in aging, if the time spent at high altitude is the same eastward as westward.]

2.10

2.10 Directional Derivatives, Gradients, and the Levi-Civita Tensor

Derivatives of vectors and tensors in Minkowski spacetime are defined in precisely the same way as in Euclidean space; see Sec. 1.7. Any reader who has not studied that section should do so now. In particular (in extreme brevity, as the explanations and justifications are the same as in Euclidean space):

directional derivative

gradient

The directional derivative of a tensor $\mathbf{T}$ along a vector $\vec{A}$ is $\nabla_{\vec{A}}\mathbf{T} \equiv \lim_{\epsilon \to 0}(1/\epsilon)[\mathbf{T}(\vec{x}_P + \epsilon \vec{A}) - \mathbf{T}(\vec{x}_P)]$; the gradient $\vec{\nabla}\mathbf{T}$ is the tensor that produces the directional derivative when one inserts $\vec{A}$ into its last slot: $\nabla_{\vec{A}}\mathbf{T} = \vec{\nabla}\mathbf{T}(\_\,,\_\,,\_\,,\vec{A})$. In slot-naming index notation (or in components on a basis), the gradient is denoted $T_{\alpha\beta\gamma;\mu}$. In a Lorentz basis (the basis vectors associated with an inertial reference frame), the components of the gradient are simply the partial derivatives of the tensor,

$T_{\alpha\beta\gamma;\mu} = \partial T_{\alpha\beta\gamma}/\partial x^\mu \equiv T_{\alpha\beta\gamma,\mu}$. (The comma means partial derivative in a Lorentz basis, as in a Cartesian basis.)

The gradient and the directional derivative obey all the familiar rules for differentiation of products, for example, $\nabla_{\vec{A}}(\boldsymbol{S} \otimes \boldsymbol{T}) = (\nabla_{\vec{A}}\boldsymbol{S}) \otimes \boldsymbol{T} + \boldsymbol{S} \otimes \nabla_{\vec{A}}\boldsymbol{T}$. The gradient of the metric vanishes, $g_{\alpha\beta;\mu} = 0$. The divergence of a vector is the contraction of its gradient, $\vec{\nabla} \cdot \vec{A} = A_{\alpha;\beta}g^{\alpha\beta} = A^\alpha{}_{;\alpha}$.

Recall that the divergence of the gradient of a tensor in Euclidean space is the Laplacian: $T_{abc;jk}g_{jk} = T_{abc,jk}\delta_{jk} = \partial^2 T_{abc}\partial x^j \partial x^j$. By contrast, in Minkowski spacetime, because $g^{00} = -1$ and $g^{jk} = \delta^{jk}$ in a Lorentz frame, the divergence of the gradient is the wave operator (also called the d'Alembertian):

d'Alembertian

$$T_{\alpha\beta\gamma;\mu\nu}g^{\mu\nu} = T_{\alpha\beta\gamma,\mu\nu}g^{\mu\nu} = -\frac{\partial^2 T_{\alpha\beta\gamma}}{\partial t^2} + \frac{\partial^2 T_{\alpha\beta\gamma}}{\partial x^j \partial x^k}\delta^{jk} = \Box T_{\alpha\beta\gamma}. \tag{2.42}$$

When one sets this to zero, one gets the wave equation.

As in Euclidean space, so also in Minkowski spacetime, there are two tensors that embody the space's geometry: the metric tensor $\boldsymbol{g}$ and the Levi-Civita tensor $\boldsymbol{\epsilon}$. The Levi-Civita tensor in Minkowski spacetime is the tensor that is completely antisymmetric in all its slots and has value $\epsilon(\vec{A}, \vec{B}, \vec{C}, \vec{D}) = +1$ when evaluated on any *right-handed set of orthonormal 4-vectors*—that is, by definition, any orthonormal set for which $\vec{A}$ is timelike and future directed, and $\{\vec{B}, \vec{C}, \vec{D}\}$ are spatial and right-handed. This means that in any right-handed Lorentz basis, the only nonzero components of $\boldsymbol{\epsilon}$ are

Levi-Civita tensor

$$\epsilon_{\alpha\beta\gamma\delta} = +1 \text{ if } \alpha, \beta, \gamma, \delta \text{ is an even permutation of } 0, 1, 2, 3;$$
$$-1 \text{ if } \alpha, \beta, \gamma, \delta \text{ is an odd permutation of } 0, 1, 2, 3;$$
$$0 \text{ if } \alpha, \beta, \gamma, \delta \text{ are not all different.} \tag{2.43}$$

By the sign-flip-if-temporal rule, $\epsilon_{0123} = +1$ implies that $\epsilon^{0123} = -1$.

2.11 Nature of Electric and Magnetic Fields; Maxwell's Equations

2.11

Now that we have introduced the gradient and the Levi-Civita tensor, we can study the relationship of the relativistic version of electrodynamics to the nonrelativistic (Newtonian) version. In doing so, we use Gaussian units (with the speed of light set to 1), as is conventional among relativity theorists, and as does Jackson (1999) in his classic textbook, switching from SI to Gaussian when he moves into the relativistic domain.

Consider a particle with charge q, rest mass m, and 4-velocity $\vec{u}$ interacting with an electromagnetic field $\boldsymbol{F}(\_, \_)$. In index notation, the electromagnetic 4-force acting on the particle [Eq. (2.20)] is

$$dp^\alpha/d\tau = q F^{\alpha\beta}u_\beta. \tag{2.44}$$

Let us examine this 4-force in some arbitrary inertial reference frame in which the particle's ordinary-velocity components are $v^j = v_j$ and its 4-velocity components are $u^0 = \gamma, u^j = \gamma v^j$ [Eqs. (2.25c)]. Anticipating the connection with the nonrelativistic viewpoint, we introduce the following notation for the contravariant components of the antisymmetric electromagnetic field tensor:

$$F^{0j} = -F^{j0} = +F_{j0} = -F_{0j} = E_j, \qquad F^{ij} = F_{ij} = \epsilon_{ijk}B_k. \qquad (2.45)$$

Inserting these components of $\boldsymbol{F}$ and $\vec{u}$ into Eq. (2.44) and using the relationship $dt/d\tau = u^0 = \gamma$ between t and τ derivatives, we obtain for the components of the 4-force $dp_j/d\tau = \gamma dp_j/dt = q(F_{j0}u^0 + F_{jk}u^k) = qu^0(F_{j0} + F_{jk}v^k) = q\gamma(E_j + \epsilon_{jki}v_kB_i)$ and $dp^0/d\tau = \gamma dp^0/dt = qF^{0j}u_j = q\gamma E_j v_j$. Dividing by γ, converting into 3-space index notation, and denoting the particle's energy by $\mathcal{E} = p^0$, we bring these into the familiar Lorentz-force form

$$d\mathbf{p}/dt = q(\mathbf{E} + \mathbf{v} \times \mathbf{B}), \qquad d\mathcal{E}/dt = q\mathbf{v} \cdot \mathbf{E}. \qquad (2.46)$$

Evidently, $\mathbf{E}$ is the electric field and $\mathbf{B}$ the magnetic field as measured in our chosen Lorentz frame.

This may be familiar from standard electrodynamics textbooks (e.g., Jackson, 1999). Not so familiar, but very important, is the following geometric interpretation of $\mathbf{E}$ and $\mathbf{B}$.

The electric and magnetic fields $\mathbf{E}$ and $\mathbf{B}$ are spatial vectors as measured in the chosen inertial frame. We can also regard them as 4-vectors that lie in a 3-surface of simultaneity $t = \text{const}$ of the chosen frame, i.e., that are orthogonal to the 4-velocity (denote it $\vec{w}$) of the frame's observers (cf. Figs. 2.5 and 2.9). We shall denote this 4-vector version of $\mathbf{E}$ and $\mathbf{B}$ by $\vec{E}_{\vec{w}}$ and $\vec{B}_{\vec{w}}$, where the subscript $\vec{w}$ identifies the 4-velocity of the observer who measures these fields. These fields are depicted in Fig. 2.9.

In the rest frame of the observer $\vec{w}$, the components of $\vec{E}_{\vec{w}}$ are $E^0_{\vec{w}} = 0, E^j_{\vec{w}} = E_j = F_{j0}$ [the E_j appearing in Eqs. (2.45)], and similarly for $\vec{B}_{\vec{w}}$; the components of $\vec{w}$ are $w^0 = 1, w^j = 0$. Therefore, in this frame Eqs. (2.45) can be rewritten as

$$E^\alpha_{\vec{w}} = F^{\alpha\beta}w_\beta, \qquad B^\beta_{\vec{w}} = \frac{1}{2}\epsilon^{\alpha\beta\gamma\delta}F_{\gamma\delta}w_\alpha. \qquad (2.47a)$$

[To verify this, insert the above components of $\boldsymbol{F}$ and $\vec{w}$ into these equations and, after some algebra, recover Eqs. (2.45) along with $E^0_{\vec{w}} = B^0_{\vec{w}} = 0$.] Equations (2.47a) say that in one special reference frame, that of the observer $\vec{w}$, the components of the 4-vectors on the left and on the right are equal. This implies that in every Lorentz frame the components of these 4-vectors will be equal; that is, Eqs. (2.47a) are true when one regards them as geometric, frame-independent equations written in slot-naming index notation. *These equations enable one to compute the electric and magnetic fields measured by an observer (viewed as 4-vectors in the observer's 3-surface of simultaneity)*

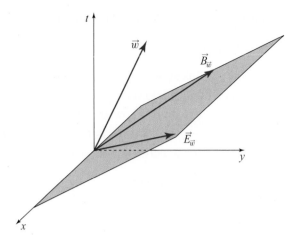

FIGURE 2.9 The electric and magnetic fields measured by an observer with 4-velocity $\vec{w}$, shown as 4-vectors $\vec{E}_{\vec{w}}$ and $\vec{B}_{\vec{w}}$ that lie in the observer's 3-surface of simultaneity (green 3-surface orthogonal to $\vec{w}$).

from the observer's 4-velocity and the electromagnetic field tensor, without the aid of any basis or reference frame.

Equations (2.47a) embody explicitly the following important fact. Although the electromagnetic field tensor **F** is a geometric, frame-independent quantity, the electric and magnetic fields $\vec{E}_{\vec{w}}$ and $\vec{B}_{\vec{w}}$ individually depend for their existence on a specific choice of observer (with 4-velocity $\vec{w}$), that is, a specific choice of inertial reference frame, or in other words, a specific choice of the split of spacetime into a 3-space (the 3-surface of simultaneity orthogonal to the observer's 4-velocity $\vec{w}$) and corresponding time (the Lorentz time of the observer's reference frame). *Only after making such an observer-dependent 3+1 split of spacetime into space plus time do the electric field and the magnetic field come into existence as separate entities.* Different observers with different 4-velocities $\vec{w}$ make this spacetime split in different ways, thereby resolving the frame-independent **F** into different electric and magnetic fields $\vec{E}_{\vec{w}}$ and $\vec{B}_{\vec{w}}$.

By the same procedure as we used to derive Eqs. (2.47a), one can derive the inverse relationship, the following expression for the electromagnetic field tensor in terms of the (4-vector) electric and magnetic fields measured by some observer:

$$F^{\alpha\beta} = w^\alpha E_{\vec{w}}^\beta - E_{\vec{w}}^\alpha w^\beta + \epsilon^{\alpha\beta}{}_{\gamma\delta} w^\gamma B_{\vec{w}}^\delta. \tag{2.47b}$$

Maxwell's equations in geometric, frame-independent form are (in Gaussian units)[5]

$$F^{\alpha\beta}{}_{;\beta} = 4\pi J^\alpha,$$

$$\epsilon^{\alpha\beta\gamma\delta} F_{\gamma\delta;\beta} = 0; \quad \text{i.e.,} \quad F_{\alpha\beta;\gamma} + F_{\beta\gamma;\alpha} + F_{\gamma\alpha;\beta} = 0. \tag{2.48}$$

Maxwell's equations

---

5. In SI units the 4π gets replaced by $\mu_0 = 1/\epsilon_0$, corresponding to the different units for the charge-current 4-vector.

Here $\vec{J}$ is the charge-current 4-vector, which in any inertial frame has components

$$J^0 = \rho_e = \text{(charge density)}, \qquad J^i = j_i = \text{(current density)}. \qquad (2.49)$$

Exercise 2.19 describes how to think about this charge density and current density as geometric objects determined by the observer's 4-velocity or 3+1 split of spacetime into space plus time. Exercise 2.20 shows how the frame-independent Maxwell's equations (2.48) reduce to the more familiar ones in terms of **E** and **B**. Exercise 2.21 explores potentials for the electromagnetic field in geometric, frame-independent language and the 3+1 split.

EXERCISES

Exercise 2.18 *Derivation and Practice: Reconstruction of* **F**
Derive Eq. (2.47b) by the same method as was used to derive Eq. (2.47a). Then show, by a geometric, frame-independent calculation, that Eq. (2.47b) implies Eq. (2.47a).

Exercise 2.19 *Problem: 3+1 Split of Charge-Current 4-Vector*
Just as the electric and magnetic fields measured by some observer can be regarded as 4-vectors $\vec{E}_{\vec{w}}$ and $\vec{B}_{\vec{w}}$ that live in the observer's 3-space of simultaneity, so also the charge density and current density that the observer measures can be regarded as a scalar $\rho_{\vec{w}}$ and 4-vector $\vec{j}_{\vec{w}}$ that live in the 3-space of simultaneity. Derive geometric, frame-independent equations for $\rho_{\vec{w}}$ and $\vec{j}_{\vec{w}}$ in terms of the charge-current 4-vector $\vec{J}$ and the observer's 4-velocity $\vec{w}$, and derive a geometric expression for $\vec{J}$ in terms of $\rho_{\vec{w}}$, $\vec{j}_{\vec{w}}$, and $\vec{w}$.

Exercise 2.20 *Problem: Frame-Dependent Version of Maxwell's Equations*
By performing a 3+1 split on the geometric version of Maxwell's equations (2.48), derive the elementary, frame-dependent version

$$\nabla \cdot \mathbf{E} = 4\pi \rho_e, \qquad \nabla \times \mathbf{B} - \frac{\partial \mathbf{E}}{\partial t} = 4\pi \mathbf{j},$$

3+1 split of Maxwell's equations

$$\nabla \cdot \mathbf{B} = 0, \qquad \nabla \times \mathbf{E} + \frac{\partial \mathbf{B}}{\partial t} = 0. \qquad (2.50)$$

Exercise 2.21 *Problem: Potentials for the Electromagnetic Field*
(a) Express the electromagnetic field tensor as an antisymmetrized gradient of a 4-vector potential: in slot-naming index notation

$$F_{\alpha\beta} = A_{\beta;\alpha} - A_{\alpha;\beta}. \qquad (2.51a)$$

Show that, whatever may be the 4-vector potential $\vec{A}$, the second of Maxwell's equations (2.48) is automatically satisfied. Show further that the electromagnetic field tensor is unaffected by a gauge change of the form

$$\vec{A}_{\text{new}} = \vec{A}_{\text{old}} + \vec{\nabla}\psi, \qquad (2.51b)$$

where ψ is a scalar field (the generator of the gauge change). Show, finally, that it is possible to find a gauge-change generator that enforces *Lorenz gauge*

$$\vec{\nabla} \cdot \vec{A} = 0 \tag{2.51c}$$

on the new 4-vector potential, and show that in this gauge, the first of Maxwell's equations (2.48) becomes (in Gaussian units)

$$\Box \vec{A} = -4\pi \vec{J}; \quad \text{i.e.,} \quad A^{\alpha;\mu}{}_{\mu} = -4\pi J^{\alpha}. \tag{2.51d}$$

(b) Introduce an inertial reference frame, and in that frame split **F** into the electric and magnetic fields **E** and **B**, split $\vec{J}$ into the charge and current densities ρ_e and **j**, and split the vector potential into a scalar potential and a 3-vector potential

$$\phi = -A_0, \quad \mathbf{A} = \text{spatial part of } \vec{A}. \tag{2.51e}$$

Deduce the 3+1 splits of Eqs. (2.51a)–(2.51d), and show that they take the form given in standard textbooks on electromagnetism.

2.12 Volumes, Integration, and Conservation Laws

2.12.1 Spacetime Volumes and Integration

In Minkowski spacetime as in Euclidean 3-space (Sec. 1.8), the Levi-Civita tensor is the tool by which one constructs volumes. The 4-dimensional parallelepiped whose legs are the four vectors $\vec{A}, \vec{B}, \vec{C}, \vec{D}$ has a 4-dimensional volume given by the analog of Eqs. (1.25) and (1.26):

$$\text{4-volume} = \epsilon_{\alpha\beta\gamma\delta} A^{\alpha} B^{\beta} C^{\gamma} D^{\delta} = \epsilon(\vec{A}, \vec{B}, \vec{C}, \vec{D}) = \det \begin{bmatrix} A^0 & B^0 & C^0 & D^0 \\ A^1 & B^1 & C^1 & D^1 \\ A^2 & B^2 & C^2 & D^2 \\ A^3 & B^3 & C^3 & D^3 \end{bmatrix}. \tag{2.52}$$

4-volume

Note that this 4-volume is positive if the set of vectors $\{\vec{A}, \vec{B}, \vec{C}, \vec{D}\}$ is right-handed and negative if left-handed [cf. Eq. (2.43)].

Equation (2.52) provides us a way to perform volume integrals over 4-dimensional Minkowski spacetime. To integrate a smooth tensor field **T** over some 4-dimensional region $\mathcal{V}$ of spacetime, we need only divide $\mathcal{V}$ up into tiny parallelepipeds, multiply the 4-volume $d\Sigma$ of each parallelepiped by the value of **T** at its center, add, and take the limit. In any right-handed Lorentz coordinate system, the 4-volume of a tiny parallelepiped whose edges are dx^{α} along the four orthogonal coordinate axes is $d\Sigma = \epsilon(dt\,\vec{e}_0, dx\,\vec{e}_x, dy\,\vec{e}_y, dz\,\vec{e}_z) = \epsilon_{0123}\,dt\,dx\,dy\,dz = dt\,dx\,dy\,dz$ (the analog of $dV = dx\,dy\,dz$). Correspondingly, the integral of **T** over $\mathcal{V}$ can be expressed as

$$\int_{\mathcal{V}} T^{\alpha\beta\gamma} d\Sigma = \int_{\mathcal{V}} T^{\alpha\beta\gamma} dt\,dx\,dy\,dz. \tag{2.53}$$

2.12 Volumes, Integration, and Conservation Laws

By analogy with the vectorial area (1.27) of a parallelogram in 3-space, any 3-dimensional parallelepiped in spacetime with legs $\vec{A}$, $\vec{B}$, $\vec{C}$ has a vectorial 3-volume $\vec{\Sigma}$ (not to be confused with the scalar 4-volume Σ) defined by

vectorial 3-volume

$$\vec{\Sigma}(\underline{}) = \boldsymbol{\epsilon}(\underline{}, \vec{A}, \vec{B}, \vec{C}); \qquad \Sigma_\mu = \epsilon_{\mu\alpha\beta\gamma} A^\alpha B^\beta C^\gamma. \tag{2.54}$$

Here we have written the 3-volume vector both in abstract notation and in slot-naming index notation. This 3-volume vector has one empty slot, ready and waiting for a fourth vector ("leg") to be inserted, so as to compute the 4-volume Σ of a 4-dimensional parallelepiped.

Notice that the 3-volume vector $\vec{\Sigma}$ is orthogonal to each of its three legs (because of the antisymmetry of $\boldsymbol{\epsilon}$), and thus (unless it is null) it can be written as $\vec{\Sigma} = V\vec{n}$, where V is the magnitude of the 3-volume, and $\vec{n}$ is the unit normal to the three legs.

Interchanging any two legs of the parallelepiped reverses the 3-volume's sign. Consequently, the 3-volume is characterized not only by its legs but also by the order of its legs, or equally well, in two other ways: (i) by the direction of the vector $\vec{\Sigma}$ (reverse the order of the legs, and the direction of $\vec{\Sigma}$ will reverse); and (ii) by the *sense* of the 3-volume, defined as follows. Just as a 2-volume (i.e., a segment of a plane) in 3-dimensional space has two sides, so a 3-volume in 4-dimensional spacetime has two

positive and negative sides, and sense of 3-volume

sides (Fig. 2.10). Every vector $\vec{D}$ for which $\vec{\Sigma} \cdot \vec{D} > 0$ points out the *positive side* of the 3-volume $\vec{\Sigma}$. Vectors $\vec{D}$ with $\vec{\Sigma} \cdot \vec{D} < 0$ point out its *negative side*. When something moves through, reaches through, or points through the 3-volume from its negative side to its positive side, we say that this thing is moving or reaching or pointing in the "positive sense;" similarly for "negative sense." The examples shown in Fig. 2.10 should make this more clear.

Figure 2.10a shows two of the three legs of the volume vector $\vec{\Sigma} = \boldsymbol{\epsilon}(\underline{}, \Delta x \vec{e}_x,$ $\Delta y \vec{e}_y, \Delta z \vec{e}_z)$, where $\{t, x, y, z\}$ are the coordinates, and $\{\vec{e}_\alpha\}$ is the corresponding right-handed basis of a specific Lorentz frame. It is easy to show that this $\vec{\Sigma}$ can also be written as $\vec{\Sigma} = -\Delta V \vec{e}_0$, where ΔV is the ordinary volume of the parallelepiped as measured by an observer in the chosen Lorentz frame, $\Delta V = \Delta x \Delta y \Delta z$. Thus, the direction of the vector $\vec{\Sigma}$ is toward the past (direction of decreasing Lorentz time t). From this, and the fact that timelike vectors have negative squared length, it is easy to infer that $\vec{\Sigma} \cdot \vec{D} > 0$ if and only if the vector $\vec{D}$ points out of the "future" side of the 3-volume (the side of increasing Lorentz time t); therefore, the positive side of $\vec{\Sigma}$ is the future side. It follows that the vector $\vec{\Sigma}$ points in the negative sense of its own 3-volume.

Figure 2.10b shows two of the three legs of the volume vector $\vec{\Sigma} = \boldsymbol{\epsilon}(\underline{}, \Delta t \vec{e}_t,$ $\Delta y \vec{e}_y, \Delta z \vec{e}_z) = -\Delta t \Delta A \vec{e}_x$ (with $\Delta A = \Delta y \Delta z$). In this case, $\vec{\Sigma}$ points in its own positive sense.

This peculiar behavior is completely general. When the normal to a 3-volume is timelike, its volume vector $\vec{\Sigma}$ points in the negative sense; when the normal is space-like, $\vec{\Sigma}$ points in the positive sense. And as it turns out, when the normal is null, $\vec{\Sigma}$

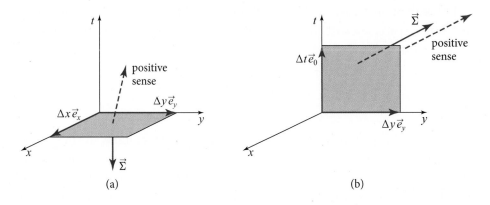

FIGURE 2.10 Spacetime diagrams depicting 3-volumes in 4-dimensional spacetime, with one spatial dimension (that along the z direction) suppressed.

lies in the 3-volume (parallel to its one null leg) and thus points neither in the positive sense nor the negative.[6]

Note the physical interpretations of the 3-volumes of Fig. 2.10. Figure 2.10a shows an instantaneous snapshot of an ordinary, spatial parallelepiped, whereas Fig. 2.10b shows the 3-dimensional region in spacetime swept out during time Δt by the parallelogram with legs $\Delta y \vec{e}_y$, $\Delta z \vec{e}_z$ and with area $\Delta A = \Delta y \Delta z$.

Vectorial 3-volume elements can be used to construct integrals over 3-dimensional volumes (also called 3-dimensional surfaces) in spacetime, for example, $\int_{\mathcal{V}_3} \vec{A} \cdot d\vec{\Sigma}$. More specifically, let (a, b, c) be (possibly curvilinear) coordinates in the 3-surface (3-volume) $\mathcal{V}_3$, and denote by $\vec{x}(a, b, c)$ the spacetime point $\mathcal{P}$ on $\mathcal{V}_3$ whose coordinate values are (a, b, c). Then $(\partial \vec{x}/\partial a)da$, $(\partial \vec{x}/\partial b)db$, $(\partial \vec{x}/\partial c)dc$ are the vectorial legs of the elementary parallelepiped whose corners are at (a, b, c), $(a + da, b, c)$, $(a, b + db, c)$, and so forth; and the spacetime components of these vectorial legs are $(\partial x^\alpha/\partial a)da$, $(\partial x^\alpha/\partial b)db$, $(\partial x^\alpha/\partial c)dc$. The 3-volume of this elementary parallelepiped is $d\vec{\Sigma} = \epsilon\big(\underline{}, (\partial \vec{x}/\partial a)da, (\partial \vec{x}/\partial b)db, (\partial \vec{x}/\partial c)dc\big)$, which has spacetime components

$$d\Sigma_\mu = \epsilon_{\mu\alpha\beta\gamma} \frac{\partial x^\alpha}{\partial a} \frac{\partial x^\beta}{\partial b} \frac{\partial x^\gamma}{\partial c} da\, db\, dc. \tag{2.55}$$

This is the integration element to be used when evaluating

$$\int_{\mathcal{V}_3} \vec{A} \cdot d\vec{\Sigma} = \int_{\mathcal{V}_3} A^\mu d\Sigma_\mu. \tag{2.56}$$

See Ex. 2.22 for an example.

---

6. This peculiar behavior gets replaced by a simpler description if one uses one-forms rather than vectors to describe 3-volumes; see, for example, Misner, Thorne, and Wheeler (1973, Box 5.2).

Just as there are Gauss's and Stokes' theorems (1.28a) and (1.28b) for integrals in Euclidean 3-space, so also there are Gauss's and Stokes' theorems in spacetime. Gauss's theorem has the obvious form

Gauss's theorem

$$\int_{\mathcal{V}_4} (\vec{\nabla} \cdot \vec{A}) d\Sigma = \int_{\partial \mathcal{V}_4} \vec{A} \cdot d\vec{\Sigma}, \tag{2.57}$$

where the first integral is over a 4-dimensional region $\mathcal{V}_4$ in spacetime, and the second is over the 3-dimensional boundary $\partial \mathcal{V}_4$ of $\mathcal{V}_4$, with the boundary's positive sense pointing outward, away from $\mathcal{V}_4$ (just as in the 3-dimensional case). We shall not write down the 4-dimensional Stokes' theorem, because it is complicated to formulate with the tools we have developed thus far; easy formulation requires *differential forms* (e.g., Flanders, 1989), which we shall not introduce in this book.

2.12.2

2.12.2 Conservation of Charge in Spacetime

In this section, we use integration over a 3-dimensional region in 4-dimensional spacetime to construct an elegant, frame-independent formulation of the law of conservation of electric charge.

We begin by examining the geometric meaning of the charge-current 4-vector $\vec{J}$. We defined $\vec{J}$ in Eq. (2.49) in terms of its components. The spatial component $J^x = J_x = J(\vec{e}_x)$ is equal to the x component of current density j_x: it is the amount Q of charge that flows across a unit surface area lying in the y-z plane in a unit time (i.e., the charge that flows across the unit 3-surface $\vec{\Sigma} = \vec{e}_x$). In other words, $\vec{J}(\vec{\Sigma}) = \vec{J}(\vec{e}_x)$ *is the total charge Q that flows across $\vec{\Sigma} = \vec{e}_x$ in $\vec{\Sigma}$'s positive sense* and similarly for the other spatial directions. The temporal component $J^0 = -J_0 = \vec{J}(-\vec{e}_0)$ is the charge density ρ_e: it is the total charge Q in a unit spatial volume. This charge is carried by particles that are traveling through spacetime from past to future and pass through the unit 3-surface (3-volume) $\vec{\Sigma} = -\vec{e}_0$. Therefore, $\vec{J}(\vec{\Sigma}) = \vec{J}(-\vec{e}_0)$ *is the total charge Q that flows through $\vec{\Sigma} = -\vec{e}_0$ in its positive sense*. This interpretation is the same one we deduced for the spatial components of $\vec{J}$.

This makes it plausible, and indeed one can show, that *for any small 3-surface $\vec{\Sigma}$,*

charge-current 4-vector

$\vec{J}(\vec{\Sigma}) \equiv J^\alpha \Sigma_\alpha$ *is the total charge Q that flows across $\vec{\Sigma}$ in its positive sense.*

This property of the charge-current 4-vector is the foundation for our frame-independent formulation of the law of charge conservation. Let $\mathcal{V}$ be a compact 4-dimensional region of spacetime and denote by $\partial \mathcal{V}$ its boundary, a closed 3-surface in 4-dimensional spacetime (Fig. 2.11). The charged media (fluids, solids, particles, etc.) present in spacetime carry electric charge through $\mathcal{V}$, from the past toward the future. The law of charge conservation says that all the charge that enters $\mathcal{V}$ through the past part of its boundary $\partial \mathcal{V}$ must exit through the future part of its boundary. If we choose the positive sense of the boundary's 3-volume element $d\vec{\Sigma}$ to point out of $\mathcal{V}$ (toward the past on the bottom boundary and toward the future on the top), then this *global law of charge conservation* can be expressed as

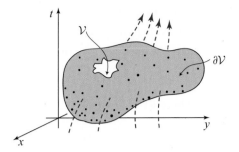

FIGURE 2.11 The 4-dimensional region $\mathcal{V}$ in spacetime and its closed 3-boundary $\partial\mathcal{V}$ (green surface), used in formulating the law of charge conservation. The dashed lines symbolize, heuristically, the flow of charge from the past toward the future.

$$\oint_{\partial\mathcal{V}} J^\alpha d\Sigma_\alpha = 0. \qquad (2.58)$$

global law of charge conservation

When each tiny charge q enters $\mathcal{V}$ through its past boundary, it contributes negatively to the integral, since it travels through $\partial\mathcal{V}$ in the negative sense (from positive side of $\partial\mathcal{V}$ toward negative side); and when that same charge exits $\mathcal{V}$ through its future boundary, it contributes positively. Therefore, its net contribution is zero, and similarly for all other charges.

In Ex. 2.23 you will show that, when this global law of charge conservation (2.58) is subjected to a 3+1 split of spacetime into space plus time, it becomes the nonrelativistic integral law of charge conservation (1.29).

This global conservation law can be converted into a *local conservation law* with the help of the 4-dimensional Gauss's theorem (2.57), $\oint_{\partial\mathcal{V}} J^\alpha d\Sigma_\alpha = \int_\mathcal{V} J^\alpha{}_{;\alpha} d\Sigma$. Since the left-hand side vanishes, so must the right-hand side; and for this 4-volume integral to vanish for every choice of $\mathcal{V}$, it is necessary that the integrand vanish everywhere in spacetime:

$$J^\alpha{}_{;\alpha} = 0; \quad \text{that is,} \quad \vec{\nabla} \cdot \vec{J} = 0. \qquad (2.59)$$

local law of charge conservation

In a specific but arbitrary Lorentz frame (i.e., in a 3+1 split of spacetime into space plus time), Eq. (2.59) becomes the standard differential law of charge conservation (1.30).

2.12.3 Conservation of Particles, Baryon Number, and Rest Mass

2.12.3

Any conserved scalar quantity obeys conservation laws of the same form as those for electric charge. For example, if the number of particles of some species (e.g., electrons, protons, or photons) is conserved, then we can introduce for that species a *number-flux 4-vector* $\vec{S}$ (analog of charge-current 4-vector $\vec{J}$): in any Lorentz frame, S^0 is the number density of particles, also designated n, and S^j is the particle flux. If $\vec{\Sigma}$ is a small

number-flux 4-vector

3-volume (3-surface) in spacetime, then $\vec{S}(\vec{\Sigma}) = S^{\alpha}\Sigma_{\alpha}$ is the number of particles that pass through Σ from its negative side to its positive side. The frame-invariant global and local conservation laws for these particles take the same form as those for electric charge:

laws of particle conservation

$$\int_{\partial\mathcal{V}} S^{\alpha}d\Sigma_{\alpha} = 0, \quad \text{where } \partial\mathcal{V} \text{ is any closed 3-surface in spacetime,} \qquad (2.60a)$$

$$S^{\alpha}{}_{;\alpha} = 0; \quad \text{that is,} \quad \vec{\nabla} \cdot \vec{S} = 0. \qquad (2.60b)$$

When fundamental particles (e.g., protons and antiprotons) are created and destroyed by quantum processes, the total baryon number (number of baryons minus number of antibaryons) is still conserved—or at least this is so to the accuracy of all experiments performed thus far. We shall assume it so in this book. This law of baryon-number conservation takes the forms of Eqs. (2.60), with $\vec{S}$ the number-flux 4-vector for baryons (with antibaryons counted negatively).

It is useful to express this baryon-number conservation law in Newtonian-like language by introducing a universally agreed-on mean rest mass per baryon $\bar{m}_B$. This $\bar{m}_B$ is often taken to be 1/56 the mass of an ^{56}Fe (iron-56) atomic nucleus, since ^{56}Fe is the nucleus with the tightest nuclear binding (i.e., the endpoint of thermonuclear evolution in stars). We multiply the baryon number-flux 4-vector $\vec{S}$ by this mean rest mass per baryon to obtain a rest-mass-flux 4-vector

rest-mass-flux 4-vector

$$\vec{S}_{\text{rm}} = \bar{m}_B \vec{S}, \qquad (2.61)$$

which (since $\bar{m}_B$ is, by definition, a constant) satisfies the same conservation laws (2.60) as baryon number.

For such media as fluids and solids, in which the particles travel only short distances between collisions or strong interactions, it is often useful to resolve the particle number-flux 4-vector and the rest-mass-flux 4-vector into a 4-velocity of the medium $\vec{u}$ (i.e., the 4-velocity of the frame in which there is a vanishing net spatial flux of particles), and the particle number density n_o or rest mass density ρ_o as measured in the medium's rest frame:

$$\vec{S} = n_o\vec{u}, \qquad \vec{S}_{\text{rm}} = \rho_o\vec{u}. \qquad (2.62)$$

See Ex. 2.24.

rest-mass conservation

We make use of the conservation laws $\vec{\nabla} \cdot \vec{S} = 0$ and $\vec{\nabla} \cdot \vec{S}_{\text{rm}} = 0$ for particles and rest mass later in this book (e.g., when studying relativistic fluids); and we shall find the expressions (2.62) for the number-flux 4-vector and rest-mass-flux 4-vector quite useful. See, for example, the discussion of relativistic shock waves in Ex. 17.9.

EXERCISES

Exercise 2.22 *Practice and Example: Evaluation of 3-Surface Integral in Spacetime*
In Minkowski spacetime, the set of all events separated from the origin by a timelike interval a^2 is a 3-surface, the hyperboloid $t^2 - x^2 - y^2 - z^2 = a^2$, where $\{t, x, y, z\}$

are Lorentz coordinates of some inertial reference frame. On this hyperboloid, introduce coordinates $\{\chi, \theta, \phi\}$ such that

$$t = a \cosh \chi, \quad x = a \sinh \chi \sin \theta \cos \phi,$$

$$y = a \sinh \chi \sin \theta \sin \phi, \quad z = a \sinh \chi \cos \theta. \quad (2.63)$$

Note that χ is a radial coordinate and (θ, ϕ) are spherical polar coordinates. Denote by $\mathcal{V}_3$ the portion of the hyperboloid with radius $\chi \leq b$.

(a) Verify that for all values of (χ, θ, ϕ), the points defined by Eqs. (2.63) do lie on the hyperboloid.

(b) On a spacetime diagram, draw a picture of $\mathcal{V}_3$, the $\{\chi, \theta, \phi\}$ coordinates, and the elementary volume element (vector field) $d\vec{\Sigma}$ [Eq. (2.55)].

(c) Set $\vec{A} \equiv \vec{e}_0$ (the temporal basis vector), and express $\int_{\mathcal{V}_3} \vec{A} \cdot d\vec{\Sigma}$ as an integral over $\{\chi, \theta, \phi\}$. Evaluate the integral.

(d) Consider a closed 3-surface consisting of the segment $\mathcal{V}_3$ of the hyperboloid as its top, the hypercylinder $\{x^2 + y^2 + z^2 = a^2 \sinh^2 b, \ 0 < t < a \cosh b\}$ as its sides, and the sphere $\{x^2 + y^2 + z^2 \leq a^2 \sinh^2 b, \ t = 0\}$ as its bottom. Draw a picture of this closed 3-surface on a spacetime diagram. Use Gauss's theorem, applied to this 3-surface, to show that $\int_{\mathcal{V}_3} \vec{A} \cdot d\vec{\Sigma}$ is equal to the 3-volume of its spherical base.

Exercise 2.23 *Derivation and Example: Global Law of Charge Conservation in an Inertial Frame*

Consider the global law of charge conservation $\int_{\partial\mathcal{V}} J^\alpha d\Sigma_\alpha = 0$ for a special choice of the closed 3-surface $\partial\mathcal{V}$: The bottom of $\partial\mathcal{V}$ is the ball $\{t = 0, x^2 + y^2 + z^2 \leq a^2\}$, where $\{t, x, y, z\}$ are the Lorentz coordinates of some inertial frame. The sides are the spherical world tube $\{0 \leq t \leq T, x^2 + y^2 + z^2 = a^2\}$. The top is the ball $\{t = T, x^2 + y^2 + z^2 \leq a^2\}$.

(a) Draw this 3-surface in a spacetime diagram.

(b) Show that for this $\partial\mathcal{V}$, $\int_{\partial\mathcal{V}} J^\alpha d\Sigma_\alpha = 0$ is a time integral of the nonrelativistic integral conservation law (1.29) for charge.

Exercise 2.24 *Example: Rest-Mass-Flux 4-Vector, Lorentz Contraction of Rest-Mass Density, and Rest-Mass Conservation for a Fluid*

Consider a fluid with 4-velocity $\vec{u}$ and rest-mass density ρ_o as measured in the fluid's rest frame.

(a) From the physical meanings of $\vec{u}$, ρ_o, and the rest-mass-flux 4-vector $\vec{S}_{rm}$, deduce Eqs. (2.62).

(b) Examine the components of $\vec{S}_{rm}$ in a reference frame where the fluid moves with ordinary velocity $\mathbf{v}$. Show that $S^0 = \rho_o \gamma$, and $S^j = \rho_o \gamma v^j$, where $\gamma = 1/\sqrt{1 - \mathbf{v}^2}$. Explain the physical interpretation of these formulas in terms of Lorentz contraction.

(c) Show that the law of conservation of rest mass $\vec{\nabla} \cdot \vec{S}_{\text{rm}} = 0$ takes the form

$$\frac{d\rho_o}{d\tau} = -\rho_o \vec{\nabla} \cdot \vec{u}, \tag{2.64}$$

where $d/d\tau$ is derivative with respect to proper time moving with the fluid.

(d) Consider a small 3-dimensional volume V of the fluid, whose walls move with the fluid (so if the fluid expands, V increases). Explain why the law of rest-mass conservation must take the form $d(\rho_o V)/d\tau = 0$. Thereby deduce that

$$\vec{\nabla} \cdot \vec{u} = (1/V)(dV/d\tau). \tag{2.65}$$

2.13

2.13.1

2.13 Stress-Energy Tensor and Conservation of 4-Momentum

2.13.1 Stress-Energy Tensor

GEOMETRIC DEFINITION

We conclude this chapter by formulating the law of 4-momentum conservation in ways analogous to our laws of conservation of charge, particles, baryon number, and rest mass. This task is not trivial, since 4-momentum is a vector in spacetime, while charge, particle number, baryon number, and rest mass are scalar quantities. Correspondingly, the density-flux of 4-momentum must have one more slot than the density-fluxes of charge, baryon number, and rest mass, $\vec{J}$, $\vec{S}$ and $\vec{S}_{\text{rm}}$, respectively; it must be a second-rank tensor. We call it the *stress-energy tensor* and denote it $\mathbf{T}(\_\,,\_\,)$. It is a generalization of the Newtonian stress tensor to 4-dimensional spacetime.

stress-energy tensor

Consider a medium or field flowing through 4-dimensional spacetime. As it crosses a tiny 3-surface $\vec{\Sigma}$, it transports a net electric charge $\vec{J}(\vec{\Sigma})$ from the negative side of $\vec{\Sigma}$ to the positive side, and net baryon number $\vec{S}(\vec{\Sigma})$ and net rest mass $\vec{S}_{\text{rm}}(\vec{\Sigma})$. Similarly, it transports a net 4-momentum $\mathbf{T}(\_\,,\vec{\Sigma})$ from the negative side to the positive side:

$$\mathbf{T}(\_\,,\vec{\Sigma}) \equiv (\text{total 4-momentum } \vec{P} \text{ that flows through } \vec{\Sigma}); \quad \text{or } T^{\alpha\beta}\Sigma_\beta = P^\alpha. \tag{2.66}$$

COMPONENTS

From this definition of the stress-energy tensor we can read off the physical meanings of its components on a specific, but arbitrary, Lorentz-coordinate basis: Making use of method (2.23b) for computing the components of a vector or tensor, we see that in a specific, but arbitrary, Lorentz frame (where $\vec{\Sigma} = -\vec{e}_0$ is a volume vector representing a parallelepiped with unit volume $\Delta V = 1$, at rest in that frame, with its positive sense toward the future):

$$-T_{\alpha 0} = \mathbf{T}(\vec{e}_\alpha, -\vec{e}_0) = \vec{P}(\vec{e}_\alpha) = \begin{pmatrix} \alpha \text{ component of 4-momentum that} \\ \text{flows from past to future across a unit} \\ \text{volume } \Delta V = 1 \text{ in the 3-space } t = \text{const} \end{pmatrix}$$

$$= (\alpha \text{ component of density of 4-momentum}). \tag{2.67a}$$

Specializing α to be a time or space component and raising indices, we obtain the specialized versions of (2.67a):

$$T^{00} = \text{(energy density as measured in the chosen Lorentz frame)},$$

$$T^{j0} = \text{(density of } j \text{ component of momentum in that frame)}. \qquad (2.67b)$$

Similarly, the αx component of the stress-energy tensor (also called the $\alpha 1$ component, since $x = x^1$ and $\vec{e}_x = \vec{e}_1$) has the meaning

$$T_{\alpha 1} \equiv T_{\alpha x} \equiv \mathbf{T}(\vec{e}_\alpha, \vec{e}_x) = \left(\begin{array}{c} \alpha \text{ component of 4-momentum that crosses} \\ \text{a unit area } \Delta y \Delta z = 1 \text{ lying in a surface of} \\ \text{constant } x, \text{ during unit time } \Delta t, \text{ crossing} \\ \text{from the } -x \text{ side toward the } +x \text{ side} \end{array} \right)$$

$$= \left(\begin{array}{c} \alpha \text{ component of flux of 4-momentum} \\ \text{across a surface lying perpendicular to } \vec{e}_x \end{array} \right). \qquad (2.67c)$$

The specific forms of this for temporal and spatial α are (after raising indices)

$$T^{0x} = \left(\begin{array}{c} \text{energy flux across a surface perpendicular to } \vec{e}_x, \\ \text{from the } -x \text{ side to the } +x \text{ side} \end{array} \right), \qquad (2.67d)$$

$$T^{jx} = \left(\begin{array}{c} \text{flux of } j \text{ component of momentum across a surface} \\ \text{perpendicular to } \vec{e}_x, \text{ from the } -x \text{ side to the } +x \text{ side} \end{array} \right)$$

$$= \left(\begin{array}{c} jx \text{ component} \\ \text{of stress} \end{array} \right). \qquad (2.67e)$$

The αy and αz components have the obvious, analogous interpretations.

These interpretations, restated much more briefly, are:

$$\boxed{\begin{array}{ll} T^{00} = \text{(energy density)}, & T^{j0} = \text{(momentum density)}, \\ T^{0j} = \text{(energy flux)}, & T^{jk} = \text{(stress)}. \end{array}} \qquad (2.67f)$$

components of stress-energy tensor

SYMMETRY

Although it might not be obvious at first sight, *the 4-dimensional stress-energy tensor is always symmetric*: in index notation (where indices can be thought of as representing the names of slots, or equally well, components on an arbitrary basis)

$$T^{\alpha\beta} = T^{\beta\alpha}. \qquad (2.68)$$

symmetry of stress-energy tensor

This symmetry can be deduced by physical arguments in a specific, but arbitrary, Lorentz frame: Consider, first, the $x0$ and $0x$ components, that is, the x components of momentum density and energy flux. A little thought, symbolized by the following heuristic equation, reveals that they must be equal:

$$T^{x0} = \left(\begin{array}{c} \text{momentum} \\ \text{density} \end{array} \right) = \frac{(\Delta\mathcal{E})dx/dt}{\Delta x \Delta y \Delta z} = \frac{\Delta\mathcal{E}}{\Delta y \Delta z \Delta t} = \left(\begin{array}{c} \text{energy} \\ \text{flux} \end{array} \right), \qquad (2.69)$$

and similarly for the other space-time and time-space components: $T^{j0} = T^{0j}$. [In the first expression of Eq. (2.69) $\Delta\mathcal{E}$ is the total energy (or equivalently mass) in the volume $\Delta x \Delta y \Delta z$, $(\Delta\mathcal{E})dx/dt$ is the total momentum, and when divided by the volume we get the momentum density. The third equality is just elementary algebra, and the resulting expression is obviously the energy flux.] The space-space components, being equal to the stress tensor, are also symmetric, $T^{jk} = T^{kj}$, by the argument embodied in Fig. 1.6. Since $T^{0j} = T^{j0}$ and $T^{jk} = T^{kj}$, all components in our chosen Lorentz frame are symmetric, $T^{\alpha\beta} = T^{\beta\alpha}$. Therefore, if we insert arbitrary vectors into the slots of $\boldsymbol{T}$ and evaluate the resulting number in our chosen Lorentz frame, we find

$$\boldsymbol{T}(\vec{A}, \vec{B}) = T^{\alpha\beta} A_\alpha B_\beta = T^{\beta\alpha} A_\alpha B_\beta = \boldsymbol{T}(\vec{B}, \vec{A}); \tag{2.70}$$

that is, $\boldsymbol{T}$ is symmetric under interchange of its slots.

Let us return to the physical meanings (2.67f) of the components of the stress-energy tensor. With the aid of $\boldsymbol{T}$'s symmetry, we can restate those meanings in the language of a 3+1 split of spacetime into space plus time: *When one chooses a specific reference frame, that choice splits the stress-energy tensor up into three parts. Its time-time part is the energy density T^{00}, its time-space part $T^{0j} = T^{j0}$ is the energy flux or equivalently the momentum density, and its space-space part T^{jk} is the stress tensor.*

2.13.2 4-Momentum Conservation

Our interpretation of $\vec{J}(\vec{\Sigma}) \equiv J^\alpha \Sigma_\alpha$ as the net charge that flows through a small 3-surface $\vec{\Sigma}$ from its negative side to its positive side gave rise to the global conservation law for charge, $\int_{\partial\mathcal{V}} J^\alpha d\Sigma_\alpha = 0$ [Eq. (2.58) and Fig. 2.11]. Similarly the role of $\boldsymbol{T}(\_\,, \vec{\Sigma})$ [$T^{\alpha\beta}\Sigma_\beta$ in slot-naming index notation] as the net 4-momentum that flows through $\vec{\Sigma}$ from its negative side to positive gives rise to the following equation for conservation of 4-momentum:

global law of 4-momentum conservation

$$\boxed{\int_{\partial\mathcal{V}} T^{\alpha\beta} d\Sigma_\beta = 0.} \tag{2.71}$$

(The time component of this equation is energy conservation; the spatial part is momentum conservation.) This equation says that all the 4-momentum that flows into the 4-volume $\mathcal{V}$ of Fig. 2.11 through its 3-surface $\partial\mathcal{V}$ must also leave $\mathcal{V}$ through $\partial\mathcal{V}$; it gets counted negatively when it enters (since it is traveling from the positive side of $\partial\mathcal{V}$ to the negative), and it gets counted positively when it leaves, so its net contribution to the integral (2.71) is zero.

This global law of 4-momentum conservation can be converted into a local law (analogous to $\vec{\nabla} \cdot \vec{J} = 0$ for charge) with the help of the 4-dimensional Gauss's theorem (2.57). Gauss's theorem, generalized in the obvious way from a vectorial integrand to a tensorial one, is:

$$\int_{\mathcal{V}} T^{\alpha\beta}{}_{;\beta}\, d\Sigma = \int_{\partial\mathcal{V}} T^{\alpha\beta} d\Sigma_\beta. \tag{2.72}$$

Since the right-hand side vanishes, so must the left-hand side; and for this 4-volume integral to vanish for every choice of $\mathcal{V}$, the integrand must vanish everywhere in spacetime:

$$T^{\alpha\beta}{}_{;\beta} = 0; \quad \text{or} \quad \vec{\nabla} \cdot \boldsymbol{T} = 0. \tag{2.73a}$$

In the second, index-free version of this local conservation law, the ambiguity about which slot the divergence is taken on is unimportant, since $\boldsymbol{T}$ is symmetric in its two slots: $T^{\alpha\beta}{}_{;\beta} = T^{\beta\alpha}{}_{;\beta}$.

In a specific but arbitrary Lorentz frame, the local conservation law (2.73a) for 4-momentum has as its temporal part

$$\frac{\partial T^{00}}{\partial t} + \frac{\partial T^{0k}}{\partial x^k} = 0, \tag{2.73b}$$

that is, the time derivative of the energy density plus the 3-divergence of the energy flux vanishes; and as its spatial part

$$\frac{\partial T^{j0}}{\partial t} + \frac{\partial T^{jk}}{\partial x^k} = 0, \tag{2.73c}$$

that is, the time derivative of the momentum density plus the 3-divergence of the stress (i.e., of momentum flux) vanishes. Thus, as one should expect, the geometric, frame-independent law of 4-momentum conservation includes as special cases both the conservation of energy and the conservation of momentum; and their differential conservation laws have the standard form that one expects both in Newtonian physics and in special relativity: time derivative of density plus divergence of flux vanishes; cf. Eq. (1.36) and associated discussion.

2.13.3 Stress-Energy Tensors for Perfect Fluids and Electromagnetic Fields

2.13.3

As an important example that illustrates the stress-energy tensor, consider a *perfect fluid*—a medium whose stress-energy tensor, evaluated in its *local rest frame* (a Lorentz frame where $T^{j0} = T^{0j} = 0$), has the form

$$T^{00} = \rho, \quad T^{jk} = P\delta^{jk} \tag{2.74a}$$

[Eq. (1.34) and associated discussion]. Here ρ is a short-hand notation for the energy density T^{00} (density of total mass-energy, including rest mass) as measured in the local rest frame, and the stress tensor T^{jk} in that frame is an isotropic pressure P. From this special form of $T^{\alpha\beta}$ in the local rest frame, one can derive the following geometric, frame-independent expression for the stress-energy tensor in terms of the 4-velocity $\vec{u}$ of the local rest frame (i.e., of the fluid itself), the metric tensor of spacetime $\boldsymbol{g}$, and the rest-frame energy density ρ and pressure P:

$$T^{\alpha\beta} = (\rho + P)u^\alpha u^\beta + Pg^{\alpha\beta}; \quad \text{i.e., } \boldsymbol{T} = (\rho + P)\vec{u} \otimes \vec{u} + P\boldsymbol{g}. \tag{2.74b}$$

See Ex. 2.26.

In Sec. 13.8, we develop and explore the laws of relativistic fluid dynamics that follow from energy-momentum conservation $\vec{\nabla} \cdot \mathbf{T} = 0$ for this stress-energy tensor and from rest-mass conservation $\vec{\nabla} \cdot \vec{S}_{\mathrm{rm}} = 0$. By constructing the Newtonian limit of the relativistic laws, we shall deduce the nonrelativistic laws of fluid mechanics, which are the central theme of Part V. Notice, in particular, that the Newtonian limit $(P \ll \rho,\ u^0 \simeq 1,\ u^j \simeq v^j)$ of the stress part of the stress-energy tensor (2.74b) is $T^{jk} = \rho v^j v^k + P \delta^{jk}$, which we met in Ex. 1.13.

Another example of a stress-energy tensor is that for the electromagnetic field, which takes the following form in Gaussian units (with $4\pi \to \mu_0 = 1/\epsilon_0$ in SI units):

electromagnetic stress-energy tensor

$$T^{\alpha\beta} = \frac{1}{4\pi} \left(F^{\alpha\mu} F^{\beta}{}_{\mu} - \frac{1}{4} g^{\alpha\beta} F^{\mu\nu} F_{\mu\nu} \right). \tag{2.75}$$

We explore this stress-energy tensor in Ex. 2.28.

EXERCISES

Exercise 2.25 *Example: Global Conservation of Energy in an Inertial Frame*
Consider the 4-dimensional parallelepiped $\mathcal{V}$ whose legs are $\Delta t \vec{e}_t$, $\Delta x \vec{e}_x$, $\Delta y \vec{e}_y$, $\Delta z \vec{e}_z$, where $(t, x, y, z) = (x^0, x^1, x^2, x^3)$ are the coordinates of some inertial frame. The boundary $\partial \mathcal{V}$ of this $\mathcal{V}$ has eight 3-dimensional "faces." Identify these faces, and write the integral $\int_{\partial \mathcal{V}} T^{0\beta} d\Sigma_\beta$ as the sum of contributions from each of them. According to the law of energy conservation, this sum must vanish. Explain the physical interpretation of each of the eight contributions to this energy conservation law. (See Ex. 2.23 for an analogous interpretation of charge conservation.)

Exercise 2.26 ***Derivation and Example: Stress-Energy Tensor and Energy-Momentum Conservation for a Perfect Fluid*
(a) Derive the frame-independent expression (2.74b) for the perfect fluid stress-energy tensor from its rest-frame components (2.74a).
(b) Explain why the projection of $\vec{\nabla} \cdot \mathbf{T} = 0$ along the fluid 4-velocity, $\vec{u} \cdot (\vec{\nabla} \cdot \mathbf{T}) = 0$, should represent energy conservation as viewed by the fluid itself. Show that this equation reduces to

$$\frac{d\rho}{d\tau} = -(\rho + P)\vec{\nabla} \cdot \vec{u}. \tag{2.76a}$$

With the aid of Eq. (2.65), bring this into the form

$$\frac{d(\rho V)}{d\tau} = -P \frac{dV}{d\tau}, \tag{2.76b}$$

where V is the 3-volume of some small fluid element as measured in the fluid's local rest frame. What are the physical interpretations of the left- and right-hand sides of this equation, and how is it related to the first law of thermodynamics?
(c) Read the discussion in Ex. 2.10 about the tensor $\mathbf{P} = \mathbf{g} + \vec{u} \otimes \vec{u}$ that projects into the 3-space of the fluid's rest frame. Explain why $P_{\mu\alpha} T^{\alpha\beta}{}_{;\beta} = 0$ should represent

the law of force balance (momentum conservation) as seen by the fluid. Show that this equation reduces to

$$(\rho + P)\vec{a} = -\boldsymbol{P} \cdot \vec{\nabla} P, \tag{2.76c}$$

where $\vec{a} = d\vec{u}/d\tau$ is the fluid's 4-acceleration. This equation is a relativistic version of Newton's $\mathbf{F} = m\mathbf{a}$. Explain the physical meanings of the left- and right-hand sides. Infer that $\rho + P$ must be the fluid's inertial mass per unit volume. It is also the enthalpy per unit volume, including the contribution of rest mass; see Ex. 5.5 and Box 13.2.

Exercise 2.27 **Example: Inertial Mass per Unit Volume*
Suppose that some medium has a rest frame (unprimed frame) in which its energy flux and momentum density vanish, $T^{0j} = T^{j0} = 0$. Suppose that the medium moves in the x direction with speed very small compared to light, $v \ll 1$, as seen in a (primed) laboratory frame, and ignore factors of order v^2. The ratio of the medium's momentum density $G_{j'} = T^{j'0'}$ (as measured in the laboratory frame) to its velocity $v_i = v\delta_{ix}$ is called its total *inertial mass per unit volume* and is denoted ρ_{ji}^{inert}:

$$T^{j'0'} = \rho_{ji}^{\text{inert}} v_i. \tag{2.77}$$

In other words, ρ_{ji}^{inert} is the 3-dimensional tensor that gives the momentum density $G_{j'}$ when the medium's small velocity is put into its second slot.

(a) Using a Lorentz transformation from the medium's (unprimed) rest frame to the (primed) laboratory frame, show that

$$\rho_{ji}^{\text{inert}} = T^{00}\delta_{ji} + T_{ji}. \tag{2.78}$$

(b) Give a physical explanation of the contribution $T_{ji}v_i$ to the momentum density.

(c) Show that for a perfect fluid [Eq. (2.74b)] the inertial mass per unit volume is isotropic and has magnitude $\rho + P$, where ρ is the mass-energy density, and P is the pressure measured in the fluid's rest frame:

$$\boxed{\rho_{ji}^{\text{inert}} = (\rho + P)\delta_{ji}.} \tag{2.79}$$

See Ex. 2.26 for this inertial-mass role of $\rho + P$ in the law of force balance.

Exercise 2.28 **Example: Stress-Energy Tensor, and Energy-Momentum
Conservation for the Electromagnetic Field*

(a) From Eqs. (2.75) and (2.45) compute the components of the electromagnetic stress-energy tensor in an inertial reference frame (in Gaussian units). Your answer should be the expressions given in electrodynamics textbooks:

$$T^{00} = \frac{\mathbf{E}^2 + \mathbf{B}^2}{8\pi}, \qquad \mathbf{G} = T^{0j}\mathbf{e}_j = T^{j0}\mathbf{e}_j = \frac{\mathbf{E} \times \mathbf{B}}{4\pi},$$

$$T^{jk} = \frac{1}{8\pi}\left[(\mathbf{E}^2 + \mathbf{B}^2)\delta_{jk} - 2(E_j E_k + B_j B_k)\right]. \tag{2.80}$$

(In SI units, $4\pi \to \mu_0 = 1/\epsilon_0$.) See also Ex. 1.14 for an alternative derivation of the stress tensor T_{jk}.

(b) Show that the divergence of the stress-energy tensor (2.75) is given by

$$T^{\mu\nu}{}_{;\nu} = \frac{1}{4\pi}(F^{\mu\alpha}{}_{;\nu}F^{\nu}{}_{\alpha} + F^{\mu\alpha}F^{\nu}{}_{\alpha;\nu} - \frac{1}{2}F_{\alpha\beta}{}^{;\mu}F^{\alpha\beta}). \tag{2.81a}$$

(c) Combine this with Maxwell's equations (2.48) to show that

$$\boldsymbol{\nabla} \cdot \boldsymbol{T} = -\mathbf{F}(\underline{}, \vec{J}); \quad \text{i.e., } T^{\alpha\beta}{}_{;\beta} = -F^{\alpha\beta}J_{\beta}. \tag{2.81b}$$

(d) The matter that carries the electric charge and current can exchange energy and momentum with the electromagnetic field. Explain why Eq. (2.81b) is the rate per unit volume at which that matter feeds 4-momentum into the electromagnetic field, and conversely, $+F^{\alpha\mu}J_{\mu}$ is the rate per unit volume at which the electromagnetic field feeds 4-momentum into the matter. Show, further, that (as viewed in any reference frame) the time and space components of this quantity are

$$\frac{d\mathcal{E}_{\text{matter}}}{dt\,dV} = F^{0j}J_j = \mathbf{E} \cdot \mathbf{j}, \qquad \frac{d\mathbf{p}_{\text{matter}}}{dt\,dV} = \rho_e\mathbf{E} + \mathbf{j} \times \mathbf{B}, \tag{2.81c}$$

where ρ_e is charge density, and $\mathbf{j}$ is current density [Eq. (2.49)]. The first of these equations describes Ohmic heating of the matter by the electric field, and the second gives the Lorentz force per unit volume on the matter (cf. Ex. 1.14b).

Bibliographic Note

For an inspiring taste of the history of special relativity, see the original papers by Einstein, Lorentz, and Minkowski, translated into English and archived in Lorentz et al. (1923).

Early relativity textbooks [see the bibliography in Jackson (1999, pp. 566–567)] emphasized the transformation properties of physical quantities, in going from one inertial frame to another, rather than their roles as frame-invariant geometric objects. Minkowski (1908) introduced geometric thinking, but only in recent decades—in large measure due to the influence of John Wheeler—has the geometric viewpoint gained ascendancy.

In our opinion, the best elementary introduction to special relativity is the first edition of Taylor and Wheeler (1966); the more ponderous second edition (Taylor and Wheeler, 1992) is also good. At an intermediate level we strongly recommend the special relativity portions of Hartle (2003).

At a more advanced level, comparable to this chapter, we recommend Goldstein, Poole, and Safko (2002) and the special relativity sections of Misner, Thorne, and Wheeler (1973), Carroll (2004), and Schutz (2009).

These all adopt the geometric viewpoint that we espouse. In this chapter, so far as possible, we have minimized the proliferation of mathematical concepts (avoiding, e.g., differential forms and dual bases). By contrast, the other advanced treatments cited above embrace the richer mathematics.

Much less geometric than these references but still good, in our view, are the special relativity sections of popular electrodynamics texts: Griffiths (1999) at an intermediate level and Jackson (1999) at a more advanced level. We recommend avoiding special relativity treatments that use imaginary time and thereby obfuscate (e.g., earlier editions of Goldstein and of Jackson, and also the more modern and otherwise excellent Zangwill (2013)).

PART II

STATISTICAL PHYSICS

In this second part of the book, we study aspects of classical statistical physics that every physicist should know but that are not usually treated in elementary thermodynamics courses. Our study lays the microphysical (particle-scale) foundations for the continuum physics of Parts III–VII, and it elucidates the intimate connections between relativistic statistical physics and the Newtonian theory, and between quantum statistical physics and the classical theory. Our treatment is both Newtonian and relativistic. Readers who prefer a solely Newtonian treatment can skip the (rather few) relativistic sections. Throughout, we presume that readers are familiar with elementary thermodynamics but not with other aspects of statistical physics.

In Chap. 3, we study *kinetic theory*—the simplest of all formalisms for analyzing systems of huge numbers of particles (e.g., molecules of air, neutrons diffusing through a nuclear reactor, or photons produced in the big-bang origin of the universe). In kinetic theory, the key concept is the distribution function, or number density of particles in phase space, $\mathcal{N}$, that is, the number of particles of some species (e.g., electrons) per unit of physical space and of momentum space. In special relativity, despite first appearances, this $\mathcal{N}$ turns out to be a geometric, reference-frame-independent entity (a scalar field in phase space). This $\mathcal{N}$ and the frame-independent laws it obeys provide us with a means for computing, from microphysics, the macroscopic quantities of continuum physics: mass density, thermal energy density, pressure, equations of state, thermal and electrical conductivities, viscosities, diffusion coefficients, ...

In Chap. 4, we develop the foundations of *statistical mechanics*. Here our statistical study is more sophisticated than in kinetic theory: we deal with ensembles of physical systems. Each ensemble is a (conceptual) collection of a huge number of physical systems that are identical in the sense that they all have the same degrees of freedom, but different in that their degrees of freedom may be in different physical states. For example, the systems in an ensemble might be balloons that are each filled with 10^{23} air molecules so each is describable by 3×10^{23} spatial coordinates (the $\{x, y, z\}$ of all the molecules) and 3×10^{23} momentum coordinates (the $\{p_x, p_y, p_z\}$ of all the molecules). The state of one of the balloons is fully described, then, by 6×10^{23} numbers. We introduce a distribution function $\mathcal{N}$ that is a function of these 6×10^{23} different coordinates (i.e., it is defined in a phase space with 6×10^{23} dimensions). This distribution function tells us how many systems (balloons) in our ensemble lie in a unit volume of that phase space. Using this distribution function, we study such issues as the statistical meaning of entropy, the relationship between entropy and information, the statistical origin of the second law of thermodynamics, the statistical meaning of "thermal equilibrium," and the evolution of ensembles into thermal equilibrium. Our applications include derivations of the Fermi-Dirac distribution for fermions in thermal equilibrium and the Bose-Einstein distribution for bosons, a study of Bose-Einstein condensation in a dilute gas, and explorations of the meaning and role of entropy in gases, black holes, and the universe as a whole.

In Chap. 5, we use the tools of statistical mechanics to study *statistical thermodynamics:* ensembles of systems that are in or near thermal equilibrium (also called statistical equilibrium). Using statistical mechanics, we derive the laws of thermodynamics, and we learn how to use thermodynamic and statistical mechanical tools, hand in hand, to study not only equilibria but also the probabilities for random, spontaneous fluctuations away from equilibrium. Among the applications we study are: (i) chemical and particle reactions, such as ionization equilibrium in a hot gas, and electron-positron pair formation in a still hotter gas and (ii) phase transitions, such as the freezing, melting, vaporization, and condensation of water. We focus special attention on a ferromagnetic phase transition, in which the magnetic moments of atoms spontaneously align with one another as iron is cooled, using it to illustrate two elegant and powerful techniques of statistical physics: the renormalization group and Monte Carlo methods.

In Chap. 6, we develop the theory of random processes (a modern, mathematical component of which is the theory of stochastic differential equations). Here we study the dynamical evolution of processes that are influenced by a huge number of factors over which we have little control and little knowledge, except of their statistical properties. One example is the Brownian motion of a dust particle being buffeted by air molecules; another is the motion of a pendulum used, say, in a gravitational-wave interferometer, where one monitors that motion so accurately that one can see the influences of seismic vibrations and of fluctuating thermal (Nyquist) forces in the pendulum's suspension wire. The position of such a dust particle or pendulum cannot be predicted as a function of time, but one can compute the probability that it will evolve in a given manner. The theory of random processes is a theory of the evolution of the position's probability distribution (and the probability distribution for any other entity driven by random, fluctuating influences). Among the random-process concepts we study are spectral densities, correlation functions, the Fokker-Planck equation (which governs the evolution of probability distributions), and the fluctuation-dissipation theorem (which says that, associated with any kind of friction, there must be fluctuating forces whose statistical properties are determined by the strength of the friction and the temperature of the entities that produce the friction).

The theory of random processes, as treated in Chap. 6, also includes the theory of signals and noise. At first sight this undeniably important topic, which lies at the heart of experimental and observational science, might seem outside the scope of this book. However, we shall discover that it is intimately connected to statistical physics and that similar principles to those used to describe, say, Brownian motion, are appropriate when thinking about, for example, how to detect the electronic signal of a rare particle event against a strong and random background. We study, for example, techniques for extracting weak signals from noisy data by filtering the data, and the limits that noise places on the accuracies of physics experiments and on the reliability of communications channels.

CHAPTER THREE

Kinetic Theory

The gaseous condition is exemplified in the soirée, where the members rush about confusedly, and the only communication is during a collision, which in some instances may be prolonged by button-holing.

JAMES CLERK MAXWELL (1873)

3.1 Overview

In this chapter, we study kinetic theory, the simplest of all branches of statistical physics. Kinetic theory deals with the statistical distribution of a "gas" made from a huge number of "particles" that travel freely, without collisions, for distances (*mean free paths*) long compared to their sizes.

Examples of particles (*italicized*) and phenomena that can be studied via kinetic theory are:

- Whether *neutrons* in a nuclear reactor can survive long enough to maintain a nuclear chain reaction and keep the reactor hot.

- How *galaxies,* formed in the early universe, congregate into clusters as the universe expands.

- How spiral structure develops in the distribution of a galaxy's *stars.*

- How, deep inside a white-dwarf star, relativistic degeneracy influences the equation of state of the star's *electrons and protons.*

- How a supernova explosion affects the evolution of the density and temperature of *interstellar molecules.*

- How anisotropies in the expansion of the universe affect the temperature distribution of the *cosmic microwave photons*—the remnants of the big bang.

- How changes of a metal's temperature affect its thermal and electrical conductivity (with the heat and current carried by *electrons*).

Most of these applications involve particle speeds small compared to that of light and so can be studied with Newtonian theory, but some involve speeds near or at the speed of light and require relativity. Accordingly, we develop both versions of the theory, Newtonian and relativistic, and demonstrate that the Newtonian theory is the low-speed limit of the relativistic theory. As is discussed in the Readers' Guide

- This chapter develops nonrelativistic (Newtonian) kinetic theory and also the relativistic theory. Sections and exercises labeled **N** are Newtonian, and those labeled **R** are relativistic. The **N** material can be read without the **R** material, but the **R** material requires the **N** material as a foundation. The **R** material is all Track Two.

- This chapter relies on the geometric viewpoint about physics developed in Chap. 1 for Newtonian physics and in Chap. 2 for relativistic physics. It especially relies on
    - Secs. 1.4 and 1.5.2 on Newtonian particle kinetics,
    - Secs. 2.4.1 and 2.6 on relativistic particle kinetics,
    - Sec. 1.8 for the Newtonian conservation laws for particles,
    - Sec. 2.12.3 for the relativistic number-flux 4-vector and its conservation law,
    - Sec. 1.9 on the Newtonian stress tensor and its role in conservation laws for momentum,
    - Sec. 2.13 on the relativistic stress-energy tensor and its role in conservation laws for 4-momentum, and
    - Sec. 1.10 on aspects of relativity theory that Newtonian readers will need.

- The Newtonian parts of this chapter are a crucial foundation for the remainder of Part II (Statistical Physics) of this book, for small portions of Part V (Fluid Dynamics; especially equations of state, the origin of viscosity, and the diffusion of heat in fluids), and for half of Part VI (Plasma Physics; Chaps. 22 and 23).

(Box 3.1), the relativistic material is all Track Two and can be skipped by readers who are focusing on the (nonrelativistic) Newtonian theory.

We begin in Sec. 3.2 by introducing the concepts of momentum space, phase space (the union of physical space and momentum space), and the distribution function (number density of particles in phase space). We meet several different versions of the distribution function, all equivalent, but each designed to optimize conceptual thinking or computations in a particular arena (e.g., photons, plasma physics, and the interface with quantum theory). In Sec. 3.3, we study the distribution functions that characterize systems of particles in thermal equilibrium. There are three such equilibrium distributions: one for quantum mechanical particles with half-integral spin (fermions), another for quantum particles with integral spin (bosons), and a third for classical particles. As special applications, we derive the Maxwell velocity distribution for low-speed, classical particles (Ex. 3.4) and its high-speed relativistic analog (Ex. 3.5 and Fig. 3.6 below), and we compute the effects of observers' motions on their measurements of the cosmic microwave radiation created in our universe's big-bang origin (Ex. 3.7). In Sec. 3.4, we learn how to compute macroscopic, physical-space quantities (particle density and flux, energy density, stress tensor, stress-energy tensor, . . .) by integrating over the momentum portion of phase space. In Sec. 3.5, we show that, if the distribution function is isotropic in momentum space, in some reference frame, then on macroscopic scales the particles constitute a perfect fluid. We use our momentum-space integrals to evaluate the equations of state of various kinds of perfect fluids: a nonrelativistic hydrogen gas in both the classical, nondegenerate regime and the regime of electron degeneracy (Sec. 3.5.2), a relativistically degenerate gas (Sec. 3.5.4), and a photon gas (Sec. 3.5.5). We use our results to discuss the physical nature of matter as a function of density and temperature (see Fig. 3.7).

In Sec. 3.6, we study the evolution of the distribution function, as described by Liouville's theorem and by the associated collisionless Boltzmann equation when collisions between particles are unimportant, and by the Boltzmann transport equation when collisions are significant. We use a simple variant of these evolution laws to study the heating of Earth by the Sun, and the key role played by the greenhouse effect (Ex. 3.15). Finally, in Sec. 3.7, we learn how to use the Boltzmann transport equation to compute the transport coefficients (diffusion coefficient, electrical conductivity, thermal conductivity, and viscosity) that describe the diffusive transport of particles, charge, energy, and momentum through a gas of particles that collide frequently. We then use the Boltzmann transport equation to study a chain reaction in a nuclear reactor (Ex. 3.21).

Readers who feel overwhelmed by the enormous amount and variety of applications in this chapter (and throughout this book) should remember the authors' goals: We want readers to learn the fundamental concepts of kinetic theory (and other topics in this book), and we want them to meet a variety of applications, so they will understand how the fundamental concepts are used. However, we do not expect readers to become expert in or even remember all these applications. To do so would require much more time and effort than most readers can afford or should expend.

3.2 Phase Space and Distribution Function

3.2.1 Newtonian Number Density in Phase Space, $\mathcal{N}$ **N**

In Newtonian, 3-dimensional space (*physical space*), consider a particle with rest mass m that moves along a path $\mathbf{x}(t)$ as universal time t passes (Fig. 3.1a). The particle's time-varying velocity and momentum are $\mathbf{v}(t) = d\mathbf{x}/dt$ and $\mathbf{p}(t) = m\mathbf{v}$. The path $\mathbf{x}(t)$ is a curve in the physical space, and the momentum $\mathbf{p}(t)$ is a time-varying, coordinate-independent vector in the physical space.

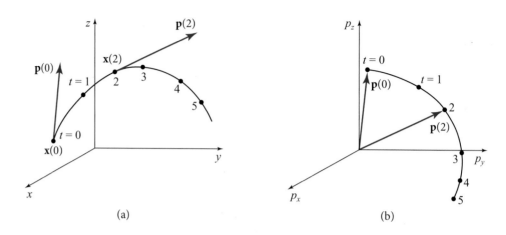

FIGURE 3.1 (a) Euclidean physical space, in which a particle moves along a curve $\mathbf{x}(t)$ that is parameterized by universal time t. In this space, the particle's momentum $\mathbf{p}(t)$ is a vector tangent to the curve. (b) Momentum space, in which the particle's momentum vector $\mathbf{p}$ is placed, unchanged, with its tail at the origin. As time passes, the momentum's tip sweeps out the indicated curve $\mathbf{p}(t)$.

momentum space

It is useful to introduce an auxiliary 3-dimensional space, called *momentum space*, in which we place the tail of $\mathbf{p}(t)$ at the origin. As time passes, the tip of $\mathbf{p}(t)$ sweeps out a curve in momentum space (Fig. 3.1b). This momentum space is "secondary" in the sense that it relies for its existence on the physical space of Fig. 3.1a. Any Cartesian coordinate system of physical space, in which the location $\mathbf{x}(t)$ of the particle has coordinates $\{x, y, z\}$, induces in momentum space a corresponding coordinate system $\{p_x, p_y, p_z\}$. The 3-dimensional physical space and 3-dimensional

phase space

momentum space together constitute a 6-dimensional *phase space*, with coordinates $\{x, y, z, p_x, p_y, p_z\}$.

In this chapter, we study a collection of a very large number of identical particles (all with the same rest mass m).[1] As tools for this study, consider a tiny 3-dimensional volume dV_x centered on some location $\mathbf{x}$ in physical space and a tiny 3-dimensional volume dV_p centered on location $\mathbf{p}$ in momentum space. Together these make up a tiny 6-dimensional volume

$$d^2V \equiv dV_x dV_p. \tag{3.1}$$

In any Cartesian coordinate system, we can think of dV_x as being a tiny cube located at (x, y, z) and having edge lengths dx, dy, dz, and similarly for dV_p. Then, as computed in this coordinate system, these tiny volumes are

$$dV_x = dx\, dy\, dz, \quad dV_p = dp_x\, dp_y\, dp_z, \quad d^2V = dx\, dy\, dz\, dp_x\, dp_y\, dp_z. \tag{3.2}$$

1. In Ex. 3.2 and Box 3.2, we extend kinetic theory to particles with a range of rest masses.

Denote by dN the number of particles (all with rest mass m) that reside inside $d^2\mathcal{V}$ in phase space (at some moment of time t). Stated more fully: dN is the number of particles that, at time t, are located in the 3-volume $d\mathcal{V}_x$ centered on the location $\mathbf{x}$ in physical space and that also have momentum vectors whose tips at time t lie in the 3-volume $d\mathcal{V}_p$ centered on location $\mathbf{p}$ in momentum space. Denote by

$$\mathcal{N}(\mathbf{x}, \mathbf{p}, t) \equiv \frac{dN}{d^2\mathcal{V}} = \frac{dN}{d\mathcal{V}_x \, d\mathcal{V}_p} \tag{3.3}$$

distribution function

the *number density of particles at location* $(\mathbf{x}, \mathbf{p})$ *in phase space at time* t. This is also called the *distribution function*.

This distribution function is kinetic theory's principal tool for describing any collection of a large number of identical particles.

In Newtonian theory, the volumes $d\mathcal{V}_x$ and $d\mathcal{V}_p$ occupied by our collection of dN particles are independent of the reference frame that we use to view them. Not so in relativity theory: $d\mathcal{V}_x$ undergoes a Lorentz contraction when one views it from a moving frame, and $d\mathcal{V}_p$ also changes; but (as we shall see in Sec. 3.2.2) their product $d^2\mathcal{V} = d\mathcal{V}_x d\mathcal{V}_p$ is the same in all frames. Therefore, in both Newtonian theory and relativity theory, the distribution function $\mathcal{N} = dN/d^2\mathcal{V}$ is independent of reference frame, and also, of course, independent of any choice of coordinates. It is a coordinate-independent scalar in phase space.

3.2.2 Relativistic Number Density in Phase Space, $\mathcal{N}$ R T2

SPACETIME

We define the special relativistic distribution function in precisely the same way as the nonrelativistic one, $\mathcal{N}(\mathbf{x}, \mathbf{p}, t) \equiv dN/d^2\mathcal{V} = dN/d\mathcal{V}_x d\mathcal{V}_p$, except that now $\mathbf{p}$ is the relativistic momentum ($\mathbf{p} = m\mathbf{v}/\sqrt{1 - v^2}$ if the particle has nonzero rest mass m). This definition of $\mathcal{N}$ appears, at first sight, to be frame dependent, since the physical 3-volume $d\mathcal{V}_x$ and momentum 3-volume $d\mathcal{V}_p$ do not even exist until we have selected a specific reference frame. In other words, this definition appears to violate our insistence that relativistic physical quantities be described by frame-independent geometric objects that live in 4-dimensional spacetime. In fact, the distribution function defined in this way *is* frame independent, though it does not look so. To elucidate this, we shall develop carefully and somewhat slowly the 4-dimensional spacetime ideas that underlie this relativistic distribution function.

Consider, as shown in Fig. 3.2a, a classical particle with rest mass m moving through spacetime along a world line $\mathcal{P}(\zeta)$, or equivalently $\vec{x}(\zeta)$, where ζ is an affine parameter related to the particle's 4-momentum by

$$\vec{p} = d\vec{x}/d\zeta \tag{3.4a}$$

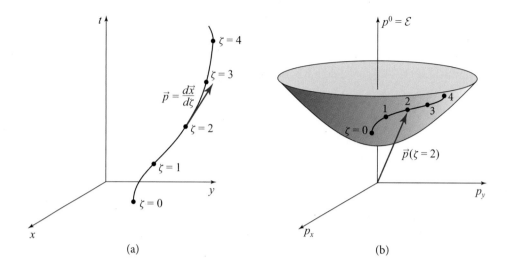

FIGURE 3.2 (a) The world line $\vec{x}(\zeta)$ of a particle in spacetime (with one spatial coordinate, z, suppressed), parameterized by a parameter ζ that is related to the particle's 4-momentum by $\vec{p} = d\vec{x}/d\zeta$. (b) The trajectory of the particle in momentum space. The particle's 4-momentum is confined to the mass hyperboloid, $\vec{p}^2 = -m^2$ (also known as the mass shell).

[as discussed following Eq. (2.10)]. If the particle has nonzero rest mass, then its 4-velocity $\vec{u}$ and proper time τ are related to its 4-momentum and affine parameter by

$$\vec{p} = m\vec{u}, \qquad \zeta = \tau/m \tag{3.4b}$$

[Eqs. (2.10) and (2.11)], and we can parameterize the world line by either τ or ζ. If the particle has zero rest mass, then its world line is null, and τ does not change along it, so we have no choice but to use ζ as the world line's parameter.

MOMENTUM SPACE AND MASS HYPERBOLOID

The particle can be thought of not only as living in 4-dimensional spacetime (Fig. 3.2a), but also as living in a 4-dimensional momentum space (Fig. 3.2b). Mo-

4-dimensional momentum space

mentum space, like spacetime, is a geometric, coordinate-independent concept: each point in momentum space corresponds to a specific 4-momentum $\vec{p}$. The tail of the vector $\vec{p}$ sits at the origin of momentum space, and its tip sits at the point representing $\vec{p}$. The momentum-space diagram drawn in Fig. 3.2b has as its coordinate axes the components $(p^0, p^1 = p_1 \equiv p_x, p^2 = p_2 \equiv p_y, p^3 = p_3 \equiv p_z)$ of the 4-momentum as measured in some arbitrary inertial frame. Because the squared length of the 4-momentum is always $-m^2$,

$$\vec{p} \cdot \vec{p} = -(p^0)^2 + (p_x)^2 + (p_y)^2 + (p_z)^2 = -m^2, \tag{3.4c}$$

the particle's 4-momentum (the tip of the 4-vector $\vec{p}$) is confined to a hyperboloid in

mass hyperboloid

momentum space. This *mass hyperboloid* requires no coordinates for its existence; it is the frame-independent set of points in momentum space for which $\vec{p} \cdot \vec{p} = -m^2$. If the particle has zero rest mass, then $\vec{p}$ is null, and the mass hyperboloid is a cone with vertex at the origin of momentum space. As in Chap. 2, we often denote the particle's energy p^0 by

$$\mathcal{E} \equiv p^0 \qquad (3.4d)$$

(with the $\mathcal{E}$ in script font to distinguish it from the energy $E = \mathcal{E} - m$ with rest mass removed and its nonrelativistic limit $E = \frac{1}{2}mv^2$), and we embody the particle's spatial momentum in the 3-vector $\mathbf{p} = p_x\mathbf{e}_x + p_y\mathbf{e}_y + p_z\mathbf{e}_z$. Therefore, we rewrite the mass-hyperboloid relation (3.4c) as

$$\mathcal{E}^2 = m^2 + |\mathbf{p}|^2. \qquad (3.4e)$$

If no forces act on the particle, then its momentum is conserved, and its location in momentum space remains fixed. A force (e.g., due to an electromagnetic field) pushes the particle's 4-momentum along some curve in momentum space that lies on the mass hyperboloid. If we parameterize that curve by the same parameter ζ as we use in spacetime, then the particle's trajectory in momentum space can be written abstractly as $\vec{p}(\zeta)$. Such a trajectory is shown in Fig. 3.2b.

Because the mass hyperboloid is 3-dimensional, we can characterize the particle's location on it by just three coordinates rather than four. We typically use as those coordinates the spatial components of the particle's 4-momentum, (p_x, p_y, p_z), or the spatial momentum vector $\mathbf{p}$ as measured in some specific (but usually arbitrary) inertial frame.

PHASE SPACE

Momentum space and spacetime, taken together, constitute the relativistic *phase space*. We can regard phase space as 8-dimensional (four spacetime dimensions plus four momentum space dimensions). Alternatively, if we think of the 4-momentum as confined to the 3-dimensional mass hyperboloid, then we can regard phase space as 7-dimensional. This 7- or 8-dimensional phase space, by contrast with the non-relativistic 6-dimensional phase space, is frame independent. No coordinates or reference frame are actually needed to define spacetime and explore its properties, and none are needed to define and explore 4-momentum space or the mass hyperboloid—though inertial (Lorentz) coordinates are often helpful in practical situations.

phase space

VOLUMES IN PHASE SPACE AND DISTRIBUTION FUNCTION

Now turn attention from an individual particle to a collection of a huge number of identical particles, each with the same rest mass m, and allow m to be finite or zero (it does not matter which). Examine those particles that pass close to a specific event $\mathcal{P}$ (also denoted $\vec{x}$) in spacetime; and *examine them from the viewpoint of a specific observer, who lives in a specific inertial reference frame.* Figure 3.3a is a spacetime diagram drawn in that observer's frame. As seen in that frame, the event $\mathcal{P}$ occurs at time t and spatial location (x, y, z).

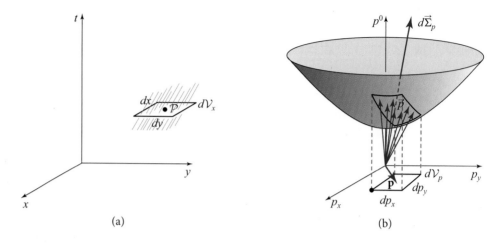

FIGURE 3.3 Definition of the distribution function from the viewpoint of a specific observer in a specific inertial reference frame, whose coordinate axes are used in these drawings. (a) At the event $\mathcal{P}$, the observer selects a 3-volume $d\mathcal{V}_x$ and focuses on the set $\mathcal{S}$ of particles that lie in $d\mathcal{V}_x$. (b) These particles have momenta lying in a region of the mass hyperboloid that is centered on $\vec{p}$ and has 3-momentum volume $d\mathcal{V}_p$. If dN is the number of particles in that set $\mathcal{S}$, then $\mathcal{N}(\mathcal{P}, \vec{p}) \equiv dN/d\mathcal{V}_x d\mathcal{V}_p$.

We ask the observer, at the time t of the chosen event, to define the distribution function $\mathcal{N}$ in identically the same way as in Newtonian theory, except that $\mathbf{p}$ is the relativistic spatial momentum $\mathbf{p} = m\mathbf{v}/\sqrt{1-v^2}$ instead of the nonrelativistic $\mathbf{p} = m\mathbf{v}$. Specifically, the observer, in her inertial frame, chooses a tiny 3-volume

$$dV_x = dx\,dy\,dz \tag{3.5a}$$

centered on location $\mathcal{P}$ (little horizontal rectangle shown in Fig. 3.3a) and a tiny 3-volume

$$dV_p = dp_x\,dp_y\,dp_z \tag{3.5b}$$

centered on $\mathbf{p}$ in momentum space (little rectangle in the p_x-p_y plane in Fig. 3.3b). Ask the observer to focus on the set $\mathcal{S}$ of particles that lie in $d\mathcal{V}_x$ and have spatial momenta in $d\mathcal{V}_p$ (Fig. 3.3). If there are dN particles in this set $\mathcal{S}$, then the observer will identify

relativistic distribution function

$$\mathcal{N} \equiv \frac{dN}{d\mathcal{V}_x d\mathcal{V}_p} \equiv \frac{dN}{d^2\mathcal{V}} \tag{3.6}$$

as the number density of particles in phase space or *distribution function*.

Notice in Fig. 3.3b that the *4-momenta* of the particles in $\mathcal{S}$ have their tails at the origin of momentum space (as by definition do all 4-momenta) and have their tips in a tiny rectangular box on the mass hyperboloid—a box centered on the 4-momentum $\vec{p}$ whose spatial part is $\mathbf{p}$ and temporal part is $p^0 = \mathcal{E} = \sqrt{m^2 + \mathbf{p}^2}$.

The momentum volume element dV_p is the projection of that mass-hyperboloid box onto the horizontal (p_x, p_y, p_z) plane in momentum space. [The mass-hyperboloid box itself can be thought of as a (frame-independent) vectorial 3-volume $d\vec{\Sigma}_p$—the momentum-space version of the vectorial 3-volume introduced in Sec. 2.12.1; see below.]

The number density $\mathcal{N}$ depends on the location $\mathcal{P}$ in spacetime of the 3-volume dV_x and on the 4-momentum $\vec{p}$ about which the momentum volume on the mass hyperboloid is centered: $\mathcal{N} = \mathcal{N}(\mathcal{P}, \vec{p})$. From the chosen observer's viewpoint, it can be regarded as a function of time t and spatial location $\mathbf{x}$ (the coordinates of $\mathcal{P}$) and of spatial momentum $\mathbf{p}$.

At first sight, one might expect $\mathcal{N}$ to depend also on the inertial reference frame used in its definition (i.e., on the 4-velocity of the observer). If this were the case (i.e., if $\mathcal{N}$ at fixed $\mathcal{P}$ and $\vec{p}$ were different when computed by the above prescription using different inertial frames), then we would feel compelled to seek some other object— one that is frame-independent—to serve as our foundation for kinetic theory. This is because the principle of relativity insists that all fundamental physical laws should be expressible in frame-independent language.

Fortunately, the distribution function (3.6) is frame independent by itself: it is a frame-independent scalar field in phase space, so we need seek no further.

PROOF OF FRAME INDEPENDENCE OF $\mathcal{N} = dN/d^2V$

To prove the frame independence of $\mathcal{N}$, we shall consider the frame dependence of the spatial 3-volume dV_x, then the frame dependence of the momentum 3-volume dV_p, and finally the frame dependence of their product $d^2V = dV_x dV_p$ and thence of the distribution function $\mathcal{N} = dN/d^2V$.

The thing that identifies the 3-volume dV_x and 3-momentum dV_p is the set of particles $\mathcal{S}$. We select that set once and for all and hold it fixed, and correspondingly, the number of particles dN in the set is fixed. Moreover, we assume that the particles' rest mass m is nonzero and shall deal with the zero-rest-mass case at the end by taking the limit $m \to 0$. Then there is a preferred frame in which to observe the particles $\mathcal{S}$: their own rest frame, which we identify by a prime.

In their rest frame and at a chosen event $\mathcal{P}$, the particles $\mathcal{S}$ occupy the interior of some box with imaginary walls that has some 3-volume $dV_{x'}$. As seen in some other "laboratory" frame, their box has a Lorentz-contracted volume $dV_x = \sqrt{1 - v^2}\, dV_{x'}$. Here v is their speed as seen in the laboratory frame. The Lorentz-contraction factor is related to the particles' energy, as measured in the laboratory frame, by $\sqrt{1 - v^2} = m/\mathcal{E}$, and therefore $\mathcal{E} dV_x = m dV_{x'}$. The right-hand side is a frame-independent constant m times a well-defined number that everyone can agree on: the particles' rest-frame volume $dV_{x'}$, i.e.,

$$\boxed{\mathcal{E} dV_x = \text{(a frame-independent quantity).}} \qquad (3.7a)$$

Thus, the spatial volume dV_x occupied by the particles is frame dependent, and their energy $\mathcal{E}$ is frame dependent, but the product of the two is independent of reference frame.

Turn now to the frame dependence of the particles' 3-volume dV_p. As one sees from Fig. 3.3b, dV_p is the projection of the frame-independent mass-hyperboloid region $d\vec{\Sigma}_p$ onto the laboratory's xyz 3-space. Equivalently, it is the time component $d\Sigma_p^0$ of $d\vec{\Sigma}_p$. Now, the 4-vector $d\vec{\Sigma}_p$, like the 4-momentum $\vec{p}$, is orthogonal to the mass hyperboloid at the common point where they intersect it, and therefore $d\vec{\Sigma}_p$ is parallel to $\vec{p}$. This means that, when one goes from one reference frame to another, the time components of these two vectors will grow or shrink in the same manner: $d\vec{\Sigma}_p^0 = dV_p$ is proportional to $p^0 = \mathcal{E}$, so their ratio must be frame independent:

$$\boxed{\frac{dV_p}{\mathcal{E}} = \text{(a frame-independent quantity).}} \tag{3.7b}$$

(If this sophisticated argument seems too slippery to you, then you can develop an alternative, more elementary proof using simpler 2-dimensional spacetime diagrams: Ex. 3.1.)

By taking the product of Eqs. (3.7a) and (3.7b) we see that for our chosen set of particles $\mathcal{S}$,

$$dV_x dV_p = d^2V = \text{(a frame-independent quantity);} \tag{3.7c}$$

and since the number of particles in the set, dN, is obviously frame-independent, we conclude that

frame independence of relativistic distribution function

$$\boxed{\mathcal{N} = \frac{dN}{dV_x dV_p} \equiv \frac{dN}{d^2V} = \text{(a frame-independent quantity).}} \tag{3.8}$$

Although we assumed nonzero rest mass ($m \neq 0$) in our derivation, the conclusions that $\mathcal{E} dV_x$ and $dV_p/\mathcal{E}$ are frame independent continue to hold if we take the limit as $m \to 0$ and the 4-momenta become null. Correspondingly, Eqs. (3.7a)–(3.8) are valid for particles with zero as well as nonzero rest mass.

EXERCISES

Exercise 3.1 *Derivation and Practice: Frame Dependences of dV_x and dV_p* R T2
Use the 2-dimensional spacetime diagrams of Fig. 3.4 to show that $\mathcal{E} dV_x$ and $dV_p/\mathcal{E}$ are frame independent [Eqs. (3.7a) and (3.7b)].

Exercise 3.2 **Example: Distribution Function for Particles with a Range of Rest Masses* R T2
A galaxy such as our Milky Way contains $\sim 10^{12}$ stars—easily enough to permit a kinetic-theory description of their distribution; each star contains so many atoms ($\sim 10^{56}$) that the masses of the stars can be regarded as continuously distributed, not

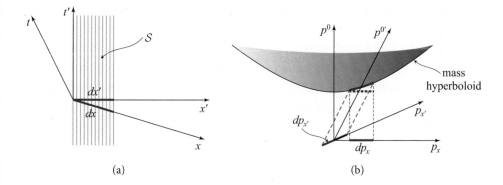

FIGURE 3.4 (a) Spacetime diagram drawn from the viewpoint of the (primed) rest frame of the particles S for the special case where the laboratory frame moves in the $-x'$ direction with respect to them. (b) Momentum-space diagram drawn from viewpoint of the unprimed observer.

discrete. Almost everywhere in a galaxy, the stars move with speeds small compared to light, but deep in the cores of most galaxies there resides a massive black hole near which the stars move with relativistic speeds. In this exercise we explore the foundations for treating such a system: "particles" with continuously distributed rest masses and relativistic speeds.

(a) For a subset S of particles like that of Fig. 3.3 and associated discussion, but with a range of rest masses dm centered on some value m, introduce the phase-space volume $d^2\mathcal{V} \equiv d\mathcal{V}_x d\mathcal{V}_p dm$ that the particles S occupy. Explain why this occupied volume is frame invariant.

(b) Show that this invariant occupied volume can be rewritten as $d^2\mathcal{V} = (d\mathcal{V}_x \, \mathcal{E}/m)(d\mathcal{V}_p d\mathcal{E}) = (d\mathcal{V}_x \, \mathcal{E}/m)(dp^0 dp^x dp^y dp^z)$. Explain the physical meaning of each term in parentheses, and show that each is frame invariant.

If the number of particles in the set S is dN, then we define the frame-invariant distribution function by

$$\mathcal{N} \equiv \frac{dN}{d^2\mathcal{V}} = \frac{dN}{d\mathcal{V}_x d\mathcal{V}_p dm}. \tag{3.9}$$

This is a function of location $\mathcal{P}$ in 4-dimensional spacetime and location $\vec{p}$ in 4-dimensional momentum space (not confined to the mass hyperboloid), and thus a function of location in 8-dimensional phase space. We explore the evolution of this distribution function in Box 3.2 (near the end of Sec. 3.6).

3.2.3 Distribution Function $f(\mathbf{x}, \mathbf{v}, t)$ for Particles in a Plasma

3.2.3

The normalization that one uses for the distribution function is arbitrary: renormalize $\mathcal{N}$ by multiplying with any constant, and $\mathcal{N}$ will still be a geometric, coordinate-independent, and frame-independent quantity and will still contain the same information as before. In this book, we use several renormalized versions of $\mathcal{N}$, depending on the situation. We now introduce them, beginning with the version used in plasma physics.

In Part VI, when dealing with nonrelativistic plasmas (collections of electrons and ions that have speeds small compared to light), we regard the distribution function as depending on time t, location $\mathbf{x}$ in Euclidean space, and velocity $\mathbf{v}$ (instead of momentum $\mathbf{p} = m\mathbf{v}$), and we denote it by[2]

plasma distribution function

$$f(t, \mathbf{x}, \mathbf{v}) \equiv \frac{dN}{d\mathcal{V}_x \, d\mathcal{V}_v} = \frac{dN}{dx\,dy\,dz \, dv_x \, dv_y \, dv_z} = m^3 \mathcal{N}. \tag{3.10}$$

(This change of viewpoint and notation when transitioning to plasma physics is typical of this textbook. When presenting any subfield of physics, we usually adopt the conventions, notation, and also the system of units that are generally used in that subfield.)

3.2.4 Distribution Function I_ν/ν^3 for Photons N R

[For readers restricting themselves to the Newtonian portions of this book: Please read Sec. 1.10, which lists a few items of special relativity that you need for the discussion here. As described there, you can deal with photons fairly easily by simply remembering that a photon has zero rest mass, energy $\mathcal{E} = h\nu$, and momentum $\mathbf{p} = (h\nu/c)\mathbf{n}$, where ν is its frequency and $\mathbf{n}$ is a unit vector tangent to the photon's spatial trajectory.]

When dealing with photons or other zero-rest-mass particles, one often expresses $\mathcal{N}$ in terms of the *specific intensity* I_ν. This quantity is defined as follows (see Fig. 3.5). An observer places a CCD (or other measuring device) perpendicular to the photons' propagation direction $\mathbf{n}$—perpendicular as measured in her reference frame. The region of the CCD that the photons hit has surface area dA as measured by her, and because the photons move at the speed of light c, the product of that surface area with c times the time dt that they take to all go through the CCD is equal to the volume they occupy at a specific moment of time:

$$d\mathcal{V}_x = dA\,c\,dt. \tag{3.11a}$$

Focus attention on a set $\mathcal{S}$ of photons in this volume that all have nearly the same frequency ν and propagation direction $\mathbf{n}$ as measured by the observer. Their energies $\mathcal{E}$ and momenta $\mathbf{p}$ are related to ν and $\mathbf{n}$ by

$$\mathcal{E} = h\nu, \quad \mathbf{p} = (h\nu/c)\mathbf{n}, \tag{3.11b}$$

where h is Planck's constant. Their frequencies lie in a range $d\nu$ centered on ν, and they come from a small solid angle $d\Omega$ centered on $-\mathbf{n}$; the volume they occupy in momentum space is related to these quantities by

$$d\mathcal{V}_p = |\mathbf{p}|^2 d\Omega d|\mathbf{p}| = (h\nu/c)^2 d\Omega (h\,d\nu/c) = (h/c)^3 \nu^2 d\Omega\,d\nu. \tag{3.11c}$$

---

2. The generalization to relativistic plasmas is straightforward; see, e.g., Ex. 23.12.

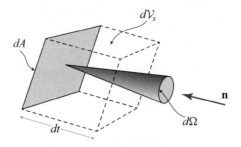

FIGURE 3.5 Geometric construction used in defining the specific intensity I_ν.

The photons' specific intensity, as measured by the observer, is defined to be the total energy

$$d\mathcal{E} = h\nu dN \qquad (3.11\text{d})$$

(where dN is the number of photons) that crosses the CCD per unit area dA, per unit time dt, per unit frequency $d\nu$, and per unit solid angle $d\Omega$ (i.e., per unit everything):

$$I_\nu \equiv \frac{d\mathcal{E}}{dA\,dt\,d\nu\,d\Omega}. \qquad (3.12)$$

specific intensity

(This I_ν is sometimes denoted $I_{\nu\Omega}$.) From Eqs. (3.8), (3.11), and (3.12) we readily deduce the following relationship between this specific intensity and the distribution function:

$$\boxed{\mathcal{N} = \frac{c^2}{h^4}\frac{I_\nu}{\nu^3}.} \qquad (3.13)$$

This relation shows that, with an appropriate renormalization, I_ν/ν^3 *is the photons'* *distribution function.*

photon distribution function

Astronomers and opticians regard the specific intensity (or equally well, I_ν/ν^3) as a function of the photon propagation direction $\mathbf{n}$, photon frequency ν, location $\mathbf{x}$ in space, and time t. By contrast, nonrelativistic physicists regard the distribution function $\mathcal{N}$ as a function of the photon momentum $\mathbf{p}$, location in space, and time; relativistic physicists regard it as a function of the photon 4-momentum $\vec{p}$ (on the photons' mass hyperboloid, which is the light cone) and of location $\mathcal{P}$ in spacetime. Clearly, the information contained in these three sets of variables, the astronomers' set and the two physicists' sets, is the same.

If two different physicists in two different reference frames at the same event in spacetime examine the same set of photons, they will measure the photons to have different frequencies ν (because of the Doppler shift between their two frames). They will also measure different specific intensities I_ν (because of Doppler shifts of frequencies, Doppler shifts of energies, dilation of times, Lorentz contraction of areas of CCDs, and aberrations of photon propagation directions and thence distortions of solid angles). However, if each physicist computes the ratio of the specific intensity that she measures to the cube of the frequency she measures, that ratio, according to Eq. (3.13), will be the same as computed by the other astronomer: the distribution function I_ν/ν^3 will be frame independent.

3.2.5

3.2.5 Mean Occupation Number η N R

Although this book is about classical physics, we cannot avoid making frequent contact with quantum theory. The reason is that modern classical physics rests on a quantum mechanical foundation. Classical physics is an approximation to quantum physics, not conversely. Classical physics is derivable from quantum physics, not conversely.

In statistical physics, the classical theory cannot fully shake itself free from its quantum roots; it must rely on them in crucial ways that we shall meet in this chapter and the next. Therefore, rather than try to free it from its roots, we expose these roots and profit from them by introducing a quantum mechanics-based normalization for the distribution function: the mean occupation number η.

As an aid in defining the mean occupation number, we introduce the concept of the *density of states:* Consider a particle of mass m, described quantum mechanically. Suppose that the particle is known to be located in a volume $d\mathcal{V}_x$ (as observed in a specific inertial reference frame) and to have a spatial momentum in the region $d\mathcal{V}_p$ centered on **p**. Suppose, further, that *the particle does not interact with any other particles or fields;* for example, ignore Coulomb interactions. (In portions of Chaps. 4 and 5, we include interactions.) Then how many single-particle quantum mechanical states[3] are available to the free particle? This question is answered most easily by constructing (in some arbitrary inertial frame) a complete set of wave functions for the particle's spatial degrees of freedom, with the wave functions (i) confined to be eigenfunctions of the momentum operator and (ii) confined to satisfy the standard periodic boundary conditions on the walls of a box with volume $d\mathcal{V}_x$. For simplicity, let the box have edge length L along each of the three spatial axes of the Cartesian spatial coordinates, so $d\mathcal{V}_x = L^3$. (This L is arbitrary and will drop out of our analysis shortly.) Then a complete set of wave functions satisfying (i) and (ii) is the set $\{\psi_{j,k,l}\}$ with

density of states

$$\psi_{j,k,l}(x, y, z) = \frac{1}{L^{3/2}} e^{i(2\pi/L)(jx+ky+lz)} e^{-i\omega t} \tag{3.14a}$$

[cf., e.g., Cohen-Tannoudji, Diu, and Laloë (1977, pp. 1440–1442), especially the Comment at the end of this page range]. Here the demand that the wave function take

---

3. A quantum mechanical state for a single particle is called an "orbital" in the chemistry literature (where the particle is an election) and in the classic thermal physics textbook by Kittel and Kroemer (1980); we shall use physicists' more conventional but cumbersome phrase "single-particle quantum state," and also, sometimes, "mode."

on the same values at the left and right faces of the box ($x = -L/2$ and $x = +L/2$), at the front and back faces, and at the top and bottom faces (the demand for periodic boundary conditions) dictates that the quantum numbers j, k, and l be integers. The basis states (3.14a) are eigenfunctions of the momentum operator $(\hbar/i)\nabla$ with momentum eigenvalues

$$p_x = \frac{2\pi\hbar}{L}j\,, \quad p_y = \frac{2\pi\hbar}{L}k\,, \quad p_z = \frac{2\pi\hbar}{L}l; \qquad (3.14b)$$

correspondingly, the wave function's frequency ω has the following values in Newtonian theory **N** and relativity **R**:

$$\boxed{\text{N}} \quad \hbar\omega = E = \frac{\mathbf{p}^2}{2m} = \frac{1}{2m}\left(\frac{2\pi\hbar}{L}\right)^2 (j^2 + k^2 + l^2); \qquad (3.14c)$$

$$\boxed{\text{R}} \quad \hbar\omega = \mathcal{E} = \sqrt{m^2 + \mathbf{p}^2} \to m + E \text{ in the Newtonian limit.} \qquad (3.14d)$$

Equations (3.14b) tell us that the allowed values of the momentum are confined to lattice sites in 3-momentum space with one site in each cube of side $2\pi\hbar/L$. Correspondingly, the total number of states in the region $d\mathcal{V}_x d\mathcal{V}_p$ of phase space is the number of cubes of side $2\pi\hbar/L$ in the region $d\mathcal{V}_p$ of momentum space:

$$dN_{\text{states}} = \frac{d\mathcal{V}_p}{(2\pi\hbar/L)^3} = \frac{L^3 d\mathcal{V}_p}{(2\pi\hbar)^3} = \frac{d\mathcal{V}_x d\mathcal{V}_p}{h^3}. \qquad (3.15)$$

This is true no matter how relativistic or nonrelativistic the particle may be.

Thus far we have considered only the particle's spatial degrees of freedom. Particles can also have an internal degree of freedom called "spin." For a particle with spin s, the number of independent spin states is

$$g_s = \begin{cases} 2s + 1 & \text{if } m \neq 0 \text{ (e.g., an electron, proton, or atomic nucleus)} \\ 2 & \text{if } m = 0 \text{ and } s > 0 \text{ [e.g., a photon } (s = 1) \text{ or graviton } (s = 2)] \\ 1 & \text{if } m = 0 \text{ and } s = 0 \text{ (i.e., a hypothetical massless scalar particle)} \end{cases}$$

$$\qquad (3.16)$$

A notable exception is each species of neutrino or antineutrino, which has nonzero rest mass and spin 1/2, but $g_s = 1$ rather than $g_s = 2s + 1 = 2$.[4] We call this number of internal spin states g_s the particle's *multiplicity*. [It will turn out to play a crucial role in computing the entropy of a system of particles (Chap. 4); thus, it places the imprint of quantum theory on the entropy of even a highly classical system.]

particle's multiplicity

Taking account of both the particle's spatial degrees of freedom and its spin degree of freedom, we conclude that the total number of independent quantum states

---

4. The reason for the exception is the particle's fixed chirality: -1 for neutrinos and $+1$ for antineutrinos; to have $g_s = 2$, a spin-1/2 particle must admit both chiralities.

available in the region $dV_x dV_p \equiv d^2V$ of phase space is $dN_{\text{states}} = (g_s/h^3)d^2V$, and correspondingly the *number density of states in phase space* is

$$\mathcal{N}_{\text{states}} \equiv \frac{dN_{\text{states}}}{d^2V} = \frac{g_s}{h^3}. \tag{3.17}$$

[Relativistic remark: Note that, although we derived this number density of states using a specific inertial frame, it is a frame-independent quantity, with a numerical value depending only on Planck's constant and (through g_s) the particle's rest mass m and spin s.]

The ratio of the number density of particles to the number density of quantum states is obviously the number of particles in each state (the state's *occupation number*) averaged over many neighboring states—but few enough that the averaging region is small by macroscopic standards. In other words, this ratio is the quantum states' *mean occupation number η*:

$$\eta = \frac{\mathcal{N}}{\mathcal{N}_{\text{states}}} = \frac{h^3}{g_s}\mathcal{N}; \quad \text{i.e.,} \quad \boxed{\mathcal{N} = \mathcal{N}_{\text{states}}\eta = \frac{g_s}{h^3}\eta.} \tag{3.18}$$

The mean occupation number η plays an important role in quantum statistical mechanics, and its quantum roots have a profound impact on classical statistical physics.

From quantum theory we learn that the allowed values of the occupation number for a quantum state depend on whether the state is that of a *fermion* (a particle with spin 1/2, 3/2, 5/2, ...) or that of a *boson* (a particle with spin 0, 1, 2, ...). For fermions, no two particles can occupy the same quantum state, so the occupation number can only take on the eigenvalues 0 and 1. For bosons, one can shove any number of particles one wishes into the same quantum state, so the occupation number can take on the eigenvalues 0, 1, 2, 3, ... Correspondingly, the mean occupation numbers must lie in the ranges

$$0 \leq \eta \leq 1 \text{ for fermions}, \qquad 0 \leq \eta < \infty \text{ for bosons}. \tag{3.19}$$

Quantum theory also teaches us that, when $\eta \ll 1$, the particles, whether fermions or bosons, behave like classical, discrete, distinguishable particles; and when $\eta \gg 1$ (possible only for bosons), the particles behave like a classical wave—if the particles are photons ($s = 1$), like a classical electromagnetic wave; and if they are gravitons ($s = 2$), like a classical gravitational wave. This role of η in revealing the particles' physical behavior will motivate us frequently to use η as our distribution function instead of $\mathcal{N}$.

Of course η, like $\mathcal{N}$, is a function of location in phase space, $\eta(\mathcal{P}, \vec{p})$ in relativity with no inertial frame chosen; or $\eta(t, \mathbf{x}, \mathbf{p})$ in both relativity and Newtonian theory when an inertial frame is in use.

Exercise 3.3 **Practice and Example: Regimes of Particulate and Wave-Like Behavior* **N** **R**

(a) Cygnus X-1 is a source of X-rays that has been studied extensively by astronomers. The observations (X-ray, optical, and radio) show that it is a distance $r \sim 6{,}000$ light-years from Earth. It consists of a very hot disk of X-ray-emitting gas that surrounds a black hole with mass $15M_\odot$, and the hole in turn is in a binary orbit with a heavy companion star. Most of the X-ray photons have energies $\mathcal{E} \sim 2$ keV, their energy flux arriving at Earth is $F \sim 10^{-10}$ W m^{-2}, and the portion of the disk that emits most of them has radius roughly 7 times that of the black hole (i.e., $R \sim 300$ km).[5] Make a rough estimate of the mean occupation number of the X-rays' photon states. Your answer should be in the region $\eta \ll 1$, so the photons behave like classical, distinguishable particles. Will the occupation number change as the photons propagate from the source to Earth?

(b) A highly nonspherical supernova in the Virgo cluster of galaxies (40 million light-years from Earth) emits a burst of gravitational radiation with frequencies spread over the band 0.5–2.0 kHz, as measured at Earth. The burst comes out in a time of about 10 ms, so it lasts only a few cycles, and it carries a total energy of roughly $10^{-3}M_\odot c^2$, where $M_\odot = 2 \times 10^{30}$ kg is the mass of the Sun. The emitting region is about the size of the newly forming neutron-star core (10 km), which is small compared to the wavelength of the waves; so if one were to try to resolve the source spatially by imaging the gravitational waves with a gravitational lens, one would see only a blur of spatial size one wavelength rather than seeing the neutron star. What is the mean occupation number of the burst's graviton states? Your answer should be in the region $\eta \gg 1$, so the gravitons behave like a classical gravitational wave.

3.3 Thermal-Equilibrium Distribution Functions **N** **R**

3.3

In Chap. 4, we introduce with care and explore in detail the concept of statistical equilibrium—also called "thermal equilibrium." That exploration will lead to a set of distribution functions for particles that are in statistical equilibrium. In this section, we summarize those equilibrium distribution functions, so as to be able to use them for examples and applications of kinetic theory.

   If a collection of many identical particles is in thermal equilibrium in the neighborhood of an event $\mathcal{P}$, then, as we shall see in Chap. 4, there is a special inertial reference frame (the *mean rest frame* of the particles near $\mathcal{P}$) in which the distribution function is isotropic, so the mean occupation number η is a function only of the magnitude $|\mathbf{p}|$

---

5. These numbers refer to what astronomers call Cygnus X-1's soft (red) state. It also, sometimes, is seen in a hard (blue) state.

of the particle momentum and does not depend on the momentum's direction. Equivalently, η is a function of the particle's energy. In the relativistic regime, we use two different energies, one denoted $\mathcal{E}$ that includes the contribution of the particle's rest mass and the other denoted E that omits the rest mass and thus represents kinetic energy (cf. Sec. 1.10):

$$E \equiv \mathcal{E} - m = \sqrt{m^2 + \mathbf{p}^2} - m \to \frac{\mathbf{p}^2}{2m} \text{ in the low-velocity, Newtonian limit.}$$

(3.20)

In the nonrelativistic, Newtonian regime we use only $E = \mathbf{p}^2/(2m)$.

Most readers already know that the details of the thermal equilibrium are fixed by two quantities: the mean density of particles and the mean energy per particle, or equivalently (as we shall see) by the *chemical potential* μ and the *temperature* T. By analogy with our treatment of relativistic energy, we use two different chemical potentials: one, $\tilde{\mu}$, that includes rest mass and the other,

$$\mu \equiv \tilde{\mu} - m,$$ (3.21)

that does not. In the Newtonian regime we use only μ.

As we prove by an elegant argument in Chap. 4, in thermal equilibrium the mean occupation number has the following form at all energies, relativistic or nonrelativistic:

Fermi-Dirac and Bose-Einstein distributions

$$\eta = \frac{1}{e^{(E-\mu)/(k_B T)} + 1} \text{ for fermions,}$$ (3.22a)

$$\eta = \frac{1}{e^{(E-\mu)/(k_B T)} - 1} \text{ for bosons.}$$ (3.22b)

Here $k_B = 1.381 \times 10^{-16} \text{ erg K}^{-1} = 1.381 \times 10^{-23} \text{ J K}^{-1}$ is Boltzmann's constant. Equation (3.22a) for fermions is the *Fermi-Dirac distribution;* Eq. (3.22b) for bosons is the *Bose-Einstein distribution.* In the relativistic regime, we can also write these distribution functions in terms of the energy $\mathcal{E}$ that includes the rest mass as

$$\boxed{\text{R}} \quad \eta = \frac{1}{e^{(E-\mu)/(k_B T)} \pm 1} = \frac{1}{e^{(\mathcal{E}-\tilde{\mu})/(k_B T)} \pm 1}.$$ (3.22c)

Notice that the equilibrium mean occupation number (3.22a) for fermions lies in the range 0–1 as required, while that (3.22b) for bosons lies in the range 0 to ∞. In the regime $\mu \ll -k_B T$, the mean occupation number is small compared to unity for all particle energies E (since E is never negative; i.e., $\mathcal{E}$ is never less than m). This is the domain of distinguishable, classical particles, and in it both the Fermi-Dirac and Bose-Einstein distributions become

$$\eta \simeq e^{-(E-\mu)/(k_B T)} = e^{-(\mathcal{E}-\tilde{\mu})/(k_B T)}$$

$$\text{when } \mu \equiv \tilde{\mu} - m \ll -k_B T \quad \text{(classical particles).}$$

(3.22d)

This limiting distribution is the *Boltzmann distribution*.[6]

By scrutinizing the distribution functions (3.22), one can deduce that the larger the temperature T at fixed μ, the larger will be the typical energies of the particles; the larger the chemical potential μ at fixed T, the larger will be the total density of particles [see Ex. 3.4 and Eqs. (3.39)]. For bosons, μ must always be negative or zero, that is, $\tilde{\mu}$ cannot exceed the particle rest mass m; otherwise, η would be negative at low energies, which is physically impossible. For bosons with μ extremely close to zero, there exist huge numbers of very-low-energy particles, leading quantum mechanically to a *Bose-Einstein condensate*; we study such condensates in Sec. 4.9.

In the special case that the particles of interest can be created and destroyed completely freely, with creation and destruction constrained only by the laws of 4-momentum conservation, the particles quickly achieve a thermal equilibrium in which the relativistic chemical potential vanishes, $\tilde{\mu} = 0$ (as we shall see in Sec. 5.5). For example, inside a box whose walls are perfectly emitting and absorbing and have temperature T, the photons acquire the mean occupation number (3.22b) with zero chemical potential, leading to the standard *blackbody (Planck)* form

$$\eta = \frac{1}{e^{h\nu/(k_B T)} - 1}, \qquad \mathcal{N} = \frac{2}{h^3} \frac{1}{e^{h\nu/(k_B T)} - 1}, \qquad I_\nu = \frac{(2h/c^2)\nu^3}{e^{h\nu/(k_B T)} - 1}. \quad (3.23)$$

(Here we have set $\mathcal{E} = h\nu$, where ν is the photon frequency as measured in the box's rest frame, and in the third expression we have inserted the factor c^{-2}, so that I_ν will be in ordinary units.)

By contrast, if one places a fixed number of photons inside a box whose walls cannot emit or absorb them but can scatter them, exchanging energy with them in the process, then the photons will acquire the Bose-Einstein distribution (3.22b) with temperature T equal to that of the walls and with nonzero chemical potential μ fixed by the number of photons present; the more photons there are, the larger will be the chemical potential.

Exercise 3.4 **Example: Maxwell Velocity Distribution** N

Consider a collection of thermalized, classical particles with nonzero rest mass, so they have the Boltzmann distribution (3.22d). Assume that the temperature is low enough ($k_B T \ll mc^2$) that they are nonrelativistic.

(a) Explain why the total number density of particles n in physical space (as measured in the particles' mean rest frame) is given by the integral $n = \int \mathcal{N} dV_p$. Show

---

6. Lynden-Bell (1967) identifies a fourth type of thermal distribution that occurs in the theory of violent relaxation of star clusters. It corresponds to individually distinguishable, classical particles (in his case stars with a range of masses) that obey the same kind of exclusion principle as fermions.

that $n \propto e^{\mu/k_B T}$, and derive the proportionality constant. [Hint: Use spherical coordinates in momentum space, so $d\mathcal{V}_p = 4\pi p^2 dp$ with $p \equiv |\mathbf{p}|$.] Your answer should be Eq. (3.39a) below.

(b) Explain why the mean energy per particle is given by $\bar{E} = n^{-1} \int (p^2/2m) \mathcal{N} d\mathcal{V}_p$. Show that $\bar{E} = \frac{3}{2} k_B T$.

(c) Show that $P(v)dv \equiv$ (probability that a randomly chosen particle will have speed $v \equiv |\mathbf{v}|$ in the range dv) is given by

$$P(v) = \frac{4}{\sqrt{\pi}} \frac{v^2}{v_o^3} e^{-v^2/v_o^2}, \quad \text{where} \quad v_o = \sqrt{\frac{2k_B T}{m}}. \tag{3.24}$$

This is called the *Maxwell velocity distribution*; it is graphed in Fig. 3.6a. Notice that the peak of the distribution is at speed v_o.

[Side remark: In the normalization of probability distributions such as this one, you will often encounter integrals of the form $\int_0^\infty x^{2n} e^{-x^2} dx$. You can evaluate this quickly via integration by parts, if you have memorized that $\int_0^\infty e^{-x^2} dx = \sqrt{\pi}/2$.]

(d) Consider particles confined to move in a plane or in one dimension (on a line). What is their speed distribution $P(v)$ and at what speed does it peak?

Exercise 3.5 *Problem: Maxwell-Jütner Velocity Distribution for Thermalized, Classical, Relativistic Particles* **R** **T2**
Show that for thermalized, classical relativistic particles the probability distribution for the speed [relativistic version of the Maxwell distribution (3.24)] is

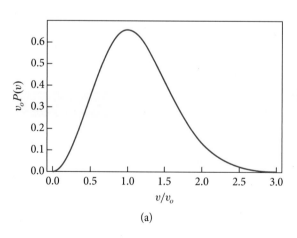

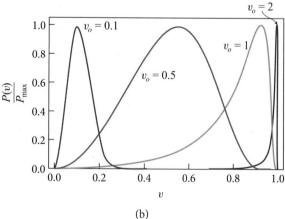

(a)

(b)

FIGURE 3.6 (a) Maxwell velocity distribution for thermalized, classical, nonrelativistic particles. (b) Extension of the Maxwell velocity distribution into the relativistic domain. In both plots $v_o = \sqrt{2k_B T/m}$.

$$P(v) = \frac{2/v_0^2}{K_2(2/v_0^2)} \frac{v^2}{(1-v^2)^{5/2}} \exp\left[-\frac{2/v_0^2}{\sqrt{1-v^2}}\right], \quad \text{where} \quad v_o = \sqrt{\frac{2k_BT}{m}}. \quad (3.25)$$

Where K_2 is the modified Bessel function of the second kind and order 2. This is sometimes called the Maxwell-Jütner distribution, and it is plotted in Fig. 3.6b for a sequence of four temperatures ranging from the nonrelativistic regime $k_BT \ll m$ toward the ultrarelativistic regime $k_BT \gg m$. In the ultrarelativistic regime the particles are (almost) all moving at very close to the speed of light, $v = 1$.

Exercise 3.6 *Example and Challenge: Radiative Processes*
We have described distribution functions for particles and photons and the forms that they have in thermodynamic equilibrium. An extension of these principles can be used to constrain the manner in which particles and photons interact, specifically, to relate the emission and absorption of radiation.

(a) Consider a two-level (two-state) electron system with energy separation $\mathcal{E} = h\nu_0$. Suppose that an electron can transition with a probability per unit time, A, from the upper level (u) to the lower level (l), creating a photon in a specific state. Use the Boltzmann distribution, ignoring degeneracy, Eq. (3.22d), and the expression for the mean photon occupation number, Eq. (3.23), to show that, when the electrons and photons are in thermal equilibrium at the same temperature T, then:

$$An_u + An_u\eta_\gamma = An_l\eta_\gamma, \quad (3.26)$$

where $n_{l,u}$ are the number densities of the electrons in the two states, and η_γ is the photon mean occupation number.

(b) The three terms in Eq. (3.26) are often called the rates per unit volume for *spontaneous emission, stimulated emission,* and *(stimulated) absorption,* respectively. (The second and third terms are sometimes expressed using Einstein B coefficients.) The expressions An_u, $An_u\eta_\gamma$, and $An_l\eta_\gamma$ for the rates of these three types of transition are commonly (and usually correctly) assumed to apply out of thermodynamic equilibrium as well as in equilibrium. Discuss briefly two conditions that need to be satisfied for this to be so: that all three types of transition proceed in a manner that is independent of anything else present, and that the stimulated transitions are time reversible.[7] What additional condition needs to be satisfied if the electron system is more complex than a two-level system? (We return to these issues in the context of cosmology, in Secs. 28.4.4 and 28.6.3.)

(c) Typical transitions have some duration τ, which implies that the photons will be emitted with a distribution of frequencies with width $\Delta\nu \sim \tau^{-1}$ about ν_0. They may also not be emitted isotropically and can carry polarization. Denote the state of a photon by its frequency ν, direction of travel $\widehat{\boldsymbol{\Omega}}$, and polarization $i = 1$ or 2,

---

7. These principles are quite classical in origin but the full justification of the heuristic argument presented here requires quantum electrodynamics.

and denote by $P^i_{\nu\widehat{\Omega}}$ the photons' probability distribution, normalized such that $\Sigma_i \int d\nu \, d\Omega P^i_{\nu\widehat{\Omega}} = 1$, with $d\Omega$ an element of solid angle around $\widehat{\Omega}$. Now define a classical emissivity and a classical absorption coefficient by

$$j^i_{\nu\widehat{\Omega}} \equiv n_u h\nu A P^i_{\nu\widehat{\Omega}}, \qquad \kappa^i \equiv n_l A(1 - e^{-h\nu_0/(k_B T_e)}) P^i_{\nu\widehat{\Omega}} c^2/\nu^2, \qquad (3.27)$$

respectively, where T_e is the electron temperature. Interpret these two quantities and prove *Kirchhoff's law,* namely, that $j^i_{\nu\widehat{\Omega}}/\kappa^i$ is the blackbody radiation intensity for one polarization mode. What happens when $h\nu_0 \ll k_B T_e$?

(d) Further generalize the results in part c by assuming that, instead of occupying just two states, the electrons have a continuous distribution of momenta and radiate throughout this distribution. Express the classical emissivity and absorption coefficient as integrals over electron momentum space.

(e) Finally, consider the weak nuclear transformation $\nu + n \to e + p$, where the neutron n and the proton p can be considered as stationary. (This is important in the early universe, as we discuss further in Sec. 28.4.2.) Explain carefully why the rate at which this transformation occurs can be expressed as

$$\frac{dn_p}{dt} = n_n \int \left(\frac{d\mathcal{V}_{p_\nu}}{h^3} \eta_\nu\right) \left(\frac{h^6 W}{m_e^5} \delta(E_\nu + m_n - E_e - m_p)\right) \left(2\frac{d\mathcal{V}_{p_e}}{h^3}(1 - \eta_e)\right)$$

(3.28)

for some W. It turns out that W is a constant (Weinberg, 2008). What is the corresponding rate for the inverse reaction?

Exercise 3.7 **Example: Observations of Cosmic Microwave Radiation from Earth** R T2

The universe is filled with cosmic microwave radiation left over from the big bang. At each event in spacetime the microwave radiation has a mean rest frame. As seen in that mean rest frame the radiation's distribution function η is almost precisely isotropic and thermal with zero chemical potential:

$$\eta = \frac{1}{e^{h\nu/(k_B T_o)} - 1}, \quad \text{with} \quad T_o = 2.725 \text{ K}. \qquad (3.29)$$

Here ν is the frequency of a photon as measured in the mean rest frame.

(a) Show that the specific intensity of the radiation as measured in its mean rest frame has the *Planck spectrum,* Eq. (3.23). Plot this specific intensity as a function of frequency, and from your plot determine the frequency of the intensity peak.

(b) Show that η can be rewritten in the frame-independent form

$$\eta = \frac{1}{e^{-\vec{p}\cdot\vec{u}_o/(k_B T_o)} - 1}, \qquad (3.30)$$

where $\vec{p}$ is the photon 4-momentum, and $\vec{u}_o$ is the 4-velocity of the mean rest frame. [Hint: See Sec. 2.6 and especially Eq. (2.29).]

(c) In actuality, Earth moves relative to the mean rest frame of the microwave background with a speed v of roughly 400 km s^{-1} toward the Hydra-Centaurus region of the sky. An observer on Earth points his microwave receiver in a direction that makes an angle θ with the direction of that motion, as measured in Earth's frame. Show that the specific intensity of the radiation received is precisely Planckian in form [Eqs. (3.23)], but with a direction-dependent *Doppler-shifted temperature*

$$T = T_o \left(\frac{\sqrt{1 - v^2}}{1 - v \cos \theta} \right) . \tag{3.31}$$

Note that this Doppler shift of T is precisely the same as the Doppler shift of the frequency of any specific photon [Eq. (2.33)]. Note also that the θ dependence corresponds to an anisotropy of the microwave radiation as seen from Earth. Show that because Earth's velocity is small compared to the speed of light, the anisotropy is very nearly dipolar in form. Measurements by the WMAP satellite give $T_o = 2.725$ K and (averaged over a year) an amplitude of 3.346×10^{-3} K for the dipolar temperature variations (Bennett et al., 2003). What, precisely, is the value of Earth's year-averaged speed v?

3.4 Macroscopic Properties of Matter as Integrals over Momentum Space

3.4.1 Particle Density n, Flux S, and Stress Tensor **T** [N]

If one knows the Newtonian distribution function $\mathcal{N} = (g_s/h^3)\eta$ as a function of momentum **p** at some location $(\mathbf{x}, t)$ in space and time, one can use it to compute various macroscopic properties of the particles.

From the definition $\mathcal{N} \equiv dN/d\mathcal{V}_x d\mathcal{V}_p$ of the distribution function, it is clear that the number density of particles $n(\mathbf{x}, t)$ in physical space is given by the integral

$$\boxed{n = \frac{dN}{d\mathcal{V}_x} = \int \frac{dN}{d\mathcal{V}_x d\mathcal{V}_p} d\mathcal{V}_p = \int \mathcal{N} d\mathcal{V}_p .} \tag{3.32a}$$

Newtonian particle density

Similarly, the number of particles crossing a unit surface in the y-z plane per unit time (i.e., the x component of the flux of particles) is

$$S_x = \frac{dN}{dydzdt} = \int \frac{dN}{dxdydzd\mathcal{V}_p} \frac{dx}{dt} d\mathcal{V}_p = \int \mathcal{N} \frac{p_x}{m} d\mathcal{V}_p,$$

where $dx/dt = p_x/m$ is the x component of the particle velocity. This and the analogous equations for S_y and S_z can be combined into a single geometric, coordinate-independent integral for the vectorial particle flux:

$$\boxed{\mathbf{S} = \int \mathcal{N} \, \mathbf{p} \, \frac{d\mathcal{V}_p}{m} .} \tag{3.32b}$$

Newtonian particle flux

Notice that, if we multiply this **S** by the particles' mass m, the integral becomes the momentum density:

$$\boxed{\mathbf{G} = m\mathbf{S} = \int \mathcal{N}\mathbf{p}\, d\mathcal{V}_p.}$$

(3.32c)

Finally, since the stress tensor **T** is the flux of momentum [Eq. (1.33)], its j-x component (j component of momentum crossing a unit area in the y-z plane per unit time) must be

$$T_{jx} = \int \frac{dN}{dydzdtd\mathcal{V}_p}\, p_j\, d\mathcal{V}_p = \int \frac{dN}{dxdydzd\mathcal{V}_p}\frac{dx}{dt}\, p_j d\mathcal{V}_p = \int \mathcal{N}p_j\frac{p_x}{m}d\mathcal{V}_p.$$

This and the corresponding equations for T_{jy} and T_{jz} can be collected together into a single geometric, coordinate-independent integral:

$$T_{jk} = \int \mathcal{N}p_j p_k\frac{d\mathcal{V}_p}{m}, \quad \text{i.e.,} \quad \boxed{\mathbf{T} = \int \mathcal{N}\mathbf{p}\otimes\mathbf{p}\frac{d\mathcal{V}_p}{m}.}$$

(3.32d)

Notice that the number density n is the zeroth moment of the distribution function in momentum space [Eq. (3.32a)], and aside from factors 1/m, the particle flux vector is the first moment [Eq. (3.32b)], and the stress tensor is the second moment [Eq. (3.32d)]. All three moments are geometric, coordinate-independent quantities, and they are the simplest such quantities that one can construct by integrating the distribution function over momentum space.

3.4.2 Relativistic Number-Flux 4-Vector $\vec{S}$ and Stress-Energy Tensor T [R] [T2]

When we switch from Newtonian theory to special relativity's 4-dimensional space-time viewpoint, we require that all physical quantities be described by geometric, frame-independent objects (scalars, vectors, tensors, …) in 4-dimensional spacetime. We can construct such objects as momentum-space integrals over the frame-independent, relativistic distribution function $\mathcal{N}(\mathcal{P}, \vec{p}) = (g_s/h^3)\eta$. The frame-independent quantities that can appear in these integrals are (i) $\mathcal{N}$ itself, (ii) the particle 4-momentum $\vec{p}$, and (iii) the frame-independent integration element $d\mathcal{V}_p/\mathcal{E}$ [Eq. (3.7b)], which takes the form $dp_x dp_y dp_z/\sqrt{m^2 + \mathbf{p}^2}$ in any inertial reference frame. By analogy with the Newtonian regime, the most interesting such integrals are the lowest three moments of the distribution function:

$$R \equiv \int \mathcal{N}\frac{d\mathcal{V}_p}{\mathcal{E}};$$

(3.33a)

$$\boxed{\vec{S} \equiv \int \mathcal{N}\vec{p}\frac{d\mathcal{V}_p}{\mathcal{E}},} \quad \text{i.e.,} \quad S^\mu \equiv \int \mathcal{N}p^\mu\frac{d\mathcal{V}_p}{\mathcal{E}};$$

(3.33b)

$$\boxed{T \equiv \int \mathcal{N}\vec{p}\otimes\vec{p}\frac{d\mathcal{V}_p}{\mathcal{E}},} \quad \text{i.e.,} \quad T^{\mu\nu} \equiv \int \mathcal{N}p^\mu p^\nu\frac{d\mathcal{V}_p}{\mathcal{E}}.$$

(3.33c)

Here and throughout this chapter, relativistic momentum-space integrals are taken over the entire mass hyperboloid unless otherwise specified.

We can learn the physical meanings of each of the momentum-space integrals (3.33) by introducing a specific but arbitrary inertial reference frame and using it to perform a 3+1 split of spacetime into space plus time [cf. the paragraph containing Eq. (2.28)]. When we do this and rewrite $\mathcal{N}$ as $dN/dV_x dV_p$, the scalar field R of Eq. (3.33a) takes the form

$$R = \int \frac{dN}{dV_x dV_p} \frac{1}{\mathcal{E}} dV_p \qquad (3.34)$$

(where of course $dV_x = dx\,dy\,dz$ and $dV_p = dp_x\,dp_y\,dp_z$). This is the sum, over all particles in a unit 3-volume, of the inverse energy. Although it is intriguing that this quantity is a frame-independent scalar, it is not a quantity that appears in any important way in the laws of physics.

By contrast, the 4-vector field $\vec{S}$ of Eq. (3.33b) plays a very important role in physics. Its time component in our chosen frame is

$$S^0 = \int \frac{dN}{dV_x dV_p} \frac{p^0}{\mathcal{E}} dV_p = \int \frac{dN}{dV_x dV_p} dV_p \qquad (3.35a)$$

(since p^0 and $\mathcal{E}$ are just different notations for the same thing—the relativistic energy $\sqrt{m^2 + \mathbf{p}^2}$ of a particle). Obviously, this S^0 is the number of particles per unit spatial volume as measured in our chosen inertial frame:

$$S^0 = n = \text{(number density of particles)}. \qquad (3.35b)$$

The x component of $\vec{S}$ is

$$S^x = \int \frac{dN}{dV_x dV_p} \frac{p^x}{\mathcal{E}} dV_p = \int \frac{dN}{dx\,dy\,dz\,dV_p} \frac{dx}{dt} dV_p = \int \frac{dN}{dt\,dy\,dz\,dV_p} dV_p, \qquad (3.35c)$$

which is the number of particles crossing a unit area in the y-z plane per unit time (i.e., the x component of the particle flux); similarly for other directions j:

$$S^j = (j \text{ component of the particle flux vector } \mathbf{S}). \qquad (3.35d)$$

[In Eq. (3.35c), the second equality follows from

$$\frac{p^j}{\mathcal{E}} = \frac{p^j}{p^0} = \frac{dx^j/d\zeta}{dt/d\zeta} = \frac{dx^j}{dt} = (j \text{ component of velocity}), \qquad (3.35e)$$

where ζ is the affine parameter such that $\vec{p} = d\vec{x}/d\zeta$.] Since S^0 is the particle number density and S^j is the particle flux, $\vec{S}$ [Eq. (3.33b)] *must be the number-flux 4-vector* introduced and studied in Sec. 2.12.3. Notice that in the Newtonian limit, where $p^0 = \mathcal{E} \to m$, the temporal and spatial parts of the formula $\vec{S} = \int \mathcal{N} \vec{p} \, (dV_p/\mathcal{E})$ reduce

number-flux 4-vector

to $S^0 = \int \mathcal{N} d\mathcal{V}_p$ and $\mathbf{S} = \int \mathcal{N}\mathbf{p}(d\mathcal{V}_p/m)$, respectively, which are the coordinate-independent expressions (3.32a) and (3.32b) for the Newtonian number density of particles and flux of particles, respectively.

Turn to the quantity T defined by the integral (3.33c). When we perform a 3+1 split of it in our chosen inertial frame, we find the following for its various parts:

$$T^{\mu 0} = \int \frac{dN}{d\mathcal{V}_x d\mathcal{V}_p} p^\mu p^0 \frac{d\mathcal{V}_p}{p^0} = \int \frac{dN}{d\mathcal{V}_x d\mathcal{V}_p} p^\mu d\mathcal{V}_p \qquad (3.36a)$$

is the μ component of 4-momentum per unit volume (i.e., T^{00} is the energy density, and T^{j0} is the momentum density). Also,

$$T^{\mu x} = \int \frac{dN}{d\mathcal{V}_x d\mathcal{V}_p} p^\mu p^x \frac{d\mathcal{V}_p}{p^0} = \int \frac{dN}{dx\,dy\,dz\,d\mathcal{V}_p} \frac{dx}{dt} p^\mu d\mathcal{V}_p = \int \frac{dN}{dt\,dy\,dz\,d\mathcal{V}_p} p^\mu d\mathcal{V}_p$$

$$(3.36b)$$

is the amount of μ component of 4-momentum that crosses a unit area in the y-z plane per unit time (i.e., it is the x component of flux of μ component of 4-momentum). More specifically, T^{0x} is the x component of energy flux (which is the same as the momentum density T^{x0}), and T^{jx} is the x component of spatial-momentum flux—or, equivalently, the jx component of the stress tensor. These and the analogous expressions and interpretations of $T^{\mu y}$ and $T^{\mu z}$ can be summarized by

$$T^{00} = \text{(energy density)}, \quad T^{j0} = \text{(momentum density)} = T^{0j} = \text{(energy flux)},$$

$$T^{jk} = \text{(stress tensor)}. \qquad (3.36c)$$

stress-energy tensor

Therefore [cf. Eq. (2.67f)], the T of Eq. (3.33c) must be the stress-energy tensor introduced and studied in Sec. 2.13. Notice that in the Newtonian limit, where $\mathcal{E} \to m$, the coordinate-independent Eq. (3.33c) for the spatial part of the stress-energy tensor (the stress) becomes $\int \mathcal{N}\mathbf{p} \otimes \mathbf{p} \, d\mathcal{V}_p/m$, which is the same as our coordinate-independent Eq. (3.32d) for the stress tensor.

3.5

3.5 Isotropic Distribution Functions and Equations of State

3.5.1

3.5.1 Newtonian Density, Pressure, Energy Density, and Equation of State N

Let us return to Newtonian theory.

If the Newtonian distribution function is isotropic in momentum space (i.e., is a function only of the magnitude $p \equiv |\mathbf{p}| = \sqrt{p_x^2 + p_y^2 + p_z^2}$ of the momentum, as is the case, e.g., when the particle distribution is thermalized), then the particle flux $\mathbf{S}$ vanishes (equal numbers of particles travel in all directions), and the stress tensor is isotropic: $\mathbf{T} = P\mathbf{g}$, or $T_{jk} = P\delta_{jk}$. Thus, it is the stress tensor of a perfect fluid. [Here P is the isotropic pressure, and $\mathbf{g}$ is the metric tensor of Euclidean 3-space, with Cartesian components equal to the Kronecker delta; Eq. (1.9f).] In this isotropic case, the pressure can be computed most easily as 1/3 the trace of the stress tensor (3.32d):

$$P = \frac{1}{3}T_{jj} = \frac{1}{3}\int \mathcal{N}(p_x^2 + p_y^2 + p_z^2)\frac{d\mathcal{V}_p}{m}$$

$$= \frac{1}{3}\int_0^\infty \mathcal{N}p^2\frac{4\pi p^2 dp}{m} = \frac{4\pi}{3m}\int_0^\infty \mathcal{N}p^4\,dp. \tag{3.37a}$$

Newtonian pressure

Here in the third step we have written the momentum-volume element in spherical polar coordinates as $d\mathcal{V}_p = p^2\sin\theta d\theta d\phi dp$ and have integrated over angles to get $4\pi p^2 dp$. Similarly, we can reexpress the number density of particles (3.32a) and the corresponding mass density as

$$n = 4\pi\int_0^\infty \mathcal{N}p^2 dp, \qquad \rho \equiv mn = 4\pi m\int_0^\infty \mathcal{N}p^2\,dp. \tag{3.37b}$$

Newtonian particle and mass density

Finally, because each particle carries an energy $E = p^2/(2m)$, the energy density in this isotropic case (which we shall denote by U) is 3/2 the pressure:

$$U = \int \frac{p^2}{2m}\mathcal{N}d\mathcal{V}_p = \frac{4\pi}{2m}\int_0^\infty \mathcal{N}p^4 dp = \frac{3}{2}P \tag{3.37c}$$

Newtonian energy density

[cf. Eq. (3.37a)].

If we know the distribution function for an isotropic collection of particles, Eqs. (3.37) give us a straightforward way of computing the collection's number density of particles n, mass density $\rho = nm$, perfect-fluid energy density U, and perfect-fluid pressure P as measured in the particles' mean rest frame. For a thermalized gas, the distribution functions (3.22a), (3.22b), and (3.22d) [with $\mathcal{N} = (g_s/h^3)\eta$] depend on two parameters: the temperature T and chemical potential μ, so this calculation gives n, U, and P in terms of μ and T. One can then invert $n(\mu, T)$ to get $\mu(n, T)$ and insert the result into the expressions for U and P to obtain *equations of state* for thermalized, nonrelativistic particles:

$$U = U(\rho, T)\,, \quad P = P(\rho, T)\,. \tag{3.38}$$

equations of state

For a gas of nonrelativistic, classical particles, the distribution function is Boltzmann [Eq. (3.22d)], $\mathcal{N} = (g_s/h^3)e^{(\mu-E)/(k_B T)}$, with $E = p^2/(2m)$, and this procedure gives, quite easily (Ex. 3.8):

thermalized, classical, nonrelativistic gas

$$n = \frac{g_s e^{\mu/(k_B T)}}{\lambda_{T\,dB}^3} = \frac{g_s}{h^3}(2\pi m k_B T)^{3/2}e^{\mu/(k_B T)}, \tag{3.39a}$$

$$U = \frac{3}{2}nk_B T\,, \quad P = nk_B T. \tag{3.39b}$$

Notice that the mean energy per particle is (cf. Ex. 3.4b)

$$\bar{E} = \frac{3}{2}k_B T\,. \tag{3.39c}$$

In Eq. (3.39a), $\lambda_{T\mathrm{dB}} \equiv h/\sqrt{2\pi m k_B T}$ is the particles' *thermal de Broglie wavelength:* the wavelength of Schrödinger wave-function oscillations for a particle with thermal kinetic energy $E = \pi k_B T$. Note that the classical regime $\eta \ll 1$ (i.e., $\mu/(k_B T) \ll -1$), in which our computation is being performed, corresponds to a mean number of particles in a thermal de Broglie wavelength small compared to 1, $n\lambda_{T\mathrm{dB}}^3 \ll 1$, which should not be surprising.

3.5.2 Equations of State for a Nonrelativistic Hydrogen Gas

As an application, consider ordinary matter. Figure 3.7 shows its physical nature as a function of density and temperature, near and above room temperature (300 K). We study solids (lower right) in Part IV, fluids (lower middle) in Part V, and plasmas (middle) in Part VI.

Our kinetic-theory tools are well suited to any situation where the particles have mean free paths large compared to their sizes. This is generally true in plasmas and sometimes in fluids (e.g., air and other gases, but not water and other liquids), and even sometimes in solids (e.g., electrons in a metal). Here we focus on a nonrelativistic

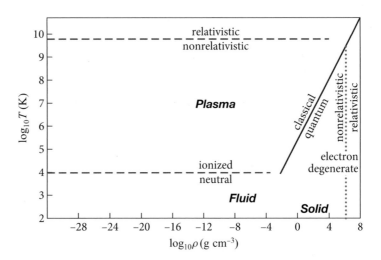

FIGURE 3.7 Physical nature of hydrogen at various densities and temperatures. The plasma regime is discussed in great detail in Part VI, and the equation of state in this regime is Eq. (3.40). The region of relativistic electron degeneracy (to the right of the vertical dotted line) is analyzed in Sec. 3.5.4, and that for the nonrelativistic regime (between slanted solid line and vertical dotted line) in the second half of Sec. 3.5.2. The boundary between the plasma regime and the electron-degenerate regime (slanted solid line) is Eq. (3.41); that between nonrelativistic degeneracy and relativistic degeneracy (vertical dotted line) is Eq. (3.46). The upper relativistic/nonrelativistic boundary is governed by electron-positron pair production (Ex. 5.9 and Fig. 5.7) and is only crudely approximated by the upper dashed line. The ionized-neutral boundary is governed by the Saha equation (Ex. 5.10 and Fig. 20.1) and is crudely approximated by the lower dashed line. For a more accurate and detailed version of this figure, including greater detail on the plasma regime and its boundaries, see Fig. 20.1.

plasma (i.e., the region of Fig. 3.7 that is bounded by the two dashed lines and the slanted solid line). For concreteness and simplicity, we regard the plasma as made solely of hydrogen.[8]

A nonrelativistic hydrogen plasma consists of a mixture of two fluids (gases): free nondegenerate hydrogen gas electrons and free protons, in equal numbers. Each fluid has a particle number density $n = \rho/m_p$, where ρ is the total mass density and m_p is the proton mass. (The electrons are so light that they do not contribute significantly to ρ.) Correspondingly, the energy density and pressure include equal contributions from the electrons and protons and are given by [cf. Eqs. (3.39b)]

$$U = 3(k_B/m_p)\rho T, \qquad P = 2(k_B/m_p)\rho T. \tag{3.40}$$

In zeroth approximation, the high-temperature boundary of validity for this equation of state is the temperature $T_{\rm rel} = m_e c^2/k_B = 6 \times 10^9$ K, at which the electrons become highly relativistic (top dashed line in Fig. 3.7). In Ex. 5.9, we compute the thermal production of electron-positron pairs in the hot plasma and thereby discover that the upper boundary is actually somewhat lower than this (Figs. 5.7 and 20.1). The bottom dashed line in Fig. 3.7 is the temperature $T_{\rm ion} \sim$ (ionization energy of hydrogen)/(a few k_B) $\sim 10^4$ K, at which electrons and protons begin to recombine and form neutral hydrogen. In Ex. 5.10 on the Saha equation, we analyze the conditions for ionization-recombination equilibrium and thereby refine this boundary (Fig. 20.1). The solid right boundary is the point at which the electrons cease to behave like classical particles, because their mean occupation number η_e ceases to be $\ll 1$. As one can see from the Fermi-Dirac distribution (3.22a), for typical electrons (which have energies $E \sim k_B T$), the regime of classical behavior ($\eta_e \ll 1$; to the left of the solid line) is $\mu_e \ll -k_B T$ and the regime of strong quantum behavior ($\eta_e \simeq 1$; *electron degeneracy;* to the right of the solid line) is $\mu_e \gg +k_B T$. The slanted solid boundary in Fig. 3.7 is thus the location $\mu_e = 0$, which translates via Eq. (3.39a) to

$$\rho = \rho_{\rm deg} \equiv 2m_p/\lambda_{\rm TdB}^3 = (2m_p/h^3)(2\pi m_e k_B T)^{3/2} = 0.00808(T/10^4 \text{ K})^{3/2} \text{ g cm}^{-3}.$$
$$\tag{3.41}$$

Although the hydrogen gas is *degenerate* to the right of this boundary, we can still degenerate hydrogen gas compute its equation of state using our kinetic-theory equations (3.37), so long as we use the quantum mechanically correct distribution function for the electrons—the Fermi-Dirac distribution (3.22a).[9] In this electron-degenerate region, $\mu_e \gg k_B T$, the electron mean occupation number $\eta_e = 1/(e^{(E-\mu_e)/(k_B T)} + 1)$ has the form shown

---

8. For both astrophysical and laboratory applications, the non-hydrogen elemental composition often matters, and involves straightforward corrections to purely hydrogen formulae given here.

9. Our kinetic-theory analysis and the Fermi-Dirac distribution ignore Coulomb interactions between the electrons, and between the electrons and the ions. They thereby miss so-called *Coulomb corrections* to the equation of state (Sec. 22.6.3) and other phenomena that are often important in condensed matter physics, but rarely important in astrophysics.

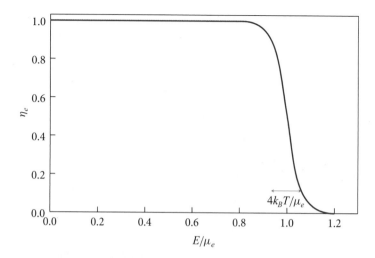

FIGURE 3.8 The Fermi-Dirac distribution function for electrons in the nonrelativistic, degenerate regime $k_B T \ll \mu_e \ll m_e$, with temperature such that $k_B T / \mu_e = 0.03$. Note that η_e drops from near 1 to near 0 over the range $\mu_e - 2k_B T \lesssim E \lesssim \mu_e + 2k_B T$. See Ex. 3.11b.

in Fig. 3.8 and thus can be well approximated by $\eta_e = 1$ for $E = p^2/(2m_e) < \mu_e$ and $\eta_e = 0$ for $E > \mu_e$; or equivalently by

$$\eta_e = 1 \text{ for } p < p_F \equiv \sqrt{2m_e \mu_e}, \qquad \eta_e = 0 \text{ for } p > p_F. \tag{3.42}$$

Fermi momentum

Here p_F is called the *Fermi momentum*. (The word "degenerate" refers to the fact that almost all the quantum states are fully occupied or are empty; i.e., η_e is everywhere nearly 1 or 0.) By inserting this degenerate distribution function [or, more precisely, $\mathcal{N}_e = (2/h^3)\eta_e$] into Eqs. (3.37) and integrating, we obtain $n_e \propto p_F^3$ and $P_e \propto p_F^5$. By then setting $n_e = n_p = \rho/m_p$ and solving for $p_F \propto n_e^{1/3} \propto \rho^{1/3}$ and inserting into the expression for P_e and evaluating the constants, we obtain (Ex. 3.9) the following equation of state for the electron pressure:

$$P_e = \frac{1}{20} \left(\frac{3}{\pi} \right)^{2/3} \frac{m_e c^2}{\lambda_c^3} \left(\frac{\rho}{m_p/\lambda_c^3} \right)^{5/3}. \tag{3.43}$$

Here

$$\lambda_c = h/(m_e c) = 2.426 \times 10^{-10} \text{ cm} \tag{3.44}$$

is the electron Compton wavelength.

The rapid growth $P_e \propto \rho^{5/3}$ of the electron pressure with increasing density is due to the degenerate electrons' being confined by the Pauli Exclusion Principle to regions of ever-shrinking size, causing their zero-point motions and associated pressure to grow. By contrast, the protons, with their far larger rest masses, remain nondegenerate [until their density becomes $(m_p/m_e)^{3/2} \sim 10^5$ times higher than Eq. (3.41)], and so their pressure is negligible compared to that of the electrons: the total pressure is

$$P = P_e = \text{Eq. (3.43)} \qquad\qquad (3.45)$$

in the regime of nonrelativistic electron degeneracy. This is the equation of state for the interior of a low-mass white-dwarf star and for the outer layers of a high-mass white dwarf—aside from tiny corrections due to Coulomb interactions. In Sec. 13.3.2 we shall see how it can be used to explore the structures of white dwarfs. It is also the equation of state for a neutron star, with m_e replaced by the rest mass of a neutron m_n (since neutron degeneracy pressure dominates over that due to the star's tiny number of electrons and protons) and ρ/m_p replaced by the number density of neutrons— except that for neutron stars there are large corrections due to the strong nuclear force (see, e.g., Shapiro and Teukolsky, 1983).

When the density of hydrogen in this degenerate regime is pushed on upward to

$$\rho_{\text{rel deg}} = \frac{8\pi m_p}{3\lambda_c^3} \simeq 9.8 \times 10^5 \text{ g cm}^{-3} \qquad\qquad (3.46)$$

(dotted vertical line in Fig. 3.7), the electrons' zero-point motions become relativistically fast (the electron chemical potential μ_e becomes of order $m_e c^2$ and the Fermi momentum p_F of order $m_e c$), so the nonrelativistic, Newtonian analysis fails, and the matter enters a domain of relativistic degeneracy (Sec. 3.5.4). Both domains, nonrelativistic degeneracy ($\mu_e \ll m_e c^2$) and relativistic degeneracy ($\mu_e \gtrsim m_e c^2$), occur for matter inside a massive white-dwarf star—the type of star that the Sun will become when it dies (see Shapiro and Teukolsky, 1983). In Sec. 26.3.5, we shall see how general relativity (spacetime curvature) modifies a star's structure. It also helps force sufficiently massive white dwarfs to collapse (Sec. 6.10 of Shapiro and Teukolsky, 1983).

The (almost) degenerate Fermi-Dirac distribution function shown in Fig. 3.8 has a thermal tail whose energy width is $4k_B T/\mu_e$. As the temperature T is increased, the number of electrons in this tail increases, thereby increasing the electrons' total energy E_{tot}. This increase is responsible for the electrons' *specific heat* (Ex. 3.11)— a quantity of importance for both the electrons in a metal (e.g., a copper wire) and the electrons in a white-dwarf star. The electrons dominate the specific heat when the temperature is sufficiently low; but at higher temperatures it is dominated by the energies of sound waves (see Ex. 3.12, where we use the kinetic theory of phonons to compute the sound waves' specific heat).

3.5.3 Relativistic Density, Pressure, Energy Density, and Equation of State R T2

Now we turn to the relativistic domain of kinetic theory, initially for a single species of particle with rest mass m and then (in the next subsection) for matter composed of electrons and protons.

The relativistic *mean rest frame* of the particles, at some event $\mathcal{P}$ in spacetime, is that frame in which the particle flux $\mathbf{S}$ vanishes. We denote by $\vec{u}_{\text{rf}}$ the 4-velocity of this mean rest frame. As in Newtonian theory (Sec. 3.5.2), we are especially interested in distribution functions $\mathcal{N}$ that are *isotropic* in the mean rest frame: distribution functions that depend on the magnitude $|\mathbf{p}| \equiv p$ of the spatial momentum of a particle but not on its direction—or equivalently, that depend solely on the particles' energy

$$\mathcal{E} = -\vec{u}_{\mathrm{rf}} \cdot \vec{p} \quad \text{expressed in frame-independent form [Eq. (2.29)]},$$

$$\mathcal{E} = p^0 = \sqrt{m^2 + p^2} \quad \text{in mean rest frame.} \tag{3.47}$$

Such isotropy is readily produced by particle collisions (Sec. 3.7).

Notice that isotropy in the mean rest frame [i.e., $\mathcal{N} = \mathcal{N}(\mathcal{P}, \mathcal{E})$] does not imply isotropy in any other inertial frame. As seen in some other (primed) frame, $\vec{u}_{\mathrm{rf}}$ will have a time component $u_{\mathrm{rf}}^{0'} = \gamma$ and a space component $\mathbf{u}'_{\mathrm{rf}} = \gamma \mathbf{V}$ [where $\mathbf{V}$ is the mean rest frame's velocity relative to the primed frame, and $\gamma = (1 - \mathbf{V}^2)^{-1/2}$]; and correspondingly, in the primed frame, $\mathcal{N}$ will be a function of

$$\mathcal{E} = -\vec{u}_{\mathrm{rf}} \cdot \vec{p} = \gamma[(m^2 + \mathbf{p}'^2)^{\frac{1}{2}} - \mathbf{V} \cdot \mathbf{p}'], \tag{3.48}$$

which is anisotropic: it depends on the direction of the spatial momentum $\mathbf{p}'$ relative to the velocity $\mathbf{V}$ of the particle's mean rest frame. An example is the cosmic microwave radiation as viewed from Earth (Ex. 3.7).

As in Newtonian theory, isotropy greatly simplifies the momentum-space integrals (3.33) that we use to compute macroscopic properties of the particles: (i) The integrands of the expressions $S^j = \int \mathcal{N} p^j (d\mathcal{V}_p/\mathcal{E})$ and $T^{j0} = T^{0j} = \int \mathcal{N} p^j p^0 (d\mathcal{V}_p/\mathcal{E})$ for the particle flux, energy flux, and momentum density are all odd in the momentum-space coordinate p^j and therefore give vanishing integrals: $S^j = T^{j0} = T^{0j} = 0$. (ii) The integral $T^{jk} = \int \mathcal{N} p^j p^k d\mathcal{V}_p/\mathcal{E}$ produces an isotropic stress tensor, $T^{jk} = P g^{jk} = P \delta^{jk}$, whose pressure is most easily computed from its trace, $P = \frac{1}{3} T^{jj}$. Using these results and the relations $|\mathbf{p}| \equiv p$ for the magnitude of the momentum, $d\mathcal{V}_p = 4\pi p^2 dp$ for the momentum-space volume element, and $\mathcal{E} = p^0 = \sqrt{m^2 + p^2}$ for the particle energy, we can easily evaluate Eqs. (3.33) for the particle number density $n = S^0$, the total density of mass-energy T^{00} (which we denote ρ— the same notation as we use for mass density in Newtonian theory), and the pressure P. The results are

$$n \equiv S^0 = \int \mathcal{N} d\mathcal{V}_p = 4\pi \int_0^\infty \mathcal{N} p^2 dp, \tag{3.49a}$$

relativistic particle density, mass-energy density, and pressure

$$\rho \equiv T^{00} = \int \mathcal{N} \mathcal{E} d\mathcal{V}_p = 4\pi \int_0^\infty \mathcal{N} \mathcal{E} p^2 dp, \tag{3.49b}$$

$$P = \frac{1}{3} \int \mathcal{N} p^2 \frac{d\mathcal{V}_p}{\mathcal{E}} = \frac{4\pi}{3} \int_0^\infty \mathcal{N} \frac{p^4 dp}{\sqrt{m^2 + p^2}}. \tag{3.49c}$$

3.5.4 Equation of State for a Relativistic Degenerate Hydrogen Gas R T2

Return to the hydrogen gas whose nonrelativistic equations of state were computed in Sec. 3.5.1. As we deduced there, at densities $\rho \gtrsim 10^5 \, \mathrm{g \, cm}^{-3}$ (near and to the right of the vertical dotted line in Fig. 3.7) the electrons are squeezed into such tiny volumes that their zero-point energies are $\gtrsim m_e c^2$, forcing us to treat them relativistically.

We can do so with the aid of the following approximation for the relativistic Fermi-Dirac mean occupation number $\eta_e = 1/[e^{(\mathcal{E}-\tilde{\mu}_e/(k_B T))} + 1]$:

$$\eta_e \simeq 1 \text{ for } \mathcal{E} < \tilde{\mu}_e \equiv \mathcal{E}_F; \text{ i.e., for } p < p_F = \sqrt{\mathcal{E}_F^2 - m^2}, \tag{3.50}$$

$$\eta_e \simeq 0 \text{ for } \mathcal{E} > \mathcal{E}_F; \text{ i.e., for } p > p_F. \tag{3.51}$$

Here $\mathcal{E}_F$ is called the relativistic *Fermi energy* and p_F the relativistic *Fermi momentum*. By inserting this η_e along with $\mathcal{N}_e = (2/h^3)\eta_e$ into the integrals (3.49) for the electron number density n_e, total density of mass-energy ρ_e, and pressure P_e, and performing the integrals (Ex. 3.10), we obtain results that are expressed most simply in terms of a parameter t (not to be confused with time) defined by

$$\mathcal{E}_F \equiv \tilde{\mu}_e \equiv m_e \cosh(t/4), \qquad p_F \equiv \sqrt{\mathcal{E}_F^2 - m_e^2} \equiv m_e \sinh(t/4). \tag{3.52a}$$

The results are

$$n_e = \frac{8\pi}{3\lambda_c^3} \left(\frac{p_F}{m_e}\right)^3 = \frac{8\pi}{3\lambda_c^3} \sinh^3(t/4), \tag{3.52b}$$

$$\rho_e = \frac{8\pi m_e}{\lambda_c^3} \int_0^{p_F/m_e} x^2\sqrt{1+x^2}\, dx = \frac{\pi m_e}{4\lambda_c^3}[\sinh(t) - t], \tag{3.52c}$$

$$P_e = \frac{8\pi m_e}{\lambda_c^3} \int_0^{p_F/m_e} \frac{x^4}{\sqrt{1+x^2}}\, dx = \frac{\pi m_e}{12\lambda_c^3}[\sinh(t) - 8\sinh(t/2) + 3t]. \tag{3.52d}$$

These parametric relationships for ρ_e and P_e as functions of the electron number density n_e are sometimes called the Anderson-Stoner equation of state, because they were first derived by Wilhelm Anderson and Edmund Stoner in 1930 (see Thorne, 1994, pp. 153–154). They are valid throughout the full range of electron degeneracy, from nonrelativistic up to ultrarelativistic.

In a white-dwarf star, the protons, with their high rest mass, are nondegenerate, the total density of mass-energy is dominated by the proton rest-mass density, and since there is one proton for each electron in the hydrogen gas, that total is

$$\rho \simeq m_p n_e = \frac{8\pi m_p}{3\lambda_c^3} \sinh^3(t/4). \tag{3.53a}$$

By contrast (as in the nonrelativistic regime), the pressure is dominated by the electrons (because of their huge zero-point motions), not the protons; and so the total pressure is

$$P = P_e = \frac{\pi m_e}{12\lambda_c^3}[\sinh(t) - 8\sinh(t/2) + 3t]. \tag{3.53b}$$

In the low-density limit, where $t \ll 1$ so $p_F \ll m_e = m_e c$, we can solve the relativistic equation (3.52b) for t as a function of $n_e = \rho/m_p$ and insert the result into the relativistic expression (3.53b); the result is the nonrelativistic equation of state (3.43).

The dividing line $\rho = \rho_{\text{rel deg}} = 8\pi m_p/(3\lambda_c^3) \simeq 1.0 \times 10^6 \text{ g cm}^{-3}$ [Eq. (3.46)] between nonrelativistic and relativistic degeneracy is the point where the electron Fermi momentum is equal to the electron rest mass [i.e., $\sinh(t/4) = 1$]. The equation of state (3.53a) and (3.53b) implies

$$P_e \propto \rho^{5/3} \quad \text{in the nonrelativistic regime, } \rho \ll \rho_{\text{rel deg}},$$

$$P_e \propto \rho^{4/3} \quad \text{in the relativistic regime, } \rho \gg \rho_{\text{rel deg}}. \tag{3.53c}$$

These asymptotic equations of state turn out to play a crucial role in the structure and stability of white dwarf stars (Secs. 13.3.2 and 26.3.5; Shapiro and Teukolsky, 1983; Thorne, 1994, Chap. 4).

3.5.5 Equation of State for Radiation N R

As was discussed at the end of Sec. 3.3, for a gas of thermalized photons in an environment where photons are readily created and absorbed, the distribution function has the blackbody (Planck) form $\eta = 1/(e^{\mathcal{E}/(k_B T)} - 1)$, which we can rewrite as $1/(e^{p/(k_B T)} - 1)$, since the energy $\mathcal{E}$ of a photon is the same as the magnitude p of its momentum. In this case, the relativistic integrals (3.49) give (see Ex. 3.13)

$$\boxed{n = bT^3, \qquad \rho = aT^4, \qquad P = \frac{1}{3}\rho,} \tag{3.54a}$$

where

$$b = 16\pi\,\zeta(3)\,\frac{k_B^3}{h^3 c^3} = 20.28 \text{ cm}^{-3}\,\text{K}^{-3}, \tag{3.54b}$$

$$a = \frac{8\pi^5}{15}\,\frac{k_B^4}{h^3 c^3} = 7.566 \times 10^{-15} \text{ erg cm}^{-3}\,\text{K}^{-4} = 7.566 \times 10^{-16} \text{ J m}^{-3}\,\text{K}^{-4} \tag{3.54c}$$

are *radiation constants*. Here $\zeta(3) = \sum_{n=1}^{\infty} n^{-3} = 1.2020569\ldots$ is the Riemann zeta function.

As we shall see in Sec. 28.4, when the universe was younger than about 100,000 years, its energy density and pressure were predominantly due to thermalized photons plus neutrinos (which contributed approximately the same as the photons), so its equation of state was given by Eq. (3.54a) with the coefficient changed by a factor of order unity. Einstein's general relativistic field equations (Sec. 25.8 and Chap. 28) relate the energy density ρ of these photons and neutrinos to the age of the universe t as measured in the photons' and neutrinos' mean rest frame:

$$\frac{3}{32\pi G t^2} = \rho \simeq aT^4. \tag{3.55a}$$

Here G is Newton's gravitation constant. Putting in numbers, we find that

$$\rho = \frac{4.5 \times 10^{-10} \text{ g cm}^{-3}}{(\tau/1 \text{ yr})^2}, \qquad T \simeq \frac{2.7 \times 10^6 \text{ K}}{(\tau/1 \text{ yr})^{1/2}}. \qquad (3.55b)$$

This implies that, when the universe was 1 minute old, its radiation density and temperature were about 100 g cm^{-3} and 2×10^9 K, respectively. These conditions and the proton density were well suited for burning hydrogen to helium; and, indeed, about 1/4 of all the mass of the universe did get burned to helium at this early epoch. We shall examine this in further detail in Sec. 28.4.2.

EXERCISES

Exercise 3.8 *Derivation and Practice: Equation of State for Nonrelativistic, Classical Gas* **N**

Consider a collection of identical, classical (i.e., with $\eta \ll 1$) particles with a distribution function $\mathcal{N}$ that is thermalized at a temperature T such that $k_B T \ll mc^2$ (nonrelativistic temperature).

(a) Show that the distribution function, expressed in terms of the particles' momenta or velocities in their mean rest frame, is

$$\mathcal{N} = \frac{g_s}{h^3} e^{\mu/(k_B T)} e^{-p^2/(2mk_B T)}, \quad \text{where } p = |\mathbf{p}| = mv, \qquad (3.56)$$

with v being the speed of a particle.

(b) Show that the number density of particles in the mean rest frame is given by Eq. (3.39a).

(c) Show that this gas satisfies the equations of state (3.39b).

Note: The following integrals, for nonnegative integral values of q, will be useful:

$$\int_0^\infty x^{2q} e^{-x^2} dx = \frac{(2q-1)!!}{2^{q+1}} \sqrt{\pi}, \qquad (3.57)$$

where $n!! \equiv n(n-2)(n-4) \ldots (2 \text{ or } 1)$; and

$$\int_0^\infty x^{2q+1} e^{-x^2} dx = \frac{1}{2} q!. \qquad (3.58)$$

Exercise 3.9 *Derivation and Practice: Equation of State for Nonrelativistic, Electron-Degenerate Hydrogen* **N**

Derive Eq. (3.43) for the electron pressure in a nonrelativistic, electron-degenerate hydrogen gas.

Exercise 3.10 *Derivation and Practice: Equation of State for Relativistic, Electron-Degenerate Hydrogen* **R** **T2**

Derive the equations of state (3.52) for an electron-degenerate hydrogen gas. (Note: It might be easiest to compute the integrals with the help of symbolic manipulation software, such as Mathematica, Matlab, or Maple.)

Exercise 3.11 *Example: Specific Heat for Nonrelativistic, Degenerate Electrons in White Dwarfs and in Metals* **N**

Consider a nonrelativistically degenerate electron gas at finite but small temperature.

(a) Show that the inequalities $k_B T \ll \mu_e \ll m_e$ are equivalent to the words "nonrelativistically degenerate."

(b) Show that the electron mean occupation number $\eta_e(E)$ has the form depicted in Fig. 3.8: It is near unity out to (nonrelativistic) energy $E \simeq \mu_e - 2k_B T$, and it then drops to nearly zero over a range of energies $\Delta E \sim 4k_B T$.

(c) If the electrons were nonrelativistic but *non*degenerate, their thermal energy density would be $U = \frac{3}{2}nk_B T$, so the total electron energy (excluding rest mass) in a volume V containing $N = nV$ electrons would be $E_{\rm tot} = \frac{3}{2}Nk_B T$, and the electron specific heat, at fixed volume, would be

$$C_V \equiv \left(\frac{\partial E_{\rm tot}}{\partial T}\right)_V = \frac{3}{2}Nk_B \quad \text{(nondegenerate, nonrelativistic).} \tag{3.59}$$

Using the semiquantitative form of η_e depicted in Fig. 3.8, show that to within a factor of order unity the specific heat of degenerate electrons is smaller than in the nondegenerate case by a factor $\sim k_B T/\mu_e$:

$$C_V \equiv \left(\frac{\partial E_{\rm tot}}{\partial T}\right)_V \sim \left(\frac{k_B T}{\mu_e}\right) Nk_B \quad \text{(degenerate, nonrelativistic).} \tag{3.60}$$

(d) Compute the multiplicative factor in Eq. (3.60) for C_V. More specifically, show that, to first order in $k_B T/\mu_e$,

$$C_V = \frac{\pi^2}{2} \left(\frac{k_B T}{\mu_e}\right) Nk_B. \tag{3.61}$$

(e) As an application, consider hydrogen inside a white dwarf with density $\rho = 10^5$ g cm^{-3} and temperature $T = 10^6$ K. (These are typical values for a white-dwarf interior). What are the numerical values of μ_e/m_e and $k_B T/\mu_e$ for the electrons? What is the numerical value of the dimensionless factor $(\pi^2/2)(k_B T/\mu_e)$ by which degeneracy reduces the electron specific heat?

(f) As a second application, consider the electrons inside a copper wire in a laboratory on Earth at room temperature. Each copper atom donates about one electron to a "gas" of freely traveling (conducting) electrons and keeps the rest of its electrons bound to itself. (We neglect interaction of this electron gas with the ions, thereby missing important condensed matter complexities, such as conduction bands and what distinguishes conducting materials from insulators.)

What are the numerical values of μ_e/m_e and $k_B T/\mu_e$ for the conducting electron gas? Verify that these are in the range corresponding to nonrelativistic degeneracy. What is the value of the factor $(\pi^2/2)(k_B T/\mu_e)$ by which degeneracy reduces the electron specific heat? At room temperature, this electron contribution to the specific heat is far smaller than the contribution from thermal vibra-tions of the copper atoms (i.e., thermal sound waves, i.e., thermal *phonons*), but at very low temperatures the electron contribution dominates, as we shall see in the next exercise.

Exercise 3.12 *Example: Specific Heat for Phonons in an Isotropic Solid* **N**
In Sec. 12.2 we will study classical sound waves propagating through an isotropic, elastic solid. As we shall see, there are two types of sound waves: *longitudinal* with frequency-independent speed C_L, and *transverse* with a somewhat smaller frequency-independent speed C_T. For each type of wave, $s = L$ or T, the material of the solid undergoes an elastic displacement $\boldsymbol{\xi} = A\mathbf{f}_s \exp(i\mathbf{k} \cdot \mathbf{x} - \omega t)$, where A is the wave amplitude, $\mathbf{f}_s$ is a unit vector (polarization vector) pointing in the direction of the displacement, $\mathbf{k}$ is the wave vector, and ω is the wave frequency. The wave speed is $C_s = \omega/|\mathbf{k}| (= C_L \text{ or } C_T)$. Associated with these waves are quanta called phonons. As for any wave, each phonon has a momentum related to its wave vector by $\mathbf{p} = \hbar\mathbf{k}$, and an energy related to its frequency by $E = \hbar\omega$. Combining these relations we learn that the relationship between a phonon's energy and the magnitude $p = |\mathbf{p}|$ of its momentum is $E = C_s p$. This is the same relationship as for photons, but with the speed of light replaced by the speed of sound! For longitudinal waves $\mathbf{f}_L$ is in the propagation direction $\mathbf{k}$, so there is just one polarization, $g_L = 1$. For transverse waves $\mathbf{f}_T$ is orthogonal to $\mathbf{k}$, so there are two orthogonal polarizations (e.g., $\mathbf{f}_T = \mathbf{e}_x$ and $\mathbf{f}_T = \mathbf{e}_y$ when $\mathbf{k}$ points in the $\mathbf{e}_z$ direction), $g_T = 2$.

(a) Phonons of both types, longitudinal and transverse, are bosons. Why? [Hint: Each normal mode of an elastic body can be described mathematically as a harmonic oscillator.]

(b) Phonons are fairly easily created, absorbed, scattered, and thermalized. A general argument that we will give for chemical reactions in Sec. 5.5 can be applied to phonon creation and absorption to deduce that, once they reach complete thermal equilibrium with their environment, the phonons will have vanishing chemical potential $\mu = 0$. What, then, will be their distribution functions η and $\mathcal{N}$?

(c) Ignoring the fact that the sound waves' wavelengths $\lambda = 2\pi/|\mathbf{k}|$ cannot be smaller than about twice the spacing between the atoms of the solid, show that the total phonon energy (wave energy) in a volume V of the solid is identical to that for blackbody photons in a volume V, but with the speed of light c replaced by the speed of sound C_s, and with the photon number of spin states, 2, replaced by $g_s = 3$ (2 for transverse waves plus 1 for longitudinal): $E_{\text{tot}} = a_s T^4 V$, with $a_s = g_s(4\pi^5/15)(k_B^4/(h^3 C_s^3))$ [cf. Eqs. (3.54)].

(d) Show that the specific heat of the phonon gas (the sound waves) is $C_V = 4a_s T^3 V$. This scales as T^3, whereas in a metal the specific heat of the degenerate electrons scales as T (previous exercise), so at sufficiently low temperatures the electron specific heat will dominate over that of the phonons.

(e) Show that in the phonon gas, only phonon modes with wavelengths longer than $\sim\lambda_T = C_s h/(k_B T)$ are excited; that is, for $\lambda \ll \lambda_T$ the mean occupation number is $\eta \ll 1$; for $\lambda \sim \lambda_T$, $\eta \sim 1$; and for $\lambda \gg \lambda_T$, $\eta \gg 1$. As T is increased, λ_T gets reduced. Ultimately it becomes of order the interatomic spacing, and our computation fails, because most of the modes that our calculation assumes are thermalized actually don't exist. What is the critical temperature (*Debye temperature*) at which our computation fails and the T^3 law for C_V changes? Show by a roughly one-line argument that above the Debye temperature, C_V is independent of temperature.

Exercise 3.13 *Derivation and Practice: Equation of State for a Photon Gas* **N** **R**

(a) Consider a collection of photons with a distribution function $\mathcal{N}$ that, in the mean rest frame of the photons, is isotropic. Show, using Eqs. (3.49b) and (3.49c), that this photon gas obeys the equation of state $P = \frac{1}{3}\rho$.

(b) Suppose the photons are thermalized with zero chemical potential (i.e., they are isotropic with a blackbody spectrum). Show that $\rho = aT^4$, where a is the radiation constant of Eq. (3.54c). [Note: Do not hesitate to use Mathematica, Matlab, or Maple, or other computer programs to evaluate integrals!]

(c) Show that for this isotropic, blackbody photon gas the number density of photons is $n = bT^3$, where b is given by Eq. (3.54b), and that the mean energy of a photon in the gas is

$$\bar{\mathcal{E}}_\gamma = \frac{\pi^4}{30\zeta(3)} k_B T \simeq 2.701\, k_B T. \tag{3.62}$$

3.6 Evolution of the Distribution Function: Liouville's Theorem, the Collisionless Boltzmann Equation, and the Boltzmann Transport Equation **N** **R**

We now turn to the issue of how the distribution function $\eta(\mathcal{P}, \vec{p})$, or equivalently, $\mathcal{N} = (g_s/h^3)\eta$, evolves from point to point in phase space. We explore the evolution under the simple assumption that between their very brief collisions, the particles all move freely, uninfluenced by any forces. It is straightforward to generalize to a situation where the particles interact with electromagnetic, gravitational, or other fields as they move, and we do so in Box 3.2, and Sec. 4.3. However, in the body of this chapter, we restrict attention to the common situation of free motion between collisions.

Initially we even rule out collisions; only at the end of this section do we restore them by inserting them as an additional term in our collision-free evolution equation for η.

The foundation for the collision-free evolution law will be *Liouville's theorem*. Consider a set $\mathcal{S}$ of particles that are initially all near some location in phase space and initially occupy an infinitesimal (frame-independent) phase-space volume $d^2\mathcal{V} = d\mathcal{V}_x d\mathcal{V}_p$. Pick a particle at the center of the set $\mathcal{S}$ and call it the "fiducial particle." Since all the particles in $\mathcal{S}$ have nearly the same initial position and velocity, they subsequently all move along nearly the same trajectory (world line): they all remain congregated around the fiducial particle. Liouville's theorem says that the phase-space volume occupied by the set of particles $\mathcal{S}$ is conserved along the trajectory of the fiducial particle:

$$\frac{d}{d\ell}(d\mathcal{V}_x d\mathcal{V}_p) = 0. \tag{3.63}$$

Liouville's theorem

Here ℓ is an arbitrary parameter along the trajectory. For example, in Newtonian theory ℓ could be universal time t or distance l traveled, and in relativity it could be proper time τ as measured by the fiducial particle (if its rest mass is nonzero) or the affine parameter ζ that is related to the fiducial particle's 4-momentum by $\vec{p} = d\vec{x}/d\zeta$.

We shall prove Liouville's theorem with the aid of the diagrams in Fig. 3.9. Assume, for simplicity, that the particles have nonzero rest mass. Consider the region in phase space occupied by the particles, as seen in the inertial reference frame (rest frame) of the fiducial particle, and choose for ℓ the time t of that inertial frame (or in Newtonian theory the universal time t). Choose the particles' region $d\mathcal{V}_x d\mathcal{V}_p$ at $t = 0$ to be a rectangular box centered on the fiducial particle (i.e., on the origin $x^j = 0$ of its inertial frame; Fig. 3.9a). Examine the evolution with time t of the 2-dimensional slice $y = p_y = z = p_z = 0$ through the occupied region. The evolution of other slices will be similar. Then, as t passes, the particle at location (x, p_x) moves with velocity $dx/dt = p_x/m$ (where the nonrelativistic approximation to the velocity is used, because all the particles are very nearly at rest in the fiducial particle's inertial frame). Because the particles move freely, each has a conserved p_x, and their motion $dx/dt = p_x/m$ (larger speeds are higher in the diagram) deforms the particles' phase space region into a skewed parallelogram as shown in Fig. 3.9b. Obviously, the area of the occupied region, $\Delta x \Delta p_x$, is conserved.

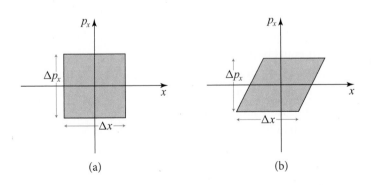

FIGURE 3.9 The phase-space region (x-p_x part) occupied by a set $\mathcal{S}$ of particles with finite rest mass, as seen in the inertial frame of the central, fiducial particle. (a) The initial region. (b) The region after a short time.

This same argument shows that the x-p_x area is conserved at *all* values of y, z, p_y, p_z; similarly for the areas in the y-p_y planes and those in the z-p_z planes. As a consequence, the total volume in phase space, $d\mathcal{V}_x d\mathcal{V}_p = \Delta x \Delta p_x \Delta y \Delta p_y \Delta z \Delta p_z$, is conserved.

Although this proof of Liouville's theorem relies on the assumption that the particles have nonzero rest mass, the theorem is also true for particles with zero rest mass—as one can deduce by taking the relativistic limit as the rest mass goes to zero and the particles' 4-momenta become null.

Since, in the absence of collisions or other nongravitational interactions, the number dN of particles in the set $\mathcal{S}$ is conserved, Liouville's theorem immediately implies the conservation of the number density in phase space, $\mathcal{N} = dN/(d\mathcal{V}_x d\mathcal{V}_p)$:

collisionless Boltzmann equation

$$\frac{d\mathcal{N}}{d\ell} = 0 \quad \text{along the trajectory of a fiducial particle.} \tag{3.64}$$

This conservation law is called the *collisionless Boltzmann equation;* in the context of plasma physics (Part VI) it is sometimes called the *Vlasov equation.* Note that it says that *not only is the distribution function $\mathcal{N}$ frame independent; $\mathcal{N}$ also is constant along the phase-space trajectory of any freely moving particle.*

The collisionless Boltzmann equation is actually far more general than is suggested by the above derivation; see Box 3.2, which is best read after finishing this section.

The collisionless Boltzmann equation is most nicely expressed in the frame-independent form Eq. (3.64). For some purposes, however, it is helpful to express the equation in a form that relies on a specific but arbitrary choice of inertial reference frame. Then $\mathcal{N}$ can be regarded as a function of the reference frame's seven phase-space coordinates, $\mathcal{N} = \mathcal{N}(t, x^j, p_k)$, and the collisionless Boltzmann equation (3.64) takes the coordinate-dependent form

$$\frac{d\mathcal{N}}{d\ell} = \frac{dt}{d\ell}\frac{\partial \mathcal{N}}{\partial t} + \frac{dx_j}{d\ell}\frac{\partial \mathcal{N}}{\partial x_j} + \frac{dp_j}{d\ell}\frac{\partial \mathcal{N}}{\partial p_j} = \frac{dt}{d\ell}\left(\frac{\partial \mathcal{N}}{\partial t} + v_j \frac{\partial \mathcal{N}}{\partial x_j}\right) = 0. \tag{3.65}$$

Here we have used the equation of straight-line motion $dp_j/dt = 0$ for the particles and have set dx_j/dt equal to the particle velocity v_j.

Since our derivation of the collisionless Boltzmann equation relies on the assumption that no particles are created or destroyed as time passes, the collisionless Boltzmann equation in turn should guarantee conservation of the number of particles, $\partial n/\partial t + \nabla \cdot \mathbf{S} = 0$ in Newtonian theory (Sec. 1.8), and $\vec{\nabla} \cdot \vec{S} = 0$ relativistically (Sec. 2.12.3). Indeed, this is so; see Ex. 3.14. Similarly, since the collisionless Boltzmann equation is based on the law of momentum (or 4-momentum) conservation for all the individual particles, it is reasonable to expect that the collisionless Boltzmann equation will guarantee the conservation of their total Newtonian momentum $[\partial \mathbf{G}/\partial t + \nabla \cdot \mathbf{T} = 0$, Eq. (1.36)] and their relativistic 4-momentum $[\vec{\nabla} \cdot \boldsymbol{T} = 0,$

Eq. (2.73a)]. And indeed, these conservation laws do follow from the collisionless Boltzmann equation; see Ex. 3.14.

Thus far we have assumed that the particles move freely through phase space with no collisions. If collisions occur, they will produce some nonconservation of $\mathcal{N}$ along the trajectory of a freely moving, noncolliding fiducial particle, and correspondingly, the collisionless Boltzmann equation will be modified to read

$$\boxed{\frac{d\mathcal{N}}{d\ell} = \left(\frac{d\mathcal{N}}{d\ell}\right)_{\text{collisions}},}$$ (3.66)

Boltzmann transport equation

where the right-hand side represents the effects of collisions. This equation, with collision terms present, is called the *Boltzmann transport equation*. The actual form of the collision terms depends, of course, on the details of the collisions. We meet some specific examples in the next section [Eqs. (3.79), (3.86a), (3.87), and Ex. 3.21] and in our study of plasmas (Chaps. 22 and 23).

When one applies the collisionless Boltzmann equation or Boltzmann transport equation to a given situation, it is helpful to simplify one's thinking in two ways: (i) Adjust the normalization of the distribution function so it is naturally tuned to the situation. For example, when dealing with photons, I_ν/ν^3 is typically best, and if—as is usually the case—the photons do not change their frequencies as they move and only a single reference frame is of any importance, then I_ν alone may do; see Ex. 3.15. (ii) Adjust the differentiation parameter ℓ so it is also naturally tuned to the situation.

Exercise 3.14 *Derivation and Problem: Collisionless Boltzmann Equation Implies Conservation of Particles and of 4-Momentum* N R

Consider a collection of freely moving, noncolliding particles that satisfy the collisionless Boltzmann equation $d\mathcal{N}/d\ell = 0$.

(a) Show that this equation guarantees that the Newtonian particle conservation law $\partial n/\partial t + \boldsymbol{\nabla} \cdot \mathbf{S} = 0$ and momentum conservation law $\partial \mathbf{G}/\partial t + \boldsymbol{\nabla} \cdot \mathbf{T} = 0$ are satisfied, where n, $\mathbf{S}$, $\mathbf{G}$, and $\mathbf{T}$ are expressed in terms of the distribution function $\mathcal{N}$ by the Newtonian momentum-space integrals (3.32).

(b) Show that the relativistic Boltzmann equation guarantees the relativistic conservation laws $\vec{\nabla} \cdot \vec{S} = 0$ and $\vec{\nabla} \cdot \boldsymbol{T} = 0$, where the number-flux 4-vector $\vec{S}$ and the stress-energy tensor $\boldsymbol{T}$ are expressed in terms of $\mathcal{N}$ by the momentum-space integrals (3.33).

Exercise 3.15 **Example: Solar Heating of Earth: The Greenhouse Effect* N

In this example we study the heating of Earth by the Sun. Along the way, we derive some important relations for blackbody radiation.

BOX 3.2. SOPHISTICATED DERIVATION OF RELATIVISTIC COLLISIONLESS BOLTZMANN EQUATION R

Denote by $\vec{X} \equiv \{\mathcal{P}, \vec{p}\}$ a point in 8-dimensional phase space. In an inertial frame the coordinates of $\vec{X}$ are $\{x^0, x^1, x^2, x^3, p_0, p_1, p_2, p_3\}$. [We use up (contravariant) indices on x and down (covariant) indices on p, because this is the form required in Hamilton's equations below; i.e., it is p_α and not p^α that is canonically conjugate to x^α.] Regard $\mathcal{N}$ as a function of location $\vec{X}$ in 8-dimensional phase space. Our particles all have the same rest mass, so $\mathcal{N}$ is nonzero only on the mass hyperboloid, which means that as a function of $\vec{X}$, $\mathcal{N}$ entails a delta function. For the following derivation, that delta function is irrelevant; the derivation is valid also for distributions of nonidentical particles, as treated in Ex. 3.2.

A particle in our distribution $\mathcal{N}$ at location $\vec{X}$ moves through phase space along a world line with tangent vector $d\vec{X}/d\zeta$, where ζ is its affine parameter. The product $\mathcal{N} d\vec{X}/d\zeta$ represents the number-flux 8-vector of particles through spacetime, as one can see by an argument analogous to Eq. (3.35c). We presume that, as the particles move through phase space, none are created or destroyed. The law of particle conservation in phase space, by analogy with $\vec{\nabla} \cdot \vec{S} = 0$ in spacetime, takes the form $\vec{\nabla} \cdot (\mathcal{N} d\vec{X}/d\zeta) = 0$. In terms of coordinates in an inertial frame, this conservation law says

$$\frac{\partial}{\partial x^\alpha}\left(\mathcal{N}\frac{dx^\alpha}{d\zeta}\right) + \frac{\partial}{\partial p_\alpha}\left(\mathcal{N}\frac{dp_\alpha}{d\zeta}\right) = 0. \tag{1}$$

The motions of individual particles in phase space are governed by Hamilton's equations

$$\frac{dx^\alpha}{d\zeta} = \frac{\partial \mathcal{H}}{\partial p_\alpha}, \quad \frac{dp_\alpha}{d\zeta} = -\frac{\partial \mathcal{H}}{\partial x^\alpha}. \tag{2}$$

For the freely moving particles of this chapter, a convenient form for the relativistic hamiltonian is [cf. Goldstein, Poole, and Safko (2002, Sec. 8.4) or Misner, Thorne, and Wheeler (1973, p. 489), who call it the super-hamiltonian]

$$\mathcal{H} = \frac{1}{2}(p_\alpha p_\beta g^{\alpha\beta} + m^2). \tag{3}$$

Our derivation of the collisionless Boltzmann equation does not depend on this specific form of the hamiltonian; it is valid for any hamiltonian and thus, for example, for particles interacting with an electromagnetic field or even a relativistic gravitational field (spacetime curvature; Part VII). By inserting Hamilton's equations (2) into the 8-dimensional law of particle conservation (1), we obtain

(continued)

BOX 3.2. (continued)

$$\frac{\partial}{\partial x^\alpha}\left(\mathcal{N}\frac{\partial \mathcal{H}}{\partial p_\alpha}\right) - \frac{\partial}{\partial p_\alpha}\left(\mathcal{N}\frac{\partial \mathcal{H}}{\partial x^\alpha}\right) = 0. \tag{4}$$

Using the rule for differentiating products and noting that the terms involving two derivatives of $\mathcal{H}$ cancel, we bring this into the form

$$0 = \frac{\partial \mathcal{N}}{\partial x^\alpha}\frac{\partial \mathcal{H}}{\partial p_\alpha} - \frac{\partial \mathcal{N}}{\partial p_\alpha}\frac{\partial \mathcal{H}}{\partial x^\alpha} = \frac{\partial \mathcal{N}}{\partial x^\alpha}\frac{dx^\alpha}{d\zeta} - \frac{\partial \mathcal{N}}{\partial p_\alpha}\frac{dp_\alpha}{d\zeta} = \frac{d\mathcal{N}}{d\zeta}, \tag{5}$$

which is the collisionless Boltzmann equation. (To get the second expression we have used Hamilton's equations, and the third follows directly from the formulas of differential calculus.) Thus, the collisionless Boltzmann equation is a consequence of just two assumptions: conservation of particles and Hamilton's equations for the motion of each particle. This implies that the Boltzmann equation has very great generality. We extend and explore this generality in the next chapter.

Since we will study photon propagation from the Sun to Earth with Doppler shifts playing a negligible role, there is a preferred inertial reference frame: that of the Sun and Earth with their relative motion neglected. We carry out our analysis in that frame. Since we are dealing with thermalized photons, the natural choice for the distribution function is I_ν/ν^3; and since we use just one unique reference frame, each photon has a fixed frequency ν, so we can forget about the ν^3 and use I_ν.

(a) Assume, as is very nearly true, that each spot on the Sun emits blackbody radiation in all outward directions with a common temperature $T_\odot = 5,800$ K. Show, by integrating over the blackbody I_ν, that the total energy flux (i.e., power per unit surface area) F emitted by the Sun is

$$F \equiv \frac{d\mathcal{E}}{dt\,dA} = \sigma T_\odot^4, \quad \text{where} \quad \sigma = \frac{ac}{4} = \frac{2\pi^5}{15}\frac{k_B^4}{h^3 c^2} = 5.67 \times 10^{-5}\frac{\text{erg}}{\text{cm}^2\,\text{s}\,\text{K}^4}.$$

$$(3.67)$$

(b) Since the distribution function I_ν is conserved along each photon's trajectory, observers on Earth, looking at the Sun, see the same blackbody specific intensity I_ν as they would if they were on the Sun's surface, and similarly for any other star. By integrating over I_ν at Earth [and not by the simpler method of using Eq. (3.67) for the flux leaving the Sun], show that the energy flux arriving at Earth is $F = \sigma T_\odot^4 (R_\odot/r)^2$, where $R_\odot = 696,000$ km is the Sun's radius and $r = 1.496 \times 10^8$ km is the mean distance from the Sun to Earth.

(c) Our goal is to compute the temperature $T_\oplus$ of Earth's surface. As a first attempt, assume that all the Sun's flux arriving at Earth is absorbed by Earth's surface, heating it to the temperature $T_\oplus$, and then is reradiated into space as blackbody radiation at temperature $T_\oplus$. Show that this leads to a surface temperature of

$$T_\oplus = T_\odot \left(\frac{R_\odot}{2r}\right)^{1/2} = 280\,\text{K} = 7\,^\circ\text{C}. \tag{3.68}$$

This is a bit cooler than the correct mean surface temperature ($287\,\text{K} = 14\,^\circ\text{C}$).

(d) Actually, Earth has an albedo of $A \simeq 0.30$, which means that 30% of the sunlight that falls onto it gets reflected back into space with an essentially unchanged spectrum, rather than being absorbed. Show that with only a fraction $1 - A = 0.70$ of the solar radiation being absorbed, the above estimate of Earth's temperature becomes

$$T_\oplus = T_\odot \left(\frac{\sqrt{1-A}\,R_\odot}{2r}\right)^{1/2} = 256\,\text{K} = -17\,^\circ\text{C}. \tag{3.69}$$

This is even farther from the correct answer.

(e) The missing piece of physics, which raises the temperature from $-17\,^\circ\text{C}$ to something much nearer the correct $14\,^\circ\text{C}$, is the *greenhouse effect:* The absorbed solar radiation has most of its energy at wavelengths $\sim 0.5\,\mu\text{m}$ (in the visual band), which pass rather easily through Earth's atmosphere. By contrast, the blackbody radiation that Earth's surface wants to radiate back into space, with its temperature $\sim 300\,\text{K}$, is concentrated in the infrared range from $\sim 8\,\mu\text{m}$ to $\sim 30\,\mu\text{m}$. Water molecules, carbon dioxide and methane in Earth's atmosphere absorb about 40% of the energy that Earth tries to reradiate at these energies (Cox, 2000, Sec. 11.22), causing the reradiated energy to be about 60% that of a blackbody at Earth's surface temperature. Show that with this (oversimplified!) greenhouse correction, $T_\oplus$ becomes about $290\,\text{K} = +17\,^\circ\text{C}$, which is within a few degrees of the true mean temperature. There is overwhelming evidence that the measured increase in the average Earth temperature in recent decades is mostly caused by the measured increase in carbon dioxide and methane, which, in turn, is mostly due to human activity. Although the atmospheric chemistry is not well enough understood to make accurate predictions, mankind is performing a very dangerous experiment.

Exercise 3.16 **Challenge: Olbers' Paradox and Solar Furnace** N

Consider a universe (not ours!) in which spacetime is flat and infinite in size and is populated throughout by stars that cluster into galaxies like our own and our neighbors, with interstellar and intergalactic distances similar to those in our neighborhood. Assume that the galaxies are *not* moving apart—there is no universal expansion. Using the collisionless Boltzmann equation for photons, show that Earth's temperature in this universe would be about the same as the surface temperatures of the universe's hotter stars, $\sim 10,000\,\text{K}$, so we would all be fried. This is Olbers' Paradox. What features of our universe protect us from this fate?

Motivated by this model universe, describe a design for a furnace that relies on sunlight for its heat and achieves a temperature nearly equal to that of the Sun's surface, 5,800 K.

3.7 Transport Coefficients N

In this section we turn to a practical application of kinetic theory: the computation of *transport coefficients*. Our primary objective is to illustrate the use of kinetic theory, but the transport coefficients themselves are also of interest: they play important roles in Parts V and VI (Fluid Dynamics and Plasma Physics) of this book.

What are transport coefficients? An example is electrical conductivity κ_e. When an electric field $\mathbf{E}$ is imposed on a sample of matter, Ohm's law tells us that the matter responds by developing a current density

electrical conductivity

$$\mathbf{j} = \kappa_e \mathbf{E}. \tag{3.70a}$$

The electrical conductivity is high if electrons can move through the material with ease; it is low if electrons have difficulty moving. The impediment to electron motion is scattering off other particles—off ions, other electrons, phonons (sound waves), plasmons (plasma waves), ... Ohm's law is valid when (as almost always) the electrons scatter many times, so they *diffuse* (random-walk their way) through the material. To compute the electrical conductivity, one must analyze, statistically, the effects of the many scatterings on the electrons' motions. The foundation for an accurate analysis is the Boltzmann transport equation (3.66).

Another example of a transport coefficient is thermal conductivity κ, which appears in the law of heat conduction

thermal conductivity

$$\mathbf{F} = -\kappa \nabla T. \tag{3.70b}$$

Here $\mathbf{F}$ is the diffusive energy flux from regions of high temperature T to low. The impediment to heat flow is scattering of the conducting particles; and, correspondingly, the foundation for accurately computing κ is the Boltzmann transport equation.

Other examples of transport coefficients are (i) the coefficient of shear viscosity η_{shear}, which determines the stress T_{ij} (diffusive flux of momentum) that arises in a shearing fluid [Eq. (13.68)]

coefficient of shear viscosity

$$T_{ij} = -2\eta_{\text{shear}}\sigma_{ij}, \tag{3.70c}$$

where σ_{ij} is the fluid's rate of shear (Ex. 3.19), and (ii) the diffusion coefficient D, which determines the diffusive flux of particles $\mathbf{S}$ from regions of high particle density n to low (Fick's law):

diffusion coefficient

$$\mathbf{S} = -D\nabla n. \tag{3.70d}$$

There is a *diffusion equation* associated with each of these transport coefficients. For example, the differential law of particle conservation $\partial n/\partial t + \nabla \cdot \mathbf{S} = 0$ [Eq. (1.30)], when applied to material in which the particles scatter many times so $\mathbf{S} = -D\nabla n$, gives the following diffusion equation for the particle number density:

diffusion equation

$$\frac{\partial n}{\partial t} = D\nabla^2 n,$$

(3.71)

where we have assumed that D is spatially constant. In Ex. 3.17, by exploring solutions to this equation, we shall see that the root mean square (rms) distance $\bar{l}$ the particles travel is proportional to the square root of their travel time, $\bar{l} = \sqrt{4Dt}$, a behavior characteristic of diffusive random walks.[10] See Sec. 6.3 for deeper insights into this.

Similarly, the law of energy conservation, when applied to diffusive heat flow $\mathbf{F} = -\kappa\nabla T$, leads to a diffusion equation for the thermal energy density U and thence for temperature [Ex. 3.18 and Eq. (18.4)]. Maxwell's equations in a magnetized fluid, when combined with Ohm's law $\mathbf{j} = \kappa_e\mathbf{E}$, lead to diffusion equation (19.6) for magnetic field lines. And the law of angular momentum conservation, when applied to a shearing fluid with $T_{ij} = -2\eta_{\text{shear}}\sigma_{ij}$, leads to diffusion equation (14.6) for vorticity.

These diffusion equations, and all other physical laws involving transport coefficients, are approximations to the real world—approximations that are valid if and only if (i) many particles are involved in the transport of the quantity of interest (e.g., charge, heat, momentum, particles) and (ii) on average each particle undergoes many scatterings in moving over the length scale of the macroscopic inhomogeneities that drive the transport. This second requirement can be expressed quantitatively in terms **mean free path** of the *mean free path* λ between scatterings (i.e., the mean distance a particle travels between scatterings, as measured in the mean rest frame of the matter) and the *macroscopic inhomogeneity scale* $\mathcal{L}$ for the quantity that drives the transport (e.g., in heat transport that scale is $\mathcal{L} \sim T/|\nabla T|$; i.e., it is the scale on which the temperature changes by an amount of order itself). In terms of these quantities, the second criterion of validity is $\lambda \ll \mathcal{L}$. These two criteria (many particles and $\lambda \ll \mathcal{L}$) together are called **diffusion criteria** *diffusion criteria,* since they guarantee that the quantity being transported (charge, heat, momentum, particles) will diffuse through the matter. If either of the two diffusion criteria fails, then the standard transport law (Ohm's law, the law of heat conduction, the Navier-Stokes equation, or the particle diffusion equation) breaks down and the corresponding transport coefficient becomes irrelevant and meaningless.

The accuracy with which one can compute a transport coefficient using the Boltzmann transport equation depends on the accuracy of one's description of the scattering. If one uses a high-accuracy collision term $(d\mathcal{N}/d\ell)_{\text{collisions}}$ in the Boltzmann

---

10. Einstein derived this diffusion law $\bar{l} = \sqrt{4Dt}$, and he used it and his formula for the diffusion coefficient D, along with observational data about the diffusion of sugar molecules in water, to demonstrate the physical reality of molecules, determine their sizes, and deduce the numerical value of Avogadro's number; see the historical discussion in Pais (1982, Chap. 5).

equation, one can derive a highly accurate transport coefficient. If one uses a very crude approximation for the collision term, the resulting transport coefficient might be accurate only to within an order of magnitude—in which case, it was probably not worth the effort to use the Boltzmann equation; a simple order-of-magnitude argument would have done just as well. If the interaction between the diffusing particles and the scatterers is electrostatic or gravitational (long-range $1/r^2$ interaction forces), then the particles cannot be idealized as moving freely between collisions, and an accurate computation of transport coefficients requires a more sophisticated analysis: the Fokker-Planck equation developed in Sec. 6.9 and discussed, for plasmas, in Secs. 20.4.3 and 20.5.

In the following three subsections, we compute the coefficient of thermal conductivity κ for hot gas inside a star (where short-range collisions hold sway and the Boltzmann transport equation is highly accurate). We do so first by an order-of-magnitude argument and then by the Boltzmann equation with an accurate collision term. In Exs. 3.19 and 3.20, readers have the opportunity to compute the coefficient of viscosity and the diffusion coefficient for particles using moderately accurate collision terms, and in Ex. 3.21 (neutron diffusion in a nuclear reactor), we will meet diffusion in momentum space, by contrast with diffusion in physical space.

Exercise 3.17 **Example: Solution of Diffusion Equation in an Infinite Homogeneous Medium** N

(a) Show that the following is a solution to the diffusion equation (3.71) for particles in a homogeneous infinite medium:

$$n = \frac{N}{(4\pi \, Dt)^{3/2}} e^{-r^2/(4Dt)}, \tag{3.72}$$

(where $r \equiv \sqrt{x^2 + y^2 + z^2}$ is radius), and that it satisfies $\int n \, d\mathcal{V}_x = N$, so N is the total number of particles. Note that this is a Gaussian distribution with width $\sigma = \sqrt{4Dt}$. Plot this solution for several values of σ. In the limit as $t \to 0$, the particles are all localized at the origin. As time passes, they random-walk (diffuse) away from the origin, traveling a mean distance $\alpha\sigma = \alpha\sqrt{4Dt}$ after time t, where α is a coefficient of order one. We will meet this square-root-of-time evolution in other random-walk situations elsewhere in this book; see especially Exs. 6.3 and 6.4, and Sec. 6.7.2.

(b) Suppose that the particles have an arbitrary initial distribution $n_o(\mathbf{x})$ at time $t = 0$. Show that their distribution at a later time t is given by the following Green's-function integral:

$$n(\mathbf{x}, t) = \int \frac{n_o(\mathbf{x}')}{(4\pi \, Dt)^{3/2}} e^{-|\mathbf{x}-\mathbf{x}'|^2/(4Dt)} d\mathcal{V}_{x'}. \tag{3.73}$$

(c) What form does the solution take in one dimension? And in two dimensions?

Exercise 3.18 **Problem: Diffusion Equation for Temperature*

Use the law of energy conservation to show that, when heat diffuses through a homogeneous medium whose pressure is being kept fixed, the evolution of the temperature perturbation $\delta T \equiv T - $ (average temperature) is governed by the diffusion equation

$$\frac{\partial T}{\partial t} = \chi \nabla^2 T, \quad \text{where} \quad \chi = \kappa/(\rho c_P) \tag{3.74}$$

is called the *thermal diffusivity*. Here c_P is the specific heat per unit mass at fixed pressure, and ρc_P is that specific heat per unit volume. For the extension of this to heat flow in a moving fluid, see Eq. (18.4).

3.7.1

3.7.1 Diffusive Heat Conduction inside a Star N

The specific transport-coefficient problem we treat here is for heat transport through hot gas deep inside a young, massive star. We confine attention to that portion of the star in which the temperature is 10^7 K $\lesssim T \lesssim 10^9$ K, the mass density is $\rho \lesssim$ 10 g cm$^{-3}(T/10^7$ K$)^2$, and heat is carried primarily by diffusing photons rather than by diffusing electrons or ions or by convection. (We study convection in Chap. 18.) In this regime the primary impediment to the photons' flow is collisions with electrons. The lower limit on temperature, 10^7 K, guarantees that the gas is almost fully ionized, so there is a plethora of electrons to do the scattering. The upper limit on density, $\rho \sim 10$ g cm$^{-3}(T/10^7$ K$)^2$, guarantees that (i) the inelastic scattering, absorption, and emission of photons by electrons accelerating in the Coulomb fields of ions (bremsstrahlung processes) are unimportant as impediments to heat flow compared to scattering off free electrons and (ii) the scattering electrons are nondegenerate (i.e., they have mean occupation numbers η small compared to unity) and thus behave like classical, free, charged particles. The upper limit on temperature, $T \sim 10^9$ K, guarantees that (i) the electrons doing the scattering are moving thermally at much less than the speed of light (the mean thermal energy $\frac{3}{2}k_B T$ of an electron is much less than its rest mass-energy $m_e c^2$) and (ii) the scattering is nearly elastic, with negligible energy exchange between photon and electron, and is describable with good accuracy by the *Thomson scattering cross section*.

In the rest frame of the electron (which to good accuracy will be the same as the mean rest frame of the gas, since the electron's speed relative to the mean rest frame is $\ll c$), the differential cross section $d\sigma$ for a photon to scatter from its initial propagation direction $\mathbf{n}'$ into a unit solid angle $d\Omega$ centered on a new propagation direction $\mathbf{n}$ is

$$\frac{d\sigma(\mathbf{n}' \rightarrow \mathbf{n})}{d\Omega} = \frac{3}{16\pi}\sigma_T[1 + (\mathbf{n} \cdot \mathbf{n}')^2]. \tag{3.75a}$$

Here σ_T is the total Thomson cross section [the integral of the differential cross section (3.75a) over solid angle]:

$$\sigma_T = \int \frac{d\sigma(\mathbf{n}' \rightarrow \mathbf{n})}{d\Omega} d\Omega = \frac{8\pi}{3}r_o^2 = 0.665 \times 10^{-24} \text{ cm}^2, \tag{3.75b}$$

where $r_o = e^2/(m_e c^2)$ is the classical electron radius. [For a derivation and discussion of the Thomson cross sections (3.75), see, e.g., Jackson (1999, Sec. 14.8).]

3.7.2 Order-of-Magnitude Analysis

Before embarking on any complicated calculation, it is always helpful to do a rough, order-of-magnitude analysis, thereby identifying the key physics and the approximate answer. The first step of a rough analysis of our heat transport problem is to identify the magnitudes of the relevant lengthscales. The inhomogeneity scale $\mathcal{L}$ for the temperature, which drives the heat flow, is the size of the hot stellar core, a moderate fraction of the Sun's radius: $\mathcal{L} \sim 10^5$ km. The mean free path of a photon can be estimated by noting that, since each electron presents a cross section σ_T to the photon and there are n_e electrons per unit volume, the probability of a photon being scattered when it travels a distance l through the gas is of order $n_e \sigma_T l$. Therefore, to build up to unit probability for scattering, the photon must travel a distance

$$\lambda \sim \frac{1}{n_e \sigma_T} \sim \frac{m_p}{\rho \sigma_T} \sim 3\,\text{cm} \left(\frac{1\,\text{g cm}^{-3}}{\rho} \right) \sim 3\,\text{cm}. \tag{3.76}$$

photon mean free path

Here m_p is the proton rest mass, $\rho \sim 1$ g cm^{-3} is the mass density in the core of a young, massive star, and we have used the fact that stellar gas is mostly hydrogen to infer that there is approximately one nucleon per electron in the gas and hence that $n_e \sim \rho/m_p$. Note that $\mathcal{L} \sim 10^5$ km is 3×10^9 times larger than $\lambda \sim 3$ cm, and the number of electrons and photons inside a cube of side $\mathcal{L}$ is enormous, so the diffusion description of heat transport is highly accurate.

In the diffusion description, the heat flux **F** as measured in the gas's rest frame is related to the temperature gradient ∇T by the law of diffusive heat conduction $\mathbf{F} = -\kappa \nabla T$. To estimate the thermal conductivity κ, orient the coordinates so the temperature gradient is in the z direction, and consider the rate of heat exchange between a gas layer located near $z = 0$ and a layer one photon mean free path away, at $z = \lambda$ (Fig. 3.10). The heat exchange is carried by photons that are emitted from one layer, propagate nearly unimpeded to the other, and then scatter. Although the individual scatterings are nearly elastic (and we thus are ignoring changes of photon

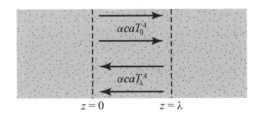

FIGURE 3.10 Heat exchange between two layers of gas separated by a distance of one photon mean free path in the direction of the gas's temperature gradient.

frequency in the Boltzmann equation), tiny changes of photon energy add up over many scatterings to keep the photons nearly in local thermal equilibrium with the gas. Thus, we approximate the photons and gas in the layer at $z = 0$ to have a common temperature T_0 and those in the layer at $z = \lambda$ to have a common temperature $T_\lambda = T_0 + \lambda dT/dz$. Then the photons propagating from the layer at $z = 0$ to that at $z = \lambda$ carry an energy flux

$$F_{0 \to \lambda} = \alpha c a (T_0)^4 \,, \tag{3.77a}$$

where a is the radiation constant of Eq. (3.54c); $a(T_0)^4$ is the photon energy density at $z = 0$; and α is a dimensionless constant of order $1/4$ that accounts for what fraction of the photons at $z = 0$ are moving rightward rather than leftward, and at what mean angle to the z direction. (Throughout this section, by contrast with early sections of this chapter, we use nongeometrized units, with the speed of light c present explicitly.) Similarly, the flux of energy from the layer at $z = \lambda$ to the layer at $z = 0$ is

$$F_{\lambda \to 0} = -\alpha c a (T_\lambda)^4 \,; \tag{3.77b}$$

and the net rightward flux, the sum of Eqs. (3.77a) and (3.77b), is

$$F = \alpha c a [(T_0)^4 - (T_\lambda)^4] = -4 \alpha c a T^3 \lambda \frac{dT}{dz}. \tag{3.77c}$$

Noting that 4α is approximately 1, inserting expression (3.76) for the photon mean free path, and comparing with the law of diffusive heat flow $\mathbf{F} = -\kappa \nabla T$, we conclude that the thermal conductivity is

thermal conductivity:
order of magnitude
estimate

$$\kappa \sim a T^3 c \lambda = \frac{a c T^3}{\sigma_T n_e} \,. \tag{3.78}$$

3.7.3

3.7.3 Analysis Using the Boltzmann Transport Equation N

With these physical insights and rough answer in hand, we turn to a Boltzmann transport analysis of the heat transfer. Our first step is to formulate the Boltzmann transport equation for the photons (including effects of Thomson scattering off the electrons) in the rest frame of the gas. Our second step will be to solve that equation to determine the influence of the heat flow on the distribution function $\mathcal{N}$, and our third step will be to compute the thermal conductivity κ by an integral over that $\mathcal{N}$.

To simplify the analysis we use, as the parameter ℓ in the Boltzmann transport equation $d\mathcal{N}/d\ell = (d\mathcal{N}/d\ell)_{\text{collisions}}$, the distance l that a fiducial photon travels, and we regard the distribution function $\mathcal{N}$ as a function of location $\mathbf{x}$ in space, the photon propagation direction (unit vector) $\mathbf{n}$, and the photon frequency ν: $\mathcal{N}(\mathbf{x}, \mathbf{n}, \nu)$. Because the photon frequency does not change during free propagation or Thomson scattering, it can be treated as a constant when solving the Boltzmann equation.

Along the trajectory of a fiducial photon, $\mathcal{N}(\mathbf{x}, \mathbf{n}, \nu)$ will change as a result of two things: (i) the scattering of photons out of the $\mathbf{n}$ direction and into other directions and (ii) the scattering of photons from other directions $\mathbf{n}'$ into the $\mathbf{n}$ direction. These effects produce the following two collision terms in the Boltzmann transport equation (3.66):

$$\frac{d\mathcal{N}(\mathbf{x}, \mathbf{n}, \nu)}{dl} = -\sigma_T n_e \mathcal{N}(\mathbf{x}, \mathbf{n}, \nu) + \int \frac{d\sigma(\mathbf{n}' \to \mathbf{n})}{d\Omega} n_e \mathcal{N}(\mathbf{x}, \mathbf{n}', \nu) d\Omega'. \quad (3.79)$$

Boltzmann transport equation for diffusing photons

(The second scattering term would be more obvious if we were to use I_ν (which is per unit solid angle) as our distribution function rather than $\mathcal{N}$; but they just differ by a constant.) Because the mean free path $\lambda = 1/(\sigma_T n_e) \sim 3$ cm is so short compared to the length scale $\mathcal{L} \sim 10^5$ km of the temperature gradient, the heat flow will show up as a tiny correction to an otherwise isotropic, perfectly thermal distribution function. Thus, we can write the photon distribution function as the sum of an unperturbed, perfectly isotropic and thermalized piece $\mathcal{N}_0$ and a tiny, anisotropic perturbation $\mathcal{N}_1$:

$$\mathcal{N} = \mathcal{N}_0 + \mathcal{N}_1, \quad \text{where } \mathcal{N}_0 = \frac{2}{h^3} \frac{1}{e^{h\nu/(k_B T)} - 1}. \quad (3.80a)$$

expand $\mathcal{N}$ in powers of $\lambda/\mathcal{L}$

Here the perfectly thermal piece $\mathcal{N}_0(\mathbf{x}, \mathbf{n}, \nu)$ has the standard blackbody form (3.23); it is independent of $\mathbf{n}$, and it depends on $\mathbf{x}$ only through the temperature $T = T(\mathbf{x})$. If the photon mean free path were vanishingly small, there would be no way for photons at different locations $\mathbf{x}$ to discover that the temperature is inhomogeneous; correspondingly, $\mathcal{N}_1$ would be vanishingly small. The finiteness of the mean free path permits $\mathcal{N}_1$ to be finite, and so it is reasonable to expect (and turns out to be true) that the magnitude of $\mathcal{N}_1$ is

$$\mathcal{N}_1 \sim \frac{\lambda}{\mathcal{L}} \mathcal{N}_0. \quad (3.80b)$$

Thus, $\mathcal{N}_0$ is the leading-order term, and $\mathcal{N}_1$ is the first-order correction in an expansion of the distribution function $\mathcal{N}$ in powers of $\lambda/\mathcal{L}$. This is called a *two-lengthscale expansion;* see Box 3.3.

Inserting $\mathcal{N} = \mathcal{N}_0 + \mathcal{N}_1$ into our Boltzmann transport equation (3.79) and using $d/dl = \mathbf{n} \cdot \nabla$ for the derivative with respect to distance along the fiducial photon trajectory, we obtain

$$n_j \frac{\partial \mathcal{N}_0}{\partial x_j} + n_j \frac{\partial \mathcal{N}_1}{\partial x_j} = \left[-\sigma_T n_e \mathcal{N}_0 + \int \frac{d\sigma(\mathbf{n}' \to \mathbf{n})}{d\Omega} n_e c \mathcal{N}_0 d\Omega' \right]$$

perturbative Boltzmann equation for $\mathcal{N}_1$

$$+ \left[-\sigma_T n_e c \mathcal{N}_1(\mathbf{n}, \nu) + \int \frac{d\sigma(\mathbf{n}' \to \mathbf{n})}{d\Omega} n_e c \mathcal{N}_1(\mathbf{n}', \nu) d\Omega' \right].$$

$$(3.80c)$$

Because $\mathcal{N}_0$ is isotropic (i.e., is independent of photon direction $\mathbf{n}'$), it can be pulled out of the integral over $\mathbf{n}'$ in the first square bracket on the right-hand side. When this is done, the first and second terms in that square bracket cancel each other. Thus, the unperturbed part of the distribution, $\mathcal{N}_0$, completely drops out of the right-hand side of Eq. (3.80c). On the left-hand side the term involving the perturbation $\mathcal{N}_1$ is tiny compared to that involving the unperturbed distribution $\mathcal{N}_0$, so we shall drop it.

BOX 3.3. TWO-LENGTHSCALE EXPANSIONS

Equation (3.80b) is indicative of the mathematical technique that underlies Boltzmann-transport computations: a perturbative expansion in the dimensionless ratio of two lengthscales, the tiny mean free path λ of the transporter particles and the far larger macroscopic scale $\mathcal{L}$ of the inhomogeneities that drive the transport. Expansions in lengthscale ratios $\lambda/\mathcal{L}$ are called *two-lengthscale expansions* and are widely used in physics and engineering. Most readers will previously have met such an expansion in quantum mechanics: the WKB approximation, where λ is the lengthscale on which the wave function changes, and $\mathcal{L}$ is the scale of changes in the potential $V(x)$ that drives the wave function. Kinetic theory itself is the result of a two-lengthscale expansion: it follows from the more sophisticated statistical-mechanics formalism of Chap. 4, in the limit where the particle sizes are small compared to their mean free paths. In this book we use two-lengthscale expansions frequently—for instance, in the geometric optics approximation to wave propagation (Chap. 7), in the study of boundary layers in fluid mechanics (Secs. 14.4, 14.5.4, and 15.5), in the quasi-linear formalism for plasma physics (Chap. 23), and in the definition of a gravitational wave (Sec. 27.4).

Because the spatial dependence of $\mathcal{N}_0$ is entirely due to the temperature gradient, we can bring the first term and the whole transport equation into the form

$$n_j \frac{\partial T}{\partial x_j} \frac{\partial \mathcal{N}_0}{\partial T} = -\sigma_T n_e \mathcal{N}_1(\mathbf{n}, \nu) + \int \frac{d\sigma(\mathbf{n}' \to \mathbf{n})}{d\Omega} n_e \mathcal{N}_1(\mathbf{n}', \nu) d\Omega'. \qquad (3.80d)$$

The left-hand side of this equation is the amount by which the temperature gradient causes $\mathcal{N}_0$ to fail to satisfy the Boltzmann equation, and the right-hand side is the manner in which the perturbation $\mathcal{N}_1$ steps into the breach and enables the Boltzmann equation to be satisfied.

Because the left-hand side is linear in the photon propagation direction n_j (i.e., it has a $\cos\theta$ dependence in coordinates where ∇T is in the z-direction; i.e., it has a dipolar, $l = 1$, angular dependence), $\mathcal{N}_1$ must also be linear in n_j (i.e., dipolar), in order to fulfill Eq. (3.80d). Thus, we write $\mathcal{N}_1$ in the dipolar form

$$\mathcal{N}_1 = K_j(\mathbf{x}, \nu) n_j, \qquad (3.80e)$$

and we shall solve the transport equation (3.80d) for the function K_j.

multipolar expansion of $\mathcal{N}$

[Side remark: This is a special case of a general situation. When solving the Boltzmann transport equation in diffusion situations, one is performing a power series expansion in $\lambda/\mathcal{L}$; see Box 3.3. The lowest-order term in the expansion, $\mathcal{N}_0$, is isotropic (i.e., it is monopolar in its dependence on the direction of motion of the diffusing particles). The first-order correction, $\mathcal{N}_1$, is down in magnitude by $\lambda/\mathcal{L}$ from $\mathcal{N}_0$ and is dipolar (or sometimes quadrupolar; see Ex. 3.19) in its dependence on the particles' direction of motion. The second-order correction, $\mathcal{N}_2$, is down in magnitude by $(\lambda/\mathcal{L})^2$ from $\mathcal{N}_0$ and its multipolar order is one higher than $\mathcal{N}_1$ (quadrupolar here; octupolar in Ex. 3.19). And so it continues to higher and higher orders.[11]]

When we insert the dipolar expression (3.80e) into the angular integral on the right-hand side of the transport equation (3.80d) and notice that the differential scattering cross section (3.75a) is unchanged under $\mathbf{n}' \rightarrow -\mathbf{n}'$, but $K_j n'_j$ changes sign, we find that the integral vanishes. As a result the transport equation (3.80d) takes the simplified form

$$n_j \frac{\partial T}{\partial x_j} \frac{\partial \mathcal{N}_0}{\partial T} = -\sigma_T n_e K_j n_j, \tag{3.80f}$$

from which we can read off the function K_j and thence $\mathcal{N}_1 = K_j n_j$:

$$\mathcal{N}_1 = -\frac{\partial \mathcal{N}_0/\partial T}{\sigma_T n_e} \frac{\partial T}{\partial x_j} n_j. \tag{3.80g}$$

solution for $\mathcal{N}_1$

Notice that, as claimed in Eq. (3.80b), the perturbation has a magnitude

$$\frac{\mathcal{N}_1}{\mathcal{N}_0} \sim \frac{1}{\sigma_T n_e} \frac{1}{T} |\boldsymbol{\nabla} T| \sim \frac{\lambda}{\mathcal{L}}. \tag{3.80h}$$

Having solved the Boltzmann transport equation to obtain $\mathcal{N}_1$, we can now evaluate the energy flux F_i carried by the diffusing photons. Relativity physicists will recognize that flux as the T^{0i} part of the stress-energy tensor and will therefore evaluate it as

$$F_i = T^{0i} = c^2 \int \mathcal{N} p^0 p^i \frac{d\mathcal{V}_p}{p^0} = c^2 \int \mathcal{N} p_i d\mathcal{V}_p \tag{3.81}$$

[cf. Eq. (3.33c) with the factors of c restored]. Newtonian physicists can deduce this formula by noticing that photons with momentum $\mathbf{p}$ in $d\mathcal{V}_p$ carry energy $E = |\mathbf{p}|c$ and move with velocity $\mathbf{v} = c\mathbf{p}/|\mathbf{p}|$, so their energy flux is $\mathcal{N} E\mathbf{v}\, d\mathcal{V}_p = c^2 \mathcal{N} \mathbf{p}\, d\mathcal{V}_p$; integrating this over momentum space gives Eq. (3.81). Inserting $\mathcal{N} = \mathcal{N}_0 + \mathcal{N}_1$ into this equation and noting that the integral over $\mathcal{N}_0$ vanishes, and inserting Eq. (3.80g) for $\mathcal{N}_1$, we obtain

$$F_i = c^2 \int \mathcal{N}_1 p_i d\mathcal{V}_p = -\frac{c}{\sigma_T n_e} \frac{\partial T}{\partial x_j} \frac{\partial}{\partial T} \int \mathcal{N}_0 c n_j p_i d\mathcal{V}_p. \tag{3.82a}$$

energy flux from $\mathcal{N}_1$

The relativity physicist will identify this integral as Eq. (3.33c) for the photons' stress tensor T_{ij} (since $n_j = p_j/p_0 = p_j/\mathcal{E}$). The Newtonian physicist, with a little thought, will recognize the integral in Eq. (3.82a) as the j component of the flux of i component

---

11. For full details of this "method-of-moments" analysis in nonrelativistic situations, see, e.g., Grad (1958); and for full relativistic details, see, e.g., Thorne (1981).

of momentum, which is precisely the stress tensor. Since this stress tensor is being computed with the isotropic, thermalized part of $\mathcal{N}$, it is isotropic, $T_{ji} = P\delta_{ji}$, and its pressure has the standard blackbody-radiation form $P = \frac{1}{3}aT^4$ [Eqs. (3.54a)]. Replacing the integral in Eq. (3.82a) by this blackbody stress tensor, we obtain our final answer for the photons' energy flux:

$$F_i = -\frac{c}{\sigma_T n_e}\frac{\partial T}{\partial x_j}\frac{d}{dT}\left(\frac{1}{3}aT^4\delta_{ji}\right) = -\frac{c}{\sigma_T n_e}\frac{4}{3}aT^3\frac{\partial T}{\partial x_i}. \tag{3.82b}$$

Thus, *from the Boltzmann transport equation we have simultaneously derived the law of diffusive heat conduction* $\mathbf{q} = -\kappa\nabla T$ *and evaluated the coefficient of thermal conductivity*

thermal conductivity from Boltzmann transport equation

$$\kappa = \frac{4}{3}\frac{acT^3}{\sigma_T n_e}. \tag{3.83}$$

Notice that this heat conductivity is 4/3 times our crude, order-of-magnitude estimate (3.78).

The above calculation, while somewhat complicated in its details, is conceptually fairly simple. The reader is encouraged to go back through the calculation and identify the main conceptual steps [expansion of distribution function in powers of $\lambda/\mathcal{L}$, insertion of zero-order plus first-order parts into the Boltzmann equation, multipolar decomposition of the zero- and first-order parts (with zero-order being monopolar and first-order being dipolar), neglect of terms in the Boltzmann equation that are smaller than the leading ones by factors $\lambda/\mathcal{L}$, solution for the coefficient of the multipolar decomposition of the first-order part, reconstruction of the first-order part from that coefficient, and insertion into a momentum-space integral to get the flux of the quantity being transported]. Precisely these same steps are used to evaluate all other transport coefficients that are governed by classical physics. [For examples of other such calculations, see, e.g., Shkarofsky, Johnston, and Bachynski (1966).]

As an application of the thermal conductivity (3.83), consider a young (main-sequence) 7-solar-mass ($7\,M_\odot$) star as modeled, for example, on pages 480 and 481 of Clayton (1968). Just outside the star's convective core, at radius $r \simeq 0.7R_\odot \simeq 5 \times 10^5$ km (where $R_\odot$ is the Sun's radius), the density and temperature are $\rho \simeq 5.5$ g cm^{-3} and $T \simeq 1.9 \times 10^7$ K, so the number density of electrons is $n_e \simeq \rho/(1.4m_p) \simeq 2.3 \times 10^{24}$ cm^{-3} where the 1.4 accounts for the star's chemical composition. For these parameters, Eq. (3.83) gives a thermal conductivity $\kappa \simeq 1.3 \times 10^{18}$ erg s^{-1} cm^{-2} K^{-1}. The lengthscale $\mathcal{L}$ on which the temperature is changing is approximately the same as the radius, so the temperature gradient is $|\nabla T| \sim T/r \sim 4 \times 10^{-4}$ K cm^{-1}. The law of diffusive heat transfer then predicts a heat flux $F = \kappa|\nabla T| \sim 5 \times 10^{14}$ erg s^{-1} cm^{-2}, and thus a total luminosity $L = 4\pi r^2 F \sim 1.5 \times 10^{37}$ erg s$^{-1} \simeq 4000L_\odot$ (4,000 solar luminosities). (This estimate is a little high. The correct value is about 3,600$L_\odot$.) What a difference the mass makes! The heavier a star, the hotter its core, the faster it burns, and the higher its luminosity will be. Increasing the mass by a factor of 7 drives the luminosity up by 4,000.

Exercise 3.19 **Example: Viscosity of a Monatomic Gas** [N]

Consider a nonrelativistic fluid that, in the neighborhood of the origin, has fluid velocity

$$v_i = \sigma_{ij} x_j, \tag{3.84}$$

with σ_{ij} symmetric and trace-free. As we shall see in Sec. 13.7.1, this represents a purely shearing flow, with no rotation or volume changes of fluid elements; σ_{ij} is called the fluid's *rate of shear*. Just as a gradient of temperature produces a diffusive flow of heat, so the gradient of velocity embodied in σ_{ij} produces a diffusive flow of momentum (i.e., a stress). In this exercise we use kinetic theory to show that, for a monatomic gas with isotropic scattering of atoms off one another, this stress is

$$\boxed{T_{ij} = -2\eta_{\text{shear}} \sigma_{ij},} \tag{3.85a}$$

with the coefficient of shear viscosity

$$\boxed{\eta_{\text{shear}} \simeq \frac{1}{3} \rho \lambda v_{th},} \tag{3.85b}$$

where ρ is the gas density, λ is the atoms' mean free path between collisions, and $v_{th} = \sqrt{3k_B T/m}$ is the atoms' rms speed. Our analysis follows the same route as the analysis of heat conduction in Secs. 3.7.2 and 3.7.3.

(a) Derive Eq. (3.85b) for the shear viscosity, to within a factor of order unity, by an order-of-magnitude analysis like that in Sec. 3.7.2.

(b) Regard the atoms' distribution function $\mathcal{N}$ as being a function of the magnitude p and direction $\mathbf{n}$ of an atom's momentum, and of location $\mathbf{x}$ in space. Show that, if the scattering is isotropic with cross section σ_s and the number density of atoms is n, then the Boltzmann transport equation can be written as

$$\frac{d\mathcal{N}}{dl} = \mathbf{n} \cdot \nabla \mathcal{N} = -\frac{1}{\lambda} \mathcal{N} + \int \frac{1}{4\pi\lambda} \mathcal{N}(p, \mathbf{n}', \mathbf{x}) d\Omega', \tag{3.86a}$$

where $\lambda = 1/n\sigma_s$ is the atomic mean free path (mean distance traveled between scatterings) and l is distance traveled by a fiducial atom.

(c) Explain why, in the limit of vanishingly small mean free path, the distribution function has the following form:

$$\mathcal{N}_0 = \frac{n}{(2\pi m k_B T)^{3/2}} \exp[-(\mathbf{p} - m\boldsymbol{\sigma} \cdot \mathbf{x})^2/(2m k_B T)]. \tag{3.86b}$$

(d) Solve the Boltzmann transport equation (3.86a) to obtain the leading-order correction $\mathcal{N}_1$ to the distribution function at $\mathbf{x} = 0$.

[Answer: $\mathcal{N}_1 = -(\lambda \, p/(k_B T)) \sigma_{ab} n_a n_b \mathcal{N}_0$.]

(e) Compute the stress at $\mathbf{x} = 0$ via a momentum-space integral. Your answer should be Eq. (3.85a) with η_{shear} given by Eq. (3.85b) to within a few tens of percent accuracy. [Hint: Along the way you will need the following angular integral:

$$\int n_a n_b n_i n_j d\Omega = \frac{4\pi}{15}(\delta_{ab}\delta_{ij} + \delta_{ai}\delta_{bj} + \delta_{aj}\delta_{bi}). \qquad (3.86c)$$

Derive this by arguing that the integral must have the above delta-function structure, and then computing the multiplicative constant by performing the integral for $a = b = i = j = z$.]

Exercise 3.20 *Example: Diffusion Coefficient in the Collision-Time Approximation* N

Consider a collection of identical test particles with rest mass $m \neq 0$ that diffuse through a collection of thermalized scattering centers. (The test particles might be molecules of one species, and the scattering centers might be molecules of a much more numerous species.) The scattering centers have a temperature T such that $k_B T \ll mc^2$, so if the test particles acquire this temperature, they will have thermal speeds small compared to the speed of light as measured in the mean rest frame of the scattering centers. We study the effects of scattering on the test particles using the following collision-time approximation for the collision terms in the Boltzmann equation, which we write in the mean rest frame of the scattering centers:

$$\left(\frac{d\mathcal{N}}{dt}\right)_{collision} = \frac{1}{\tau}(\mathcal{N}_0 - \mathcal{N}), \quad \text{where} \quad \mathcal{N}_0 \equiv \frac{e^{-p^2/(2mk_B T)}}{(2\pi m k_B T)^{3/2}}n. \qquad (3.87)$$

Here the time derivative d/dt is taken moving with a fiducial test particle along its unscattered trajectory, $p = |\mathbf{p}|$ is the magnitude of the test particles' spatial momentum, $n = \int \mathcal{N} dV_p$ is the number density of test particles, and τ is a constant to be discussed below.

(a) Show that this collision term preserves test particles in the sense that

$$\left(\frac{dn}{dt}\right)_{collision} \equiv \int \left(\frac{d\mathcal{N}}{dt}\right)_{collision} dp_x dp_y dp_z = 0. \qquad (3.88)$$

(b) Explain why this collision term corresponds to the following physical picture: Each test particle has a probability $1/\tau$ per unit time of scattering; when it scatters, its direction of motion is randomized and its energy is thermalized at the scattering centers' temperature.

(c) Suppose that the temperature T is homogeneous (spatially constant), but the test particles are distributed inhomogeneously, $n = n(\mathbf{x}) \neq$ const. Let $\mathcal{L}$ be the lengthscale on which their number density n varies. What condition must $\mathcal{L}, \tau, T$, and m satisfy for the diffusion approximation to be reasonably accurate? Assume that this condition is satisfied.

(d) Compute, in order of magnitude, the particle flux $\mathbf{S} = -D\nabla n$ produced by the gradient of the number density n, and thereby evaluate the diffusion coefficient D.

(e) Show that the Boltzmann transport equation takes the form (sometimes known as the BKG or Crook model)

$$\frac{\partial \mathcal{N}}{\partial t} + \frac{p_j}{m}\frac{\partial \mathcal{N}}{\partial x_j} = \frac{1}{\tau}(\mathcal{N}_0 - \mathcal{N}). \tag{3.89a}$$

(f) Show that to first order in a small diffusion-approximation parameter, the solution of this equation is $\mathcal{N} = \mathcal{N}_0 + \mathcal{N}_1$, where $\mathcal{N}_0$ is as defined in Eq. (3.87), and

$$\mathcal{N}_1 = -\frac{p_j \tau}{m}\frac{\partial n}{\partial x_j}\frac{e^{-p^2/(2mk_BT)}}{(2\pi m k_B T)^{3/2}}. \tag{3.89b}$$

Note that $\mathcal{N}_0$ is monopolar (independent of the direction of $\mathbf{p}$), while $\mathcal{N}_1$ is dipolar (linear in $\mathbf{p}$).

(g) Show that the perturbation $\mathcal{N}_1$ gives rise to a particle flux given by Eq. (3.70d), with the diffusion coefficient

$$D = \frac{k_B T}{m}\tau. \tag{3.90}$$

How does this compare with your order-of-magnitude estimate in part (d)?

Exercise 3.21 **Example: Neutron Diffusion in a Nuclear Reactor** [N]
A simplified version of a commercial nuclear reactor involves *fissile material* such as enriched uranium[12] and a *moderator* such as graphite, both of which will be assumed in this exercise. Slow (thermalized) neutrons, with kinetic energies ~ 0.1 eV, are captured by the ^{235}U nuclei and cause them to undergo fission, releasing ~ 170 MeV of kinetic energy per fission which appears as heat. Some of this energy is then converted into electric power. The fission releases an average of two or three (assume two) fast neutrons with kinetic energies ~ 1 MeV. (This is an underestimate.) The fast neutrons must be slowed to thermal speeds where they can be captured by ^{235}U atoms and induce further fissions. The slowing is achieved by scattering off the moderator atoms—a scattering in which the crucial effect, energy loss, occurs in momentum space. The momentum-space scattering is elastic and isotropic in the center-of-mass frame, with total cross section (to scatter off one of the moderator's carbon atoms) $\sigma_s \simeq 4.8 \times 10^{-24}$ cm$^2 \equiv 4.8$ barns. Using the fact that in the moderator's rest frame, the incoming neutron has a much higher kinetic energy than the moderator carbon atoms, and using energy and momentum conservation and the isotropy of the scattering, one can show that in the moderator's rest frame, the logarithm of the neutron's

---

12. Natural uranium contains ~ 0.007 of the fissile isotope ^{235}U; the fraction is increased to ~ 0.01–0.05 in enriched uranium.

energy is reduced in each scattering by an average amount ξ that is independent of energy and is given by

$$\xi \equiv -\overline{\Delta \ln E} = 1 + \frac{(A-1)^2}{2A} \ln \left(\frac{A-1}{A+1} \right) \simeq 0.16, \tag{3.91}$$

a quantity sometimes known as *lethargy*. Here $A \simeq 12$ is the ratio of the mass of the scattering atom to that of the scattered neutron.

There is a dangerous hurdle that the diffusing neutrons must overcome during their slowdown: as the neutrons pass through a critical energy region of about ~ 7 to ~ 6 eV, the ^{238}U atoms can absorb them. The absorption cross section has a huge resonance there, with width ~ 1 eV and resonance integral $\int \sigma_a d \ln E \simeq 240$ barns. For simplicity, we approximate the cross section in this absorption resonance by $\sigma_a \simeq 1600$ barns at 6 eV $< E < 7$ eV, and zero outside this range. To achieve a viable fission chain reaction and keep the reactor hot requires about half of the neutrons (one per original ^{235}U fission) to slow down through this resonant energy without getting absorbed. Those that make it through will thermalize and trigger new ^{235}U fissions (about one per original fission), maintaining the chain reaction.

We idealize the uranium and moderator atoms as homogeneously mixed on lengthscales small compared to the neutron mean free path, $\lambda_s = 1/(\sigma_s n_s) \simeq 2$ cm, where n_s is the number density of moderator (carbon) atoms. Then the neutrons' distribution function $\mathcal{N}$, as they slow down, is isotropic in direction and independent of position; and in our steady-state situation, it is independent of time. It therefore depends only on the magnitude p of the neutron momentum or equivalently, on the neutron kinetic energy $E = p^2/(2m)$: $\mathcal{N} = \mathcal{N}(E)$.

Use the Boltzmann transport equation or other considerations to develop the theory of the slowing of the neutrons in momentum space and of their struggle to pass through the ^{238}U resonance region without getting absorbed. More specifically, do the following.

(a) Use as the distribution function not $\mathcal{N}(E)$ but rather $n_E(E) \equiv dN/dV_x dE$ = (number of neutrons per unit volume and per unit kinetic energy), and denote by $q(E)$ the number of neutrons per unit volume that slow down through energy E per unit time. Show that outside the resonant absorption region these two quantities are related by

$$q = \sigma_s n_s \xi E n_E v, \quad \text{where } v = \sqrt{2mE} \tag{3.92}$$

is the neutron speed, so q contains the same information as the distribution function n_E. Explain why the steady-state operation of the nuclear reactor requires q to be independent of energy in this nonabsorption region, and infer that $n_E \propto E^{-3/2}$.

(b) Show further that inside the resonant absorption region, 6 eV $< E < 7$ eV, the relationship between q and E is modified:

$$q = (\sigma_s n_s + \sigma_a n_a) \xi E n_E v. \tag{3.93}$$

Here n_s is the number density of scattering (carbon) atoms, and n_a is the number density of absorbing (^{238}U) atoms. [Hint: Require that the rate at which neutrons scatter into a tiny interval of energy $\delta E \ll \xi E$ is equal to the rate at which they leave that tiny interval.] Then show that the absorption causes q to vary with energy according to the following differential equation:

$$\frac{d \ln q}{d \ln E} = \frac{\sigma_a n_a}{(\sigma_s n_s + \sigma_a n_a)\xi}. \tag{3.94}$$

(c) By solving this differential equation in our idealization of constant σ_a over the range 7 to 6 eV, show that the condition to maintain the chain reaction is

$$\frac{n_s}{n_a} \simeq \frac{\sigma_a}{\sigma_s} \left(\frac{\ln(7/6)}{\xi \ln 2} - 1 \right) \simeq 0.41 \frac{\sigma_a}{\sigma_s} \simeq 140. \tag{3.95}$$

Thus, to maintain the reaction in the presence of the huge ^{238}U absorption resonance for neutrons, it is necessary that approximately 99% of the reactor volume be taken up by moderator atoms and 1% by uranium atoms.

This is a rather idealized version of what happens inside a nuclear reactor, but it provides insight into some of the important processes and the magnitudes of various relevant quantities. For a graphic example of an additional complexity, see the description of "xenon poisoning" of the chain reaction in the first production-scale nuclear reactor (built during World War II to make plutonium for the first American atomic bombs) in John Archibald Wheeler's autobiography (Wheeler, 2000).

Bibliographic Note

Newtonian kinetic theory is treated in many textbooks on statistical physics. At an elementary level, Kittel and Kroemer (1980, Chap. 14) is rather good. Texts at a more advanced level include Kardar (2007, Chap. 3), Reif (2008, Secs. 7.9–7.13 and Chaps. 12–14), and Reichl (2009, Chap. 11). For a very advanced treatment with extensive applications to electrons and ions in plasmas, and electrons, phonons, and quasi-particles in liquids and solids, see Lifshitz and Pitaevskii (1981).

Relativistic kinetic theory is rarely touched on in statistical-physics textbooks but should be. It is well known to astrophysicists. The treatment in this chapter is easily lifted into general relativity theory (see, e.g., Misner, Thorne, and Wheeler, 1973, Sec. 22.6).

Statistical Mechanics

Willard Gibbs did for statistical mechanics and for thermodynamics what Laplace did for celestial mechanics and Maxwell did for electrodynamics, namely, made his field a well-nigh finished theoretical structure.

ROBERT A. MILLIKAN (1938)

4.1 Overview

4.1

While kinetic theory (Chap. 3) gives a powerful description of some statistical features of matter, other features are outside its realm and must be treated using the more sophisticated tools of statistical mechanics. Examples include:

- *Correlations:* Kinetic theory's distribution function $\mathcal{N}$ tells us, on average, how many particles will occupy a given phase-space volume, but it says nothing about whether the particles like to clump or prefer to avoid one another. It is therefore inadequate to describe, for example, the distributions of galaxies and stars, which cluster under their mutual gravitational attraction (Sec. 4.10.1), or that of electrons in a plasma, which are mutually repulsive and thus are spatially anticorrelated (Sec. 22.6).

- *Fluctuations:* In experiments to measure a very weak mechanical force (e.g., tests of the equivalence principle and searches for gravitational waves), one typically monitors the motion of a pendulum's test mass, on which the force acts. Molecules of gas hitting the test mass also make it move. Kinetic theory predicts how many molecules will hit in 1 ms, on average, and how strong is the resulting pressure acting in all directions; but kinetic theory's distribution function $\mathcal{N}$ cannot tell us the probability that in 1 ms more molecules will hit one side of the test mass than the other, mimicking the force one is trying to measure. The probability distribution for fluctuations is an essential tool for analyzing the noise in this and any other physical experiment, and it falls in the domain of statistical mechanics, not kinetic theory (Sec. 5.6).

- *Strongly interacting particles:* As should be familiar, the thermal motions of an ionic crystal are best described not in terms of individual atoms (as in the Einstein theory), but instead by decomposing the atoms' motion into normal modes (phonons; Debye theory). The thermal excitation of phonons is governed by statistical mechanics [Eq. (4.26)].

- *Microscopic origin of thermodynamic laws:* The laws of classical thermodynamics can be (and often are) derived from a few elementary, macroscopic postulates without any reference to the microscopic, atomic nature of matter. Kinetic theory provides a microscopic foundation for some of thermodynamics' abstract macroscopic ideas (e.g., the first law of thermodynamics) and permits the computation of equations of state. However, a full appreciation of entropy and the second law of thermodynamics (Sec. 4.7), and of behavior at phase transitions (Secs. 4.9, 5.5.2, 5.8.2, and 5.8.4) requires the machinery of statistical mechanics.

In this chapter, we develop the conceptual foundations for classical statistical mechanics and its interface with quantum physics, and we also delve deeply enough into the quantum world to be able to treat a few simple quantum problems. More specifically, in Sec. 4.2, we introduce the concepts of systems, ensembles of systems, and the distribution function for an ensemble. In Sec. 4.3, we use Hamiltonian dynamics to study the evolution of an ensemble's distribution function and derive the statistical mechanical version of Liouville's theorem. In Sec. 4.4, we develop the concept of statistical equilibrium and derive the general form of distribution functions for ensembles of systems that have reached statistical equilibrium (Sec. 4.4.2) and specific forms that depend on what additive macroscopic quantities the systems can exchange with their thermalized environment: energy exchange (canonical distribution, Sec. 4.4.1), energy and volume exchange (Gibbs distribution, Sec. 4.4.2), energy and particle exchange (grand canonical distribution, Sec. 4.4.2), and nothing exchanged (microcanonical distribution, Sec. 4.5).

In Chap. 5, we study these equilibrium distributions in considerable detail, especially their relationship to the laws of thermodynamics. Here in Chap. 4, we use them to explore some fundamental statistical mechanics issues:

1. a derivation of the Bose-Einstein and Fermi-Dirac distributions for the mean occupation number of a single-particle quantum state, which we studied in depth in Chap. 3 (Sec. 4.4.3);

2. a discussion and proof of the equipartition theorem for classical, quadratic degrees of freedom (Sec. 4.4.4);

3. the relationship between the microcanonical ensemble and *ergodicity* (the ergodic evolution of a single, isolated system; Sec. 4.6);

4. the concept of the entropy of an arbitrary ensemble of systems, and the increase of entropy (second law of thermodynamics) as an ensemble of isolated ("closed") systems evolves into its equilibrium, microcanonical form via *phase mixing, coarse graining,* and (quantum mechanically) via *discarding quantum correlations*—it's the physicist's fault!—(Sec. 4.7);

5. the entropy increase when two gases are mixed, and how quantum mechanics resolved the highly classical Gibbs Paradox (Ex. 4.8);

6. the power of entropy per particle as a tool for studying the evolution of physical systems (Sec. 4.8 and Ex. 4.10); and

7. Bose-Einstein condensation of a dilute gas of bosonic atoms (Sec. 4.9).

We conclude with a discussion of statistical mechanics in the presence of gravity (applications to galaxies, black holes, and the universe as a whole; Sec. 4.10) and a brief introduction to the concept of *information* and its connection to entropy (Sec. 4.11).

4.2 Systems, Ensembles, and Distribution Functions

4.2.1 Systems

Systems play the same role in statistical mechanics as is played by particles in kinetic theory. A system is any physical entity. (Obviously, this is an exceedingly general concept!) Examples are a galaxy, the Sun, a sapphire crystal, the fundamental mode of vibration of that crystal, an aluminum atom in that crystal, an electron from that aluminum atom, a quantum state in which that electron could reside, ...

SEMICLOSED SYSTEMS

Statistical mechanics focuses special attention on systems that couple only weakly to the rest of the universe. Stated more precisely, we are interested in systems whose **relevant** internal evolution timescales, τ_{int}, are short compared with the external timescales, τ_{ext}, on which they exchange energy, entropy, particles, and so forth, with their surroundings. Such systems are said to be *semiclosed,* and in the idealized limit where one completely ignores their external interactions, they are said to be *closed.* The statistical mechanics formalism for dealing with them relies on the assumption $\tau_{int}/\tau_{ext} \ll 1$; in this sense, it is a variant of a two-lengthscale expansion (Box 3.3).

Some examples will elucidate these concepts. For a galaxy of, say, 10^{11} stars, τ_{int} is the time it takes a star to cross the galaxy, so $\tau_{int} \sim 10^8$ yr. The external timescale is the time since the galaxy's last collison with a neighboring galaxy or the time since it was born by separating from the material that formed neighboring galaxies; both these times are $\tau_{ext} \sim 10^{10}$ yr, so $\tau_{int}/\tau_{ext} \sim 1/100$, and the galaxy is semiclosed. For a small volume of gas inside the Sun (say, 1 m on a side), τ_{int} is the timescale for the constituent electrons, ions, and photons to interact through collisions, $\tau_{int} \lesssim 10^{-10}$ s; this is much smaller than the time for external heat or particles to diffuse from the cube's surface to its center, $\tau_{ext} \gtrsim 10^{-5}$ s, so the cube is semiclosed. An individual atom in a crystal is so strongly coupled to its neighboring atoms by electrostatic forces that $\tau_{int} \sim \tau_{ext}$, which means the atom is not semiclosed. By contrast, for a vibrational mode of the crystal, τ_{int} is the mode's vibration period, and τ_{ext} is the time to exchange energy with other modes and thereby damp the chosen mode's vibrations; quite generally, the damping time is far longer than the period, so the mode is semiclosed. (For a highly polished, cold sapphire crystal weighing several kilograms, τ_{ext} can be $\sim 10^9 \tau_{int}$.) Therefore, it is the crystal's vibrational normal modes and not its atoms that are amenable to the statistical mechanical tools we shall develop.

CLOSED SYSTEMS AND THEIR HAMILTONIANS

When a semiclosed classical system is idealized as closed, so its interactions with the external universe are ignored, then its evolution can be described using Hamiltonian dynamics (see, e.g., Marion and Thornton, 1995; Landau and Lifshitz, 1976; Goldstein, Poole, and Safko, 2002). The system's classical state is described by *generalized coordinates* $\mathbf{q} \equiv \{q_j\}$ and *generalized momenta* $\mathbf{p} \equiv \{p_j\}$, where the index j runs from 1 to $W =$ (the system's number of degrees of freedom). The evolution of $\mathbf{q}, \mathbf{p}$ is governed by *Hamilton's equations*

$$\boxed{\frac{dq_j}{dt} = \frac{\partial H}{\partial p_j}, \quad \frac{dp_j}{dt} = -\frac{\partial H}{\partial q_j},} \tag{4.1}$$

where $H(\mathbf{q}, \mathbf{p})$ is the *hamiltonian,* and each equation is really W separate equations. Note that, because the system is idealized as closed, there is no explicit time dependence in the hamiltonian. Of course, not all physical systems (e.g., not those with strong internal dissipation) are describable by Hamiltonian dynamics, though in principle this restriction can usually be circumvented by increasing the number of degrees of freedom to include the cause of the dissipation.[1]

EXAMPLES

Let us return to our examples. For an individual star inside a galaxy, there are three degrees of freedom ($W = 3$), which we might choose to be the motion along three mutually orthogonal Cartesian directions, so $q_1 = x, q_2 = y, q_3 = z$. Because the star's speed is small compared to that of light, its hamiltonian has the standard form for a nonrelativistic particle:

$$H(\mathbf{q}, \mathbf{p}) = \frac{1}{2m}(p_1{}^2 + p_2{}^2 + p_3{}^2) + m\Phi(q_1, q_2, q_3). \tag{4.2}$$

Here m is the stellar mass, and $\Phi(q_1, q_2, q_3)$ is the galaxy's Newtonian gravitational potential (whose sign we take to be negative). Now, consider not just one star, but $K \sim 10^{11}$ of them in a galaxy. There are now $W = 3K$ degrees of freedom, and the hamiltonian is simply the sum of the hamiltonians for each individual star, so long as we ignore interactions between pairs of stars. The great power of the principles that we will develop is that they do not depend on whether $W = 1$ or $W = 3 \times 10^{11}$.

If our system is the fundamental mode of a sapphire crystal, then the number of degrees of freedom is only $W = 1$, and we can take the single generalized coordinate q to be the displacement of one end of the crystal from equilibrium. There will be an effective mass M for the mode (approximately equal to the actual mass of the crystal) such that the mode's generalized momentum is $p = M dq/dt$. The hamiltonian will be the standard one for a harmonic oscillator:

$$H(p, q) = \frac{p^2}{2M} + \frac{1}{2}M\omega^2 q^2, \tag{4.3a}$$

where ω is the mode's angular frequency of oscillation.

If we want to describe a whole crystal weighing several kilograms and having $K \sim 10^{26}$ atoms, then we obtain H by summing over $W = 3K$ oscillator hamiltonians for the crystal's W normal modes and adding an interaction potential H_{int} that accounts for the very weak interactions among modes:

$$H = \sum_{j=1}^{W}\left\{\frac{p_j^2}{2M_j} + \frac{1}{2}M_j\omega_j{}^2 q_j{}^2\right\} + H_{\text{int}}(q_1, \ldots, q_W, p_1, \ldots, p_W). \tag{4.3b}$$

Here M_j is the effective mass of mode j, and ω_j is the mode's angular frequency. This description of the crystal is preferable to one in which we use, as our generalized coordinates and momenta, the coordinate locations and momentum components of each of the 10^{26} atoms. Why? Because the normal modes are so weakly coupled to one

---

1. For example, if we add damping to a simple harmonic oscillator, we can either treat the system as (in principle) hamiltonian by allowing for all the internal degrees of freedom associated with the damping, for example friction, or as (in practice) non-hamiltonian with an external heat sink.

another that they are semiclosed subsystems of the crystal, whereas the atoms are so strongly coupled that they are not, individually, semiclosed. As we shall see, there is great power in decomposing a complicated system into semiclosed subsystems.

4.2.2 Ensembles

In kinetic theory, we study statistically a collection of a huge number of particles. Similarly, in statistical mechanics, we study statistically a collection or *ensemble* of a huge number of systems. This ensemble is actually only a conceptual device, a foundation for statistical arguments that take the form of thought experiments. As we shall see, there are many different ways that one can imagine forming an ensemble, and this freedom can be used to solve many different types of problems.

In some applications, we require that all the systems in the ensemble be closed and be identical in the sense that they all have the same number of degrees of freedom, W; are governed by hamiltonians with the same functional forms $H(\mathbf{q}, \mathbf{p})$; and have the same volume V and total internal energy E (or $\mathcal{E}$, including rest masses). However, the values of the generalized coordinates and momenta at a specific time t, $\{\mathbf{q}(t), \mathbf{p}(t)\}$, need not be the same (i.e., the systems need not be in the same state at time t). If such a conceptual ensemble of identical closed systems (first studied by Boltzmann) evolves until it reaches statistical equilibrium (Sec. 4.5), it then is called *microcanonical*; see Table 4.1.

Sometimes we deal with an ensemble of systems that can exchange energy (heat) with their identical surroundings, so the internal energy of each system can fluctuate. If the surroundings (sometimes called *heat baths*) have far greater heat capacity than the individual systems, and if statistical equilibrium has been reached, then we call this sort of ensemble (introduced by Gibbs) *canonical*.

At the next level of generality, the systems can also expand (i.e., they can exchange volume as well as energy with their identical surroundings). This case was also studied by Gibbs—his text (Gibbs, 1902) still repays reading—and in equilibrium it is known as the *Gibbs ensemble*; its environment is a heat and volume bath. A fourth ensemble in common use is Pauli's *grand canonical ensemble*, in which each system can exchange

---

TABLE 4.1: Statistical-equilibrium ensembles used in this chapter

Ensemble	Quantities exchanged with surroundings
Microcanonical	Nothing
Canonical	Energy E
Gibbs	Energy E and volume V
Grand canonical	Energy E and number of particles N_I of various species I

Note: For relativistic systems we usually include the rest masses of all particles in the energy, so E is replaced by $\mathcal{E}$ in the above formulas.

energy and particles (but not volume) with its surroundings; see Table 4.1. We study these equilibrium ensembles and their baths in Secs. 4.4 and 4.5.

4.2.3 Distribution Function

DISTRIBUTION FUNCTION AS A PROBABILITY

In kinetic theory (Chap. 3), we described the statistical properties of a collection of identical particles by a distribution function, and we found it useful to tie that distribution function's normalization to quantum theory: $\eta(t; \mathbf{x}, \mathbf{p}) =$ (mean number of particles that occupy a quantum state at location $\{\mathbf{x}, \mathbf{p}\}$ in 6-dimensional phase space at time t). In statistical mechanics, we use the obvious generalization of this: $\eta =$ (mean number of systems that occupy a quantum state at location $\{\mathbf{q}, \mathbf{p}\}$ in an ensemble's $2W$-dimensional phase space, at time t)—except that we need two modifications:

1. This generalized η is proportional to the number of systems N_{sys} in our ensemble. (If we double N_{sys}, then η will double.) Because our ensemble is only a conceptual device, we don't really care how many systems it contains, so we divide η by N_{sys} to get a renormalized, N_{sys}-independent distribution function, $\rho = \eta/N_{\text{sys}}$, whose physical interpretation is

$$\rho(t; \mathbf{q}, \mathbf{p}) = \begin{pmatrix} \text{probability that a system, drawn randomly} \\ \text{from our ensemble, will be in a quantum state} \\ \text{at location } (\mathbf{q}, \mathbf{p}) \text{ in phase space at time } t \end{pmatrix}. \quad (4.4)$$

probabilistic distribution function

2. If the systems of our ensemble can exchange particles with the external universe (as is the case, for example, in the grand canonical ensemble of Table 4.1), then their number of degrees of freedom, W, can change, so ρ may depend on W as well as on location in the $2W$-dimensional phase space: $\rho(t; W, \mathbf{q}, \mathbf{p})$.

In the sector of the system's phase space with W degrees of freedom, denote the number density of quantum states by

$$\mathcal{N}_{\text{states}}(W, \mathbf{q}, \mathbf{p}) = \frac{dN_{\text{states}}}{d^W q \, d^W p} \equiv \frac{dN_{\text{states}}}{d\Gamma_W}. \quad (4.5)$$

density of states

Here we have used

$$d^W q \equiv dq_1 dq_2 \cdots dq_W, \quad d^W p \equiv dp_1 dp_2 \cdots dp_W, \quad d\Gamma_W \equiv d^W q \, d^W p. \quad (4.6)$$

Then the sum of the occupation probability ρ over all quantum states, which must (by the meaning of probability) be unity, takes the form

$$\sum_n \rho_n = \sum_W \int \rho \mathcal{N}_{\text{states}} d\Gamma_W = 1. \quad (4.7)$$

normalization of distribution function

On the left-hand side n is a formal index that labels the various quantum states $|n\rangle$ available to the ensemble's systems; on the right-hand side the sum is over all possible values of the system's dimensionality W, and the integral is over all of the $2W$-dimensional phase space, with $d\Gamma_W$ a short-hand notation for the phase-space integration element $d^W q \, d^W p$.

GEOMETRICAL VIEWPOINT

Equations (4.4)–(4.7) require some discussion. Just as the events and 4-momenta in relativistic kinetic theory are geometric, frame-independent objects, similarly *location in phase space* in statistical mechanics is a geometric, coordinate-independent concept (though our notation does not emphasize it). The quantities $\{\mathbf{q}, \mathbf{p}\} \equiv \{q_1, q_2, \ldots, q_W, p_1, p_2, \ldots, p_W\}$ are the coordinates of that phase-space location. When one makes a canonical transformation from one set of generalized coordinates and momenta to another (Ex. 4.1), the qs and ps change, but the geometric location in phase space does not. Moreover, just as the individual spatial and momentum volumes $d\mathcal{V}_x$ and $d\mathcal{V}_p$ occupied by a set of relativistic particles in kinetic theory are frame dependent, but their product $d\mathcal{V}_x d\mathcal{V}_p$ is frame-independent [cf. Eqs. (3.7a)–(3.7c)], so also in statistical mechanics the volumes $d^W q$ and $d^W p$ occupied by some chosen set of systems are dependent on the choice of canonical coordinates (they change under a canonical transformation), but the product $d^W q \, d^W p \equiv d\Gamma_W$ (the systems' total volume in phase space) is independent of the choice of canonical coordinates and is unchanged by a canonical transformation. Correspondingly, *the number density of*

ρ and $\mathcal{N}_{\text{states}}$ as geometric objects

states in phase space $\mathcal{N}_{\text{states}} = d N_{\text{states}}/d\Gamma_W$ *and the statistical mechanical distribution function* $\rho(t; W, \mathbf{q}, \mathbf{p})$, *like their kinetic-theory counterparts, are geometric, coordinate-independent quantities: they are unchanged by a canonical transformation.* See Ex. 4.1 and references cited there.

DENSITY OF STATES

Classical thermodynamics was one of the crowning achievements of nineteenth-century science. However, thermodynamics was inevitably incomplete and had to remain so until the development of quantum theory. A major difficulty, one that we have already confronted in Chap. 3, was how to count the number of states available to a system. As we saw in Chap. 3, the number density of quantum mechanical states in the 6-dimensional, single-particle phase space of kinetic theory is (ignoring particle spin) $\mathcal{N}_{\text{states}} = 1/h^3$, where h is Planck's constant. Generalizing to the $2W$-dimensional phase space of statistical mechanics, the number density of states turns out to be $1/h^W$ [one factor of $1/h$ for each of the canonical pairs $(q_1, p_1), (q_2, p_2), \cdots, (q_W, p_W)$]. Formally, this follows from the canonical quantization procedure of elementary quantum mechanics.

DISTINGUISHABILITY AND MULTIPLICITY

distinguishability of particles

There was a second problem in nineteenth-century classical thermodynamics, that of *distinguishability*: If we swap two similar atoms in phase space, do we have a new state or not? If we mix two containers of the same gas at the same temperature and pressure, does the entropy increase (Ex. 4.8)? This problem was recognized classically but was not resolved in a completely satisfactory classical manner. When the laws of quantum mechanics were developed, it became clear that all identical particles are indistinguishable, so having particle 1 at location $\mathcal{A}$ in phase space and an identical particle 2 at location $\mathcal{B}$ must be counted as the same state as particle 1 at $\mathcal{B}$ and particle 2 at $\mathcal{A}$. Correspondingly, if we attribute half the quantum state to the classical phase-space location {1 at $\mathcal{A}$, 2 at $\mathcal{B}$} and the other half to {1 at $\mathcal{B}$, 2 at $\mathcal{A}$}, then the classical number density of states per unit volume of phase space must be reduced by a factor of 2—and more generally by some *multiplicity factor* $\mathcal{M}$. In general, therefore, we can write the actual number density of states in phase space as

multiplicity

$$\mathcal{N}_{\text{states}} = \frac{dN_{\text{states}}}{d\Gamma_W} = \frac{1}{\mathcal{M}h^W}, \tag{4.8a}$$

density of states

and correspondingly, we can rewrite the normalization condition (4.7) for our probabilistic distribution function as

$$\sum_n \rho_n \equiv \sum_W \int \rho \mathcal{N}_{\text{states}} d\Gamma_W = \sum_W \int \rho \frac{d\Gamma_W}{\mathcal{M}h^W} = 1. \tag{4.8b}$$

This equation can be regarded, in the classical domain, as defining the meaning of the sum over states n. We shall make extensive use of such sums over states.

For N identical and indistinguishable particles with zero spin, it is not hard to see that $\mathcal{M} = N!$. If we include the effects of quantum mechanical spin (and the spin states can be regarded as degenerate), then there are g_s [Eq. (3.16)] more states present in the phase space of each particle than we thought, so an individual state's multiplicity $\mathcal{M}$ (the number of different phase-space locations to be attributed to the state) is reduced to

$$\mathcal{M} = \frac{N!}{g_s^N} \quad \text{for a system of N identical particles with spin } s. \tag{4.8c}$$

This is the quantity that appears in the denominator of the sum over states [Eq. (4.8b)].

Occasionally, for conceptual purposes it is useful to introduce a renormalized distribution function $\mathcal{N}_{\text{sys}}$, analogous to kinetic theory's number density of particles in phase space:

$$\mathcal{N}_{\text{sys}} = N_{\text{sys}} \mathcal{N}_{\text{states}} \rho = \frac{d(\text{number of systems})}{d(\text{volume in } 2W\text{-dimensional phase space})}. \tag{4.9}$$

number density of systems in phase space

However, this version of the distribution function will rarely if ever be useful computationally.

ENSEMBLE AVERAGE

Each system in an ensemble is endowed with a total energy that is equal to its hamiltonian, $E = H(\mathbf{q}, \mathbf{p})$ [or relativistically, $\mathcal{E} = H(\mathbf{q}, \mathbf{p})$]. Because different systems reside at different locations $(\mathbf{q}, \mathbf{p})$ in phase space, they typically will have different energies.

A quantity of much interest is the *ensemble-averaged energy,* which is the average value of E over all systems in the ensemble:

$$\langle E \rangle = \sum_n \rho_n \, E_n = \sum_W \int \rho \, E \mathcal{N}_{\text{states}} d\Gamma_W = \sum_W \int \rho \, E \frac{d\Gamma_W}{\mathcal{M}h^W}. \qquad (4.10a)$$

For any other function $A(\mathbf{q}, \mathbf{p})$ defined on the phase space of a system (e.g., the linear momentum or the angular momentum), one can compute an ensemble average by the obvious analog of Eq. (4.10a):

ensemble average

$$\boxed{\langle A \rangle = \sum_n \rho_n A_n.} \qquad (4.10b)$$

Our probabilistic distribution function $\rho_n = \rho(t; W, \mathbf{q}, \mathbf{p})$ has deeper connections to quantum theory than the above discussion reveals. In the quantum domain, even if we start with a system whose wave function ψ is in a *pure state* (ordinary, everyday type of quantum state), the system may evolve into a *mixed state* as a result of (i) interaction with the rest of the universe and (ii) our choice not to keep track of correlations between the universe and the system (Box 4.2 and Sec. 4.7.2). The system's initial, pure state can be described in geometric, basis-independent quantum language by a state vector ("ket") $|\psi\rangle$; but its final, mixed state requires a different kind of quantum description: a *density operator* $\hat{\rho}$. In the classical limit, the quantum mechanical density operator $\hat{\rho}$ becomes our classical probabilistic distribution function $\rho(t, W, \mathbf{q}, \mathbf{p})$; see Box 4.2 for some details.

EXERCISES

Exercise 4.1 *Example: Canonical Transformation*
Canonical transformations are treated in advanced textbooks on mechanics, such as Goldstein, Poole, and Safko (2002, Chap. 9) or, more concisely, Landau and Lifshitz (1976). This exercise gives a brief introduction. For simplicity we assume the hamiltionian is time independent.

Let (q_j, p_k) be one set of generalized coordinates and momenta for a given system. We can transform to another set (Q_j, P_k), which may be more convenient, using a *generating function* that connects the old and new sets. One type of generating function is $F(q_j, P_k)$, which depends on the old coordinates $\{q_j\}$ and new momenta $\{P_k\}$, such that

$$p_j = \frac{\partial F}{\partial q_j}, \quad Q_j = \frac{\partial F}{\partial P_j}. \qquad (4.11)$$

(a) As an example, what are the new coordinates and momenta in terms of the old that result from

$$F = \sum_{i=1}^{W} f_i(q_1, q_2, \ldots, q_W) P_i, \qquad (4.12)$$

where f_i are arbitrary functions of the old coordinates?

BOX 4.2. DENSITY OPERATOR AND QUANTUM STATISTICAL MECHANICS T2

Here we describe briefly the connection of our probabilistic distribution function ρ to the full quantum statistical theory as laid out, for example, in Sethna (2006, Sec. 7.1.1) and Cohen-Tannoudji, Diu, and Laloë (1977, complement E_{III}), or in much greater detail in Pathria and Beale (2011, Chap. 5) and Feynman (1972).

Consider a single quantum mechanical system that is in a pure state $|\psi\rangle$. One can formulate the theory of such a pure state equally well in terms of $|\psi\rangle$ or the density operator $\hat{\varrho} \equiv |\psi\rangle\langle\psi|$. For example, the expectation value of some observable, described by a Hermitian operator $\hat{A}$, can be expressed equally well as $\langle A \rangle = \langle \psi | \hat{A} | \psi \rangle$ or as $\langle A \rangle = \text{Trace}(\hat{\varrho}\hat{A})$.[2]

If our chosen system interacts with the external universe and we have no knowledge of the correlations that the interaction creates between the system and the universe, then the interaction drives the system into a mixed state, which is describable by a density operator $\hat{\varrho}$ but not by a ket vector $|\psi\rangle$. This $\hat{\varrho}$ can be regarded as a classical-type average of $|\psi\rangle\langle\psi|$ over an ensemble of systems, each of which has interacted with the external universe and then has been driven into a pure state $|\psi\rangle$ by a measurement of the universe. Equivalently, $\hat{\varrho}$ can be constructed from the pure state of universe plus system by "tracing over the universe's degrees of freedom."

If the systems in the ensemble behave nearly classically, then it turns out that in the basis $|\phi_n\rangle$, whose states are labeled by the classical variables $n = \{W, \mathbf{q}, \mathbf{p}\}$, the density matrix $\varrho_{nm} \equiv \langle \phi_n | \hat{\varrho} | \phi_m \rangle$ is very nearly diagonal. The classical probability ρ_n of classical statistical mechanics (and of this book when dealing with classical or quantum systems) is then equal to the diagonal value of this density matrix: $\rho_n = \varrho_{nn}$.

It can be demonstrated that the equation of motion for the density operator $\hat{\varrho}$, when the systems in the quantum mechanical ensemble are all evolving freely (no significant interactions with the external universe), is

$$\frac{\partial \hat{\varrho}}{\partial t} + \frac{1}{i\hbar}[\hat{\varrho}, \hat{H}] = 0. \tag{1}$$

This is the quantum statistical analog of the Liouville equation (4.15), and the quantum mechanical commutator $[\hat{\varrho}, \hat{H}]$ appearing here is the quantum

---

2. In any basis $|\phi_i\rangle$, "Trace" is just the trace of the matrix product $\sum_j \varrho_{ij} A_{jk}$, where $A_{jk} \equiv \langle \phi_j | \hat{A} | \phi_k \rangle$, and $\varrho_{ij} \equiv \langle \phi_i | \hat{\varrho} | \phi_j \rangle$ is called the *density matrix* in that basis. Sometimes $\hat{\rho}$ itself is called the density matrix, even though it is an operator.

(continued)

BOX 4.2. (continued)

mechanical analog of the Poisson bracket $[\rho, H]_{\mathbf{q},\mathbf{p}}$, which appears in the Liouville equation. If the ensemble's quantum systems are in eigenstates of their hamiltonians, then $\hat{\varrho}$ commutes with $\hat{H}$, so the density matrix is constant in time and there will be no transitions. This is the quantum analog of the classical ρ being constant in time and thusa constant of the motion (Sec. 4.4).

(b) The canonical transformation generated by Eq. (4.11) for arbitrary $F(q_j, P_k)$ leaves unchanged the value, but not the functional form, of the hamiltonian at each point in phase space. In other words, H is a geometric, coordinate-independent function (scalar field) of location in phase space. Show, for the special case of a system with one degree of freedom (one q, one p, one Q, and one P), that if Hamilton's equations (4.1) are satisfied in the old variables (q, p), then they will be satisfied in the new variables (Q, P).

(c) Show, for a system with one degree of freedom, that although $dq \neq dQ$ and $dp \neq dP$, the volume in phase space is unaffected by the canonical transformation: $dp\,dq = dP\,dQ$.

(d) Hence show that for any closed path in phase space, $\oint p\,dq = \oint P\,dQ$. These results are readily generalized to more than one degree of freedom.

4.3

4.3 Liouville's Theorem and the Evolution of the Distribution Function

In kinetic theory, the distribution function $\mathcal{N}$ is not only a frame-independent entity, it is also a constant along the trajectory of any freely moving particle, so long as collisions between particles are negligible. Similarly, in statistical mechanics, the probability ρ is not only coordinate-independent (unaffected by canonical transformations); ρ *is also a constant along the phase-space trajectory of any freely evolving system,* so long as the systems in the ensemble are not interacting significantly with the external universe (i.e., so long as they can be idealized as closed). This is the statistical mechanical version of Liouville's theorem, and its proof is a simple exercise in Hamiltonian mechanics, analogous to the "sophisticated" proof of the collisionless Boltzmann equation in Box 3.2.

Liouville's theorem

Since the ensemble's systems are closed, no system changes its dimensionality W during its evolution. This permits us to fix W in the proof. Since no systems are created or destroyed during their evolution, the number density of systems in phase space, $\mathcal{N}_{\text{sys}} = N_{\text{sys}}\,\mathcal{N}_{\text{states}}\,\rho$ [Eq. (4.9)], must obey the same kind of conservation law as we encountered in Eq. (1.30) for electric charge and particle number in Newtonian physics. For particle number in a fluid, the conservation law is $\partial n/\partial t + \nabla \cdot (n\mathbf{v}) = 0$, where n is the number density of particles in physical space, $\mathbf{v}$ is their common velocity (the fluid's velocity) in physical space, and $n\mathbf{v}$ is their flux. Our ensemble's systems have velocity $dq_j/dt = \partial H/\partial p_j$ in physical space and "velocity" $dp_j/dt = -\partial H/\partial q_j$ in momentum space, so by analogy with particle number in a fluid, the conservation law (valid for ρ as well as for $\mathcal{N}_{\text{sys}}$, since they are proportional to each other) is

$$\frac{\partial \rho}{\partial t} + \frac{\partial}{\partial q_j}\left(\rho \frac{dq_j}{dt}\right) + \frac{\partial}{\partial p_j}\left(\rho \frac{dp_j}{dt}\right) = 0. \qquad (4.13)$$

conservation law for systems

Equation (4.13) has an implicit sum, from 1 to W, over the repeated index j (recall the Einstein summation convention, Sec. 1.5). Using Hamilton's equations, we can rewrite this as

$$\begin{aligned}
0 &= \frac{\partial \rho}{\partial t} + \frac{\partial}{\partial q_j}\left(\rho \frac{\partial H}{\partial p_j}\right) - \frac{\partial}{\partial p_j}\left(\rho \frac{\partial H}{\partial q_j}\right) \\
&= \frac{\partial \rho}{\partial t} + \frac{\partial \rho}{\partial q_j}\frac{\partial H}{\partial p_j} - \frac{\partial \rho}{\partial p_j}\frac{\partial H}{\partial q_j} = \frac{\partial \rho}{\partial t} + [\rho, H]_{\mathbf{q},\mathbf{p}},
\end{aligned} \qquad (4.14)$$

where $[\rho, H]_{\mathbf{q},\mathbf{p}}$ is the *Poisson bracket* (e.g., Landau and Lifshitz, 1976; Marion and Thornton, 1995; Goldstein, Poole, and Safko, 2002). By using Hamilton's equations once again in the second expression, we discover that this is the time derivative of ρ moving with a fiducial system through the $2W$-dimensional phase space:

$$\boxed{\left(\frac{d\rho}{dt}\right)_{\substack{\text{moving with a} \\ \text{fiducial system}}} \equiv \frac{\partial \rho}{\partial t} + \frac{dq_j}{dt}\frac{\partial \rho}{\partial q_j} + \frac{dp_j}{dt}\frac{\partial \rho}{\partial p_j} = \frac{\partial \rho}{\partial t} + [\rho, H]_{\mathbf{q},\mathbf{p}} = 0.}$$

Liouville equation (collisionless Boltzmann equation)

$$(4.15)$$

Therefore, the probability ρ is constant along the system's phase space trajectory, as was to be proved.

We call Eq. (4.15), which embodies this Liouville theorem, the statistical mechanical *Liouville equation* or *collisionless Boltzmann equation*.

As a simple, qualitative example, consider a system consisting of hot gas expanding adiabatically, so its large random kinetic energy is converted into ordered radial motion. If we examine a set $\mathcal{S}$ of such systems very close to one another in phase space, then it is apparent that, as the expansion proceeds, the size of $\mathcal{S}$'s physical-space volume $d^W q$ increases and the size of its momentum-space volume $d^W p$ diminishes, so that the product $d^W q\, d^W p$ remains constant (Fig. 4.1), and correspondingly $\rho \propto \mathcal{N}_{\text{sys}} = dN_{\text{sys}}/d^W q\, d^W p$ is constant.

What happens if the systems being studied interact weakly with their surroundings? We must then include an interaction term on the right-hand side of Eq. (4.15),

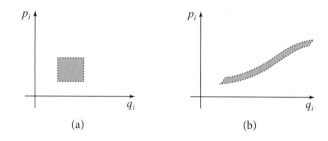

FIGURE 4.1 Liouville's theorem. (a) The region in the q_i-p_i part of phase space (with i fixed) occupied by a set $\mathcal{S}$ of identical, closed systems at time $t = 0$. (b) The region occupied by the same set of systems a short time later, $t > 0$. The hamiltonian-generated evolution of the individual systems has moved them in such a manner as to skew the region they occupy, but the volume $\int dp_i dq_i$ is unchanged.

thereby converting it into the statistical mechanical version of the Boltzmann transport equation:

Boltzmann transport equation

$$\left(\frac{d\rho}{dt} \right)_{\substack{\text{moving with a} \\ \text{fiducial system}}} = \left(\frac{d\rho}{dt} \right)_{\text{interactions}}. \tag{4.16}$$

The time derivative on the left is now taken moving through phase space with a fiducial system that does not interact with the external universe.

4.4

4.4 Statistical Equilibrium

In this section's discussion of statistical equilibrium, we begin by writing formulas in the relativistic form, where rest masses are included in the energy $\mathcal{E}$; afterward we take the Newtonian limit.

the meaning of temperature

Before we start, we should clarify what we mean by the temperature of a system. The first point to make is that a rigorous definition presumes thermodynamic equilibrium for the system in question, which can be anything from a two-level atom to the early universe. We can then say that two systems that are in equilibrium and whose average properties do not change when they are brought into contact (so they can exchange heat) have the same temperature and that this attribute is transitive. (This is known as the zeroth law of thermodynamics.) This allows us to define the temperature of a reference system as a monotonic function that increases as heat is added to it. This reference system can then be used to assign a temperature to every other body with which it can be brought into equilibrium; cf. Sec. 5.2.3. There are several ways to specify this function. The one that we shall follow [Eqs. (4.19) and following text] is surprisingly general and depends on simple ideas from probability.[3]

---

3. Other approaches to the definition of temperature involve the engineering-inspired Carnot cycle, the mathematical Carathéodory approach, and the fluctuation-dissipation theorem expressed, for example,

STATISTICAL EQUILIBRIUM AND JEANS' THEOREM

Consider an ensemble of identical systems, all of which have the same huge number of degrees of freedom (dimensionality $W \gg 1$). Put all the systems initially in the same state, and then let them exchange heat (but not particles, volume, or anything else) with an external *thermal bath* that has a huge heat capacity and is in thermodynamic equilibrium at some temperature T. (For example, the systems might be impermeable cubes of gas 1 km on a side near the center of the Sun, and the thermal bath might be all the surrounding gas near the Sun's center; or the systems might be identical sapphire crystals inside a huge cryostat, and the thermal bath might be the cryostat's huge store of liquid helium.) After a sufficiently long time, $t \gg \tau_{ext}$, the ensemble will settle down into equilibrium with the bath (i.e., it will become the canonical ensemble mentioned in Table 4.1 above). In this final, canonical equilibrium state, the probability $\rho(t, \mathbf{q}, \mathbf{p})$ is independent of time t, and it no longer is affected by interactions with the external environment. In other words, the interaction terms in the evolution equation (4.16) have ceased to have any net effect: on average, for each interaction event that feeds energy into a system, there is an interaction event that takes away an equal amount of energy. The distribution function, therefore, satisfies the interaction-free, collisionless Boltzmann equation (4.15) with the time derivative $\partial \rho / \partial t$ removed:

$$[\rho, H]_{\mathbf{q}, \mathbf{p}} \equiv \frac{\partial \rho}{\partial q_j} \frac{\partial H}{\partial p_j} - \frac{\partial \rho}{\partial p_j} \frac{\partial H}{\partial q_j} = 0. \qquad (4.17)$$

statistical equilibrium

We use the phrase *statistical equilibrium* to refer to any ensemble whose distribution function has attained such a state and thus satisfies Eq. (4.17).

Equation (4.17) is a well-known equation in Hamiltonian mechanics. It says that ρ is a function solely of constants of the individual systems' hamiltonian-induced motions (e.g., Landau and Lifshitz, 1976; Marion and Thornton, 1995; Goldstein, Poole, and Safko, 2002); in other words, ρ can depend on location $(\mathbf{q}, \mathbf{p})$ in phase space only through those constants of the motion. Sometimes this goes by the name *Jeans' theorem*. Among the constants of motion in typical situations (for typical hamiltonians) are the system's energy $\mathcal{E}$, its linear momentum $\mathbf{P}$, its angular momentum $\mathbf{J}$, its number N_I of conserved particles of various types I (e.g., electrons, protons), and its volume V. Notice that these constants of motion $\mathcal{E}, \mathbf{P}, \mathbf{J}, N_I$, and V are all additive: if we double the size of a system, they each double. We call such additive constants of the hamiltonian-induced motion *extensive variables* (a term borrowed from thermodynamics) and denote them by an enumerated list $K_1, K_2, \ldots$.

Jeans' theorem

extensive variables

Now, the systems we are studying have exchanged energy $\mathcal{E}$ with their environment (the thermal bath) and thereby have acquired some range of $\mathcal{E}$ values; therefore, ρ

as Johnson noise (Sec. 6.8.1). These are covered in many thermodynamics texts (e.g., Reif, 2008) and lead to the same temperature scale.

can depend on $\mathcal{E}$. However, the systems have not exchanged anything else with their environment, and they all thus have retained their original (identical) values of the other extensive variables K_A; therefore, ρ must be a delta function in these other variables. We write

$$\rho = \rho(\mathcal{E}) \tag{4.18a}$$

and do not write down the delta functions explicitly.

As an aid in discovering the form of the function $\rho(\mathcal{E})$, let us decompose each system in the ensemble into a huge number of subsystems. For example, each system might be a cube 1 km on a side inside the Sun, and its subsystems might be the 10^9 1-m cubes into which the system can be divided, or the systems might be identical sapphire crystals each containing 10^{26} atoms, and the subsystems might be the crystals' 3×10^{26} normal modes of vibration. We label the subsystems of each system by an integer a in such a way that subsystem a in one system has the same hamiltonian as subsystem a in any other system. (For the sapphire crystals, $a = 1$ could be the fundamental mode of vibration, $a = 2$ the first harmonic, $a = 3$ the second harmonic, etc.) The subsystems with fixed a make up a *subensemble* because of their relationship to the original ensemble.

Because the full ensemble is in statistical equilibrium, the subensembles will also be in statistical equilibrium; therefore, their probabilities must be functions of those extensive variables $\mathcal{E}, K_A$ that they can exchange with one another:

$$\rho_a = \rho_a(\mathcal{E}_a, K_{1a}, K_{2a}, \ldots). \tag{4.18b}$$

(Although each system can exchange only energy $\mathcal{E}$ with its heat bath, the subsystems may be able to exchange other quantities with one another; for example, if subsystem a is a 1-m cube inside the Sun with permeable walls, then it can exchange energy $\mathcal{E}_a$ and particles of all species I, so $K_{Ia} = N_{Ia}$.)

Since there is such a huge number of subsystems in each system, it is reasonable to expect that in statistical equilibrium there will be no significant correlations at any given time between the actual state of subsystem a and the state of any other subsystem. In other words, the probability $\rho_a(W_a, \mathbf{q}_a, \mathbf{p}_a)$ that subsystem a is in a quantum state with W_a degrees of freedom and with its generalized coordinates and momenta near the values $(\mathbf{q}_a, \mathbf{p}_a)$ is independent of the state of any other subsystem. This lack of correlations, which can be written as

$$\rho(\mathcal{E}) = \prod_a \rho_a, \tag{4.18c}$$

is called *statistical independence*.[4]

---

4. Statistical independence is actually a consequence of a two-lengthscale approximation [Box 3.3]. The size of each subsystem is far smaller than that of the full system, and precise statistical independence arises in the limit as the ratio of these sizes goes to zero.

Statistical independence places a severe constraint on the functional forms of ρ and ρ_a, as the following argument shows. By taking the logarithm of Eq. (4.18c), we obtain

$$\ln \rho(\mathcal{E}) = \sum_a \ln \rho_a(\mathcal{E}_a, K_{1a}, \ldots). \qquad (4.18d)$$

We also know, since energy is a linearly additive quantity, that

$$\mathcal{E} = \sum_a \mathcal{E}_a. \qquad (4.18e)$$

Now, we have not stipulated the way in which the systems are decomposed into subsystems. For our solar example, the subsystems might have been 2-m or 7-m cubes rather than 1-m cubes. Exploiting this freedom, one can deduce that Eqs. (4.18d) and (4.18e) can be satisfied simultaneously if and only if $\ln \rho$ and $\ln \rho_a$ depend linearly on the energies $\mathcal{E}$ and $\mathcal{E}_a$, with the same proportionality constant $-\beta$:

$$\ln \rho_a = -\beta \mathcal{E}_a + (\text{some function of } K_{1a}, K_{2a}, \ldots), \qquad (4.19a)$$

$$\ln \rho = -\beta \mathcal{E} + \text{constant}. \qquad (4.19b)$$

definition of temperature

The reader presumably will identify β with $1/(k_B T)$, where T is the temperature of the thermal bath, as we have implicitly done in Chap. 3. However, what does this temperature actually mean and how do we define it? One approach is to choose as our subsystem a single identified atom in a classical monatomic perfect gas, with atomic rest mass high enough that the gas is nonrelativistic. The energy $\mathcal{E}$ is then just the atom's kinetic energy (plus rest mass) and the pressure is $n k_B T$. We then repeat this for every other atom. Measuring the pressure—mechanically—in this simple system then defines $k_B T$. The choice of k_B is a mere convention; it is most commonly chosen to set the temperature of the triple point of water as 273.16 K. Much more complex systems can then be assigned the same temperature as the perfect gas with which they would be in equilibrium as outlined above. If the reader protests that gases are not perfect at low or high temperature, then we can substitute free electrons or photons and adopt their distribution functions as we have described in Sec. 3.3 and will derive in Sec. 4.4.3.

To summarize, an ensemble of identical systems with many degrees of freedom $W \gg 1$, which have reached statistical equilibrium by exchanging energy but nothing else with a huge thermal bath, has the following canonical distribution function:

canonical distribution

$$\boxed{\rho_{\text{canonical}} = C \exp(-\mathcal{E}/k_B T),} \quad \boxed{\rho_{\text{canonical}} = C' \exp(-E/k_B T) \text{ nonrelativistically.}}$$

$$(4.20)$$

Here $\mathcal{E}(\mathbf{q}, \mathbf{p})$ is the energy of a system at location $\{\mathbf{q}, \mathbf{p}\}$ in phase space, k_B is Boltzmann's constant, T is the temperature of the heat bath, and C is whatever normalization constant is required to guarantee that $\sum_n \rho_n = 1$. The nonrelativistic expression is obtained by removing all the particle rest masses from the total energy $\mathcal{E}$ and then taking the low-temperature, low-thermal-velocities limit.

Actually, we have proved more than Eq. (4.20). Not only must the ensemble of huge systems ($W \gg 1$) have the energy dependence $\rho \propto \exp(-\mathcal{E}/(k_B T))$, so must each subensemble of smaller systems, $\rho_a \propto \exp(-\mathcal{E}_a/(k_B T))$, even if (for example) the subensemble's identical subsystems have only one degree of freedom, $W_a = 1$. Thus, if the subsystems exchanged only heat with their parent systems, then they must have the same canonical distribution (4.20) as the parents. This shows that *the canonical distribution is the equilibrium state independently of the number of degrees of freedom W.*

4.4.2 General Equilibrium Ensemble and Distribution; Gibbs Ensemble; Grand Canonical Ensemble

GENERAL EQUILIBRIUM ENSEMBLE

We can easily generalize the canonical distribution to an ensemble of systems that exchange other additive conserved quantities (extensive variables) K_1, K_2, ..., in addition to energy $\mathcal{E}$, with a huge, thermalized bath. By an obvious generalization of the argument in Sec. 4.4.1, the resulting statistical equilibrium distribution function must have the form

general equilibrium distribution

$$\rho = C \exp\left(-\beta\mathcal{E} - \sum_A \beta_A K_A\right). \tag{4.21}$$

extensive variables: energy, volume, particle numbers, momentum, angular momentum

When the extensive variables K_A that are exchanged with the bath (and thus appear explicitly in the distribution function ρ) are energy $\mathcal{E}$, momentum $\mathbf{P}$, angular momentum $\mathbf{J}$, the number N_I of the species I of conserved particles, volume V, or any combination of these quantities, it is conventional to rename the multiplicative factors β and β_A so that ρ takes on the form

$$\rho = C \exp\left[\frac{-\mathcal{E} + \mathbf{U} \cdot \mathbf{P} + \boldsymbol{\Omega} \cdot \mathbf{J} + \sum_I \tilde{\mu}_I N_I - PV}{k_B T}\right]. \tag{4.22}$$

bath's intensive variables: temperature, pressure, chemical potentials, velocity, angular velocity

Here T, $\mathbf{U}$, $\boldsymbol{\Omega}$, $\tilde{\mu}_I$, and P are constants (called *intensive* variables) that are the same for all systems and subsystems (i.e., that characterize the full ensemble and all its subensembles and therefore must have been acquired from the bath); *any extensive variable that is not exchanged with the bath must be omitted from the exponential and be replaced by an implicit delta function.*

As we have seen in Sec. 3.3, T is the temperature that the ensemble and subensembles acquired from the bath (i.e., it is the bath temperature). From the Lorentz transformation law for energy and momentum [$\mathcal{E}' = \gamma(\mathcal{E} - \mathbf{U} \cdot \mathbf{P})$; Eqs. (2.37) and Ex. 2.13] we see that, if we were to transform to a reference frame that moves with velocity $\mathbf{U}$ with respect to our original frame, then the $\exp(\mathbf{U} \cdot \mathbf{P}/(k_B T))$ term in ρ would disappear, and the distribution function would be isotropic in $\mathbf{P}$. Thus $\mathbf{U}$ is the velocity of the bath with respect to our chosen reference frame. By a similar argument, $\boldsymbol{\Omega}$ is the bath's angular velocity with respect to an inertial frame. By comparison with

Eq. (3.22d), we see that $\tilde{\mu}_I$ is the chemical potential of the conserved species I. Finally, experience with elementary thermodynamics suggests (and it turns out to be true) that P is the bath's pressure.[5] Note that, by contrast with the corresponding extensive variables $\mathcal{E}$, $\mathbf{P}$, $\mathbf{J}$, N_I, and V, the intensive variables T, $\mathbf{U}$, $\boldsymbol{\Omega}$, $\tilde{\mu}_I$, and P do not double when the size of a system is doubled (i.e., they are not additive); rather, they are properties of the ensemble as a whole and thus are independent of the systems' sizes.

By removing the rest masses m_I of all particles from each system's energy and similarly removing the particle rest mass from each chemical potential,

$$\boxed{E \equiv \mathcal{E} - \sum_I N_I m_I,} \qquad \boxed{\mu_I \equiv \tilde{\mu}_I - m_I} \qquad (4.23)$$

nonrelativistic energy and chemical potentials obtained by removing rest masses

[Eqs. (3.20) and (3.21)], we bring the distribution function into a form that is identical to Eq. (4.22) but with $\mathcal{E} \to E$ and $\tilde{\mu}_I \to \mu_I$:

$$\boxed{\rho = C \exp\left[\frac{-E + \mathbf{U} \cdot \mathbf{P} + \boldsymbol{\Omega} \cdot \mathbf{J} + \sum_I \mu_I N_I - PV}{k_B T}\right].} \qquad (4.24)$$

general equilibrium distribution

This is the form used in Newtonian theory, but it is also valid relativistically.

SPECIAL EQUILIBRIUM ENSEMBLES

Henceforth (except in Sec. 4.10.2, when discussing black-hole atmospheres), we restrict our baths always to be at rest in our chosen reference frame and to be nonrotating with respect to inertial frames, so that $\mathbf{U} = \boldsymbol{\Omega} = 0$. The distribution function ρ can then either be a delta function in the system momentum $\mathbf{P}$ and angular momentum $\mathbf{J}$ (if momentum and angular momentum are not exchanged with the bath), or it can involve no explicit dependence on $\mathbf{P}$ and $\mathbf{J}$ (if momentum and angular momentum are exchanged with the bath; cf. Eq. (4.22) with $\mathbf{U} = \boldsymbol{\Omega} = 0$). In either case, if energy is the only other quantity exchanged with the bath, then the distribution function is the canonical one [Eq. (4.20)]:

$$\boxed{\rho_{\text{canonical}} = C \exp\left[\frac{-\mathcal{E}}{k_B T}\right] = C' \exp\left[\frac{-E}{k_B T}\right],} \qquad (4.25a)$$

where (obviously) the constants C and C' are related by

$$C' = C \exp\left[-\sum_I N_I m_I / k_B T\right].$$

If, in addition to energy, volume can also be exchanged with the bath (e.g., if the systems are floppy bags of gas whose volumes can change and through which heat can

---

5. One can also identify these physical interpretations of T, $\tilde{\mu}_I$, and P by analyzing idealized measuring devices; cf. Sec. 5.2.2.

flow),[6] then the equilibrium is the *Gibbs ensemble,* which has the distribution function

$$\rho_{\text{Gibbs}} = C \exp\left[\frac{-(\mathcal{E} + PV)}{k_B T}\right] = C' \exp\left[\frac{-(E + PV)}{k_B T}\right] \qquad (4.25\text{b})$$

(with an implicit delta function in N_I and possibly in **J** and **P**). The combination

$\mathcal{E} + PV$ is known as the *enthalpy H.* If the exchanged quantities are energy and particles but not volume (e.g., if the systems are 1-m cubes inside the Sun with totally imaginary walls through which particles and heat can flow), then the equilibrium is the *grand canonical ensemble,* with

$$\rho_{\text{grand canonical}} = C \exp\left[\frac{-\mathcal{E} + \sum_I \tilde{\mu}_I N_I}{k_B T}\right] = C \exp\left[\frac{-E + \sum_I \mu_I N_I}{k_B T}\right]$$

$$(4.25\text{c})$$

(with an implicit delta function in V and perhaps in **J** and **P**).

We mention, as a preview of an issue to be addressed in Chap. 5, that an individual system, picked randomly from the ensemble and then viewed as a bath for its own tiny subsystems, will not have the same temperature T, and/or chemical potential $\tilde{\mu}_I$, and/or pressure P as the huge bath with which the ensemble has equilibrated; rather, the individual system's T, $\tilde{\mu}_I$, and/or P can fluctuate a tiny bit around the huge bath's values (around the values that appear in the above probabilities), just as its $\mathcal{E}$, N_I, and/or V fluctuate. We study these fluctuations in Sec. 5.6.

4.4.3 Fermi-Dirac and Bose-Einstein Distributions

The concepts and results developed in this chapter have enormous generality. They are valid (when handled with sufficient care) for quantum systems as well as classical ones, and they are valid for semiclosed or closed systems of any type. The systems need not resemble the examples we have met in the text. They can be radically different, but so long as they are closed or semiclosed, our concepts and results will apply.

SINGLE-PARTICLE QUANTUM STATES (MODES)

As an important example, let each system be a single-particle quantum state of some field. These quantum states can exchange particles (quanta) with one another. As we shall see, in this case the above considerations imply that, in statistical equilibrium at temperature T, the mean number of particles in a state, whose individual particle energies are $\mathcal{E}$, is given by the Fermi-Dirac formula (for fermions) $\eta = 1/(e^{(\mathcal{E} - \tilde{\mu})/(k_B T)} + 1)$ and Bose-Einstein formula (for bosons) $\eta = 1/(e^{(\mathcal{E} - \tilde{\mu})/(k_B T)} - 1)$, which we used in our kinetic-theory studies in the last chapter [Eqs. (3.22a), (3.22b)]. Our derivation of these mean occupation numbers will

---

6. For example, the huge helium-filled balloons made of thin plastic that are used to lift scientific payloads into Earth's upper atmosphere.

illustrate the closeness of classical statistical mechanics and quantum statistical mechanics: the proof is fundamentally quantum mechanical, because the regime $\eta \sim 1$ is quantum mechanical (it violates the classical condition $\eta \ll 1$); nevertheless, the proof makes use of precisely the same concepts and techniques as we have developed for our classical studies.

As a conceptual aid in the derivation, consider an ensemble of complex systems in statistical equilibrium. Each system can be regarded as made up of a large number of fermions (electrons, protons, neutrons, neutrinos, . . .) and/or bosons (photons, gravitons, alpha particles, phonons, . . .). We analyze each system by identifying a complete set of single-particle quantum states (which we call *modes*) that the particles can occupy[7] (see, e.g., Chandler, 1987, chap. 4). A complete enumeration of modes is the starting point for the *second quantization* formulation of quantum field theory and is also the starting point for our far simpler analysis.

Choose one specific mode S [e.g., a nonrelativistic electron plane-wave mode in a box of side L with spin up and momentum $\mathbf{p} = (5, 3, 17)h/L$]. There is one such mode S in each of the systems in our ensemble, and these modes (all identical in their properties) form a subensemble of our original ensemble. Our derivation focuses on this subensemble of identical modes S. Because each of these modes can exchange energy and particles with all the other modes in its system, the subensemble is grand canonically distributed.

The *(many-particle) quantum states* allowed for mode S are states in which S contains a finite number n of particles (quanta). Denote by $\mathcal{E}_S$ the energy of one particle residing in mode S. Then the mode's total energy when it is in the state $|n\rangle$ (when it contains n quanta) is $\mathcal{E}_n = n\mathcal{E}_S$. [For a freely traveling, relativistic electron mode, $\mathcal{E}_S = \sqrt{m^2 + \mathbf{p}^2}$, Eq. (1.40), where $\mathbf{p}$ is the mode's momentum, $p_x = jh/L$ for some integer j and similarly for p_y and p_z; for a phonon mode with angular eigenfrequency of vibration ω, $\mathcal{E}_S = \hbar\omega$.] Since the distribution of the ensemble's modes among the allowed quantum states is grand canonical, the probability ρ_n of being in state $|n\rangle$ is [Eq. (4.25c)]

$$\rho_n = C \, \exp\left(\frac{-\mathcal{E}_n + \tilde{\mu} n}{k_B T}\right) = C \, \exp\left(\frac{n(\tilde{\mu} - \mathcal{E}_S)}{k_B T}\right), \tag{4.26}$$

where $\tilde{\mu}$ and T are the chemical potential and temperature of the bath of other modes, with which the mode S interacts.[8]

---

7. For photons, these modes are the normal modes of the classical electromagnetic field; for phonons in a crystal, they are the normal modes of the crystal's vibrations; for nonrelativistic electrons or protons or alpha particles, they are energy eigenstates of the nonrelativistic Schrödinger equation; for relativistic electrons, they are energy eigenstates of the Dirac equation.

8. Here and throughout Chaps. 4 and 5 we ignore quantum zero point energies, since they are unobservable in this context (though they are observed in the fluctuational forces discussed in Sec. 6.8). Equally well we could include the zero point energies in $\mathcal{E}_n$, $\mathcal{E}_s$, and $\tilde{\mu}$, and they would cancel out in the combinations that appear, e.g., in Eq. (4.26): $\tilde{\mu} - \mathcal{E}_s$, and so forth.

modes

second quantization

many-particle quantum states

FERMION MODES: FERMI-DIRAC DISTRIBUTION

Suppose that S is a fermion mode (i.e., its particles have half-integral spin). Then the Pauli exclusion principle dictates that S cannot contain more than one particle: n can take on only the values 0 and 1. In this case, the normalization constant in the distribution function (4.26) is determined by $\rho_0 + \rho_1 = 1$, which implies that

Fermi-Dirac distribution

$$\rho_0 = \frac{1}{1 + \exp[(\tilde{\mu} - \mathcal{E}_S)/(k_B T)]}, \quad \rho_1 = \frac{\exp[(\tilde{\mu} - \mathcal{E}_S)/(k_B T)]}{1 + \exp[(\tilde{\mu} - \mathcal{E}_S)/(k_B T)]}. \quad (4.27a)$$

This is the explicit form of the grand canonical distribution for a fermion mode. For many purposes (including all those in Chap. 3), this full probability distribution is more than one needs. Quite sufficient instead is the mode's mean occupation number

Fermi-Dirac mean occupation number

$$\boxed{\eta_S \equiv \langle n \rangle = \sum_{n=0}^{1} n\rho_n = \frac{1}{\exp[(\mathcal{E}_S - \tilde{\mu})/(k_B T)] + 1} = \frac{1}{\exp[(E_S - \mu)/(k_B T)] + 1}.}$$

$$(4.27b)$$

Here $E_S = \mathcal{E}_S - m$ is the energy of a particle in the mode with rest mass removed, and $\mu = \tilde{\mu} - m$ is the chemical potential with rest mass removed—the quantities used in the nonrelativistic (Newtonian) regime.

Equation (4.27b) is the *Fermi-Dirac mean occupation number* asserted in Chap. 3 [Eq. (3.22a)] and studied there for the special case of a gas of freely moving, non-interacting fermions. Because our derivation is completely general, we conclude that this mean occupation number and the underlying grand canonical distribution (4.27a) are valid for any mode of a fermion field—for example, the modes for an electron trapped in an external potential well or a magnetic bottle, and the (single-particle) quantum states of an electron in a hydrogen atom.

BOSON MODES: BOSE-EINSTEIN DISTRIBUTION

Suppose that S is a boson mode (i.e., its particles have integral spin), so it can contain any nonnegative number of quanta; that is, n can assume the values $0, 1, 2, 3, \ldots$. Then the normalization condition $\sum_{n=0}^{\infty} \rho_n = 1$ fixes the constant in the grand canonical distribution (4.26), resulting in

Bose-Einstein distribution

$$\rho_n = \left[1 - \exp\left(\frac{\tilde{\mu} - \mathcal{E}_S}{k_B T} \right) \right] \exp\left(\frac{n(\tilde{\mu} - \mathcal{E}_S)}{k_B T} \right). \quad (4.28a)$$

From this grand canonical distribution we can deduce the mean number of bosons in mode S:

Bose-Einstein mean occupation number

$$\boxed{\eta_S \equiv \langle n \rangle = \sum_{n=1}^{\infty} n\rho_n = \frac{1}{\exp[(\mathcal{E}_S - \tilde{\mu})/(k_B T)] - 1} = \frac{1}{\exp[(E_S - \mu)/(k_B T)] - 1},}$$

$$(4.28b)$$

in accord with Eq. (3.22b). As for fermions, this *Bose-Einstein mean occupation number* and underlying grand canonical distribution (4.28a) are valid generally, and not solely for the freely moving bosons of Chap. 3.

When the mean occupation number is small, $\eta_S \ll 1$, both the bosonic and the fermionic distribution functions are well approximated by the classical *Boltzmann mean occupation number*

$$\eta_S = \exp[-(\mathcal{E}_S - \tilde{\mu})/(k_B T)].$$
(4.29)

Boltzmann mean occupation number

In Sec. 4.9 we explore an important modern application of the Bose-Einstein mean occupation number (4.28b): *Bose-Einstein condensation* of bosonic atoms in a magnetic trap.

4.4.4 Equipartition Theorem for Quadratic, Classical Degrees of Freedom

4.4.4

As a second example of statistical equilibrium distribution functions, we derive the classical equipartition theorem using statistical methods.

To motivate this theorem, consider a diatomic molecule of nitrogen, N_2. To a good approximation, its energy (its hamiltonian) can be written as

$$E = \frac{p_x^2}{2M} + \frac{p_y^2}{2M} + \frac{p_z^2}{2M} + \frac{P_\ell^2}{2M_\ell} + \frac{1}{2} M_\ell \, \omega_v^2 \, \ell^2 + \frac{J_x^2}{2I} + \frac{J_y^2}{2I}.$$
(4.30)

Here M is the molecule's mass; p_x, p_y, and p_z are the components of its translational momentum; and the first three terms are the molecule's kinetic energy of translation. The next two terms are the molecule's longitudinal vibration energy, with ℓ the change of the molecule's length (change of the separation of its two nuclei) from equilibrium, P_ℓ the generalized momentum conjugate to that length change, ω_v the vibration frequency, and M_ℓ the generalized mass associated with that vibration. The last two terms are the molecule's energy of end-over-end rotation, with J_x and J_y the components of angular momentum associated with this two-dimensional rotator and I its moment of inertia.

Notice that every term in this hamiltonian is quadratic in a generalized coordinate or generalized momentum! Moreover, each of these coordinates and momenta appears only in its single quadratic term and nowhere else, and the density of states is independent of the value of that coordinate or momentum. We refer to such a coordinate or momentum as a *quadratic degree of freedom*.

quadratic degree of freedom

In some cases (e.g., the vibrations and rotations but not the translations), the energy $E_\xi = \alpha \xi^2$ of a quadratic degree of freedom ξ is quantized, with some energy separation ε_0 between the ground state and first excited state (and with energy separations to higher states that are $\lesssim \varepsilon_0$). If (and only if) the thermal energy $k_B T$ is significantly larger than ε_0, then the quadratic degree of freedom ξ will be excited far above its ground state and will behave classically. The equipartition theorem applies only at these high temperatures. For diatomic nitrogen, the rotational degrees of freedom J_x and J_y have $\varepsilon_0 \sim 10^{-4}$ eV and $\varepsilon_0/k_B \sim 1$ K, so temperatures big compared to 1 K are required for J_x and J_y to behave classically. By contrast, the vibrational degrees of freedom ℓ and P_ℓ have $\varepsilon_0 \sim 0.1$ eV and $\varepsilon_0/k_B \sim 1{,}000$ K, so temperatures of a few thousand Kelvins are required for them to behave classically. Above $\sim 10^4$ K, the hamiltonian (4.30) fails: electrons around the nuclei are driven into excited states, and the molecule breaks apart (dissociates into two free atoms of nitrogen).

The equipartition theorem holds for any classical, quadratic degree of freedom [i.e., at temperatures somewhat higher than $T_o = \varepsilon_o/(k_B T)$]. We derive this theorem using the canonical distribution (4.25a). We write the molecule's total energy as $E = \alpha\xi^2 + E'$, where E' does not involve ξ. Then the mean energy associated with ξ is

$$\langle E_\xi \rangle = \frac{\int \alpha\xi^2 \, e^{-\beta(\alpha\xi^2 + E')} d\xi \, d(\text{other degrees of freedom})}{\int e^{-\beta(\alpha\xi^2 + E')} d\xi \, d(\text{other degrees of freedom})}. \tag{4.31}$$

Here the exponential is that of the canonical distribution function (4.25a), the denominator is the normalizing factor, and we have set $\beta \equiv 1/(k_B T)$. Because ξ does not appear in the portion E' of the energy, its integral separates out from the others in both numerator and denominator, and the integrals over E' in numerator and denominator cancel. Rewriting $\int \alpha\xi^2 \exp(-\beta\alpha\xi^2) \, d\xi$ as $-d/d\beta[\int \exp(-\beta\alpha\xi^2) \, d\xi]$, Eq. (4.31) becomes

$$\langle E_\xi \rangle = -\frac{d}{d\beta} \ln\left[\int \exp(-\beta\alpha\xi^2) \, d\xi \right]$$

$$= -\frac{d}{d\beta} \ln\left[\frac{1}{\sqrt{\beta\alpha}} \int du \, e^{-u^2} du \right] = \frac{1}{2\beta} = \frac{1}{2} k_B T. \tag{4.32}$$

Therefore, *in statistical equilibrium, the mean energy associated with any classical,*

quadratic degree of freedom is $\frac{1}{2} k_B T$. This is the equipartition theorem. Note that the factor $\frac{1}{2}$ follows from the quadratic nature of the degrees of freedom.

For our diatomic molecule, at room temperature there are three translational and two rotational classical, quadratic degrees of freedom (p_x, p_y, p_z, J_x, J_y), so the mean total energy of the molecule is $\frac{5}{2} k_B T$. At a temperature of several thousand Kelvins, the two vibrational degrees of freedom, ℓ and P_ℓ, become classical and the molecule's mean total energy is $\frac{7}{2} k_B T$. Above $\sim 10^4$ K the molecule dissociates, and its two parts (the two nitrogen atoms) have only translational quadratic degrees of freedom, so the mean energy per atom is $\frac{3}{2} k_B T$.

The equipartition theorem is valid for any classical, quadratic degree of freedom, whether it is part of a molecule or not. For example, it applies to the generalized coordinate and momentum of any harmonic-oscillator mode of any system: a vibrational mode of Earth or of a crystal, or a mode of an electromagnetic field.

4.5 The Microcanonical Ensemble

Let us now turn from ensembles of systems that interact with an external, thermal bath (as discussed in Sec. 4.4.1) to an ensemble of identical, precisely closed systems (i.e., systems that have no interactions with the external universe). By "identical" we mean that every system in the ensemble has (i) the same set of degrees of freedom, and thus (ii) the same number of degrees of freedom W, (iii) the same hamiltonian, and (iv) the same values for all the additive constants of motion ($\mathcal{E}$, K_1, K_2, ...) except perhaps total momentum $\mathbf{P}$ and total angular momentum $\mathbf{J}$.[9]

Suppose that these systems begin with values of $(\mathbf{q}, \mathbf{p})$ that are spread out in some (arbitrary) manner over a hypersurface in phase space that has $H(\mathbf{q}, \mathbf{p})$ equal to the common value of energy $\mathcal{E}$. Of course, we cannot choose systems whose energy is precisely equal to $\mathcal{E}$. For most $\mathcal{E}$ this would be a set of measure zero (see Ex. 4.7). Instead we let the systems occupy a tiny range of energy between $\mathcal{E}$ and $\mathcal{E} + \delta\mathcal{E}$ and then discover (in Ex. 4.7) that our results are highly insensitive to $\delta\mathcal{E}$ as long as it is extremely small compared with $\mathcal{E}$.

It seems reasonable to expect that this ensemble, after evolving for a time much longer than its longest internal dynamical time scale $t \gg \tau_{\mathrm{int}}$, will achieve statistical equilibrium (i.e., will evolve into a state with $\partial\rho/\partial t = 0$). (In the next section we justify this expectation.) The distribution function ρ then satisfies the collisionless Boltzmann equation (4.15) with vanishing time derivative, and therefore is a function only of the hamiltonian's additive constants of the motion $\mathcal{E}$ and K_A. However, we already know that ρ is a delta function in K_A and almost a delta function with a tiny but finite spread in $\mathcal{E}$; and the fact that it cannot depend on any other phase-space quantities then implies *ρ is a constant over the hypersurface in phase space that has the prescribed values of K_A and $\mathcal{E}$, and it is zero everywhere else in phase space.* This equilibrium ensemble is called "microcanonical."

microcanonical distribution

There is a subtle aspect of this microcanonical ensemble that deserves discussion. Suppose we split each system in the ensemble up into a huge number of subsystems that can exchange energy (but for concreteness, nothing else) with one another. We thereby obtain a huge number of subensembles, in the manner of Sec. 4.4.1. The original systems can be regarded as a thermal bath for the subsystems, and correspondingly, the subensembles will have canonical distribution functions, $\rho_a = Ce^{-\mathcal{E}_a/(k_B T)}$. One might also expect the subensembles to be statistically independent, so that $\rho = \prod_a \rho_a$. However, such independence is not possible, since together with additivity of energy $\mathcal{E} = \sum_a \mathcal{E}_a$, it would imply that $\rho = Ce^{-\mathcal{E}/(k_B T)}$ (i.e., that the full ensemble is canonically distributed rather than microcanonical). What is wrong here?

The answer is that there is a tiny correlation between the subensembles: if, at some moment of time, subsystem $a = 1$ happens to have an unusually large energy, then the other subsystems must correspondingly have a little less energy than usual. This very slightly invalidates the statistical-independence relation $\rho = \prod_a \rho_a$, thereby

correlations in the microcanonical ensemble

---

9. Exercise 4.7 below is an example of a microcanonical ensemble where $\mathbf{P}$ and $\mathbf{J}$ are not precisely fixed, though we do not discuss this in the exercise. The gas atoms in that example are contained inside an impermeable box whose walls cannot exchange energy or atoms with the gas, but obviously can and do exchange momentum and angular momentum when atoms collide with the walls. Because the walls are at rest in our chosen reference frame, the distribution function has $\mathbf{U} = \mathbf{\Omega} = 0$ and so is independent of $\mathbf{P}$ and $\mathbf{J}$ [Eq. (4.24)] rather than having precisely defined values of them.

enabling the full ensemble to be microcanonical, even though all its subensembles are canonical. In the language of two-lengthscale expansions, where one expands in the dimensionless ratio (size of subsystems)/(size of full system) [Box 3.3], this correlation is a higher-order correction to statistical independence.

We are now in a position to understand more deeply the nature of the thermalized bath that we have invoked to drive ensembles into statistical equilibrium. That bath can be any huge system that contains the systems we are studying as subsystems; the bath's thermal equilibrium can be either a microcanonical statistical equilibrium or a statistical equilibrium involving exponentials of its extensive variables.

Exercise 4.7 gives a concrete illustration of the microcanonical ensemble, but we delay presenting it until we have developed some additional concepts that it also illustrates.

4.6

4.6 The Ergodic Hypothesis

The ensembles we have been studying are almost always just conceptual ones that do not exist in the real universe. We have introduced them and paid so much attention to them not for their own sakes, but because, in the case of statistical-equilibrium ensembles, they can be powerful tools for studying the properties of a single, individual system that really does exist in the universe or in our laboratory.

This power comes about because a sequence of snapshots of the single system, taken at times separated by sufficiently large intervals Δt, has a probability distribution ρ (for the snapshots' instantaneous locations $\{\mathbf{q}, \mathbf{p}\}$ in phase space) that is the same as the distribution function ρ of some conceptual, statistical-equilibrium ensemble. If the single system is closed, so its evolution is driven solely by its own hamiltonian, then the time between snapshots should be $\Delta t \gg \tau_{\mathrm{int}}$, and its snapshots will be (very nearly) microcanonically distributed. If the single system exchanges energy, and only energy, with a thermal bath on a timescale τ_{ext}, then the time between snapshots should be $\Delta t \gg \tau_{\mathrm{ext}}$, and its snapshots will be canonically distributed; similarly for the other types of bath interactions. This property of snapshots is equivalent to the statement that *for the individual system, the long-term time average*[10] *of any function of the system's location in phase space is equal to the statistical-equilibrium ensemble average:*

ergodicity: equality of time average and ensemble average

$$\bar{A} \equiv \lim_{T \to \infty} \frac{1}{T} \int_{-T/2}^{+T/2} A(\mathbf{q}(t), \mathbf{p}(t)) = \langle A \rangle \equiv \sum_n A_n \rho_n. \tag{4.33}$$

ergodicity

This property comes about because of *ergodicity:* the individual system, as it evolves, visits each accessible quantum state n for a fraction of the time that is equal to the equilibrium ensemble's probability ρ_n. Or, stated more carefully, the system comes

---

10. Physicists often study a system's evolution for too short a time to perform this average, and therefore the system does not reveal itself to be ergodic.

sufficiently close to each state n for a sufficient length of time that, for practical purposes, we can approximate it as spending a fraction ρ_n of its time at n.

At first sight, ergodicity may seem trivially obvious. However, it is not a universal property of all systems. One can easily devise idealized examples of nonergodic behavior (e.g., a perfectly elastic billiard ball bouncing around a square billiard table), and some few-body systems that occur in Nature are nonergodic (e.g., fortunately, planetary systems, such as the Sun's). Moreover, one can devise realistic, nonergodic models of some many-body systems. For examples and a clear discussion see Sethna (2006, Chap. 4). On the other hand, generic closed and semiclosed systems in Nature, whose properties and parameters are not carefully fine tuned, do typically behave ergodically, though to prove so is one of the most difficult problems in statistical mechanics.[11]

We assume throughout this book's discussion of statistical physics that all the systems we study are indeed ergodic; this is called the *ergodic hypothesis.* Correspondingly, **ergodic hypothesis** sometimes (for ease of notation) we denote the ensemble average with a bar.

One must be cautious in practical applications of the ergodic hypothesis: it can sometimes require much longer than one might naively expect for a system to wander sufficiently close to accessible states that $\bar{A} = \langle A \rangle$ for observables A of interest.

4.7 Entropy and Evolution toward Statistical Equilibrium

4.7.1 Entropy and the Second Law of Thermodynamics

For any ensemble of systems, whether it is in statistical equilibrium or not, and also whether it is quantum mechanical or not, the ensemble's *entropy* S is defined, in words, by the following awful sentence: S is the mean value (ensemble average) of the natural logarithm of the probability that a random system in the ensemble occupies a given quantum state, summed over states and multiplied by $-k_B$. More specifically, denoting the probability that a system is in state n by ρ_n, the ensemble's entropy S is the following sum over quantum states (or the equivalent integral over phase space):

$$S \equiv -k_B \sum_n \rho_n \ln \rho_n.$$

(4.34)   **entropy of an ensemble**

If all the systems are in the same quantum state, for example, in the state $n = 17$, then $\rho_n = \delta_{n,17}$ so we know precisely the state of any system pulled at random from the ensemble, and Eq. (4.34) dictates that the entropy vanish. Vanishing entropy thus corresponds to perfect knowledge of the system's quantum state; it corresponds to the quantum state being pure.

Entropy is a measure of our lack of information about the state of any system chosen at random from an ensemble (see Sec. 4.11). In this sense, *the entropy can be*

---

11. The ergodic hypothesis was introduced in 1871 by Boltzmann. For detailed analyses of it from physics viewpoints, see, e.g., ter Haar (1955) and Farquhar (1964). Attempts to understand it rigorously, and the types of systems that do or do not satisfy it, have spawned a branch of mathematical physics called "ergodic theory."

regarded as a property of a random individual system in the ensemble, as well as of the ensemble itself.

By contrast, consider a system in microcanonical statistical equilibrium. In this case, all states are equally likely (ρ is constant), so if there are N_{states} states available to the system, then $\rho_n = 1/N_{\text{states}}$, and the entropy (4.34) takes the form[12]

entropy for microcanonical ensemble

$$\boxed{S = k_B \ln N_{\text{states}}.} \tag{4.35}$$

The entropy, so defined, has some important properties. One is that when the ensemble can be broken up into statistically independent subensembles of subsystems (as is generally the case for big systems in statistical equilibrium), so that $\rho = \prod_a \rho_a$, then the entropy is additive: $S = \sum_a S_a$ (see Ex. 4.3). This permits us to regard the entropy, like the systems' additive constants of motion, as an extensive variable.

second law of thermodynamics

A second very important property is that, as an ensemble of systems evolves, its entropy cannot decrease, and it generally tends to increase. This is the statistical mechanical version of the second law of thermodynamics.

As an example of this second law, consider two different gases (e.g., nitrogen and oxygen) in a container, separated by a thin membrane. One set of gas molecules is constrained to lie on one side of the membrane; the other set lies on the opposite side. The total number of available states N_{states} is less than if the membrane is ruptured and the two gases are allowed to mix. The mixed state is readily accessible from the partitioned state and not vice versa. When the membrane is removed, the entropy begins to increase in accord with the second law of thermodynamics (cf. Ex. 4.8).

Since any ensemble of identical, closed systems will ultimately, after a time $t \gg \tau_{\text{int}}$, evolve into microcanonical statistical equilibrium, it must be that the microcanonical distribution function $\rho = $ constant has a larger entropy than any other distribution function that the ensemble could acquire. That this is indeed so can be demonstrated formally as follows.

Consider the class of all distribution functions ρ that: (i) vanish unless the constants of motion have the prescribed values $\mathcal{E}$ (in the tiny range $\delta\mathcal{E}$) and K_A; (ii) can be nonzero anywhere in the region of phase space, which we call $\mathcal{Y}_o$, where the prescribed values $\mathcal{E}$, K_A are taken; and (iii) are correctly normalized so that

$$\sum_n \rho_n \equiv \int_{\mathcal{Y}_o} \rho N_{\text{states}} d\Gamma = 1 \tag{4.36a}$$

[Eq. (4.8b)]. We ask which ρ in this class gives the largest entropy

$$S = -k_B \sum_n \rho_n \ln \rho_n.$$

The requirement that the entropy be extremal (stationary) under variations $\delta\rho$ of ρ that preserve the normalization (4.36a) is embodied in the variational principle (see, e.g., Boas, 2006, Chap. 9):

---

12. This formula, with slightly different notation, can be found on Boltzmann's tomb.

$$\delta S = \delta \int_{\mathcal{Y}_o} (-k_B \rho \ln \rho - \Lambda \rho) \mathcal{N}_{\text{states}} d\Gamma = 0. \qquad (4.36b)$$

Here Λ is a Lagrange multiplier that enforces the normalization (4.36a). Performing the variation, we find that

$$\int_{\mathcal{Y}_o} (-k_B \ln \rho - k_B - \Lambda) \delta \rho \mathcal{N}_{\text{states}} d\Gamma = 0, \qquad (4.36c)$$

which is satisfied if and only if ρ is a constant, $\rho = e^{-1-\Lambda/k_B}$, independent of location in the allowed region $\mathcal{Y}_o$ of phase space (i.e., if and only if ρ is that of the microcanonical ensemble). This calculation actually only shows that the microcanonical ensemble has stationary entropy. To show it is a maximum, one must perform the second variation (i.e., compute the second-order contribution of $\delta \rho$ to $\delta S = \delta \int (-k_B \rho \ln \rho) \mathcal{N}_{\text{states}} d\Gamma$). That second-order contribution is easily seen to be

$$\delta^2 S = \int_{\mathcal{Y}_o} \left(-k_B \frac{(\delta \rho)^2}{2\rho} \right) \mathcal{N}_{\text{states}} d\Gamma < 0. \qquad (4.36d)$$

Thus, the microcanonical distribution does maximize the entropy, as claimed.

entropy maximized when ρ is constant (microcanonical distribution)

4.7.2 What Causes the Entropy to Increase?

4.7.2

There is an apparent paradox at the heart of statistical mechanics, and, at various stages in the development of the subject it has led to confusion and even despair. It still creates controversy (see, e.g., Hawking and Penrose, 2010; Penrose, 1999). Its simplest and most direct expression is to ask: how can the time-reversible, microscopic laws, encoded in a time-independent hamiltonian, lead to the remorseless increase of entropy?

COARSE-GRAINING

For insight, first consider a classical, microcanonical ensemble of precisely closed systems (no interaction at all with the external universe). Assume, for simplicity, that at time $t = 0$ all the systems are concentrated in a small but finite region of phase space with volume $\Delta \Gamma$, as shown in Fig. 4.2a, with $\rho = 1/(\mathcal{N}_{\text{states}} \Delta \Gamma)$ in the occupied region and $\rho = 0$ everywhere else. As time passes each system evolves under the action of the systems' common hamiltonian. As depicted in Fig. 4.2b, this evolution distorts the occupied region of phase space; but Liouville's theorem dictates that the occupied region's volume $\Delta \Gamma$ remain unchanged and, correspondingly, that the ensemble's entropy

$$S = -k_B \int (\rho \ln \rho) \mathcal{N}_{\text{states}} d\Gamma = k_B \ln(\mathcal{N}_{\text{states}} \Delta \Gamma) \qquad (4.37)$$

remain unchanged.

How can this be so? The ensemble is supposed to evolve into statistical equilibrium, with its distribution function uniformly spread out over that entire portion of

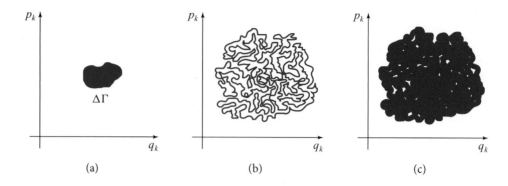

FIGURE 4.2 Evolution of a classical ensemble at $t = 0$ (a) toward statistical equilibrium by means of phase mixing (b) (cf. Fig. 4.1) followed by coarse-graining of one's viewpoint (c).

phase space allowed by the hamiltonian's constants of motion—a portion of phase space far, far larger than $\Delta\Gamma$—and in the process the entropy is supposed to increase.

Figure 4.2b,c resolves the paradox. As time passes, the occupied region becomes more and more distorted. It retains its phase-space volume, but gets strung out into a winding, contorted surface (Fig. 4.2b), which (by virtue of the ergodic hypothesis) ultimately passes arbitrarily close to any given point in the region allowed by the constants of motion. This ergodic wandering is called *phase mixing*. Ultimately, the physicist gets tired of keeping track (or ceases to be able to keep track) of all these contortions of the occupied region and chooses instead to take a *coarse-grained* viewpoint that averages over scales larger than the distance between adjacent portions of the occupied surface, and thereby regards the ensemble as having become spread over the entire allowed region (Fig. 4.2c). More typically, the physicist will perform a coarse-grained smearing out on some given, constant scale at all times. Once the transverse scale of the ensemble's lengthening and narrowing phase-space region drops below the smearing scale, its smeared volume and its entropy start to increase. Thus, *for an ensemble of closed systems it is the physicist's choice (though often a practical necessity) to perform coarse-grain averaging that causes entropy to increase and causes the ensemble to evolve into statistical equilibrium.*

The situation is a bit more subtle for an ensemble of systems interacting with a thermal bath. The evolution toward statistical equilibrium is driven by the interactions. Thus, it might appear at first sight that the physicist is not, this time, to blame for the entropy increase and the achievement of statistical equilibrium. However, a deeper examination reveals the physicist's ultimate culpability. If the physicist were willing to keep track of all those dynamical degrees of freedom of the bath that are influenced by and influence the systems in the ensemble, then the physicist could incorporate those degrees of freedom into the description of the systems and define a phase-space volume that obeys Liouville's theorem and thus does not increase, and an entropy that correspondingly remains constant. However, physicists instead generally choose to ignore the microscopic details of the bath, and that choice forces them to

phase mixing

coarse graining causes entropy increase

attribute a growing entropy to the ensemble of systems bathed by the bath and regard the ensemble as approaching statistical equilibrium.

DISCARDING CORRELATIONS

When one reexamines these issues in quantum mechanical language, one discovers that the entropy increase is caused by the physicist's discarding the quantum mechanical correlations (the off-diagonal terms in the density matrix of Box 4.2) that get built up through the systems' interaction with the rest of the universe. This discarding of correlations is accomplished through a trace over the external universe's basis states (Box 4.2), and if the state of system plus universe was originally pure, this tracing (discarding of correlations) makes it mixed. From this viewpoint, then, *it is the physicist's choice to discard correlations with the external universe that causes the entropy increase and the evolution toward statistical equilibrium.* Heuristically, we can say that the entropy does not increase until the physicist actually (or figuratively) chooses to let it increase by ignoring the rest of the universe. For a simple example, see Box 4.3 and Ex. 4.9.

> discarding quantum correlations (decoherence) causes entropy increase

This viewpoint then raises a most intriguing question. What if we regard the universe as the ultimate microcanonical system? In this case, we might expect that the entropy of the universe will remain identically zero for all time, unless physicists (or other intelligent beings) perform some sort of coarse-graining or discard some sort of correlations. However, such coarse-graining or discarding are made deeply subtle by the fact that the physicists (or other intelligent beings) are themselves part of the system being studied. Further discussion of these questions introduces fascinating, though ill-understood, quantum mechanical and cosmological considerations, which will reappear in a different context in Sec. 28.7.

EXERCISES

Exercise 4.2 *Practice: Estimating Entropy*
Make rough estimates of the entropy of the following systems, assuming they are in statistical equilibrium.

(a) An electron in a hydrogen atom at room temperature.

(b) A glass of wine.

(c) The Pacific ocean.

(d) An ice cube.

(e) The observable universe. [Its entropy is mostly contained in the 3-K microwave background radiation and in black holes (Sec. 4.10). Why?]

Exercise 4.3 *Derivation and Practice: Additivity of Entropy for Statistically Independent Systems*
Consider an ensemble of classical systems with each system made up of a large number of statistically independent subsystems, so $\rho = \prod_a \rho_a$. Show that the entropy of the full ensemble is equal to the sum of the entropies of the subensembles a: $S = \sum_a S_a$.

BOX 4.3. ENTROPY INCREASE DUE TO DISCARDING
QUANTUM CORRELATIONS

As an idealized, pedagogical example of entropy increase due to physicists'
discarding quantum correlations, consider an electron that interacts with a
photon. The electron's initial quantum state is $|\psi_e\rangle = \alpha|\uparrow\rangle + \beta|\downarrow\rangle$, where
$|\uparrow\rangle$ is the state with spin up, $|\downarrow\rangle$ is that with spin down, and α and β are
complex probability amplitudes with $|\alpha|^2 + |\beta|^2 = 1$. The interaction is so
arranged that if the electron spin is up, then the photon is put into a positive
helicity state $|+\rangle$, and if down, the photon is put into a negative helicity state
$|-\rangle$. Therefore, after the interaction the combined system of electron plus
photon is in the state $|\Psi\rangle = \alpha|\uparrow\rangle \otimes |+\rangle + \beta|\downarrow\rangle \otimes |-\rangle$, where $\otimes$ is the tensor
product [Eq. (1.5a) generalized to the vector space of quantum states].

The photon flies off into the universe leaving the electron isolated. Suppose
that we measure some electron observable $\hat{A}_e$. The expectation values for the
measurement before and after the interaction with the photon are

$$\text{Before:} \quad \langle\psi_e|\hat{A}_e|\psi_e\rangle = |\alpha|^2\langle\uparrow|\hat{A}_e|\uparrow\rangle + |\beta|^2\langle\downarrow|\hat{A}_e|\downarrow\rangle$$

$$+ \alpha^*\beta\langle\uparrow|\hat{A}_e|\downarrow\rangle + \beta^*\alpha\langle\downarrow|\hat{A}_e|\uparrow\rangle; \quad (1)$$

$$\text{After:} \quad \langle\Psi|\hat{A}_e|\Psi\rangle = |\alpha|^2\langle\uparrow|\hat{A}_e|\uparrow\rangle \underbrace{\langle+|+\rangle}_{1} + |\beta|^2\langle\downarrow|\hat{A}_e|\downarrow\rangle \underbrace{\langle-|-\rangle}_{1}$$

$$+ \alpha^*\beta\langle\uparrow|\hat{A}_e|\downarrow\rangle \underbrace{\langle+|-\rangle}_{0} + \beta^*\alpha\langle\downarrow|\hat{A}_e|\uparrow\rangle \underbrace{\langle-|+\rangle}_{0}$$

$$= |\alpha|^2\langle\uparrow|\hat{A}_e|\uparrow\rangle + |\beta|^2\langle\downarrow|\hat{A}_e|\downarrow\rangle. \quad (2)$$

Comparing Eqs. (1) and (2), we see that *the correlations with the photon
have removed the $\alpha^*\beta$ and $\beta^*\alpha$ quantum interference terms from the expectation
value.* The two pieces $\alpha|\uparrow\rangle$ and $\beta|\downarrow\rangle$ of the electron's original quantum state
$|\psi_e\rangle$ are said to have *decohered*. Since the outcomes of all measurements
can be expressed in terms of expectation values, this quantum decoherence
is complete in the sense that no quantum interference between the $\alpha|\uparrow\rangle$
and $\beta|\downarrow\rangle$ pieces of the electron state $|\psi_e\rangle$ will ever be seen again in any
measurement on the electron, unless the photon returns, interacts with the
electron, and thereby removes its correlations with the electron state.

If physicists are confident the photon will never return and the correlations will never be removed, then they are free to change their mathematical
description of the electron state. Instead of describing the postinterac-
tion state as $|\Psi\rangle = \alpha|\uparrow\rangle \otimes |+\rangle + \beta|\downarrow\rangle \otimes |-\rangle$, the physicists can discard the

(continued)

correlations with the photon and regard the electron as having *classical* probabilities $\rho_\uparrow = |\alpha|^2$ for spin up and $\rho_\downarrow = |\beta|^2$ for spin down (i.e., as being in a *mixed* state). This new, mixed-state viewpoint leads to the same expectation value (2) for all physical measurements as the old, correlated, pure-state viewpoint $|\Psi\rangle$.

The important point for us is that, when discarding the quantum correlations with the photon (with the external universe), the physicist changes the entropy from zero (the value for any pure state including $|\Psi\rangle$) to $S = -k_B(\rho_\uparrow \ln \rho_\uparrow + \rho_\downarrow \ln \rho_\downarrow) = -k_B(|\alpha|^2 \ln |\alpha|^2 + |\beta|^2 \ln |\beta|^2) > 0$. The physicist's change of viewpoint has increased the entropy.

In Ex. 4.9, this pedagogical example is reexpressed in terms of the density operator discussed in Box 4.2.

Exercise 4.4 **Example: Entropy of a Thermalized Mode of a Field*
Consider a mode S of a fermionic or bosonic field, as discussed in Sec. 4.4.3. Suppose that an ensemble of identical such modes is in statistical equilibrium with a heat and particle bath and thus is grand canonically distributed.

(a) Show that if S is fermionic, then the ensemble's entropy is

$$S_S = -k_B[\eta \ln \eta + (1 - \eta) \ln(1 - \eta)]$$
$$\simeq -k_B \eta (\ln \eta - 1) \quad \text{in the classical regime } \eta \ll 1, \qquad (4.38a)$$

where η is the mode's fermionic mean occupation number (4.27b).

(b) Show that if the mode is bosonic, then the entropy is

$$S_S = k_B[(\eta + 1) \ln(\eta + 1) - \eta \ln \eta]$$
$$\simeq -k_B \eta (\ln \eta - 1) \quad \text{in the classical regime } \eta \ll 1, \qquad (4.38b)$$

where η is the bosonic mean occupation number (4.28b). Note that in the classical regime, $\eta \simeq e^{-(\mathcal{E}-\tilde{\mu})/(k_B T)} \ll 1$, the entropy is insensitive to whether the mode is bosonic or fermionic.

(c) Explain why the entropy per particle in units of Boltzmann's constant is $\sigma = S_S/(\eta k_B)$. Plot σ as a function of η for fermions and for bosons. Show analytically that for degenerate fermions ($\eta \simeq 1$) and for the bosons' classical-wave regime ($\eta \gg 1$) the entropy per particle is small compared to unity. See Sec. 4.8 for the importance of the entropy per particle.

Exercise 4.5 *Example: Entropy of Thermalized Radiation Deduced from Entropy per Mode*

Consider fully thermalized electromagnetic radiation at temperature T, for which the mean occupation number has the standard Planck (blackbody) form $\eta = 1/(e^x - 1)$ with $x = h\nu/(k_B T)$.

(a) Show that the entropy per mode of this radiation is

$$S_S = k_B[x/(e^x - 1) - \ln(1 - e^{-x})].$$

(b) Show that the radiation's entropy per unit volume can be written as the following integral over the magnitude of the photon momentum:

$$S/V = (8\pi/h^3) \int_0^\infty S_S \, p^2 dp.$$

(c) By performing the integral (e.g., using Mathematica), show that

$$\frac{S}{V} = \frac{4}{3}\frac{U}{T} = \frac{4}{3}aT^3, \tag{4.39}$$

where $U = aT^4$ is the radiation energy density, and $a = (8\pi^5 k_B^4/15)/(ch)^3$ is the radiation constant [Eqs. (3.54)].

(d) Verify Eq. (4.39) for the entropy density by using the first law of thermodynamics $dE = TdS - PdV$ (which you are presumed to know before reading this book, and which we discuss below and study in the next chapter).

Exercise 4.6 *Problem: Entropy of a Classical, Nonrelativistic, Perfect Gas, Deduced from Entropy per Mode*

Consider a classical, nonrelativistic gas whose particles do not interact and have no excited internal degrees of freedom (a *perfect gas*—not to be confused with perfect fluid). Let the gas be contained in a volume V and be thermalized at temperature T and chemical potential μ. Using the gas's entropy per mode, Ex. 4.4, show that the total entropy in the volume V is

$$S = \left(\frac{5}{2} - \frac{\mu}{k_B T}\right) k_B N, \tag{4.40}$$

where $N = g_s(2\pi m k_B T/h^2)^{3/2} e^{\mu/(k_B T)} V$ is the number of particles in the volume V [Eq. (3.39a), derived in Chap. 3 using kinetic theory], and g_s is each particle's number of spin states.

Exercise 4.7 *Example: Entropy of a Classical, Nonrelativistic, Perfect Gas in a Microcanonical Ensemble*

Consider a microcanonical ensemble of closed cubical cells with volume V. Let each cell contain precisely N particles of a classical, nonrelativistic, perfect gas and contain a nonrelativistic total energy $E \equiv \mathcal{E} - Nmc^2$. For the moment (by contrast with the text's discussion of the microcanonical ensemble), assume that E is precisely fixed instead of being spread over some tiny but finite range.

(a) Explain why the region $\mathcal{Y}_o$ of phase space accessible to each system is

$$|x_A| < L/2, \quad |y_A| < L/2, \quad |z_A| < L/2, \quad \sum_{A=1}^{N} \frac{1}{2m} |\mathbf{p}_A|^2 = E, \quad (4.41a)$$

where A labels the particles, and $L \equiv V^{1/3}$ is the side of the cell.

(b) To compute the entropy of the microcanonical ensemble, we compute the volume $\Delta\Gamma$ in phase space that it occupies, multiply by the number density of states in phase space (which is independent of location in phase space), and then take the logarithm. Explain why

$$\Delta\Gamma \equiv \prod_{A=1}^{N} \int_{\mathcal{Y}_o} dx_A dy_A dz_A dp_A^x dp_A^y dp_A^z \quad (4.41b)$$

vanishes. This illustrates the "set of measure zero" statement in the text (second paragraph of Sec. 4.5), which we used to assert that we must allow the systems' energies to be spread over some tiny but finite range.

(c) Now permit the energies of our ensemble's cells to lie in the tiny but finite range $E_o - \delta E_o < E < E_o$. Show that

$$\Delta\Gamma = V^N [\mathcal{V}_\nu(a) - \mathcal{V}_\nu(a - \delta a)], \quad (4.41c)$$

where $\mathcal{V}_\nu(a)$ is the volume of a sphere of radius a in a Euclidean space with $\nu \gg 1$ dimensions, and where

$$a \equiv \sqrt{2mE_o}, \quad \frac{\delta a}{a} = \frac{1}{2} \frac{\delta E_o}{E_o}, \quad \nu \equiv 3N. \quad (4.41d)$$

It can be shown (and you might want to try to show it) that

$$\mathcal{V}_\nu(a) = \frac{\pi^{\nu/2}}{(\nu/2)!} a^\nu \quad \text{for } \nu \gg 1. \quad (4.41e)$$

(d) Show that, so long as $1 \gg \delta E_o/E_o \gg 1/N$ (where N in practice is an exceedingly huge number),

$$\mathcal{V}_\nu(a) - \mathcal{V}_\nu(a - \delta a) \simeq \mathcal{V}_\nu(a)[1 - e^{-\nu\delta a/a}] \simeq \mathcal{V}_\nu(a), \quad (4.41f)$$

which is independent of δE_o and thus will produce a value for $\Delta\Gamma$ and thence N_{states} and S independent of δE_o, as desired. From this and with the aid of Stirling's approximation (Reif, 2008, Appendix A.6) $n! \simeq (2\pi n)^{1/2} (n/e)^n$ for large n, and taking account of the multiplicity $\mathcal{M} = N!$, show that the entropy of the microcanonically distributed cells is given by

$$S(V, E, N) = Nk_B \ln\left[\frac{V}{N} \left(\frac{E}{N}\right)^{3/2} g_s \left(\frac{4\pi m}{3h^2}\right)^{3/2} e^{5/2} \right]. \quad (4.42)$$

This is known as the *Sackur-Tetrode* equation.

(e) Using Eqs. (3.39), show that this is equivalent to Eq. (4.40) for the entropy, which we derived by a very different method.

Exercise 4.8 **Example: Entropy of Mixing, Indistinguishability of Atoms, and the Gibbs Paradox*

(a) Consider two identical chambers, each with volume V, separated by an impermeable membrane. Into one chamber put energy E and N atoms of helium, and into the other, energy E and N atoms of xenon, with E/N and N/V small enough that the gases are nonrelativistic and nondegenerate. The membrane is ruptured, and the gases mix. Show that this mixing drives the entropy up by an amount $\Delta S = 2Nk_B \ln 2$. [Hint: Use the Sackur-Tetrode equation (4.42).]

(b) Suppose that energy E and N atoms of helium are put into both chambers (no xenon). Show that, when the membrane is ruptured and the gases mix, there is no increase of entropy. Explain why this result is reasonable, and explain its relationship to entropy being an extensive variable.

(c) Suppose that the N helium atoms were distinguishable instead of indistinguishable. Show this would mean that, in the microcanonical ensemble, they have $N!$ more states available to themselves, and their entropy would be larger by $k_B \ln N! \simeq k_B(N \ln N - N)$; and as a result, the Sackur-Tetrode formula (4.42) would be

Distinguishable particles:

$$S(V, E, N) = Nk_B \ln \left[V \left(\frac{E}{N} \right)^{3/2} \left(\frac{4\pi m}{3h^2} \right)^{3/2} e^{3/2} \right]. \tag{4.43}$$

Before the advent of quantum theory, physicists thought that atoms were distinguishable, and up to an additive multiple of N (which they could not compute), they deduced this entropy.

(d) Show that if, as prequantum physicists believed, atoms were distinguishable, then when the membrane between two identical helium-filled chambers is ruptured, there would be an entropy increase identical to that when the membrane between helium and xenon is ruptured: $\Delta S = 2Nk_B \ln 2$ [cf. parts (a) and (b)]. This result, which made prequantum physicists rather uncomfortable, is called the *Gibbs paradox*.

Exercise 4.9 *Problem: Quantum Decoherence and Entropy Increase in Terms of the Density Operator* T2

Reexpress Box 4.3's pedagogical example of quantum decoherence and entropy increase in the language of the quantum mechanical density operator $\hat{\rho}$ (Box 4.2). Use

this example to explain the meaning of the various statements made in the next-to-last
paragraph of Sec. 4.7.2.

4.8 Entropy per Particle

The entropy per particle in units of Boltzmann's constant,

$$\sigma \equiv S/(Nk_B),$$ (4.44)

is a very useful concept in both quantitative and order-of-magnitude analyses. (See,
e.g., Exs. 4.10, 4.11, and 4.4c, and the discussion of entropy in the expanding universe
in Sec. 4.10.3.) One reason is the second law of thermodynamics. Another is that in
the real universe σ generally lies somewhere between 0 and 100 and thus is a natural
quantity in terms of which to think about and remember the magnitudes of various
quantities.

For example, for ionized hydrogen gas in the nonrelativistic, classical domain,
the Sackur-Tetrode equation (4.42) for the entropy (derived from the microcanonical
ensemble in Ex. 4.7), when specialized either to the gas's protons or to its electrons
(both of which, with their spin 1/2, have $g_s = 2$), gives

$$\sigma_p = \frac{5}{2} + \ln\left[\frac{2m_p}{\rho}\left(\frac{2\pi mk_BT}{h^2}\right)^{3/2}\right].$$ (4.45)

Here we have set the number density of particles N/V (either protons or electrons)
to ρ/m_p, since almost all the mass density ρ is in the protons, and we have set the
thermal energy per particle E/N to $\frac{3}{2}k_BT$ [see Eq. (3.39b) derived in Chap. 3 using
kinetic theory]. The only way this formula depends on which particle species we are
considering, the gas's protons or its electrons, is through the particle mass m; and this
factor in Eq. (4.45) tells us that the protons' entropy per proton is a factor $\simeq 10$ higher
than the electrons': $\sigma_p - \sigma_e = \frac{3}{2}\ln(m_p/m_e) = 11.27 \simeq 10$. Therefore, for an ionized
hydrogen gas most of the entropy is in the protons.

The protons' entropy per proton is plotted as a function of density and temperature
in Fig. 4.3. It grows logarithmically as the density ρ decreases, and it ranges from $\sigma \ll 1$
in the regime of extreme proton degeneracy (lower right of Fig. 4.3; see Ex. 4.4) to $\sigma \sim$
1 near the onset of proton degeneracy (the boundary of the classical approximation),
to $\sigma \sim 100$ at the lowest density that occurs in the universe, $\rho \sim 10^{-29}$ g cm^{-3}. This
range is an example of the fact that the logarithms of almost all dimensionless numbers
that occur in Nature lie between approximately -100 and $+100$.

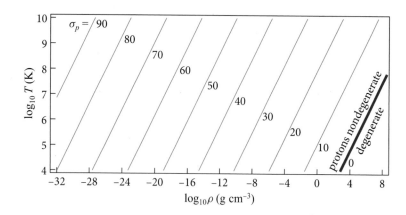

FIGURE 4.3 Proton entropy per proton σ_p for an ionized hydrogen gas. Each line is labeled with its value of σ_p. The electron entropy per electron σ_e is a factor $\simeq 10$ smaller; the electrons are degenerate when $\sigma_e \simeq \sigma_p - 10 \lesssim 1$. The protons are degenerate when $\sigma_p \lesssim 1$.

EXERCISES

Exercise 4.10 **Problem: Primordial Element Formation*

(a) As we shall see in Sec. 28.4.1, when the early universe was $\sim$200 s old, its principal constituents were photons, protons, neutrons, electrons, positrons, and (thermodynamically isolated) neutrinos and gravitons. The photon temperature was $\sim 9 \times 10^8$ K, and the baryon density was $\sim$0.02 kg m^{-3}. The photons, protons, electrons, and positrons were undergoing rapid electromagnetic interactions that kept them in thermodynamic equilibrium. Use the neutron-proton mass difference of 1.3 MeV to argue that, if the neutrons had also been in thermodynamic equilibrium, then their density would have been negligible at this time.

(b) However, the universe expanded too rapidly for weak interactions to keep the neutrons in equilibrium with the other particles, and so the fraction of the baryons at 200 s, in the form of neutrons that did not subsequently decay, was $\sim$0.14. At this time (Sec. 28.4.2) the neutrons began rapidly combining with protons to form alpha particles through a short chain of reactions, of which the first step—the one that was hardest to make go—was $n + p \rightarrow d + 2.22$ MeV, with the 2.22 MeV going into heat. Only about 10^{-5} of the baryons needed to go into deuterium in order to start and then maintain the full reaction chain. Show, using entropy per particle and the first law of thermodynamics, that at time $t \sim 200$ s after the big bang, this reaction was just barely entropically favorable.

(c) During this epoch, roughly how do you expect the baryon density to have decreased as a function of decreasing photon temperature? (For this you can neglect the heat being released by the nuclear burning and the role of positrons.) Show that before $t \sim 200$ s, the deuterium-formation reaction could not take place, and after $t \sim 200$ s, it rapidly became strongly entropically favorable.

(d) The nuclear reactions shut down when the neutrons had all cycled through deuterium and almost all had been incoporated into α particles, leaving only $\sim 10^{-5}$ of them in deuterium. About what fraction of all the baryons wound up in α particles (Helium 4)? (See Fig. 28.7 for details from a more complete analysis sketched in Sec. 28.4.2.)

Exercise 4.11 *Problem: Reionization of the Universe*
Following the epoch of primordial element formation (Ex. 4.10), the universe continued to expand and cool. Eventually when the temperature of the photons was $\sim$3,000 K, the free electrons and protons combined to form atomic hydrogen; this was the *epoch of recombinaton*. Later, when the photon temperature had fallen to $\sim$30 K, some hot stars and quasars formed and their ultraviolet radiation dissociated the hydrogen; this was the *epoch of reionization*. The details are poorly understood. (For more discussion, see Ex. 28.19.) Making the simplifying assumption that reionization happened rapidly and homogeneously, show that the increase in the entropy per baryon was $\sim 60 k_B$, depending weakly on the temperature of the atomic hydrogen, which you can assume to be $\sim$100 K.

4.9 Bose-Einstein Condensate

In this section, we explore an important modern application of the Bose-Einstein mean occupation number for bosons in statistical equilibrium. Our objectives are (i) to exhibit, in action, the tools developed in this chapter and (ii) to give a nice example of the connections between quantum statistical mechanics (which we use in the first 3/4 of this section) and classical statistical mechanics (which we use in the last 1/4).

For bosons in statistical equilibrium, the mean occupation number $\eta = 1/[e^{(E-\mu)/(k_B T)} - 1]$ diverges as $E \to 0$, if the chemical potential μ vanishes. This divergence is intimately connected to *Bose-Einstein condensation*.

Consider a dilute atomic gas in the form of a large number N of bosonic atoms, spatially confined by a magnetic trap. When the gas is cooled below some critical temperature T_c, μ is negative but gets very close to zero [see Eq. (4.47d)], causing η to become huge near zero energy. This huge η manifests physically as a large number N_0 of atoms collecting into the trap's mode of lowest (vanishing) energy, the Schrödinger equation's ground state [see Eq. (4.49a) and Fig. 4.4a later in this section].

This condensation was predicted by Einstein (1925), but an experimental demonstration was not technologically feasible until 1995, when two research groups independently exhibited it: one at JILA (University of Colorado) led by Eric Cornell and Carl Wieman; the other at MIT led by Wolfgang Ketterle. For these experiments, Cornell, Ketterle, and Wieman were awarded the 2001 Nobel Prize. Bose-Einstein condensates have great promise as tools for precision-measurement technology and nanotechnology.

Bose-Einstein condensation

As a concrete example of Bose-Einstein condensation, we analyze an idealized version of one of the early experiments by the JILA group (Ensher et al., 1996): a gas of 40,000 ^{87}Rb atoms placed in a magnetic trap that we approximate as a spherically symmetric, harmonic oscillator potential:[13]

$$V(r) = \frac{1}{2}m\omega_o^2 r^2 = \frac{1}{2}m\omega_o^2(x^2 + y^2 + z^2). \qquad (4.46a)$$

Here x, y, z are Cartesian coordinates, and r is radius. The harmonic-oscillator frequency ω_o and associated temperature $\hbar\omega_o/k_B$, the number N of rubidium atoms trapped in the potential, and the atoms' rest mass m are

$$\omega_o/(2\pi) = 181\,\text{Hz}, \quad \hbar\omega_o/k_B = 8.7\,\text{nK}, \quad N = 40,000, \quad m = 1.444 \times 10^{-25}\,\text{kg}. \qquad (4.46b)$$

Our analysis is adapted, in part, from a review article by Dalfovo et al. (1999).

Each ^{87}Rb atom is made from an even number of fermions [$Z = 37$ electrons, $Z = 37$ protons, and $(A - Z) = 50$ neutrons], and the many-particle wave function Ψ for the system of $N = 40,000$ atoms is antisymmetric (changes sign) under interchange of each pair of electrons, each pair of protons, and each pair of neutrons. Therefore, when any pair of atoms is interchanged (entailing interchange of an even number of fermion pairs), there is an even number of sign flips in Ψ. Thus Ψ is symmetric (no sign change) under interchange of atoms (i.e., the atoms behave like bosons and must obey Bose-Einstein statistics).

Repulsive forces between the atoms have a moderate influence on the experiment, but only a tiny influence on the quantities that we compute (see, e.g., Dalfovo et al., 1999). We ignore those forces and treat the atoms as noninteracting.

To make contact with our derivation, in Sec. 4.4.3, of the Bose-Einstein distribu-

modes: single-atom quantum states

tion, we must identify the modes (single-atom quantum states) S available to the atoms. Those modes are the energy eigenstates of the Schrödinger equation for a ^{87}Rb atom in the harmonic-oscillator potential $V(r)$. Solution of the Schrödinger equation (e.g., Cohen-Tannoudji, Diu, and Laloë, 1977, complement B_{VII}) reveals that the energy eigenstates can be labeled by the number of quanta of energy $\{n_x, n_y, n_z\}$ associated with an atom's motion along the x, y, and z directions; the energy of the mode $\{n_x, n_y, n_z\}$ is $E_{n_x,n_y,n_z} = \hbar\omega_o[(n_x + 1/2) + (n_y + 1/2) + (n_z + 1/2)]$. We simplify subsequent formulas by subtracting $\frac{3}{2}\hbar\omega_o$ from all energies and all chemical potentials. This is merely a change in what energy we regard as zero (renormalization), a change under which our statistical formalism is invariant (cf. foot-

13. In the actual experiment, the potential was harmonic but prolate spheroidal rather than spherical, i.e., in Cartesian coordinates $V(x, y, z) = \frac{1}{2}m[\omega_\varpi^2(x^2 + y^2) + \omega_z^2 z^2]$, with ω_z somewhat smaller than ω_ϖ. For pedagogical simplicity we treat the potential as spherical, with ω_o set to the geometric mean of the actual frequencies along the three Cartesian axes, $\omega_o = (\omega_\varpi^2 \omega_z)^{1/3}$. This choice of ω_o gives good agreement between our model's predictions and the prolate-spheroidal predictions for the quantities that we compute.

note 8 on page 175). Correspondingly, we attribute to the mode $\{n_x, n_y, n_z\}$ the energy $E_{n_x,n_y,n_z} = \hbar\omega_o(n_x + n_y + n_z)$. Our calculations will be simplified by lumping together all modes that have the same energy, so we switch from $\{n_x, n_y, n_z\}$ to $q \equiv n_x + n_y + n_z =$ (the mode's total number of quanta) as our fundamental quantum number, and we write the mode's energy as

$$E_q = q\hbar\omega_o. \tag{4.47a}$$

It is straightforward to verify that the number of independent modes with q quanta (the number of independent ways to choose $\{n_x, n_y, n_z\}$ such that their sum is q) is $\frac{1}{2}(q + 1)(q + 2)$. (Of course, one can also derive this same formula in spherical polar coordinates.)

Of special interest is the ground-state mode of the potential, $\{n_x, n_y, n_z\} = \{0, 0, 0\}$. This mode has $q = 0$, and it is unique: $(q + 1)(q + 2)/2 = 1$. Its energy is $E_0 = 0$, and its Schrödinger wave function is $\psi_o = (\pi\sigma_o^2)^{-3/4} \exp(-r^2/(2\sigma_o^2))$, so for any atom that happens to be in this ground-state mode, the probability distribution for its location is the Gaussian

$$|\psi_o(r)|^2 = \left(\frac{1}{\pi\sigma_o^2}\right)^{3/2} \exp\left(-\frac{r^2}{\sigma_o^2}\right), \quad \text{where} \quad \sigma_o = \sqrt{\frac{\hbar}{m\omega_o}} = 0.800\,\mu\text{m}. \tag{4.47b}$$

The entire collection of N atoms in the magnetic trap is a system; it interacts with its environment, which is at temperature T, exchanging energy but nothing else, so a conceptual ensemble consisting of this N-atom system and a huge number of identical systems has the canonical distribution $\rho = C\,\exp(-E_{\text{tot}}/(k_B T))$, where E_{tot} is the total energy of all the atoms. Each mode (labeled by $\{n_x, n_y, n_z\}$ or by q and two degeneracy parameters that we have not specified) is a subsystem of this system. Because the modes can exchange atoms as well as energy with one another, a conceptual ensemble consisting of any chosen mode and its clones is grand-canonically distributed [Eq. (4.28a)], so its mean occupation number is given by the Bose-Einstein formula (4.28b) with $E_S = q\hbar\omega_o$:

$$\eta_q = \frac{1}{\exp[(q\hbar\omega_o - \mu)/(k_B T)] - 1}. \tag{4.47c}$$

The temperature T is inherited from the environment (heat bath) in which the atoms live. The chemical potential μ is common to all the modes and takes on whatever value is required to guarantee that the total number of atoms in the trap is N—or, equivalently, that the sum of the mean occupation number (4.47c) over all the modes is N.

For the temperatures of interest to us:

1. The number $N_0 \equiv \eta_0$ of atoms in the ground-state mode $q = 0$ will be large, $N_0 \gg 1$, which permits us to expand the exponential in Eq. (4.47c) with $q = 0$ to obtain $N_0 = 1/[e^{-\mu/(k_B T)} - 1] \simeq -k_B T/\mu$, that is,

$$\mu/(k_B T) = -1/N_0. \tag{4.47d}$$

4.9 Bose-Einstein Condensate **195**

2. The atoms in excited modes will be spread out smoothly over many values of q, so we can approximate the total number of excited-mode atoms by an integral:

$$N - N_0 = \sum_{q=1}^{\infty} \frac{(q+1)(q+2)}{2} \eta_q \simeq \int_0^{\infty} \frac{\frac{1}{2}(q^2 + 3q + 2)dq}{\exp[(q\hbar\omega_o/(k_B T) + 1/N_0)] - 1}.$$

(4.48a)

The integral is dominated by large qs, so it is a rather good approximation to keep only the q^2 term in the numerator and to neglect the $1/N_0$ in the exponential:

$$N - N_0 \simeq \int_0^{\infty} \frac{q^2/2}{\exp(q\hbar\omega_o/(k_B T)) - 1} dq = \zeta(3) \left(\frac{k_B T}{\hbar\omega_o}\right)^3.$$

(4.48b)

Here $\zeta(3) \simeq 1.202$ is the Riemann zeta function [which also appeared in our study of the equation of state of thermalized radiation, Eq. (3.54b)]. It is useful to rewrite Eq. (4.48b) as

$$N_0 = N \left[1 - \left(\frac{T}{T_c^0}\right)^3 \right],$$

(4.49a)

where

$$T_c^0 = \frac{\hbar\omega_o}{k_B} \left(\frac{N}{\zeta(3)}\right)^{1/3} = 280 \text{ nK} \gg \hbar\omega_o/k_B = 8.7 \text{ nK}$$

(4.49b)

is our leading-order approximation to the critical temperature T_c. Obviously, we cannot have a negative number of atoms in the ground-state mode, so Eq. (4.49a) must fail for $T > T_c^0$. Presumably, N_0 becomes so small there that our approximation (4.47d) fails.

Figure 4.4a, adapted from the review article by Dalfovo et al. (1999), compares our simple prediction (4.49a) for $N_0(T)$ (dashed curve) with the experimental measurements by the JILA group (Ensher et al., 1996). Both theory and experiment show that, when one lowers the temperature T through a critical temperature T_c, the atoms suddenly begin to accumulate in large numbers in the ground-state mode. At $T \simeq 0.8T_c$,

half the atoms have condensed into the ground state (Bose-Einstein condensation); at $T \simeq 0.2T_c$ almost all are in the ground state. The simple formula (4.49a) is remarkably good at $0 < T < T_c$; evidently, T_c^0 [Eq. (4.49b)] is a rather good leading-order approximation to the critical temperature T_c at which the Bose-Einstein condensate begins to form.

Exercise 4.12 and Fig. 4.4b use more accurate approximations to Eq. (4.48a) to explore the onset of condensation as T is gradually lowered through the critical temperature. The onset is actually continuous when viewed on a sufficiently fine temperature scale; but on scales $0.01T_c$ or greater, it appears to be discontinuous.

The onset of Bose-Einstein condensation is an example of a *phase transition*: a sudden (nearly) discontinuous change in the properties of a thermalized system. Among

Chapter 4. Statistical Mechanics

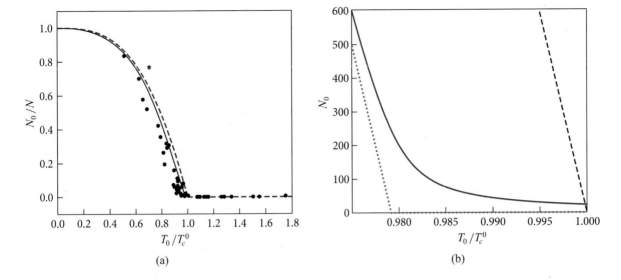

FIGURE 4.4 The number N_0 of atoms in the Bose-Einstein condensate at the center of a magnetic trap as a function of temperature T. (a) Low resolution; (b) High resolution. The dashed curve in each panel is the prediction (4.49a) of the simple theory presented in the text, using the parameters shown in Eq. (4.46b). The dotted curve in panel b is the prediction derived in Ex. 4.12c. The solid curves are our most accurate prediction (4.54) [Ex. 4.12d], including details of the condensation turning on. The large dots are experimental data from Ensher et al. (1996). Panel a is adapted from Dalfovo et al. (1999).

the sudden changes accompanying this phase transition is a (nearly) discontinuous change of the atoms' specific heat (Ex. 4.13). We study some other phase transitions in Chap. 5.

Notice that the critical temperature T_c^0 is larger, by a factor $\sim N^{1/3} = 34$, than the temperature $\hbar\omega_o/k_B = 8.7$ nK associated with the harmonic-oscillator potential. Correspondingly, at the critical temperature there are significant numbers of atoms in modes with q as large as ~ 34—which means that nearly all of the atoms are actually behaving rather classically at $T \simeq T_c^0$, despite our use of quantum mechanical concepts to analyze them!

It is illuminating to compute the spatial distribution of these atoms at the critical temperature using classical techniques. (This distribution could equally well be deduced using the above quantum techniques.) In the near classical, outer region of the trapping potential, the atoms' number density in phase space $\mathcal{N} = (g_s/h^3)\eta$ must have the classical, Boltzmann-distribution form (4.29): $dN/d\mathcal{V}_x d\mathcal{V}_p \propto \exp[-E/(k_B T)] = \exp\{-[V(r) + p^2/(2m)]/(k_B T_c)\}$, where $V(r)$ is the harmonic-oscillator potential (4.46a). Integrating over momentum space, $d\mathcal{V}_p = 4\pi p^2 dp$, we obtain for the number density of atoms $n = dN/d\mathcal{V}_x$

$$n(r) \propto \exp\left(\frac{-V(r)}{k_B T_c}\right) = \exp\left(\frac{-r^2}{a_o^2}\right), \tag{4.50a}$$

where [using Eqs. (4.49b) for T_c and (4.47b) for ω_o]

$$a_o = \sqrt{\frac{2k_B T_c}{m\omega_o^2}} = \frac{\sqrt{2}}{[\zeta(3)]^{1/6}} N^{1/6}\sigma_o = 1.371 N^{1/6}\sigma_o = 8.02\sigma_o = 6.4\ \mu\text{m}. \quad (4.50b)$$

Thus, at the critical temperature, the atoms have an approximately Gaussian spatial distribution with radius a_o eight times larger than the trap's 0.80 μm ground-state Gaussian distribution. This size of the distribution gives insight into the origin of the Bose-Einstein condensation: The mean inter-atom spacing at the critical temperature T_c^0 is $a_o/N^{1/3}$. It is easy to verify that this is approximately equal to the typical atom's de Broglie wavelength $\lambda_{T\,\text{dB}} = h/\sqrt{2\pi m k_B T} = h/(\text{typical momentum})$—which is the size of the region that we can think of each atom as occupying. The atomic separation is smaller than this in the core of the atomic cloud, so the atoms there are beginning to overlap and feel one another's presence, and thereby want to accumulate into the same quantum state (i.e., want to begin condensing). By contrast, the mean separation is larger than λ_{dB} in the outer portion of the cloud, so the atoms there continue to behave classically.

At temperatures below T_c, the N_0 condensed, ground-state-mode atoms have a spatial Gaussian distribution with radius $\sigma_o \sim a_o/8$ (i.e., eight times smaller than the region occupied by the classical, excited-state atoms). Therefore, the condensation is visually manifest by the growth of a sharply peaked core of atoms at the center of the larger, classical, thermalized cloud. In momentum space, the condensed atoms and classical cloud also have Gaussian distributions, with rms momenta $p_{\text{cloud}} \sim 8p_{\text{condensate}}$ (Ex. 4.14) or equivalently, rms speeds $v_{\text{cloud}} \sim 8v_{\text{condensate}}$. In early experiments, the existence of the condensate was observed by suddenly shutting off the trap and letting the condensate and cloud expand ballistically to sizes $v_{\text{condensate}}t$ and $v_{\text{cloud}}t$, and then observing them visually. The condensate was revealed as a sharp Gaussian peak, sticking out of the roughly eight times larger, classical cloud (Fig. 4.5).

EXERCISES

Exercise 4.12 **Example: Onset of Bose-Einstein Condensation*
By using more accurate approximations to Eq. (4.48a), explore the onset of the condensation near $T = T_c^0$. More specifically, do the following.

(a) Approximate the numerator in Eq. (4.48a) by $q^2 + 3q$, and keep the $1/N_0$ term in the exponential. Thereby obtain

$$N - N_0 = \left(\frac{k_B T}{\hbar\omega_o}\right)^3 \text{Li}_3(e^{-1/N_0}) + \frac{3}{2}\left(\frac{k_B T}{\hbar\omega_o}\right)^2 \text{Li}_2(e^{-1/N_0}). \quad (4.51)$$

Here

$$\text{Li}_n(u) = \sum_{p=1}^{\infty} \frac{u^p}{p^n} \quad (4.52a)$$

is a special function called the *polylogarithm* (Lewin, 1981), which is known to Mathematica and other symbolic manipulation software and has the properties

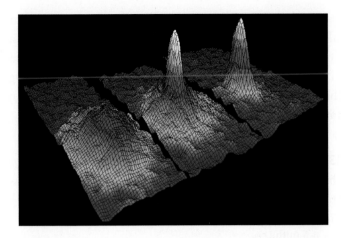

FIGURE 4.5 Velocity distribution of rubidium atoms in a Bose-Einstein condensate experiment by Anderson et al. (1995), as observed by the ballistic expansion method described in the text. In the left frame T is slightly higher than T_c, and there is only the classical cloud. In the center frame T is a bit below T_c, and the condensate sticks up sharply above the cloud. The right frame, at still lower T, shows almost pure condensate. Figure from Cornell (1996).

$$\text{Li}_n(1) = \zeta(n), \qquad \frac{d\text{Li}_n(u)}{du} = \frac{\text{Li}_{n-1}(u)}{u}, \tag{4.52b}$$

where $\zeta(n)$ is the Riemann zeta function.

(b) Show that by setting $e^{-1/N_0} = 1$ and ignoring the second polylogarithm in Eq. (4.51), one obtains the leading-order description of the condensation discussed in the text: Eqs. (4.48b) and (4.49).

(c) By continuing to set $e^{-1/N_0} = 1$ but keeping the second polylogarithm, obtain an improved equation for $N_0(T)$. Your answer should continue to show a discontinuous turn on of the condensation, but at a more accurate, slightly lower critical temperature

$$T_c^1 = T_c^0 \left[1 - \frac{\zeta(2)}{2\zeta(3)^{2/3}} \frac{1}{N^{1/3}} \right] = 0.979 T_c^0. \tag{4.53}$$

This equation illustrates the fact that our approximations are a *large-N expansion* (i.e., an expansion in powers of $1/N$).

(d) By keeping all details of Eq. (4.51) but rewriting it in terms of T_c^0, show that

$$N_0 = N \left[1 - \left(\frac{T}{T_c^0} \right)^3 \frac{\text{Li}_3(e^{-1/N_0})}{\zeta(3)} - \frac{3}{2\zeta(3)^{2/3}} \frac{1}{N^{1/3}} \left(\frac{T}{T_c^0} \right)^2 \text{Li}_2(e^{-1/N_0}) \right]. \tag{4.54}$$

Solve this numerically to obtain $N_0(T/T_c^0)$ for $N = 40{,}000$, and plot your result graphically. It should take the form of the solid curves in Fig. 4.4: a continuous turn on of the condensation over the narrow temperature range $0.98T_c^0 \lesssim T \lesssim 0.99T_c^0$ (i.e., a range $\Delta T \sim T_c^0/N^{1/3}$). In the limit of an arbitrarily large number of atoms, the turn on is instantaneous, as described by Eq. (4.49a)—an instantaneous phase transition.

Exercise 4.13 ****Problem: Discontinuous Change of Specific Heat*
Analyze the behavior of the atoms' total energy near the onset of condensation, in the limit of arbitrarily large N (i.e., keeping only the leading order in our $1/N^{1/3}$ expansion and approximating the condensation as turning on discontinuously at $T = T_c^0$). More specifically, do the following.

(a) Show that the total energy of the atoms in the magnetic trap is

$$E_{\text{total}} = \frac{3\zeta(4)}{\zeta(3)} N k_B T_c^0 \left(\frac{T}{T_c^0}\right)^4 \quad \text{when } T < T_c^0,$$

(4.55a)

$$E_{\text{total}} = \frac{3\text{Li}_4(e^{\mu/(k_B T)})}{\zeta(3)} N k_B T_c^0 \left(\frac{T}{T_c}\right)^4 \quad \text{when } T > T_c^0,$$

where (at $T > T_c$) $e^{\mu/(k_B T)}$ is a function of N and T determined by $N = \left(k_B T/(\hbar\omega_o)\right)^3 \text{Li}_3(e^{\mu/(k_B T)})$, so $\mu = 0$ at $T = T_c^0$ (see Ex. 4.12). Because $\text{Li}_n(1) = \zeta(n)$, this energy is continuous across the critical temperature T_c.

(b) Show that the specific heat $C = (\partial E_{\text{total}}/\partial T)_N$ is discontinuous across the critical temperature T_c^0:

$$C = \frac{12\zeta(4)}{\zeta(3)} N k_B = 10.80 N k_B \quad \text{as } T \to T_c \text{ from below,}$$

(4.55b)

$$C = \left(\frac{12\zeta(4)}{\zeta(3)} - \frac{9\zeta(3)}{\zeta(2)}\right) N k_B = 4.228 N k_B \quad \text{as } T \to T_c \text{ from above.}$$

Note that for gas contained within the walls of a box, there are two specific heats: $C_V = (\partial E/\partial T)_{N,V}$ when the box volume is held fixed, and $C_P = (\partial E/\partial T)_{N,P}$ when the pressure exerted by the box walls is held fixed. For our trapped atoms, there are no physical walls; the quantity held fixed in place of V or P is the trapping potential $V(r)$.

Exercise 4.14 *Derivation: Momentum Distributions in Condensate Experiments*
Show that in the Bose-Einstein condensate discussed in the text, the momentum distribution for the ground-state-mode atoms is Gaussian with rms momentum $p_{\text{condensate}} = \sqrt{3/2}\hbar/\sigma_o = \sqrt{3\hbar m\omega_o/2}$ and that for the classical cloud it is Gaussian with rms momentum $p_{\text{cloud}} = \sqrt{3mk_B T_c} \simeq \sqrt{3(N/\zeta(3))^{1/3}\hbar m\omega_o} \simeq 8p_{\text{condensate}}$.

Exercise 4.15 *Problem: Bose-Einstein Condensation in a Cubical Box*

Analyze Bose-Einstein condensation in a cubical box with edge lengths L [i.e., for a potential $V(x, y, z)$ that is zero inside the box and infinite outside it]. In particular, using the analog of the text's simplest approximation, show that the critical temperature at which condensation begins is

$$T_c^0 = \frac{1}{2\pi m k_B} \left[\frac{2\pi\hbar}{L} \left(\frac{N}{\zeta(3/2)} \right)^{1/3} \right]^2, \tag{4.56a}$$

and the number of atoms in the ground-state condensate, when $T < T_c^0$, is

$$N_0 = N \left[1 - \left(\frac{T}{T_c^0} \right)^{3/2} \right]. \tag{4.56b}$$

4.10 Statistical Mechanics in the Presence of Gravity T2

Systems with significant gravity behave quite differently in terms of their statistical mechanics than do systems without gravity. This has led to much controversy as to whether statistical mechanics can really be applied to gravitating systems. Despite that controversy, statistical mechanics has been applied in the presence of gravity in a variety of ways, with great success, resulting in important, fundamental conclusions. In this section, we sketch some of those applications: to galaxies, black holes, the universe as a whole, and the formation of structure in the universe. Our discussion is intended to give just the flavor of these subjects and not full details, so we state some things without derivation. This is necessary in part because many of the phenomena we describe rely for their justification on general relativity (Part VII) and/or quantum field theory in curved spacetime (see, e.g., Parker and Toms, 2009).

4.10.1 Galaxies T2

A galaxy is dominated by a roughly spherical distribution of dark matter (believed to comprise elementary particles with negligible collision cross section) with radius $R_D \sim 3 \times 10^{21}$ m and mass $M_D \sim 10^{42}$ kg. The dark matter and roughly $N \sim 10^{11}$ stars, each with fiducial mass $m \sim 10^{30}$ kg, move in a common gravitational potential well. (As we discuss in Chap. 28, the ratio of regular, or baryonic, matter to dark matter is roughly 1:5 by mass.) The baryons (stars plus gas) are mostly contained within a radius $R \sim 3 \times 10^{20}$ m. The characteristic speed of the dark matter and the stars and gas is $v \sim (GM_D/R_D)^{1/2} \sim (GNm/R)^{1/2} \sim 200$ km s^{-1}. For the moment, focus on the stars, with total mass $M = Nm$, ignoring the dark matter and gas, whose presence does not change our conclusions.

The time it takes stars moving in the dark matter's gravitational potential to cross the baryonic galaxy is $\tau_{\text{int}} \sim 2R/v \sim 10^8$ yr.[14] This time is short compared with the

---

14. $1\,\text{yr} \simeq \pi \times 10^7$ s.

age of a galaxy, $\sim 10^{10}$ yr. Galaxies have distant encounters with their neighbors on timescales that can be smaller than their ages but still much longer than τ_{int}; in this sense, they can be thought of as semiclosed systems weakly coupled to their environments. In this subsection, we idealize our chosen galaxy as fully closed (no interaction with its environment). Direct collisions between stars are exceedingly rare, and strong two-star gravitational encounters, which happen when the impact parameter[15] is smaller than $\sim Gm/v^2 \sim R/N$, are also negligibly rare except, sometimes, near the center of a galaxy (which we ignore until the last paragraph of this subsection). We can therefore regard each of the galaxy's stars as moving in a gravitational potential determined by the smoothed-out mass of the dark matter and all the other stars, and can use Hamiltonian dynamics to describe their motions.

Imagine that we have an ensemble of such galaxies, all with the same number of stars N, the same mass M, and the same energy E (in a tiny range δE). We begin our study of that ensemble by making an order-of-magnitude estimate of the probability ρ of finding a chosen galaxy from the ensemble in some chosen quantum state. We compute that probability from the corresponding probabilities for its subsystems, individual stars. The phase-space volume available to each star in the galaxy is $\sim R^3(mv)^3$, the density of single-particle quantum states (modes) in each star's phase space is $1/h^3$, the number of available modes is the product of these, $\sim (Rmv/h)^3$, and the probability of the star occupying the chosen mode, or any other mode, is the reciprocal of this product, $\sim [h/(Rmv)]^3$. The probability of the galaxy occupying a state in its phase space is the product of the probabilities for each of its N stars [Eq. (4.18c)]:

distribution function for
stars in a galaxy

$$\rho \sim \left(\frac{h}{Rmv}\right)^{3N} \sim 10^{-2.7 \times 10^{13}}. \tag{4.57}$$

This very small number suggests that it is somewhat silly of us to use quantum mechanics to normalize the distribution function (i.e., silly to use the probabilistic distribution function ρ) when dealing with a system as classical as a whole galaxy. Silly, perhaps; but dangerous, no. The key point is that, so far as classical statistical mechanics is concerned, the only important feature of ρ is that it is proportional to the classical distribution function $\mathcal{N}_{\text{sys}}$; its absolute normalization is usually not important, classically. It was this fact that permitted so much progress to be made in statistical mechanics prior to the advent of quantum mechanics.

Are real galaxies in statistical equilibrium? To gain insight into this question, we estimate the entropy of a galaxy in our ensemble and then ask whether that entropy has any chance of being the maximum value allowed to the galaxy's stars (as it must be if the galaxy is in statistical equilibrium).

Obviously, the stars (by contrast with electrons) are distinguishable, so we can assume multiplicity $\mathcal{M} = 1$ when estimating the galaxy's entropy. Ignoring the (neg-

---

15. The impact parameter is the closest distance between the two stars along their undeflected trajectories.

ligible) correlations among stars, the entropy computed by integating $\rho \ln \rho$ over the galaxy's full $6N$-dimensional phase space is just N times the entropy associated with a single star, which is $S \sim N k_B \ln(\Delta\Gamma/h^3)$ [Eqs. (4.37) and (4.8a)], where $\Delta\Gamma$ is the phase-space volume over which the star wanders in its ergodic, hamiltonian-induced motion (i.e., the phase space volume available to the star). We express this entropy in terms of the galaxy's total mass M and its total nonrelativistic energy $E \sim -GM^2/(2R)$ as follows. Since the characteristic stellar speed is $v \sim (GM/R)^{1/2}$, the volume of phase space over which the star wanders is $\Delta\Gamma \sim (mv)^3 R^3 \sim (GMm^2R)^{3/2} \sim (-G^2M^3m^2/(2E))^{3/2}$, and the entropy is therefore

$$S_{\text{Galaxy}} \sim (M/m)k_B \ln(\Delta\Gamma/h^3) \sim (3M/(2m))k_B \ln(-G^2M^3m^2/(2Eh^2)). \quad (4.58)$$

Is this the maximum possible entropy available to the galaxy, given the constraints that its mass be M and its nonrelativistic energy be E? No. Its entropy can be made larger by removing a single star from the galaxy to radius $r \gg R$, where the star's energy is negligible. The entropy of the remaining stars will decrease slightly, since the mass M diminishes by m at constant E. However, the entropy associated with the removed star, $\sim(3/2) \ln(GMm^2r/h^2)$, can be made arbitrarily large by making its orbital radius r arbitrarily large. By this thought experiment, we discover that galaxies cannot be in a state of maximum entropy at fixed E and M; they therefore cannot be in a true state of statistical equilibrium.[16] (One might wonder whether there is entropy associated with the galaxy's gravitational field, some of which is due to the stars, and whether that entropy invalidates our analysis. The answer is no. The gravitational field has no randomness, beyond that of the stars themselves, and thus no entropy; its structure is uniquely determined, via Newton's gravitational field equation, by the stars' spatial distribution.)

galaxy never in statistical equilibrium

In a real galaxy or other star cluster, rare near-encounters between stars in the cluster core (ignored in the above discussion) cause individual stars to be ejected from the core into distant orbits or to be ejected from the cluster altogether. These ejections increase the entropy of the cluster plus ejected stars in just the manner of our thought experiment. The core of the galaxy shrinks, a diffuse halo grows, and the total number of stars in the galaxy gradually decreases. This evolution to ever-larger entropy is demanded by the laws of statistical mechanics, but by contrast with systems without gravity, it does not bring the cluster to statistical equilibrium. The long-range influence of gravity prevents a true equilibrium from being reached. Ultimately, the cluster's or galaxy's core may collapse to form a black hole—and, indeed, most large galaxies are observed to have massive black holes in their cores. Despite this somewhat negative conclusion, the techniques of statistical mechanics can be used to understand

to increase entropy, galaxy core shrinks and halo grows

gravity as key to galaxy's behavior

---

16. A true equilibrium can be achieved if the galaxy is enclosed in an idealized spherical box whose walls prevent stars from escaping, or if the galaxy lives in an infinite thermalized bath of stars so that, on average, when one star is ejected into a distant orbit in the bath, another gets injected into the galaxy (see, e.g., Ogorodnikov, 1965; Lynden-Bell, 1967). However, in the real universe galaxies are not surrounded by walls or by thermalized star baths.

galactic dynamics over the comparatively short timescales of interest to astronomers (e.g., Binney and Tremaine, 2003). For a complementary description of a dark matter galaxy in which the stars are ignored, see Ex. 28.7. Further discussion of stellar and dark matter distributions in galaxies is presented in Chap. 28.

4.10.2 Black Holes T2

Quantum field theory predicts that, near the horizon of a black hole, the vacuum fluctuations of quantized fields behave thermally, as seen by stationary (non-infalling) observers. More specifically, such observers see the horizon surrounded by an atmosphere that is in statistical equilibrium (a thermalized atmosphere) and that rotates with the same angular velocity $\mathbf{\Omega}_H$ as the hole's horizon. This remarkable conclusion, due to Stephen Hawking (1976), William Unruh (1976), and Paul Davies (1977), is discussed pedagogically in books by Thorne, Price, and MacDonald (1986) and Frolov and Zelnikov (2011), and more rigorously in a book by Wald (1994). The atmosphere contains all types of particles that can exist in Nature. Very few of the particles manage to escape from the hole's gravitational pull; most emerge from the horizon, fly up to some maximum height, then fall back down to the horizon. Only if they start out moving almost vertically upward (i.e., with nearly zero angular momentum) do they have any hope of escaping. The few that do escape make up a tiny trickle of *Hawking radiation* (Hawking, 1975) that will ultimately cause the black hole to evaporate, unless it grows more rapidly due to infall of material from the external universe (which it will unless the black hole is far less massive than the Sun).

In discussing the distribution function for the hole's thermalized, rotating atmosphere, one must take account of the fact that the locally measured energy of a particle decreases as it climbs out of the hole's gravitational field (Ex. 26.4). One does so by attributing to the particle the energy that it would ultimately have if it were to escape from the hole's gravitational grip. This is called the particle's "redshifted energy" and is denoted by $\mathcal{E}_\infty$. This $\mathcal{E}_\infty$ is conserved along the particle's world line, as is the projection $\mathbf{j} \cdot \hat{\mathbf{\Omega}}_H$ of the particle's orbital angular momentum $\mathbf{j}$ along the hole's spin axis (unit direction $\hat{\mathbf{\Omega}}_H$).

The hole's horizon behaves like the wall of a blackbody cavity. Into each upgoing mode (single-particle quantum state) a of any and every quantum field that can exist in Nature, it deposits particles that are thermalized with (redshifted) temperature T_H, vanishing chemical potential, and angular velocity $\mathbf{\Omega}_H$. As a result, the mode's distribution function—which is the probability of finding N_a particles in it with net redshifted energy $\mathcal{E}_{a\infty} = N_a \times$ (redshifted energy of one quantum in the mode) and with net axial component of angular momentum $\mathbf{j}_a \cdot \hat{\mathbf{\Omega}}_H = N_a \times$ (angular momentum of one quantum in the mode)—is

$$\rho_a = C \exp\left[\frac{-\mathcal{E}_{a\infty} + \mathbf{\Omega}_H \cdot \mathbf{j}_a}{k_B T_H}\right] \tag{4.59}$$

[see Eq. (4.22) and note that $\mathbf{\Omega}_H = \Omega_H \hat{\mathbf{\Omega}}_H$]. The distribution function for the entire thermalized atmosphere (made of all modes that emerge from the horizon) is, of course, $\rho = \prod_a \rho_a$. (Ingoing modes, which originate at infinity—i.e., far from the black hole—are not thermalized; they contain whatever the universe chooses to send toward the hole.) Because $\mathcal{E}_{a\,\infty}$ is the redshifted energy in mode a, T_H is similarly a redshifted temperature: it is the temperature that the Hawking radiation exhibits when it has escaped from the hole's gravitational grip. Near the horizon, the locally measured atmospheric temperature is gravitationally blue-shifted to much higher values than T_H.

The temperature T_H and angular velocity $\mathbf{\Omega}_H$, like all properties of a black hole, are determined completely by the hole's spin angular momentum $\mathbf{J}_H$ and its mass M_H. To within factors of order unity, they have magnitudes [Ex. 26.16 and Eq. (26.77)]

$$T_H \sim \frac{\hbar}{8\pi k_B G M_H/c^3} \sim \frac{6 \times 10^{-8}\,\text{K}}{M_H/M_\odot}, \quad \Omega_H \sim \frac{J_H}{M_H(2GM_H/c^2)^2}. \qquad (4.60)$$

black hole temperature and angular velocity

For a very slowly rotating hole the "$\sim$" becomes an "$=$" in both equations. Notice how small the hole's temperature is, if its mass is greater than or of order $M_\odot$. For such holes the thermal atmosphere is of no practical interest, though it has deep implications for fundamental physics. Only for tiny black holes (that might conceivably have been formed in the big bang) is T_H high enough to be physically interesting.

Suppose that the black hole evolves much more rapidly by accreting matter than by emitting Hawking radiation. Then the evolution of its entropy can be deduced from the first law of thermodynamics for its atmosphere. By techniques analogous to some developed in the next chapter, one can argue that the atmosphere's equilibrium distribution (4.59) implies the following form for the first law (where we set $c = 1$):

$$dM_H = T_H dS_H + \mathbf{\Omega}_H \cdot d\mathbf{J}_H \qquad (4.61)$$

first law of thermodynamics for black hole

[cf. Eq. (26.92)]. Here dM_H is the change of the hole's mass due to the accretion (with each infalling particle contributing its $\mathcal{E}_\infty$ to dM_H), $d\mathbf{J}_H$ is the change of the hole's spin angular momentum due to the accretion (with each infalling particle contributing its $\mathbf{j}$), and dS_H is the increase of the black hole's entropy.

Because this first law can be deduced using the techniques of statistical mechanics (Chap. 5), it can be argued (e.g., Zurek and Thorne, 1985) that the hole's entropy increase has the standard statistical mechanical origin and interpretation: if N_{states} is the total number of quantum states that the infalling material could have been in (subject only to the requirement that the total infalling mass-energy be dM_H and total infalling angular momentum be $d\mathbf{J}_H$), then $dS_H = k_B \log N_{\text{states}}$ [cf. Eq. (4.35)]. In other words, the hole's entropy increases by k_B times the logarithm of the number of quantum mechanically different ways that we could have produced its changes of mass and angular momentum, dM_H and $d\mathbf{J}_H$. Correspondingly, we can regard the hole's total entropy as k_B times the logarithm of the number of ways in which it could have been made. That number of ways is enormous, and correspondingly, the hole's entropy is enormous. This analysis, when carried out in full detail (Zurek and Thorne, 1985), reveals that the entropy is [Eq. (26.93)]

black hole's entropy

$$S_H = k_B \frac{A_H}{4L_P{}^2} \sim 1 \times 10^{77} k_B \left(\frac{M_H}{M_\odot} \right)^2, \tag{4.62}$$

where $A_H \sim 4\pi (2GM_H/c^2)$ is the surface area of the hole's horizon, and $L_P = \sqrt{G\hbar/c^3} = 1.616 \times 10^{-33}$ cm is the Planck length—a result first proposed by Bekenstein (1972) and first proved by Hawking (1975).

What is it about a black hole that leads to this peculiar thermal behavior and enormous entropy? Why is a hole so different from a star or galaxy? The answer lies in the black-hole horizon and the fact that things that fall inward through the horizon cannot get back out. From the perspective of quantum field theory, the horizon produces the thermal behavior. From that of statistical mechanics, the horizon produces the loss of information about how the black hole was made and the corresponding entropy increase. In this sense, the horizon for a black hole plays a role analogous to coarse-graining in conventional classical statistical mechanics.[17]

horizon as key to hole's thermal behavior

The above statistical mechanical description of a black hole's atmosphere and thermal behavior is based on the laws of quantum field theory in curved spacetime—laws in which the atmosphere's fields (electromagnetic, neutrino, etc.) are quantized, but the hole itself is not governed by the laws of quantum mechanics. For detailed but fairly brief analyses along the lines of this section, see Zurek and Thorne (1985); Frolov and Page (1993). For a review of the literature on black-hole thermodynamics and of conundrums swept under the rug in our simple-minded discussion, see Wald (2001).[18]

EXERCISES

Exercise 4.16 *Example: Thermal Equilibria for a Black Hole inside a Box* T2
This problem was posed and partially solved by Hawking (1976), and fully solved by Page et al. (1977). Place a nonrotating black hole with mass M at the center of a spherical box with volume V that is far larger than the black hole: $V^{1/3} \gg 2GM/c^2$. Put into the box, and outside the black hole, thermalized radiation with energy $\mathcal{E} - Mc^2$ (so the total energy in the box is $\mathcal{E}$). The black hole will emit Hawking

---

17. It seems likely, as of 2017, that the information about what fell into the hole gets retained, in some manner, in the black hole's atmosphere, and we have coarse-grained it away by our faulty understanding of the relevant black-hole quantum mechanics.

18. A much deeper understanding involves string theory—the most promising approach to quantum gravity and to quantization of black holes. Indeed, the thermal properties of black holes, and most especially their entropy, are a powerful testing ground for candidate theories of quantum gravity. A recent, lively debate around the possibility that a classical event horizon might be accompanied by a *firewall* (Almheiri et al., 2013) is a testament to the challenge of these questions, and to the relevance that they have to the relationship between gravity and the other fundamental interactions.

radiation and will accrete thermal radiation very slowly, causing the hole's mass M and the radiation energy $\mathcal{E} - Mc^2$ to evolve. Assume that the radiation always thermalizes during the evolution, so its temperature T is always such that $aT^4V = \mathcal{E} - Mc^2$, where a is the radiation constant. (For simplicity assume that the only form of radiation that the black hole emits and that resides in the box is photons; the analysis is easily extended to the more complicated case where other kinds of particles are present and are emitted.) Discuss the evolution of this system using the second law of thermodynamics. More specifically, do the following.

(a) To simplify your calculations, use so-called *natural units* in which not only is c set equal to unity (geometrized units), but also $G = \hbar = k_B = 1$. If you have never used natural units before, verify that no dimensionless quantities can be constructed from G, c, $\hbar$, and k_B; from this, show that you can always return to cgs or SI units by inserting the appropriate factors of G, c, $\hbar$, and k_B to get the desired units. Show that in natural units the radiation constant is $a = \pi^2/15$.

(b) Show that in natural units the total entropy inside the box is $S = \frac{4}{3}aVT^3 + 4\pi M^2$. The mass M and radiation temperature T will evolve so as to increase this S subject to the constraint that $\mathcal{E} = M + aVT^4$, with fixed $\mathcal{E}$ and V.

(c) Find a rescaling of variables that enables $\mathcal{E}$ to drop out of the problem. [Answer: Set $m = M/\mathcal{E}$, $\Sigma = S/(4\pi\mathcal{E}^2)$, and $\nu = Va/[(3\pi)^4\mathcal{E}^5]$, so m, Σ, and ν are rescaled black-hole mass, entropy in the box, and box volume.] Show that the rescaled entropy is given by

$$\Sigma = m^2 + \nu^{1/4}(1-m)^{3/4}. \tag{4.63}$$

This entropy is plotted as a function of the black hole mass m in Fig. 4.6 for various ν [i.e., for various $V = (3\pi)^4(E^5/a)\nu$].

(d) Show that

$$\frac{d\Sigma}{dm} = \frac{3\nu^{1/4}}{4(1-m)^{1/4}}(\tau - 1), \tag{4.64}$$

where $\tau = T/T_H = 8\pi MT$ is the ratio of the radiation temperature to the black hole's Hawking temperature. Thereby show that (i) when $\tau > 1$, the black hole accretes more energy than it emits, and its mass M grows, thereby increasing the entropy in the box and (ii) when $\tau < 1$, the black hole emits more energy than it accretes, thereby decreasing its mass and again increasing the total entropy.

(e) From the shapes of the curves in Fig. 4.6, it is evident that there are two critical values of the box's volume V (or equivalently, of the rescaled volume ν): V_h and V_g. For $V > V_h$, the only state of maximum entropy is $m = 0$: no black hole. For these large volumes, the black hole will gradually evaporate and disappear. For $V < V_h$, there are two local maxima: one with a black hole whose mass is somewhere between $m = M/\mathcal{E} = 4/5$ and 1; the other with no black hole, $m = 0$.

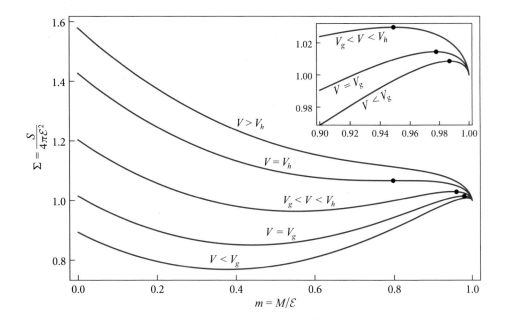

FIGURE 4.6 Total entropy S inside a box of volume V that contains a black hole with mass M and thermalized radiation with energy $\mathcal{E} - Mc^2$ and temperature $T = [(\mathcal{E} - Mc^2)/(aV)]^{1/4}$. For $V < V_h$ there is a statistical equilibrium that contains a black hole, at the local maximum of entropy (large filled circles). At that equilibrium, the radiation temperature T is equal to the black hole's Hawking temperature T_H. See Exercise 4.16e for the definitions of V_g and V_h.

If $V = V_g$, the two states have the same entropy; if $V < V_g$, the state with a black hole has the larger entropy; if $V_g < V < V_h$, the state without a black hole has the larger entropy. Show that

$$V_h = \frac{2^{20}\pi^4}{5^5 a}\mathcal{E}^5 = 4.97 \times 10^4 \left(\frac{\mathcal{E}}{M_P c^2}\right)^2 \left(\frac{G\mathcal{E}}{c^4}\right)^3, \quad V_g = 0.256 V_h, \quad (4.65)$$

where the first expression is in natural units. Here $M_P = \sqrt{\hbar c/G} = 2.177 \times 10^{-5}$ g is the Planck mass, and $G\mathcal{E}/c^4$ is half the radius of a black hole with mass $\mathcal{E}/c^2$. Show that, if $V_g < V < V_h$, and if the box's volume V is made as large as possible—the size of the universe—then the mass of the equilibrium black hole will be roughly 1/100 the mass of the Sun.

When there are two entropy maxima, that is, two states of statistical equilibrium (i.e., for $V < V_h$), an entropy barrier exists between the two equilibria. In principle, quantum fluctuations should be able to carry the system through that barrier, from the equilibrium of lower entropy to that of higher entropy, though the rate will be extremely small for $\mathcal{E} \gg M_P c^2$. Those fluctuations will be governed by the laws of quantum gravity, so studying them is a useful thought experiment in the quest to understand quantum gravity.

4.10.3 The Universe T2

Observations and theory agree that the universe, when far younger than 1 s old, settled into a very hot, highly thermalized state. All particles except gravitons were in statistical equilibrium at a common, declining temperature, until the dark matter and the neutrinos dropped out of equilibrium and (like the gravitons) became thermodynamically isolated.

During this early relativistic era, the equations of relativistic cosmology imply (as discussed in Sec. 28.4.1) that the temperature T of the universe at age t satisfied $T/T_P \sim (t/t_P)^{-1/2}$. Here $T_P \equiv [\hbar c^5/(G k_B^2)]^{1/2} \sim 10^{32}$ K is the Planck temperature, and $t_P \equiv (\hbar G/c^5)^{1/2} \sim 10^{-43}$ s is the Planck time. (This approximate T/T_P relationship can be justified on dimensional grounds.) Now the region that was in causal contact at time t (i.e., that was contained within a mutual cosmological horizon) had a volume $\sim (ct)^3$, and thermodynamic considerations imply that the number of relativistic particles that were in causal contact at time t was $N \sim (k_B T t/\hbar)^3 \sim (t/t_P)^{3/2}$. (This remains roughly true today when N has grown to $\sim 10^{91}$, essentially all in microwave background photons.) The associated entropy was then $S \sim N k_B$ (cf. Sec. 4.8).

number of particles and entropy inside cosmological horizon

Although this seems like an enormous entropy, gravity can do even better. The most efficient way to create entropy, as described in Sec. 4.10.2, is to form massive black holes. Suppose that out of all the relativistic particle mass within the horizon, $M \sim N k_B T/c^2$, a fraction f has collapsed into black holes of mass M_H. Then, with the aid of Sec. 4.10.2, we estimate that the associated entropy is $S_H \sim f(M_H/M)(t/t_P)^{1/2}S$. If we use the observation that every galaxy has a central black hole with mass in the $\sim 10^6$–10^9 solar mass range, we find that $f \sim 10^{-4}$ and $S_H \sim 10^{11}S$ today!

entropy in black holes

Now it might be claimed that massive black holes are thermodynamically isolated from the rest of the universe because they will take so long to evaporate. That may be so as a practical matter, but more modest gravitational condensations that create stars and starlight can produce large local departures from thermodynamic equilibrium, accompanied by (indeed, driven by) a net increase of entropy and can produce the conditions necessary for life to develop.

Given these considerations, it should not come as a surprise to learn that the behavior of entropy in the expanding universe has provided a foundation for the standard model of cosmology, which we outline in Chap. 28. It has also provided a stimulus for many imaginative proposals addressing fascinating questions that lie beyond the standard model, some of which connect with the nature of life in the universe.[19]

4.10.4 Structure Formation in the Expanding Universe: Violent Relaxation and Phase Mixing T2

The formation of stars and galaxies ("structure") by gravitational condensation provides a nice illustration of the phase mixing and coarse-graining that underlie the second law of thermodynamics (Sec. 4.7.2).

It is believed that galaxies formed when slight overdensities in the dark matter and gas (presumably irregular in shape) stopped expanding and began to contract under their mutual gravitational attraction. Much of the gas was quickly converted into stars. The dark-matter particles and the stars had very little random motion at this stage relative to their random motions today, $v \sim 200$ km s^{-1}. Correspondingly, although their physical volume $\mathcal{V}_x$ was initially only moderately larger than today, their momentum-space volume $\mathcal{V}_p$ was far smaller than it is today. Translated into the language of an ensemble of N such galaxies, the initial coordinate-space volume $\int d^{3N}x \sim \mathcal{V}_x{}^N$ occupied by each of the ensemble's galaxies was moderately larger than it is today, while its momentum-space volume $\int d^{3N}p \sim \mathcal{V}_p{}^N$ was far smaller. The phase-space volume $\mathcal{V}_x^N \mathcal{V}_p^N$ must therefore have increased considerably during the galaxy formation—with the increase due to a big increase in the relative momenta of neighboring stars. For this to occur, it was necessary that the stars changed their relative energies during the contraction, which requires a time-dependent hamiltonian. In other words, the gravitational potential Φ felt by the stars must have varied rapidly, so that the individual stellar energies would vary according to

$$\frac{dE}{dt} = \frac{\partial H}{\partial t} = m\frac{\partial \Phi}{\partial t}. \tag{4.66}$$

violent relaxation of star distribution

The largest changes of energy occurred when the galaxy was contracting dynamically (collapsing), so the potential changed significantly on the timescale it took stars to cross the galaxy, $\tau_{\text{int}} \sim 2R/v$. Numerical simulations show that this energy transfer was highly efficient. This process is known as *violent relaxation*. Although violent relaxation could create the observed stellar distribution functions, it was not by itself a means of diluting the phase-space density, since Liouville's theorem still applied.

phase mixing and coarse-graining of star distribution

The mechanism that changed the phase-space density was phase mixing and coarse-graining (Sec. 4.7.2 above). During the initial collapse, the particles and newly formed stars could be thought of as following highly perturbed radial orbits. The orbits of nearby stars were somewhat similar, though not identical. Therefore small elements of occupied phase space became highly contorted as the particles and stars moved along their phase-space paths.

---

19. For an early and highly influential analysis, see Schrödinger (1944), and for a more recent discussion, see Penrose (2016).

Let us make a simple model of this process by assuming the individual particles and stars initially populate a fraction $f \ll 1$ of the final occupied phase-space volume $\mathcal{V}_{\text{final}}$. After one dynamical timescale $\tau_{int} \sim R/v$, this small volume $f\mathcal{V}_{\text{final}}$ is (presumably) deformed into a convoluted surface that folds back on itself once or twice like dough being kneaded by a baker, while still occupying the same volume $f\mathcal{V}_{\text{final}}$. After n dynamical timescales, there are $\sim 2^n$ such folds (cf. Fig. 4.2b above). After $n \sim -\log_2 f$ dynamical timescales, the spacing between folds becomes comparable with the characteristic thickness of this convoluted surface, and it is no longer practical to distinguish the original distribution function. We expect that coarse-graining has been accomplished for all practical purposes; only a pathological physicist would resist it and insist on trying to continue keeping track of which contorted phase-space regions have the original high density and which do not. For a galaxy we might expect that $f \sim 10^{-3}$ and so this natural coarse-graining can occur in a time approximately equal to $-\log_2 10^{-3}\tau_{int} \sim 10\, \tau_{int} \sim 10^9$ yr, which is 10 times shorter than the present age of galaxies. Therefore it need not be a surprise that the galaxy we know best, our own Milky Way, exhibits little obvious vestigial trace of its initial high-density (low phase-space-volume) distribution function.[20]

4.11 Entropy and Information T2

4.11.1 Information Gained When Measuring the State of a System in a Microcanonical Ensemble T2

In Sec. 4.7, we said that entropy is a measure of our lack of information about the state of any system chosen at random from an ensemble. In this section we make this heuristic statement useful by introducing a precise definition of *information*.

Consider a microcanonical ensemble of identical systems. Each system can reside in any one of a finite number, N_{states}, of quantum states, which we label by integers $n = 1, 2, 3, \ldots, N_{\text{states}}$. Because the ensemble is microcanonical, all N_{states} states are equally probable; they have probabilities $\rho_n = 1/N_{\text{states}}$. Therefore the entropy of any system chosen at random from this ensemble is $S = -k_B \sum_n \rho_n \ln \rho_n = k_B \ln N_{\text{states}}$ [Eqs. (4.34) and (4.35)].

Now suppose that we measure the state of our chosen system and find it to be (for example) state number 238 out of the N_{states} equally probable states. How much information have we gained? For this thought experiment, and more generally (see Sec. 4.11.2 below), *the amount of information gained, expressed in bits, is defined to be the minimum number of binary digits required to distinguish the measured state from all the other N_{states} states that the system could have been in.* To evaluate this information gain, we label each state n by the number $n-1$ written in binary code (state $n = 1$ is labeled by {000}, state $n = 2$ is labeled by {001}, 3 is {010}, 4 is {011}, 5 is {100},

amount of information gained in a measurement for microcanonical ensemble

20. However, the dark-matter particles may very well not be fully coarse-grained at the present time, and this influences strategies for detecting it.

6 is {101}, 7 is {110}, 8 is {111}, etc.). If $N_\text{states} = 4$, then the number of binary digits needed is 2 (the leading 0 in the enumeration above can be dropped), so in measuring the system's state we gain 2 bits of information. If $N_\text{states} = 8$, the number of binary digits needed is 3, so our measurement gives us 3 bits of information. In general, we need $\log_2 N_\text{states}$ binary digits to distinguish the states from one another, so *the amount of information gained in measuring the system's state is the base-2 logarithm of the number of states the system could have been in:*

$$I = \log_2 N_\text{states} = (1/\ln 2) \ln N_\text{states} = 1.4427 \ln N_\text{states}. \tag{4.67a}$$

Notice that this information gain is proportional to the entropy $S = k_B \ln N_\text{states}$ of the system before the measurement was made:

information gain related to entropy decrease

$$I = S/(k_B \ln 2). \tag{4.67b}$$

The measurement reduces the system's entropy from $S = -k_B \ln N_\text{states}$ to zero (and increases the entropy of the rest of the universe by at least this amount), and it gives us $I = S/(k_B \ln 2)$ bits of information about the system. We shall discover below that this entropy/information relationship is true of measurements made on a system drawn from any ensemble, not just a microcanonical ensemble. But first we must develop a more complete understanding of information.

4.11.2

4.11.2 Information in Communication Theory T2

The definition of "the amount of information I gained in a measurement" was formulated by Claude Shannon (1948) in the context of his laying the foundations of *communication theory.* Communication theory deals (among other things) with the problem of how to encode most efficiently a message as a binary string (a string of 0s and 1s) in order to transmit it across a communication channel that transports binary signals. Shannon defined the information in a message as *the number of bits required, in the most compressed such encoding, to distinguish this message from all other messages that might be transmitted.*

communication theory

general definition of information

symbols and messages

Shannon focused on messages that, before encoding, consist of a sequence of symbols. For an English-language message, each symbol might be a single character (a letter A, B, C, . . . , Z or a space; $N = 27$ distinct symbols in all), and a specific message might be the following sequence of length $L = 19$ characters: "I DO NOT UNDERSTAND". Suppose, for simplicity, that in the possible messages, all N distinct symbols appear with equal frequency (this, of course, is not the case for English-language messages), and suppose that the length of some specific message (its number of symbols) is L. Then the number of bits needed to encode this message and distinguish it from all other possible messages of length L is

$$I = \log_2 N^L = L \log_2 N. \tag{4.68a}$$

In other words, the average number of bits per symbol (the average amount of information per symbol) is

$$\bar{I} = \log_2 N. \tag{4.68b}$$

If there are only two possible symbols, we have one bit per symbol in our message. If there are four possible (equally likely) symbols, we have two bits per symbol, and so forth.

It is usually the case that not all symbols occur with the same frequency in the allowed messages. For example, in English messages the letter "A" occurs with a frequency $p_A \simeq 0.07$, while the letter "Z" occurs with the much smaller frequency $p_Z \simeq 0.001$. All English messages, of character length $L \gg N = 27$, constructed by a typical English speaker, will have these frequencies of occurrence for "A" and "Z". Any purported message with frequencies for "A" and "Z" differing substantially from 0.07 and 0.001 will not be real English messages, and thus need not be included in the binary encoding of messages. As a result, it turns out that the most efficient binary encoding of English messages (the most *compressed* encoding) will use an average number of bits per character somewhat less than $\log_2 N = \log_2 27 = 4.755$. In other words, the average information per character in English language messages is somewhat less than $\log_2 27$.

To deduce the average information per character when the characters do not all occur with the same frequency, we begin with a simple example: the number of distinct characters to be used in the message is just $N = 2$ (the characters "B" and "E"), and their frequencies of occurrence in very long allowed messages are $p_B = 3/5$ and $p_E = 2/5$. For example, in the case of messages with length $L = 100$, the message

$$\begin{aligned}
&\text{EBBEEBBBBEBBBBBBBBBEBEBBBEEEEBBBEB} \\
&\text{BEEEEEBEEBEEEEEEBBEBBBBBBEBBBBBEBBE} \\
&\text{BBBEBBBEEBBBEBBBBBBBEBBBBEBBEEBEB}
\end{aligned} \tag{4.69a}$$

contains 63 Bs and 37 Es, and thus (to within statistical variations) has the correct frequencies $p_B \simeq 0.6$, $p_E \simeq 0.4$ to be an allowed message. By contrast, the message

$$\begin{aligned}
&\text{BBBBBBBBBBBBBBBEBBBBBBBBBBBBBBBBBBBB} \\
&\text{BBBBBBBBBBBEBBBBBBBBBBBBBBBBBBBBBBBB} \\
&\text{BBBBBBBBBBBBBBBEBBBBBBBBBBBBBBBBB}
\end{aligned} \tag{4.69b}$$

contains 97 Bs and 3 Es and thus is not an allowed message. To deduce the number of allowed messages and thence the number of bits required to encode them distinguishably, we map this problem of 60% probable Bs and 40% probable Es onto the problem of messages with five equally probable symbols as follows. Let the set of distinct symbols be the letters "a", "b", "c", "y", and "z", all occurring in allowed messages equally frequently: $p_a = p_b = p_c = p_y = p_z = 1/5$. An example of an allowed message is

$$\begin{aligned}
&\text{zcczzcaabzccbabcccczaybacyzyzcbbyc} \\
&\text{ayyyyyayzcyzzzzzcczacbabybbbcczabz} \\
&\text{bbbybaazybccybaccabazacbzbayycyc}
\end{aligned} \tag{4.69c}$$

We map each such message from our new message set into one from the previous message set by identifying "a", "b", and "c" as from the beginning of the alphabet (and thus converting them to "B"), and identifying "y" and "z" as from the end of the alphabet (and thus converting them to "E"). Our message (4.69c) from the new message set then maps into the message (4.69a) from the old set. This mapping enables us to deduce the number of bits required to encode the old messages, with their unequal frequencies $p_B = 3/5$ and $p_E = 2/5$, from the number required for the new messages, with their equal frequencies $p_a = p_b = \ldots = p_z = 1/5$.

The number of bits needed to encode the new messages, with length $L \gg N_{new} = 5$, is $I = L \log_2 5$. The characters "a," "b," and "c" from the beginning of the alphabet occur $\frac{3}{5}L$ times in each new message (in the limit that L is arbitrarily large). When converting the new message to an old one, we no longer need to distinguish between "a", "b", and "c", so we no longer need the $\frac{3}{5}L \log_2 3$ bits that were being used to make those distinctions. Similarly, the number of bits we no longer need when we drop the distinction between our two end-of-alphabet characters "y" and "z" is $\frac{2}{5}L \log_2 2$. As a result, the number of bits still needed to distinguish between old messages (messages with "B" occurring 3/5 of the time and "E" occurring 2/5 of the time) is

$$I = L \log_2 5 - \frac{3}{5}L \log_2 3 - \frac{2}{5} \log_2 2 = L \left[-\frac{3}{5} \log_2 \frac{3}{5} - \frac{2}{5} \log_2 \frac{2}{5} \right]$$

$$= L(-p_B \log_2 p_B - p_E \log_2 p_E). \tag{4.69d}$$

A straightforward generalization of this argument (Ex. 4.17) shows that, *when one constructs messages with very large length $L \gg N$ from a pool of N symbols that occur with frequencies $p_1, p_2, \ldots, p_N$, the minimum number of bits required to distinguish all the allowed messages from one another (i.e., the amount of information in each message) is*

$$I = L \sum_{n=1}^{N} -p_n \log_2 p_n; \tag{4.70}$$

so *the average information per symbol in the message is*

information per symbol in a message

$$\boxed{\bar{I} = \sum_{n=1}^{N} -p_n \log_2 p_n = (1/\ln 2) \sum_{n=1}^{N} -p_n \ln p_n.} \tag{4.71}$$

4.11.3 Examples of Information Content [T2]

Notice the similarity of the general information formula (4.70) to the general formula (4.34) for the entropy of an arbitrary ensemble. This similarity has a deep consequence.

INFORMATION FROM MEASURING THE STATE OF ONE SYSTEM IN AN ENSEMBLE

Consider an arbitrary ensemble of systems in statistical mechanics. As usual, label the quantum states available to each system by the integer $n = 1, 2, \ldots, N_{states}$, and denote by p_n the probability that any chosen system in the ensemble will turn out to be in state n. Now select one system out of the ensemble and measure its quantum state n_1;

select a second system and measure its state, n_2. Continue this process until some large number $L \gg N_{states}$ of systems have been measured. The sequence of measurement results $\{n_1, n_2, \ldots, n_L\}$ can be regarded as a message. The minimum number of bits needed to distinguish this message from all other possible such messages is given by the general information formula (4.70). This is the total information in the L system measurements. Correspondingly, *the amount of information we get from measuring the state of one system (the average information per measurement) is given by Eq. (4.71).* This acquired information is related to the entropy of the system before measurement [Eq. (4.34)] by the same standard formula (4.67b) as we obtained earlier for the special case of the microcanonical ensemble:

information from measurement for a general ensemble

$$\bar{I} = S/(k_B \ln 2). \tag{4.72}$$

INFORMATION IN AN ENGLISH-LANGUAGE MESSAGE

For another example of information content, we return to English-language messages (Shannon, 1948). Evaluating the information content of a long English message is a very difficult task, since it requires figuring out how to compress the message most compactly. We shall make a series of estimates.

A crude initial estimate of the information per character is that obtained by idealizing all the characters as occurring equally frequently: $\bar{I} = \log_2 27 \simeq 4.76$ bits per character [Eq. (4.68b)]. This is an overestimate, because the 27 characters actually occur with very different frequencies. We could get a better estimate by evaluating $\bar{I} = \sum_{n=1}^{27} -p_n \log_2 p_n$, taking account of the characters' varying frequencies p_n (the result is about $\bar{I} = 4.1$), but we can do even better by converting from characters as our symbols to words. The average number of characters in an English word is about 4.5 letters plus 1 space, or 5.5 characters per word. We can use this number to convert from characters as our symbols to words. The number of words in a typical English speaker's vocabulary is roughly 12,000. If we idealize these 12,000 words as occurring with the same frequencies, then the information per word is $\log_2 12{,}000 \simeq 13.6$, so the information per character is $\bar{I} = (1/5.5) \log_2 12{,}000 \simeq 2.46$. This is much smaller than our previous estimates. A still better estimate is obtained by using Zipf's (1935) approximation $p_n = 0.1/n$ of the frequencies of occurrence of the words in English messages.[21] To ensure that $\sum_{n=1}^{N} p_n = 1$ for Zipf's approximation, we require that the number of words be $N = 12{,}367$. We then obtain, as our improved estimate of the information per word, $\sum_{n=1}^{12,367}(-0.1/n) \log_2(0.1/n) = 9.72$, corresponding to a value of information per character $\bar{I} \simeq 9.72/5.5 = 1.77$. This is substantially smaller than our initial, crudest estimate of 4.76 and is close to more careful estimates $\bar{I} \simeq 1.0$ to 1.5 (Schneier, 1997, Sec. 11.1).

information per character in English message

---

21. The most frequently occurring word is "THE", and its frequency is about 0.1 (1 in 10 words is "THE" in a long message). The next most frequent words are "OF", "AND", and "TO"; their frequencies are about 0.1/2, 0.1/3, and 0.1/4, respectively; and so forth.

4.11.4 Some Properties of Information `T2`

Because of the similarity of the general formulas for information and entropy (both proportional to $\sum_n -p_n \ln p_n$), information has very similar properties to entropy. In particular (Ex. 4.18):

1. Information is additive (just as entropy is additive). The information in two successive, independent messages is the sum of the information in each message.

2. If the frequencies of occurrence of the symbols in a message are $p_n = 0$ for all symbols except one, which has $p_n = 1$, then the message contains zero information. This is analogous to the vanishing entropy when all states have zero probability except for one, which has unit probability.

3. For a message L symbols long, whose symbols are drawn from a pool of N distinct symbols, the information content is maximized if the probabilities of the symbols are all equal ($p_n = 1/N$), and the maximal value of the information is $I = L \log_2 N$. This is analogous to the microcanonical ensemble having maximal entropy.

4.11.5 Capacity of Communication Channels; Erasing Information from Computer Memories `T2`

NOISELESS COMMUNICATION

A noiseless communication channel has a maximum rate (number of bits per second) at which it can transmit information. This rate is called the *channel capacity* and is denoted C. When one subscribes to a cable internet connection in the United States, one typically pays a monthly fee that depends on the connection's channel capacity; for example, in Pasadena, California, in summer 2011 the fee was \$29.99 per month for a connection with capacity $C = 12$ megabytes/s $= 96$ megabits/s, and \$39.99 for $C = 144$ megabits/s. (This was a 30-fold increase in capacity per dollar since 2003!)

> **channel capacity of noiseless communication**

It should be obvious from the way we have defined the information I in a message that the maximum rate at which we can transmit optimally encoded messages, each with information content I, is C/I messages per second.

NOISY COMMUNICATION

When a communication channel is noisy (as all channels actually are), for high-confidence transmission of messages one must put some specially designed redundancy into one's encoding. With cleverness, one can thereby identify and correct errors in a received message, caused by the noise (error-correcting code); see, for example, Shannon (1948), Raisbeck (1963), and Pierce (2012).[22] The redundancy needed for such error identification and correction reduces the channel's capacity. As an example, consider a *symmetric binary channel:* one that carries messages made of 0s and

---

22. A common form of error-correcting code is based on parity checks.

1s, with equal frequency $p_0 = p_1 = 0.5$, and whose noise randomly converts a 0 into a 1 with some small error probability p_e, and randomly converts a 1 into 0 with that same probability p_e. Then one can show (e.g., Pierce, 2012; Raisbeck, 1963) that the channel capacity is reduced—by the need to find and correct for these errors—by a factor

<aside>reduction of channel capacity due to error correction</aside>

$$C = C_{\text{noiseless}}[1 - \bar{I}(p_e)], \quad \text{where } \bar{I}(p_e) = -p_e \log_2 p_e - (1 - p_e) \log_2(1 - p_e).$$

$$(4.73)$$

Note that the fractional reduction of capacity is by the amount of information per symbol in messages made from symbols with frequencies equal to the probabilities p_e of making an error and $1 - p_e$ of not making an error—a remarkable and nontrivial conclusion! This is one of many important results in communication theory.

MEMORY AND ENTROPY

Information is also a key concept in the theory of computation. As an important example of the relationship of information to entropy, we cite Landauer's (1961, 1991) theorem: In a computer, when one erases L bits of information from memory, one necessarily increases the entropy of the memory and its environment by at least $\Delta S = L k_B \ln 2$ and correspondingly, one increases the thermal energy (heat) of the memory and environment by at least $\Delta Q = T \Delta S = L k_B T \ln 2$ (Ex. 4.21).

<aside>Landauer's theorem: entropy increase due to erasure</aside>

EXERCISES

Exercise 4.17 *Derivation: Information per Symbol When Symbols Are Not Equally Probable* **T2**

Derive Eq. (4.70) for the average number of bits per symbol in a long message constructed from N distinct symbols, where the frequency of occurrence of symbol n is p_n. [Hint: Generalize the text's derivation of Eq. (4.69d).]

Exercise 4.18 *Derivation: Properties of Information* **T2**

Prove the properties of entropy enumerated in Sec. 4.11.4.

Exercise 4.19 *Problem: Information per Symbol for Messages Built from Two Symbols* **T2**

Consider messages of length $L \gg 2$ constructed from just two symbols ($N = 2$), which occur with frequencies p and $(1 - p)$. Plot the average information per symbol $\bar{I}(p)$ in such messages, as a function of p. Explain why your plot has a maximum $\bar{I} = 1$ when $p = 1/2$, and has $\bar{I} = 0$ when $p = 0$ and when $p = 1$. (Relate these properties to the general properties of information.)

Exercise 4.20 *Problem: Information in a Sequence of Dice Throws* **T2**

Two dice are thrown randomly, and the sum of the dots showing on the upper faces is computed. This sum (an integer n in the range $2 \le n \le 12$) constitutes a symbol, and the sequence of results of $L \gg 12$ throws is a message. Show that the amount of information per symbol in this message is $\bar{I} \simeq 3.2744$.

Exercise 4.21 *Derivation: Landauer's Theorem* T2
Derive, or at least give a plausibility argument for, Landauer's theorem (stated at the end of Sec. 4.11.5).

Bibliographic Note

Statistical mechanics has inspired a variety of readable and innovative texts. The classic treatment is Tolman (1938). Classic elementary texts are Kittel (2004) and Kittel and Kroemer (1980). Among more modern approaches that deal in much greater depth with the topics covered in this chapter are Lifshitz and Pitaevskii (1980), Chandler (1987), Sethna (2006), Kardar (2007), Reif (2008), Reichl (2009), and Pathria and Beale (2011). The Landau-Lifshitz textbooks (including Lifshitz and Pitaevskii, 1980) are generally excellent after one has already learned the subject at a more elementary level. A highly individual and advanced treatment, emphasizing quantum statistical mechanics, is Feynman (1972). A particularly readable account in which statistical mechanics is used heavily to describe the properties of solids, liquids, and gases is Goodstein (2002). Readable, elementary introductions to information theory are Raisbeck (1963) and Pierce (2012); an advanced text is McEliece (2002).

5

Statistical Thermodynamics

One of the principal objects of theoretical research is to find the point of view from which the subject appears in the greatest simplicty.

J. WILLARD GIBBS (1881)

5.1 Overview

In Chap. 4, we introduced the concept of statistical equilibrium and briefly studied some properties of equilibrated systems. In this chapter, we develop the theory of statistical equilibrium in a more thorough way.

The title of this chapter, "Statistical Thermodynamics," emphasizes two aspects of the theory of statistical equilibrium. The term *thermodynamics* is an ancient one that predates statistical mechanics. It refers to a study of the macroscopic properties of systems that are in or near equilibrium, such as their energy and entropy. Despite paying no attention to the microphysics, classical thermodynamics is a very powerful theory for deriving general relationships among macroscopic properties. Microphysics influences the macroscopic world in a statistical manner, so in the late nineteenth century, Willard Gibbs and others developed statistical mechanics and showed that it provides a powerful conceptual underpinning for classical thermodynamics. The resulting synthesis, *statistical thermodynamics,* adds greater power to thermodynamics by augmenting it with the statistical tools of ensembles and distribution functions.

In our study of statistical thermodynamics, we restrict attention to an ensemble of large systems that are in statistical equilibrium. By "large" is meant a system that can be broken into a large number N_{ss} of subsystems that are all macroscopically identical to the full system except for having $1/N_{ss}$ as many particles, $1/N_{ss}$ as much volume, $1/N_{ss}$ as much energy, $1/N_{ss}$ as much entropy, and so forth. (Note that this definition constrains the energy of interaction between the subsystems to be negligible.) Examples are 1 kg of plasma in the center of the Sun and a 1-kg sapphire crystal.

The equilibrium thermodynamic properties of any type of large system (e.g., an ideal gas)[1] can be derived using any one of the statistical equilibrium ensembles of the last chapter (microcanonical, canonical, grand canonical, or Gibbs). For example,

---

1. An *ideal gas* is one with negligible interactions among its particles.

each of these ensembles will predict the same equation of state $P = (N/V)k_B T$ for an ideal gas, even though in one ensemble each system's number of particles N is precisely fixed, while in another ensemble N can fluctuate so that strictly speaking, one should write the equation of state as $P = (\overline{N}/V)k_B T$, with $\overline{N}$ the ensemble average of N. (Here and throughout this chapter, for compactness we use bars rather than brackets to denote ensemble averages, i.e., $\overline{N}$ rather than $\langle N \rangle$.) The equations of state are the same to very high accuracy because the fractional fluctuations of N are so extremely small: $\Delta N/N \sim 1/\sqrt{\overline{N}}$ (cf. Ex. 5.11).

Although the thermodynamic properties are independent of the equilibrium ensemble, specific properties are often derived most quickly, and the most insight usually accrues, from the ensemble that most closely matches the physical situation being studied. In Secs. 5.2–5.5, we use the microcanonical, grand canonical, canonical, and Gibbs ensembles to derive many useful results from statistical thermodynamics: fundamental potentials expressed as statistical sums over quantum states, variants of the first law of thermodynamics, equations of state, Maxwell relations, Euler's equation, and others. Table 5.1 summarizes the most important of those statistical-equilibrium results and some generalizations of them. Readers are advised to delay studying this table until they have read further into the chapter.

As we saw in Chap. 4, when systems are out of statistical equilibrium, their evolution toward equilibrium is driven by the law of entropy increase—the second law of thermodynamics. In Sec. 5.5, we formulate the fundamental potential (Gibbs potential) for an out-of-equilibrium ensemble that interacts with a heat and volume bath, and we discover a simple relationship between that fundamental potential and the entropy of system plus bath. From that relationship, we learn that in this case the second law is equivalent to a law of decrease of the Gibbs potential. As applications, we learn how chemical potentials drive chemical reactions and phase transitions. In Sec. 5.6, we discover how the Gibbs potential can be used to study spontaneous fluctuations of a system away from equilibrium, when it is coupled to a heat and particle bath.

TABLE 5.1: Representations and ensembles for systems in statistical equilibrium, in relativistic notation

Representation and ensemble	First law	Quantities exchanged with bath	Distribution function ρ
Energy and microcanonical (Secs. 4.5 and 5.2)	$d\mathcal{E} = T dS + \tilde{\mu} dN - P dV$	None	$\text{const} = e^{-S/k_B}$ $\mathcal{E}$ const in $\delta\mathcal{E}$
Enthalpy (Exs. 5.5 and 5.13)	$dH = T dS + \tilde{\mu} dN + V dP$	V and $\mathcal{E}$ $d\mathcal{E} = -P dV$	$\text{const} = e^{-S/k_B}$ H const
Physical free energy and canonical (Secs. 4.4.1 and 5.4)	$dF = -S dT + \tilde{\mu} dN - P dV$	$\mathcal{E}$	$e^{(F-\mathcal{E})/(k_B T)}$
Gibbs (Secs. 4.4.2 and 5.5)	$dG = -S dT + \tilde{\mu} dN + V dP$	$\mathcal{E}$ and V	$e^{(G-\mathcal{E}-PV)/(k_B T)}$
Grand canonical (Secs. 4.4.2 and 5.3)	$d\Omega = -S dT - N d\tilde{\mu} - P dV$	$\mathcal{E}$ and N	$e^{(\Omega-\mathcal{E}+\tilde{\mu}N)/(k_B T)}$

Notes: The nonrelativistic formulas are the same but with the rest masses of particles removed from the chemical potentials ($\tilde{\mu} \to \mu$) and from all fundamental potentials except Ω ($\mathcal{E} \to E$, but no change of notation for H, F, and G). This table will be hard to understand until after reading the sections referenced in column one.

In Sec. 5.7, we employ these tools to explore fluctuations and the gas-to-liquid phase transition for a model of a real gas due to the Dutch physicist Johannes van der Waals. Out-of-equilibrium aspects of statistical mechanics (evolution toward equilibrium and fluctuations away from equilibrium) are summarized in Table 5.2 and discussed in Secs. 5.5.1 and 5.6, not just for heat and volume baths, but for a variety of baths.

Deriving the macroscopic properties of real materials by statistical sums over their quantum states can be formidably difficult. Fortunately, in recent years some powerful approximation techniques have been devised for performing the statistical sums. In Secs. 5.8.3 and 5.8.4, we give the reader the flavor of two of these techniques: the *renormalization group* and *Monte Carlo methods*. We illustrate and compare these techniques by using them to study a phase transition in a simple model for ferromagnetism called the *Ising model*.

5.2 Microcanonical Ensemble and the Energy Representation of Thermodynamics

5.2

5.2.1 Extensive and Intensive Variables; Fundamental Potential

5.2.1

Consider a microcanonical ensemble of large, closed systems that have attained statistical equilibrium. We can describe the ensemble macroscopically using a set of thermodynamic variables. These variables can be divided into two classes: *extensive variables* (Sec. 4.4.1), which double if one doubles the system's size, and *intensive variables* (Sec. 4.4.2), whose magnitudes are independent of the system's size. Familiar examples of extensive variables are a system's total energy $\mathcal{E}$, entropy S, volume V, and number of conserved particles of various species N_I. Corresponding examples of

extensive and intensive variables

intensive variables are temperature T, pressure P, and the chemical potentials $\tilde{\mu}_I$ for various species of particles.

complete set of extensive variables

For a large, closed system, there is a *complete set of extensive variables that we can specify independently*—usually its volume V, total energy $\mathcal{E}$ or entropy S, and number N_I of particles of each species I. The values of the other extensive variables and all the intensive variables are determined in terms of this complete set by methods that we shall derive.

The particle species I in the complete set must only include those species whose particles are conserved on the timescales of interest. For example, if photons can be emitted and absorbed, then one must not specify N_γ, the number of photons; rather, N_γ will come to an equilibrium value that is governed by the values of the other extensive variables. Also, one must omit from the set $\{I\}$ any conserved particle species whose numbers are automatically determined by the numbers of other, included species. For example, gas inside the Sun is always charge neutral to very high precision, and therefore (neglecting all elements except hydrogen and helium), the number of electrons N_e in a sample of gas is determined by the number of protons N_p and the number of helium nuclei (alpha particles) N_α: $N_e = N_p + 2N_\alpha$. Therefore, if one includes N_p and N_α in the complete set of extensive variables being used, one must omit N_e.

As in Chap. 4, we formulate the theory relativistically correctly but formulate it solely in the mean rest frames of the systems and baths being studied. Correspondingly, in our formulation we generally include the particle rest masses m_I in the total energy $\mathcal{E}$ and in the chemical potentials $\tilde{\mu}_I$. For very nonrelativistic systems, however, we usually replace $\mathcal{E}$ by the nonrelativistic energy $E \equiv \mathcal{E} - \sum_I N_I m_I c^2$ and $\tilde{\mu}_I$ by the nonrelativistic chemical potential $\mu_I \equiv \tilde{\mu}_I - m_I c^2$ (though, as we shall see in Sec. 5.5 when studying chemical reactions, the identification of the appropriate rest mass m_I to subtract is a delicate issue).

5.2.2 Energy as a Fundamental Potential

For simplicity, we temporarily specialize to a microcanonical ensemble of one-species systems, which all have the same values of a complete set of three extensive variables: the energy $\mathcal{E}$,[2] number of particles N, and volume V. Suppose that the microscopic nature of the ensemble's systems is known. Then, at least in principle and often in practice, one can identify from that microscopic nature the quantum states that are available to the system (given its constrained values of $\mathcal{E}$, N, and V), one can count those quantum states, and from their total number N_{states} one can compute the ensemble's total entropy $S = k_B \ln N_{\text{states}}$ [Eq. (4.35)]. The resulting entropy can be regarded as a function of the complete set of extensive variables,

$$S = S(\mathcal{E}, N, V), \tag{5.1}$$

---

2. In practice, as illustrated in Ex. 4.7, one must allow $\mathcal{E}$ to fall in some tiny but finite range $\delta\mathcal{E}$ rather than constraining it precisely, and one must then check to be sure that the results of the analysis are independent of $\delta\mathcal{E}$.

and this equation can then be inverted to give the total energy in terms of the entropy and the other extensive variables:

$$\boxed{\mathcal{E} = \mathcal{E}(S, N, V).}$$ (5.2)

We call the energy $\mathcal{E}$, viewed as a function of S, N, and V, the *fundamental thermo-dynamic potential for the microcanonical ensemble*. When using this fundamental potential, we regard S, N, and V as our complete set of extensive variables rather than $\mathcal{E}$, N, and V. From the fundamental potential, as we shall see, one can deduce all other thermodynamic properties of the system.

energy as fundamental thermodynamic potential for microcanonical ensemble

5.2.3 Intensive Variables Identified Using Measuring Devices; First Law of Thermodynamics

5.2.3

TEMPERATURE

In Sec. 4.4.1, we used kinetic-theory considerations to identify the thermodynamic temperature T of the canonical ensemble [Eq. (4.20)]. It is instructive to discuss how this temperature arises in the microcanonical ensemble. Our discussion makes use of an idealized *thermometer* consisting of an idealized atom that has only two quantum states, $|0\rangle$ and $|1\rangle$, with energies $\mathcal{E}_0$ and $\mathcal{E}_1 = \mathcal{E}_0 + \Delta\mathcal{E}$. The atom, initially in its ground state, is brought into thermal contact with one of the large systems of our microcanonical ensemble and then monitored over time as it is stochastically excited and deexcited. The ergodic hypothesis (Sec. 4.6) guarantees that the atom traces out a history of excitation and deexcitation that is governed statistically by the canonical ensemble for a collection of such atoms exchanging energy (heat) with our large system (the heat bath). More specifically, if T is the (unknown) temperature of our system, then the fraction of time the atom spends in its excited state, divided by the fraction spent in its ground state, is equal to the canonical distribution's probability ratio:

idealized thermometer

$$\frac{\rho_1}{\rho_0} = \frac{e^{-\mathcal{E}_1/(k_BT)}}{e^{-\mathcal{E}_0/(k_BT)}} = e^{-\Delta\mathcal{E}/(k_BT)}$$ (5.3a)

[cf. Eq. (4.20)].

This ratio can also be computed from the properties of the full system augmented by the two-state atom. This augmented system is microcanonical with a total energy $\mathcal{E} + \mathcal{E}_0$, since the atom was in the ground state when first attached to the full system. Of all the quantum states available to this augmented system, the ones in which the atom is in the ground state constitute a total number $N_0 = e^{S(\mathcal{E}, N, V)/k_B}$; and those with the atom in the excited state constitute a total number $N_1 = e^{S(\mathcal{E}-\Delta\mathcal{E}, N, V)/k_B}$. Here we have used the fact that the number of states available to the augmented system is equal to that of the original, huge system with energy $\mathcal{E}$ or $\mathcal{E} - \Delta\mathcal{E}$ (since the atom, in each of the two cases, is forced to be in a unique state), and we have expressed that number of states of the original system, for each of the two cases, in terms of the original system's entropy function [Eq. (5.1)]. The ratio of the number of states N_1/N_0 is (by the ergodic hypothesis) the ratio of the time that the augmented system spends with the atom excited to the time spent with the atom in its ground state (i.e., it is equal to ρ_1/ρ_0):

$$\frac{\rho_1}{\rho_0} = \frac{N_1}{N_0} = \frac{e^{S(\mathcal{E}-\Delta\mathcal{E},N,V)/k_B}}{e^{S(\mathcal{E},N,V)/k_B}} = \exp\left[-\frac{\Delta\mathcal{E}}{k_B}\left(\frac{\partial S}{\partial \mathcal{E}}\right)_{N,V}\right]. \tag{5.3b}$$

By equating Eqs. (5.3a) and (5.3b), we obtain an expression for the original system's temperature T in terms of the partial derivative $(\partial\mathcal{E}/\partial S)_{N,V}$ of its fundamental potential $\mathcal{E}(S, N, V)$:

temperature from energy potential $\mathcal{E}(S, N, V)$

$$T = \frac{1}{(\partial S/\partial \mathcal{E})_{N,V}} = \left(\frac{\partial \mathcal{E}}{\partial S}\right)_{N,V}, \tag{5.3c}$$

where we have used Eq. (3) of Box 5.2.

CHEMICAL POTENTIAL AND PRESSURE

A similar thought experiment—using a highly idealized measuring device that can exchange one particle ($\Delta N = 1$) with the system but cannot exchange any energy with it—gives for the fraction of the time spent with the extra particle in the measuring device ("state 1") and in the system ("state 0"):

$$\frac{\rho_1}{\rho_0} = e^{\tilde{\mu}\Delta N/k_B T}$$

$$= \frac{e^{S(\mathcal{E},N-\Delta N,V)/k_B}}{e^{S(\mathcal{E},N,V)/k_B}} = \exp\left[-\frac{\Delta N}{k_B}\left(\frac{\partial S}{\partial N}\right)_{\mathcal{E},V}\right]. \tag{5.4a}$$

Here the first expression is computed from the viewpoint of the measuring device's equilibrium ensemble,[3] and the second from the viewpoint of the combined system's microcanonical ensemble. Equating these two expressions, we obtain

chemical potential from energy potential

$$\tilde{\mu} = -T\left(\frac{\partial S}{\partial N}\right)_{\mathcal{E},V} = \left(\frac{\partial \mathcal{E}}{\partial N}\right)_{S,V}. \tag{5.4b}$$

In the last step we use Eq. (5.3c) and Eq. (4) of Box 5.2. The reader should be able to construct a similar thought experiment involving an idealized pressure transducer (Ex. 5.1), which yields the following expression for the system's pressure:

pressure from energy potential

$$P = -\left(\frac{\partial \mathcal{E}}{\partial V}\right)_{S,N}. \tag{5.5}$$

FIRST LAW OF THERMODYNAMICS

Having identifed the three intensive variables T, $\tilde{\mu}$, and P as partial derivatives [Eqs. (5.3c), (5.4b), and (5.5)], we now see that the fundamental potential's differential relation

$$d\mathcal{E}(S, N, V) = \left(\frac{\partial \mathcal{E}}{\partial S}\right)_{N,V} dS + \left(\frac{\partial \mathcal{E}}{\partial N}\right)_{S,V} dN + \left(\frac{\partial \mathcal{E}}{\partial V}\right)_{S,N} dV \tag{5.6}$$

---

3. This ensemble has $\rho = \text{constant } e^{-\tilde{\mu}N/(k_B T)}$, since only particles can be exchanged with the device's heat bath (our system).

BOX 5.2. TWO USEFUL RELATIONS BETWEEN PARTIAL DERIVATIVES

Expand a differential increment in the energy $\mathcal{E}(S, N, V)$ in terms of differentials of its arguments S, N, and V:

$$d\mathcal{E}(S, N, V) = \left(\frac{\partial \mathcal{E}}{\partial S}\right)_{N,V} dS + \left(\frac{\partial \mathcal{E}}{\partial N}\right)_{S,V} dN + \left(\frac{\partial \mathcal{E}}{\partial V}\right)_{S,N} dV. \quad (1)$$

Next expand the entropy $S(\mathcal{E}, N, V)$ similarly, and substitute the resulting expression for dS into the above equation to obtain

$$d\mathcal{E} = \left(\frac{\partial \mathcal{E}}{\partial S}\right)_{N,V} \left(\frac{\partial S}{\partial \mathcal{E}}\right)_{N,V} d\mathcal{E}$$

$$+ \left[\left(\frac{\partial \mathcal{E}}{\partial S}\right)_{N,V} \left(\frac{\partial S}{\partial N}\right)_{E,V} + \left(\frac{\partial \mathcal{E}}{\partial N}\right)_{S,V}\right] dN$$

$$+ \left[\left(\frac{\partial \mathcal{E}}{\partial S}\right)_{N,V} \left(\frac{\partial S}{\partial V}\right)_{N,E} + \left(\frac{\partial \mathcal{E}}{\partial V}\right)_{S,N}\right] dV. \quad (2)$$

Noting that this relation must be satisfied for all values of $d\mathcal{E}$, dN, and dV, we conclude that

$$\left(\frac{\partial \mathcal{E}}{\partial S}\right)_{N,V} = \frac{1}{(\partial S/\partial \mathcal{E})_{N,V}}, \quad (3)$$

$$\left(\frac{\partial \mathcal{E}}{\partial N}\right)_{S,V} = -\left(\frac{\partial \mathcal{E}}{\partial S}\right)_{N,V} \left(\frac{\partial S}{\partial N}\right)_{\mathcal{E},V}, \quad (4)$$

and so forth; similarly for other pairs and triples of partial derivatives.

These equations, and their generalization to other variables, are useful in manipulations of thermodynamic equations.

is nothing more nor less than the ordinary first law of thermodynamics

$$\boxed{d\mathcal{E} = TdS + \tilde{\mu}dN - PdV} \quad (5.7)$$

first law of thermo-dynamics from energy potential

(cf. Table 5.1 above).

Notice the pairing of intensive and extensive variables in this first law: temperature T is paired with entropy S; chemical potential $\tilde{\mu}$ is paired with number of particles N; and pressure P is paired with volume V. We can think of each intensive variable as a "generalized force" acting on its corresponding extensive variable to change the energy of the system. We can add additional pairs of intensive and extensive variables if appropriate, calling them X_A, Y_A (e.g., an externally imposed magnetic field **B** and a material's magnetization **M**; Sec. 5.8). We can also generalize to a multicomponent system (i.e., one that has several types of conserved particles with numbers N_I and associated chemical potentials $\tilde{\mu}_I$). We can convert to nonrelativistic language by subtracting off the rest-mass contributions (switching from $\mathcal{E}$ to $E \equiv \mathcal{E} - \sum N_I m_I c^2$ and from $\tilde{\mu}_I$ to $\mu_I = \tilde{\mu}_I - m_I c^2$). The result is the nonrelativistic, extended first law:

extended first law

$$dE = TdS + \sum_I \mu_I dN_I - PdV + \sum_A X_A dY_A. \qquad (5.8)$$

EXERCISES

Exercise 5.1 *Problem: Pressure-Measuring Device*
For the microcanonical ensemble considered in this section, derive Eq. (5.5) for the pressure using a thought experiment involving a pressure-measuring device.

5.2.4

5.2.4 Euler's Equation and Form of the Fundamental Potential

We can integrate the differential form of the first law to obtain a remarkable—though essentially trivial—relation known as *Euler's equation*. We discuss this for the one-species system whose first law is $d\mathcal{E} = TdS + \tilde{\mu}dN - PdV$. The generalization to other systems should be obvious.

We decompose our system into a large number of subsystems in equilibrium with one another. As they are in equilibrium, they will all have the same values of the intensive variables T, $\tilde{\mu}$, and P; therefore, if we add up all their energies $d\mathcal{E}$ to obtain $\mathcal{E}$, their entropies dS to obtain S, and so forth, we obtain from the first law (5.7)[4]

Euler's equation for energy

$$\mathcal{E} = TS + \tilde{\mu}N - PV. \qquad (5.9a)$$

Since the energy $\mathcal{E}$ is itself extensive, Euler's equation (5.9a) must be expressible as

$$\mathcal{E} = Nf(V/N, S/N) \qquad (5.9b)$$

for some function f. This is a useful functional form for the fundamental potential $\mathcal{E}(N, V, S)$. For example, for a nonrelativistic, classical, monatomic perfect gas,[5] the

---

4. There are a few (but very few!) systems for which some of the thermodynamic laws, including Euler's equation, take on forms different from those presented in this chapter. A black hole is an example (cf. Sec. 4.10.2). A black hole cannot be divided up into subsystems, so the above derivation of Euler's equation fails. Instead of increasing linearly with the mass $M_H = \mathcal{E}/c^2$ of the hole, the hole's extensive variables $S_H =$ (entropy) and $J_H =$ (spin angular momentum) increase quadratically with M_H. And instead of being independent of the hole's mass, the intensive variables $T_H =$ (temperature) and $\Omega_H =$ (angular velocity) scale as $1/M_H$. See, e.g., Tranah and Landsberg (1980) and see Sec. 4.10.2 for some other aspects of black-hole thermodynamics.
5. Recall (Ex. 4.6) that a perfect gas is one that is ideal (i.e., has negligible interactions among its particles) and whose particles have no excited internal degrees of freedom. The phrase "perfect gas" must not be confused with "perfect fluid" (a fluid whose viscosity is negligible so its stress tensor, in its rest frame, consists solely of an isotropic pressure).

Sackur-Tetrode equation (4.42) can be solved for $E = \mathcal{E} - Nmc^2$ to get the following form of the fundamental potential:

energy potential for nonrelativistic, classical, perfect gas

$$E(V, S, N) = N \left(\frac{3h^2}{4\pi m g_s^{2/3}} \right) \left(\frac{V}{N} \right)^{-2/3} \exp\left(\frac{2}{3k_B} \frac{S}{N} - \frac{5}{3} \right). \qquad (5.9c)$$

Here m is the mass of each of the gas's particles, and h is Planck's constant.

5.2.5 Everything Deducible from First Law; Maxwell Relations

5.2.5

There is no need to memorize a lot of thermodynamic relations; *nearly all relations can be deduced almost trivially from the functional form of the first law of thermodynamics,* the main formula shown on the first line of Table 5.1.

For example, in the case of our simple one-species system, the first law $d\mathcal{E} = TdS + \tilde{\mu}dN - PdV$ tells us that the system energy $\mathcal{E}$ should be regarded as a function of the things that appear as differentials on the right-hand side: S, N, and V; that is, the fundamental potential must have the form $\mathcal{E} = \mathcal{E}(S, N, V)$. By thinking about building up our system from smaller systems by adding entropy dS, particles dN, and volume dV at fixed values of the intensive variables, we immediately deduce, from the first law, the Euler equation $\mathcal{E} = TS + \tilde{\mu}N - PV$. By writing out the differential relation (5.6)—which is just elementary calculus—and comparing with the first law, we immediately read off the intensive variables in terms of partial derivatives of the fundamental potential:

$$T = \left(\frac{\partial \mathcal{E}}{\partial S} \right)_{V,N}, \qquad \tilde{\mu} = \left(\frac{\partial \mathcal{E}}{\partial N} \right)_{V,S}, \qquad P = -\left(\frac{\partial \mathcal{E}}{\partial V} \right)_{S,N}. \qquad (5.10a)$$

We can then go on to notice that the resulting $P(V, S, N)$, $T(V, S, N)$, and $\tilde{\mu}(V, S, N)$ are not all independent. The equality of mixed partial derivatives (e.g., $\partial^2 \mathcal{E}/\partial V \partial S = \partial^2 \mathcal{E}/\partial S \partial V$) together with Eqs. (5.10a) implies that they must satisfy the following *Maxwell relations:*

$$\left(\frac{\partial T}{\partial N} \right)_{S,V} = \left(\frac{\partial \tilde{\mu}}{\partial S} \right)_{N,V}, \quad -\left(\frac{\partial P}{\partial S} \right)_{V,N} = \left(\frac{\partial T}{\partial V} \right)_{S,N}, \quad \left(\frac{\partial \tilde{\mu}}{\partial V} \right)_{N,S} = -\left(\frac{\partial P}{\partial N} \right)_{V,S}.$$

Maxwell relations from energy potential

$$(5.10b)$$

Additional relations can be generated using the types of identities proved in Box 5.2—or they can be generated more easily by applying the above procedure to the fundamental potentials associated with other ensembles; see Secs. 5.3–5.5. All equations of state [i.e., all such relations as Eqs. (5.11) between intensive and extensive variables] must satisfy the Maxwell relations. For our simple example of a nonrelativistic, classical, perfect gas, we can substitute the fundamental potential E [Eq. (5.9c)] into Eqs. (5.10a) to obtain

$$T(V, S, N) = \left(\frac{h^2}{2\pi m k_B g_s^{2/3}} \right) \left(\frac{N}{V} \right)^{2/3} \exp \left(\frac{2S}{3k_B N} - \frac{5}{3} \right),$$

$$\mu(V, S, N) = \left(\frac{h^2}{4\pi m g_s^{2/3}} \right) \left(\frac{N}{V} \right)^{2/3} \left(5 - 2\frac{S}{k_B N} \right) \exp \left(\frac{2S}{3k_B N} - \frac{5}{3} \right)$$

$$P(V, S, N) = \left(\frac{h^2}{2\pi m g_s^{2/3}} \right) \left(\frac{N}{V} \right)^{5/3} \exp \left(\frac{2S}{3k_B N} - \frac{5}{3} \right), \tag{5.11}$$

(Ex. 5.2). These clearly do satisfy the Maxwell relations.

EXERCISES

Exercise 5.2 *Derivation: Energy Representation for a Nonrelativistic, Classical, Perfect Gas*

(a) Use the fundamental potential $E(V, S, N)$ for the nonrelativistic, classical, perfect gas [Eq. (5.9c)] to derive Eqs. (5.11) for the gas pressure, temperature, and chemical potential.

(b) Show that these equations of state satisfy Maxwell relations (5.10b).

(c) Combine these equations of state to obtain the ideal-gas equation of state

$$P = \frac{N}{V} k_B T, \tag{5.12}$$

which we derived in Ex. 3.8 using kinetic theory.

5.2.6

5.2.6 Representations of Thermodynamics

The treatment of thermodynamics given in this section is called the *energy representation*, because it is based on the fundamental potential $\mathcal{E}(S, V, N)$ in which the energy is expressed as a function of the complete set of extensive variables $\{S, V, N\}$. As we have seen, this energy representation is intimately related to the microcanonical ensemble. In Sec. 5.3, we meet the grand-potential representation for thermodynamics, which is intimately related to the grand canonical ensemble for systems of volume V in equilibrium with a heat and particle bath that has temperature T and chemical potential $\tilde{\mu}$. Then in Secs. 5.4 and 5.5, we meet the two representations of thermodynamics that are intimately related to the canonical and Gibbs ensembles, and discover their power to handle certain special issues. And in Ex. 5.5, we consider a representation and ensemble based on enthalpy. These five representations and their ensembles are summarized in Table 5.1 above.

5.3 Grand Canonical Ensemble and the Grand-Potential Representation of Thermodynamics

We now turn to the grand canonical ensemble, and its grand-potential representation of thermodynamics, for a semiclosed system that exchanges heat and particles with a thermalized bath. For simplicity, we assume that all particles are identical (just one particle species), but we allow them to be relativistic (speeds comparable to the speed of light) or not and allow them to have nontrivial internal degrees of freedom (e.g., vibrations and rotations; Sec. 5.4.2), and allow them to exert forces on one another via an interaction potential that appears in their hamiltonian (e.g., van der Waals forces; Secs. 5.3.2 and 5.7). We refer to these particles as a gas, though our analysis is more general than gases. The nonrelativistic limit of all our fundamental equations is trivially obtained by removing particle rest masses from the energy ($\mathcal{E}$ gets replaced by $E = \mathcal{E} - mc^2$) and from the chemical potential ($\tilde{\mu}$ gets replaced by $\mu = \tilde{\mu} - mc^2$), but not from the grand potential, as it never has rest masses in it [see, e.g., Eq. (5.18) below].

We begin in Sec. 5.3.1 by deducing the grand potential representation of thermodynamics from the grand canonical ensemble, and by deducing a method for computing the thermodynamic properties of our gas from a grand canonical sum over the quantum states available to the system. In Ex. 5.3, the reader will apply this grand canonical formalism to a relativistic perfect gas. In Sec. 5.3.2, we apply the formalism to a nonrelativistic gas of particles that interact via van der Waals forces, thereby deriving the van der Waals equation of state, which is surprisingly accurate for many nonionized gases.

5.3.1 The Grand-Potential Representation, and Computation of Thermodynamic Properties as a Grand Canonical Sum

Figure 5.1 illustrates the ensemble of systems that we are studying and its bath. Each system is a cell of fixed volume V, with imaginary walls, inside a huge thermal bath of identical particles. Since the cells' walls are imaginary, the cells can and do exchange energy and particles with the bath. The bath is characterized by chemical potential $\tilde{\mu}$ for these particles and by temperature T. Since we allow the particles to be relativistic, we include the rest mass in the chemical potential $\tilde{\mu}$.

We presume that our ensemble of cells has reached statistical equilibrium with the bath, so its probabilistic distribution function has the grand canonical form (4.25c):

$$\rho_n = \frac{1}{Z} \exp\left(\frac{-\mathcal{E}_n + \tilde{\mu} N_n}{k_B T}\right) = \exp\left(\frac{\Omega - \mathcal{E}_n + \tilde{\mu} N_n}{k_B T}\right). \tag{5.13}$$

grand canonical distribution function

Here the index n labels the quantum state $|n\rangle$ of a cell, N_n is the number of particles in that quantum state, $\mathcal{E}_n$ is the total energy of that quantum state (including each particle's rest mass; its energy of translational motion; its internal energy if it has internal vibrations, rotations, or other internal excitations; and its energy of interaction with

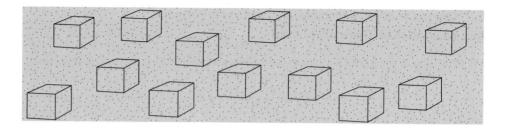

FIGURE 5.1 An ensemble of cells, each with volume V and imaginary walls, inside a heat and particle bath.

other particles), and $1/Z \equiv e^{\Omega/(k_B T)}$ is the normalization constant that guarantees $\sum_n \rho_n = 1$:

grand partition function and grand potential

$$Z \equiv \exp\left(\frac{-\Omega}{k_B T}\right) \equiv \sum_n \exp\left(\frac{-\mathcal{E}_n + \tilde{\mu} N_n}{k_B T}\right). \tag{5.14}$$

This normalization constant, whether embodied in Z or in Ω, is a function of the bath's temperature T and chemical potential $\tilde{\mu}$, and also of the cells' common volume V (which influences the set of available states $|n\rangle$). When regarded as a function of T, $\tilde{\mu}$, and V, the quantity $Z(V, \tilde{\mu}, T)$ is called the gas's *grand partition function*, and $\Omega(T, \tilde{\mu}, V)$ is called its *grand potential*. The following general argument shows that *once one has computed the explicit functional form for the grand potential*

$$\Omega(V, \tilde{\mu}, T) \tag{5.15}$$

[or for the grand partition function $Z(V, \tilde{\mu}, T)$], one can then derive from it all the thermodynamic properties of the thermally equilibrated system. The argument is so general that it applies to every grand canonical ensemble of systems, not just to our chosen gas of identical particles.

As key quantities in the argument, we introduce the mean energy and mean number of particles in the ensemble's systems (i.e., cells of Fig. 5.1):

$$\bar{\mathcal{E}} \equiv \sum_n \rho_n \mathcal{E}_n, \quad \bar{N} \equiv \sum_n \rho_n N_n. \tag{5.16}$$

(We denote these quantities with bars $\bar{\mathcal{E}}$ rather than brackets $\langle \mathcal{E} \rangle$ for ease of notation.) Inserting expression (5.13) for ρ_n into the log term in the definition of entropy $S = -k_B \sum_n \rho_n \ln \rho_n$ and using Eqs. (5.16), we obtain

$$S = -k_B \sum_n \rho_n \ln \rho_n = -k_B \sum_n \rho_n \left(\frac{\Omega - \mathcal{E}_n + \tilde{\mu} N_n}{k_B T}\right) = -\frac{\Omega - \bar{\mathcal{E}} + \tilde{\mu} \bar{N}}{T}; \tag{5.17}$$

or, equivalently,

Legendre transformation between thermodynamic representations

$$\Omega = \bar{\mathcal{E}} - T S - \tilde{\mu} \bar{N}. \tag{5.18}$$

This can be regarded as a *Legendre transformation* that leads from the energy representation of thermodynamics to the grand-potential representation. Legendre transformations are a common tool, for example, in classical mechanics (e.g., Landau and Lifshitz, 1976; Marion and Thornton, 1995; Goldstein, Poole, and Safko, 2002), for switching from one set of independent variables to another. Note that removing rest masses from $\overline{\mathcal{E}} = \bar{E} + Nmc^2$ and from $\tilde{\mu} = \mu + mc^2$ to get a nonrelativistic formula leaves Ω unchanged.

We now ask how the grand potential will change if the temperature T and chemical potential $\tilde{\mu}$ of the bath (and therefore of the ensemble) are slowly altered, and the volumes V of all the ensemble's boxes are slowly altered. Differentiating Eq. (5.18), we obtain $d\Omega = d\overline{\mathcal{E}} - T dS - S dT - \tilde{\mu} d\overline{N} - \overline{N} d\tilde{\mu}$. Expressing $d\mathcal{E}$ in terms of the energy representation's first law of thermodynamics (5.7) (with $\mathcal{E}$ replaced by $\overline{\mathcal{E}}$ and N replaced by $\overline{N}$), we bring this expression into the form

$$d\Omega = -P dV - \overline{N} d\tilde{\mu} - S dT.$$ (5.19)

first law in grand potential representation

This is the *grand-potential representation* of the first law of thermodynamics. The quantities P, $\overline{N}$, and S paired with the independent variables V, $\tilde{\mu}$, and T, respectively, can be thought of as generalized forces that push on the independent variables as they change, to produce changes in the grand potential.

generalized forces

From this version of the first law (the key grand canonical equation listed in the last line of Table 5.1), we can easily deduce almost all other equations of the grand-potential representation of thermodynamics. We just follow the same procedure as we used for the energy representation (Sec. 5.2.5).

The grand-potential representation's complete set of independent variables consists of those that appear as differentials on the right-hand side of the first law (5.19): V, $\tilde{\mu}$, and T. From the form (5.19) of the first law we see that Ω is being regarded as a function of these three independent variables: $\Omega = \Omega(V, \tilde{\mu}, T)$. This is the fundamental potential.

The Euler equation of this representation is deduced by building up a system from small pieces that all have the same values of the intensive variables $\tilde{\mu}$, T, and P. The first law (5.19) tells us that this buildup will produce

Euler equation for grand potential

$$\Omega = -PV.$$ (5.20)

Thus, if we happen to know P as a function of this representation's independent intensive variables—$P(\tilde{\mu}, T)$ (it must be independent of the extensive variable V)—then we can simply multiply by V to get the functional form of the grand potential: $\Omega(V, \tilde{\mu}, T) = P(\tilde{\mu}, T)V$; see Eqs. (5.24a) and (5.25) as a concrete example.

By comparing the grand-potential version of the first law (5.19) with the elementary calculus equation $d\Omega = (\partial\Omega/\partial V)dV + (\partial\Omega/\partial\tilde{\mu})d\tilde{\mu} + (\partial\Omega/\partial T)dT$, we infer equations for the system's "generalized forces," the pressure P, mean number of particles $\overline{N}$, and entropy S:

$$\overline{N} = -\left(\frac{\partial \Omega}{\partial \tilde{\mu}}\right)_{V,T}, \quad S = -\left(\frac{\partial \Omega}{\partial T}\right)_{V,\tilde{\mu}}, \quad P = -\left(\frac{\partial \Omega}{\partial V}\right)_{\tilde{\mu},T}. \tag{5.21}$$

Maxwell relations from grand potential

By differentiating these relations and equating mixed partial derivatives, we can derive Maxwell relations analogous to those [Eqs. (5.10b)] of the energy representation; for example, $(\partial \overline{N}/\partial T)_{V,\tilde{\mu}} = (\partial S/\partial \tilde{\mu})_{V,T}$. Equations of state are constrained by these Maxwell relations.

If we had begun with a specific functional form of the grand potential as a function of this representation's complete set of independent variables $\Omega(V, T, \tilde{\mu})$ [e.g., Eq. (5.24a) below], then Eqs. (5.21) would give us the functional forms of almost all the other dependent thermodynamic variables. The only one we are missing is the mean energy $\overline{\mathcal{E}}(V, \tilde{\mu}, T)$ in a cell. If we have forgotten Eq. (5.18) (the Legendre transformation) for that quantity, we can easily rederive it from the grand canonical distribution function $\rho = \exp[(\Omega - \mathcal{E} + \tilde{\mu}N)/(k_B T)]$ (the other key equation, besides the first law, on the last line of Table 5.1), via the definition of entropy as $S = -k_B \sum_n \rho_n \ln \rho_n = -k_B \overline{\ln \rho}$, as we did in Eq. (5.17).

This illustrates the power of the sparse information in Table 5.1. From it and little else we can deduce all the thermodynamic equations for each representation of thermodynamics.

It should be easy to convince oneself that *the nonrelativistic versions of all the above equations in this section can be obtained by the simple replacements $\mathcal{E} \to E$ (removal of rest masses from total energy) and $\tilde{\mu} \to \mu$ (removal of rest mass from chemical potential).*

5.3.2 Nonrelativistic van der Waals Gas

5.3.2

computing the grand potential from a statistical sum

The statistical sum $Z \equiv e^{-\Omega/(k_B T)} = \sum_n e^{(-\mathcal{E}_n + \tilde{\mu} N_n)/(k_B T)}$ [Eq. (5.14)] is a powerful method for computing the grand potential $\Omega(V, \tilde{\mu}, T)$, a method often used in condensed matter physics. Here we present a nontrivial example: a nonrelativistic, monatomic gas made of atoms or molecules (we call them "particles") that interact with so-called "van der Waals forces." In Ex. 5.3, the reader will explore a simpler example: a relativistic, perfect gas.

We assume that the heat and particle bath that bathes the cells of Fig. 5.1 has (i) sufficiently low temperature that the gas's particles are not ionized (therefore they are also nonrelativistic, $k_B T \ll mc^2$) and (ii) sufficiently low chemical potential that the mean occupation number η of the particles' quantum states is small compared to unity, so they behave classically, $\mu \equiv \tilde{\mu} - mc^2 \ll -k_B T$ [Eq. (3.22d)].

The orbital electron clouds attached to each particle repel one another when the distance r between the particles' centers of mass is smaller than about the diameter r_o of the particles. At larger separations, the particles' electric dipoles (intrinsic or induced) attract one another weakly. The interaction energy (potential energy) $u(r)$

associated with these forces has a form well approximated by the Lennard-Jones potential

$$u(r) = \varepsilon_o \left[\left(\frac{r_o}{r} \right)^{12} - \left(\frac{r_o}{r} \right)^6 \right],$$

(5.22a)

where ε_o is a constant energy. When a gradient is taken, the first term gives rise to the small-r repulsive force and the second to the larger-r attractive force. For simplicity of analytic calculations, we use the cruder approximation

$$u(r) = \infty \text{ for } r < r_o, \qquad u(r) = -\varepsilon_o(r_o/r)^6 \text{ for } r > r_o,$$

(5.22b)

approximate interaction energy for van der Waals gas

which has an infinitely sharp repulsion at $r = r_o$ (a *hard wall*). For simplicity, we assume that the mean interparticle separation is much larger than r_o (*dilute gas*), so it is highly unlikely that three or more particles are close enough together simultaneously, $r \sim r_o$, to interact (i.e., we confine ourselves to two-particle interactions).[6]

We compute the grand potential $\Omega(V, \mu, T)$ for an ensemble of cells embedded in a bath of these particles (Fig. 5.1), and from $\Omega(V, \mu, T)$ we compute how the particles' interaction energy $u(r)$ alters the gas's equation of state from the form $P = (\overline{N}/V)k_B T$ for an ideal gas [Eqs. (3.39b,c)]. Since this is our objective, any internal and spin degrees of freedom that the particles might have are irrelevant, and we ignore them.

For this ensemble, the nonrelativistic grand partition function $Z = \sum_n \exp[(-E_n + \mu N_n)/(k_B T)]$ is

$$Z = e^{-\Omega/(k_B T)}$$

$$= \sum_{N=0}^{\infty} \frac{e^{\mu N/(k_B T)}}{N!} \int \frac{d^{3N}x \, d^{3N}p}{h^{3N}} \exp\left[-\sum_{i=1}^{N} \frac{\mathbf{p}_i^2}{2mk_B T} - \sum_{i=1}^{N} \sum_{j=i+1}^{N} \frac{u_{ij}}{k_B T} \right].$$

(5.23a)

Here we have used Eq. (4.8b) for the sum over states $\sum_n$ [with $\mathcal{M} = N!$ (the multiplicity factor of Eqs. (4.8)), $W = 3N$, and $d\Gamma_W = d^{3N}x \, d^{3N}p$; cf. Ex. 5.3], and we have written E_n as the sum over the kinetic energies of the N particles in the cell and the interaction energies

$$u_{ij} \equiv u(r_{ij}), \qquad r_{ij} \equiv |\mathbf{x}_i - \mathbf{x}_j|$$

(5.23b)

of the $\frac{1}{2}N(N-1)$ pairs of particles.

Evaluation of the integrals and sum in Eq. (5.23a), with the particles' interaction energies given by Eqs. (5.23b) and (5.22b), is a rather complex task, which we relegate to the Track-Two Box 5.3. The result for the grand potential, $\Omega = -k_B T \ln Z$, accurate

---

6. Finding effective ways to tackle many-body problems is a major challenge in many areas of modern theoretical physics.

to first order in the particles' interactions (in the parameters a and b below), is

$$\Omega = -k_B T V \frac{(2\pi m k_B T)^{3/2}}{h^3} e^{\mu/(k_B T)} \left[1 + \frac{(2\pi m k_B T)^{3/2}}{h^3} e^{\mu/(k_B T)} \left(\frac{a}{k_B T} - b\right)\right].$$

(5.24a)

Here b is four times the volume of each hard-sphere particle, and a is that volume times the interaction energy ε_o of two hard-sphere particles when they are touching:

$$b = \frac{16\pi}{3} \left(\frac{r_o}{2}\right)^3, \qquad a = b\varepsilon_o.$$

(5.24b)

By differentiating this grand potential, we obtain the following expressions for the pressure P and mean number of particles $\overline{N}$ in a volume-V cell:

$$P = -\left(\frac{\partial\Omega}{\partial V}\right)_{\mu,T}$$

$$= k_B T \frac{(2\pi m k_B T)^{3/2}}{h^3} e^{\mu/(k_B T)} \left[1 + \frac{(2\pi m k_B T)^{3/2}}{h^3} e^{\mu/(k_B T)} \left(\frac{a}{k_B T} - b\right)\right],$$

$$\overline{N} = -\left(\frac{\partial\Omega}{\partial \mu}\right)_{V,T}$$

$$= V \frac{(2\pi m k_B T)^{3/2}}{h^3} e^{\mu/(k_B T)} \left[1 + 2\frac{(2\pi m k_B T)^{3/2}}{h^3} e^{\mu/(k_B T)} \left(\frac{a}{k_B T} - b\right)\right].$$

(5.25)

Notice that, when the interaction energy is turned off so $a = b = 0$, the second equation gives the standard ideal-gas particle density $\overline{N}/V = (2\pi m k_B T)^{3/2} e^{\mu/(k_B T)}/h^3 = \zeta/\lambda_{T\,\mathrm{dB}}^3$ where ζ and $\lambda_{T\,\mathrm{dB}}$ are defined in Eq. (2) of Box 5.3. Inserting this into the square bracketed expression in Eqs. (5.25), taking the ratio of expressions (5.25) and multiplying by V and expanding to first order in $a/(k_B T) - b$, we obtain $PV/\overline{N} = k_B T[1 + (\overline{N}/V)(b - a/(k_B T))]$—accurate to first order in $(b - a/(k_B T))$. Bringing the a term to the left-hand side, multiplying both sides by $[1 - (\overline{N}/V)b]$, and linearizing in b, we obtain the standard *van der Waals equation of state*:

$$\boxed{\left(P + \frac{a}{(V/\overline{N})^2}\right)(V/\overline{N} - b) = k_B T.}$$

(5.26)

The quantity $V/\overline{N}$ is the *specific volume* (volume per particle).

A few comments are in order.

1. The factor $(V/\overline{N} - b)$ in the equation of state corresponds to an excluded volume $b = (16\pi/3)(r_o/2)^3$ that is four times larger than the actual volume of each hard-sphere particle (whose radius is $r_o/2$).

2. The term linear in a, $P = -a\overline{N}/V = -b\varepsilon_o\overline{N}/V$, is a pressure reduction due to the attractive force between particles.

BOX 5.3.   DERIVATION OF VAN DER WAALS GRAND POTENTIAL T2

T2

In Eq. (5.23a) the momentum integrals and the space integrals separate, and the N momentum integrals are identical, so Z takes the form

$$Z = \sum_{N=0}^{\infty} \frac{e^{\mu N/(k_B T)}}{N! \, h^{3N}} \left[\int_0^{\infty} 4\pi p^2 dp \, \exp\left(\frac{-p^2}{2mk_B T} \right) \right]^N J_N$$

$$= \sum_{N=0}^{\infty} \frac{(\zeta/\lambda_{T\,dB}{}^3)^N}{N!} J_N,$$

(1)

where

$$\zeta \equiv e^{\mu/(k_B T)}, \quad \lambda_{T\,dB} \equiv \frac{h}{(2\pi mk_B T)^{1/2}},$$

$$J_N = \int d^{3N}x \, \exp\left[-\sum_{i=1}^{N} \sum_{j=i+1}^{N} \frac{u_{ij}}{k_B T} \right].$$

(2)

Note that λ is the particles' thermal deBroglie wavelength. The Boltzmann factor $e^{-u_{ij}/(k_B T)}$ is unity for large interparticle separations $r_{ij} \gg r_o$, so we write

$$e^{-u_{ij}/(k_B T)} \equiv 1 + f_{ij},$$

(3)

where f_{ij} is zero except when $r_{ij} \lesssim r_o$. Using this definition and rewriting the exponential of a sum as the products of exponentials, we bring J_N into the form

$$J_N = \int d^{3N}x \prod_{i=1}^{N} \prod_{j=i+1}^{N} (1 + f_{ij}).$$

(4)

The product contains (i) terms linear in f_{ij} that represent the influence of pairs of particles that are close enough ($r_{ij} \lesssim r_o$) to interact, plus (ii) quadratic terms, such as $f_{14} f_{27}$, that are nonzero only if particles 1 and 4 are near each other and 2 and 7 are near each other (there are so many of these terms that we cannot neglect them!), plus (iii) quadratic terms such as $f_{14} f_{47}$ that are nonzero only if particles 1, 4, and 7 are all within a distance $\sim r_o$ of one another (because our gas is dilute, it turns out these three-particle terms can be neglected), plus (iv) cubic and higher-order terms. At all orders ℓ (linear, quadratic, cubic, quartic, etc.) for our dilute gas, we can ignore terms that require three or more particles to be near one another, so we focus only on terms $f_{ij} f_{mn} \cdots f_{pq}$ where all indices are different. Eq. (4) then becomes

(continued)

BOX 5.3.   (continued)

$$J_N = \int d^{3N}x [1 + \underbrace{(f_{12} + f_{13} + \cdots)}_{n_1 \text{ terms}} + \underbrace{(f_{12}f_{34} + f_{13}f_{24} + \cdots)}_{n_2 \text{ terms}}$$

$$+ \underbrace{(f_{12}f_{34}f_{56} + f_{13}f_{24}f_{56} + \cdots)}_{n_3 \text{ terms}} \cdots], \qquad (5)$$

where n_ℓ is the number of terms of order ℓ with all 2ℓ particles different. Denoting

$$V_o \equiv \int f(r)4\pi r^2 dr, \qquad (6)$$

and performing the integrals, we bring Eq. (5) into the form

$$J_N = \sum_{\ell=0}^{\infty} n_\ell V^{N-\ell} V_o^\ell. \qquad (7)$$

At order ℓ the number of unordered sets of 2ℓ particles that are all different is $N(N-1)\cdots(N-2\ell+1)/\ell!$. The number of ways that these 2ℓ particles can be assembled into unordered pairs is $(2\ell-1)(2\ell-3)(2\ell-5)\cdots 1 \equiv (2\ell-1)!!$. Therefore, the number of terms of order ℓ that appear in Eq. (7) is

$$n_\ell = \frac{N(N-1)\cdots(N-2\ell+1)}{\ell!}(2\ell-1)!!$$

$$= \frac{N(N-1)\cdots(N-2\ell+1)}{2^\ell \ell!}. \qquad (8)$$

Inserting Eqs. (7) and (8) into Eq. (1) for the partition function, we obtain

$$Z = \sum_{N=0}^{\infty} \frac{(\zeta/\lambda^3)^N}{N!} \sum_{\ell=0}^{[N/2]} \frac{N(N-1)\cdots(N-2\ell+1)}{2^\ell \ell!} V^{N-\ell} V_o^\ell, \qquad (9)$$

where $[N/2]$ means the largest integer less than or equal to $N/2$. Performing a little algebra and then reversing the order of the summations, we obtain

$$Z = \sum_{\ell=0}^{\infty} \sum_{N=2\ell}^{\infty} \frac{1}{(N-2\ell)!} \left(\frac{\zeta V}{\lambda^3}\right)^{N-2\ell} \frac{1}{\ell!} \left(\frac{\zeta V}{\lambda^3}\frac{\zeta V_o}{2\lambda^3}\right)^\ell. \qquad (10)$$

By changing the summation index from N to $N' = N - 2\ell$, we decouple the two summations. Each of the sums is equal to an exponential, giving

(continued)

BOX 5.3. (continued)

T2

$$Z = e^{-\Omega/(k_B T)} = \exp\left(\frac{\zeta V}{\lambda^3}\right)\exp\left(\frac{\zeta V}{\lambda^3}\frac{\zeta V_o}{2\lambda^3}\right)$$

$$= \exp\left[\frac{\zeta V}{\lambda^3}\left(1 + \frac{\zeta V_o}{2\lambda^3}\right)\right]. \tag{11}$$

Therefore, the grand potential for our van der Waals gas is

$$\Omega = \frac{-k_B T \zeta V}{\lambda^3}\left(1 + \frac{\zeta V_o}{2\lambda^3}\right). \tag{12}$$

From kinetic theory [Eq. (3.39a)], we know that for an ideal gas, the mean number density is $\overline{N}/V = \zeta/\lambda^3$; this is also a good first approximation for our van der Waals gas, which differs from an ideal gas only by the weakly perturbative interaction energy $u(r)$. Thus $\zeta V_o/(2\lambda^3)$ is equal to $\frac{1}{2}V_o/$(mean volume per particle), which is $\ll 1$ by our dilute-gas assumption. If we had kept three-particle interaction terms, such as $f_{14}f_{47}$, they would have given rise to fractional corrections of order $(\zeta V_o/\lambda^3)^2$, which are much smaller than the leading-order fractional correction $\zeta V_o/(2\lambda^3)$ that we have computed [Eq. (12)]. The higher-order corrections are derived in statistical mechanics textbooks, such as Pathria and Beale (2011, Chap. 10) and Kardar (2007, Chap. 5) using a technique called the "cluster expansion."

For the hard-wall potential (5.22b), f is -1 at $r < r_o$, and assuming that the temperature is high enough that $\varepsilon_o/(k_B T) \ll 1$, then at $r > r_o$, f is very nearly $-u/(k_B T) = \left(\varepsilon_o/(k_B T)\right)(r_o/r)^6$; therefore we have

$$\frac{V_o}{2} \equiv \frac{1}{2}\int f(r)4\pi r^2 dr = \frac{a}{k_B T} - b, \quad \text{where } b = \frac{2\pi r_o^3}{3}, \quad a = b\varepsilon_o. \tag{13}$$

Inserting this expression for $V_o/2$ and Eqs. (2) for ζ and λ into Eq. (12), we obtain Eqs. (5.24) for the grand potential of a van der Waals gas.

3. Our derivation is actually only accurate to first order in a and b, so it does not justify the quadratic term $P = ab(\overline{N}/V)^2$ in the equation of state (5.26). However, that quadratic term does correspond to the behavior of real gases: a sharp rise in pressure at high densities due to the short-distance repulsion between particles.

We study this van der Waals equation of state in Sec. 5.7, focusing on the gas-to-liquid phase transition that it predicts and on fluctuations of thermodynamic quantities associated with that phase transition.

In this section we have presented the grand canonical analysis for a van der Waals gas not because such a gas is important (though it is), but rather as a concrete example of how one uses the formalism of statistical mechanics and introduces ingenious approximations to explore the behavior of realistic systems made of interacting particles.

EXERCISES

Exercise 5.3 *Derivation and Example: Grand Canonical Ensemble for a Classical, Relativistic, Perfect Gas*

Consider cells that reside in a heat and particle bath of a classical, relativistic, perfect gas (Fig. 5.1). Each cell has the same volume V and imaginary walls. Assume that the bath's temperature T has an arbitrary magnitude relative to the rest mass-energy mc^2 of the particles (so the thermalized particles might have relativistic velocities), but require $k_B T \ll -\mu$ (so all the particles behave classically). Ignore the particles' spin degrees of freedom, if any. For ease of notation use geometrized units (Sec. 1.10) with $c = 1$.

(a) The number of particles in a chosen cell can be anything from $N = 0$ to $N = \infty$. Restrict attention, for the moment, to a situation in which the cell contains a precise number of particles, N. Explain why the multiplicity is $\mathcal{M} = N!$ even though the density is so low that the particles' wave functions do not overlap, and they are behaving classically (cf. Ex. 4.8).

(b) Still holding fixed the number of particles in the cell, show that the number of degrees of freedom W, the number density of states in phase space $\mathcal{N}_{\text{states}}$, and the energy $\mathcal{E}_N$ in the cell are

$$W = 3N \, , \quad \mathcal{N}_{\text{states}} = \frac{1}{N! h^{3N}} \, , \quad \mathcal{E}_N = \sum_{j=1}^{N} (\mathbf{p}_j{}^2 + m^2)^{\frac{1}{2}} \, , \qquad (5.27a)$$

where $\mathbf{p}_j$ is the momentum of classical particle number j.

(c) Using Eq. (4.8b) to translate from the formal sum over states $\sum_n$ to a sum over $W = 3N$ and an integral over phase space, show that the sum over states (5.14) for the grand partition function becomes

$$Z = e^{-\Omega/(k_B T)} = \sum_{N=0}^{\infty} \frac{V^N}{N! \, h^{3N}} e^{\tilde{\mu} N/(k_B T)} \left[\int_0^\infty \exp\left(-\frac{(p^2 + m^2)^{\frac{1}{2}}}{k_B T} \right) 4\pi p^2 dp \right]^N .$$

$$(5.27b)$$

(d) Evaluate the momentum integral in the nonrelativistic limit $k_B T \ll m$, and thereby show that

$$\Omega(T, \mu, V) = -k_B T V \frac{(2\pi m k_B T)^{3/2}}{h^3} e^{\mu/(k_B T)}, \qquad (5.28a)$$

where $\mu = \tilde{\mu} - m$ is the nonrelativistic chemical potential. This is the interaction-free limit $V_o = a = b = 0$ of our grand potential (5.24a) for a van der Waals gas.

(e) Show that in the extreme relativistic limit $k_B T \gg m$, Eq. (5.27b) gives

$$\Omega(T, \tilde{\mu}, V) = -\frac{8\pi V (k_B T)^4}{h^3} e^{\tilde{\mu}/(k_B T)}. \qquad (5.29)$$

(f) For the extreme relativistic limit use your result (5.29) for the grand potential $\Omega(V, T, \tilde{\mu})$ to derive the mean number of particles $\overline{N}$, the pressure P, the entropy S, and the mean energy $\overline{\mathcal{E}}$ as functions of V, $\tilde{\mu}$, and T. Note that for a photon gas, because of the spin degree of freedom, the correct values of $\overline{N}$, $\overline{\mathcal{E}}$, and S will be twice as large as what you obtain in this calculation. Show that the energy density is $\overline{\mathcal{E}}/V = 3P$ (a relation valid for any ultrarelativistic gas); and that $\overline{\mathcal{E}}/\overline{N} = 3 k_B T$ (which is higher than the $2.7011780\ldots k_B T$ for blackbody radiation, as derived in Ex. 3.13, because in the classical regime of $\eta \ll 1$, photons don't cluster in the same states at low frequency; that clustering lowers the mean photon energy for blackbody radiation).

5.4 Canonical Ensemble and the Physical-Free-Energy Representation of Thermodynamics

In this section, we turn to an ensemble of single-species systems that can exchange energy but nothing else with a heat bath at temperature T. The systems thus have variable total energy $\mathcal{E}$, but they all have the same, fixed values of the two remaining extensive variables N and V. We presume that the ensemble has reached statistical equilibrium, so it is canonical with a distribution function (probability of occupying any quantum state of energy $\mathcal{E}$) given by Eq. (4.20):

$$\rho_n = \frac{1}{z} e^{-\mathcal{E}_n/(k_B T)} \equiv e^{(F - \mathcal{E}_n)/(k_B T)}. \qquad (5.30)$$

canonical distribution function

Here, as in the grand canonical ensemble [Eq. (5.13)], we have introduced special notations for the normalization constant: $1/z = e^{F/(k_B T)}$, where z (the *partition function*) and F (the *physical free energy* or *Helmholtz free energy*) are functions of the systems' fixed N and V and the bath temperature T. Once the systems' quantum states $|n\rangle$ (with fixed N and V but variable $\mathcal{E}$) have been identified, the functions $z(T, N, V)$ and $F(T, N, V)$ can be computed from the normalization relation $\sum_n \rho_n = 1$:

$$e^{-F/(k_B T)} \equiv z(T, N, V) = \sum_n e^{-\mathcal{E}_n/(k_B T)}. \qquad (5.31)$$

partition function and physical free energy

This canonical sum over states, like the grand canonical sum (5.14) that we used for the van der Waals gas, is a powerful tool in statistical mechanics. As an example, in Secs. 5.8.2 and 5.8.3 we use the canonical sum to evaluate the physical free energy F for a model of ferromagnetism and use the resulting F to explore a ferromagnetic phase transition.

Having evaluated $z(T, N, V)$ [or equivalently, $F(T, N, V)$], one can then proceed as follows to determine other thermodynamic properties of the ensemble's systems. The entropy S can be computed from the standard expression $S = -k_B \sum_n \rho_n \ln \rho_n = -k_B \overline{\ln \rho}$, which, with Eq. (5.30) for ρ_n, implies $S = (\overline{\mathcal{E}} - F)/T$. It is helpful to rewrite this as an equation for the physical free energy F:

Legendre transformation

$$F = \overline{\mathcal{E}} - TS. \tag{5.32}$$

This is the Legendre transformation that leads from the energy representation of thermodynamics to the physical-free-energy representation.

Suppose that the canonical ensemble's parameters T, N, and V are changed slightly. By how much will the physical free energy change? Equation (5.32) tells us that $dF = d\overline{\mathcal{E}} - TdS - SdT$. Because macroscopic thermodynamics is independent of the statistical ensemble being studied, we can evaluate $d\overline{\mathcal{E}}$ using the first law of thermodynamics (5.7) with the microcanonical exact energy $\mathcal{E}$ replaced by the canonical mean energy $\overline{\mathcal{E}}$. The result is

first law in physical-free-energy representation

$$dF = -SdT + \tilde{\mu}dN - PdV. \tag{5.33}$$

Equation (5.33) contains the same information as the first law of thermodynamics and can be thought of as the first law rewritten in the physical-free-energy representation. From this form of the first law, we can deduce the other equations of the physical-free-energy representation by the same procedure we used for the energy representation in Sec. 5.2.5 and the grand-potential representation in Sec. 5.3.1.

If we have forgotten our representation's independent variables, we read them off the first law (5.33); they appear as differentials on the right-hand side: T, N, and V. The fundamental potential is the quantity that appears on the left-hand side of the first law: $F(T, N, V)$. By building up a full system from small subsystems that all have the same intensive variables T, $\tilde{\mu}$, and P, we deduce from the first law the Euler relation for this representation:

Euler relation for physical free energy

$$F = \tilde{\mu}N - PV. \tag{5.34}$$

From the first law (5.33) we read off equations of state for this representation's generalized forces [e.g., $-S = (\partial F/\partial T)_{N,V}$]. Maxwell relations can be derived from the equality of mixed partial derivatives.

Thus, as for the energy and grand-potential representations, all the equations of the physical-free-energy representation are easily deducible from the minimal information in Table 5.1.

And as for those representations, *the Newtonian version of this representation's fundamental equations (5.30)–(5.34) is obtained by simply removing rest masses from $\tilde{\mu}$ (which becomes μ), $\mathcal{E}$ (which becomes E), and F (whose notation does not change).*

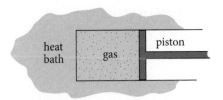

FIGURE 5.2 Origin of the name "physical free energy" for $F(V, T, N)$.

5.4.1 Experimental Meaning of Physical Free Energy

The name "physical free energy" for F can be understood using the idealized experiment shown in Fig. 5.2. Gas is placed in a chamber, one wall of which is a piston, and the chamber comes into thermal equilibrium with a heat bath, with which it can exchange heat but not particles. The volume of the chamber has some initial value V_i; and correspondingly, the gas has some initial physical free energy $F(V_i, T, N)$. The gas is then allowed to push the piston to the right sufficiently slowly for the gas to remain always in thermal equilibrium with the heat bath, at the bath's temperature T. When the chamber has reached its final volume V_f, the total work done on the piston by the gas (i.e., the total energy extracted by the piston from this "engine") is

$$\mathcal{E}_{\text{extracted}} = \int_{V_i}^{V_f} -P dV. \qquad (5.35a)$$

Using the first law $dF = -SdT + \tilde{\mu} dN - P dV$ and remembering that T and N are kept constant, Eq. (5.35a) becomes

$$\mathcal{E}_{\text{extracted}} = F(V_f, T, N) - F(V_i, T, N) \equiv \Delta F. \qquad (5.35b)$$

physical free energy extracted in isothermal expansion

Thus, *F is the energy that is free to be extracted in an isothermal, physical expansion of the gas.*[7]

If the expansion had been done in a chamber that was perfectly thermally insulated, so no heat could flow in or out of it, then there would have been no entropy change. Correspondingly, with S and N held fixed but V changing during the expansion, the natural way to analyze the expansion would have been in the energy representation; that representation's first law $d\mathcal{E} = -P dV + T dS + \tilde{\mu} dN$ would have told us that the total energy extracted, $\int -P dV$, was the change $\Delta \mathcal{E}$ of the gas's total energy. Such a process, which occurs without any heat flow or entropy increase, is called *adiabatic.* Thus, *the energy $\mathcal{E}$ (or in the nonrelativistic regime, E) measures the amount of energy that can be extracted from an adiabatic engine, by contrast with F, which measures the energy extracted from an isothermal engine.*

---

7. More generally, the phrase "free energy" means the energy that can be extracted in a process that occurs in contact with some sort of environment. The nature of the free energy depends on the nature of the contact. We will meet chemical free energy in Sec. 5.5, and the free energy of a body on which a steady force is acting in Sec. 11.6.1.

5.4.2 Ideal Gas with Internal Degrees of Freedom

As an example of the canonical distribution, we explore the influence of internal molecular degrees of freedom on the properties of a nonrelativistic, ideal gas.[8] This example is complementary to the van der Waals gas that we analyzed in Sec. 5.3.2 using the grand canonical distribution. There we assumed no internal degrees of freedom, but we allowed each pair of particles to interact via an interaction potential $u(r)$ that depended on the particles' separation r. Here, because the gas is ideal, there are no interactions, but we allow for internal degrees of freedom—rotational, vibrational, and electron excitations.

(We have previously studied internal degrees of freedom in Sec. 4.4.4, where we proved the equipartition theorem for those whose generalized coordinates and/or momenta are quadratic in the hamiltonian and are classically excited, such as the vibrations and rotations of a diatomic molecule. Here we allow the internal degrees of freedom to have any form whatsoever and to be arbitrarily excited or nonexcited.)

Our gas is confined to a fixed volume V, it has a fixed number of molecules N, it is in contact with a heat bath with temperature T, and its equilibrium distribution is therefore canonical, $\rho_n = e^{(F-E_n)/(k_BT)}$. The quantum states $|n\rangle$ available to the gas can be characterized by the locations $\{\mathbf{x}_i, \mathbf{p}_i\}$ in phase space of each of the molecules $i = 1, \ldots, N$, and by the state $|K_i\rangle$ of each molecule's internal degrees of freedom. Correspondingly, the partition function and physical free energy are given by [Eq. (5.31)]

$$z = e^{-F/(k_BT)} = \frac{g_s{}^N}{N!} \int \frac{d^{3N}x\, d^{3N}p}{h^{3N}} \sum_{K_1, K_2, \ldots, K_N} \exp\left[-\sum_{i=1}^{N}\left(\frac{\mathbf{p}_i^2}{2mk_BT} + \frac{E_{K_i}}{k_BT}\right)\right].$$

(5.36a)

It is instructive to compare this with Eq. (5.23a) for the grand partition function of the van der Waals gas. In Eq. (5.36a) there is no interaction energy u_{ij} between molecules, no sum over N, and no $e^{\mu N/(k_BT)}$ (because N is fixed and there is no particle bath). However, we now have sums over the internal states K_i of each of the molecules and a factor g_s in the multiplicity to allow for the molecules' g_s different spin states [cf. Eq. (4.8c)].

Because there are no interactions between molecules, the partition function can be split up into products of independent contributions from each of the molecules. Because there are no interactions between a molecule's internal and translational degrees of freedom, the partition function can be split into a product of translational and internal terms; and because the molecules are all identical, their contributions are all identical, leading to

$$z = e^{-F/(k_BT)} = \frac{1}{N!}\left[g_s \int \frac{d^3x\, d^3p}{h^3} e^{-\mathbf{p}^2/(k_BT)}\right]^N \left[\sum_K e^{-E_K/(k_BT)}\right]^N. \quad (5.36b)$$

---

8. See footnote 5 of this chapter on p. 226 for the meaning of "ideal gas."

The $\int d^3x d^3p\, h^{-3} e^{-\mathbf{p}^2/(k_B T)}$ integral is the same as we encountered in the grand canonical analysis; it gives V/λ^3, where $\lambda = h/(2\pi m k_B T)^{1/2}$ [cf. Eq. (1) in Box 5.3]. The sum over internal states gives a contribution that is some function of temperature:

contribution of internal states to physical free energy of an ideal gas

$$f(T) \equiv \sum_K e^{-E_K/(k_B T)}. \qquad (5.37)$$

Correspondingly [using Stirling's approximation, $N! \simeq (N/e)^N$, to the accuracy needed here], the physical free energy becomes

$$F(N, V, T) = N k_B T \ln \left[\frac{N}{e} \frac{h^3/g_s}{(2\pi m k_B T)^{3/2} V} \right] - N k_B T \ln[f(T)]. \qquad (5.38)$$

Note that because the molecules' translational and internal degrees of freedom are decoupled, their contributions to the free energy are additive. We could have computed them separately and then simply added their free energies.

Because the contribution of the internal degrees of freedom depends only on temperature and not on volume, the ideal gas's pressure

$$P = -(\partial F/\partial V)_{N,T} = (N/V)k_B T \qquad (5.39)$$

is unaffected by the internal degrees of freedom. By contrast, the entropy and the total energy in the box do have internal contributions, which depend on temperature but not on the gas's volume and thence not on its density N/V:

$$S = -(\partial F/\partial T)_{N,V} = S_{\text{translational}} + N k_B (\ln f + d \ln f/d \ln T), \qquad (5.40)$$

where the entropy $S_{\text{translational}}$ can straightforwardly be shown to be equivalent to the Sackur-Tetrode formula (4.42) and

$$\bar{E} = F + TS = N k_B T \left(\frac{3}{2} + \frac{d \ln f}{d \ln T} \right). \qquad (5.41)$$

For degrees of freedom that are classical and quadratic, the internal contribution $N k_B T\, d \ln f/d \ln T$ gives $\frac{1}{2} k_B T$ for each quadratic term in the hamiltonian, in accord with the equipartition theorem (Sec. 4.4.4).

If there is more than one particle species present (e.g., electrons and protons at high temperatures, so hydrogen is ionized), then the contributions of the species to F, P, S, and E simply add, just as the contributions of internal and translational degrees of freedom added in Eq. (5.38).

EXERCISES

Exercise 5.4 *Example and Derivation: Adiabatic Index for Ideal Gas*

In Part V, when studying fluid dynamics, we shall encounter an *adiabatic index*

$$\Gamma \equiv -\left(\frac{\partial \ln P}{\partial \ln V} \right)_S \qquad (5.42)$$

[Eq. (13.2)] that describes how the pressure P of a fluid changes when it is compressed adiabatically (i.e., compressed at fixed entropy, with no heat being added or removed).

Derive an expression for Γ for an ideal gas that may have internal degrees of freedom (e.g., Earth's atmosphere). More specifically, do the following.

(a) Consider a fluid element (a small sample of the fluid) that contains N molecules. These molecules can be of various species; all species contribute equally to the ideal gas's pressure $P = (N/V)k_B T$ and contribute additively to its energy. Define the fluid element's specific heat at fixed volume to be the amount of heat TdS that must be inserted to raise its temperature by an amount dT while the volume V is held fixed:

$$C_V \equiv T(\partial S/\partial T)_{V,N} = (\partial E/\partial T)_{V,N}. \tag{5.43}$$

Deduce the second equality from the first law of thermodynamics. Show that in an adiabatic expansion the temperature T drops at a rate given by $C_V dT = -PdV$. [Hint: Use the first law of thermodynamics and the fact that for an ideal gas the energy of a fluid element depends only on its temperature and not on its volume (or density); Eq. (5.41).]

(b) Combine the temperature change $dT = (-P/C_V)dV$ for an adiabatic expansion with the equation of state $PV = Nk_B T$ to obtain $\Gamma = (C_V + Nk_B)/C_V$.

(c) To interpret the numerator $C_V + Nk_B$, imagine adding heat to a fluid element while holding its pressure fixed (which requires a simultaneous volume change). Show that in this case the ratio of heat added to temperature change is

$$C_P \equiv T(\partial S/\partial T)_{P,N} = C_V + Nk_B. \tag{5.44}$$

Combining with part (b), conclude that the adiabatic index for an ideal gas is given by

$$\Gamma = \gamma \equiv C_P/C_V, \tag{5.45}$$

a standard result in elementary thermodynamics.

Exercise 5.5 *Example: The Enthalpy Representation of Thermodynamics* T2

(a) Enthalpy H is a macroscopic thermodynamic variable defined by

$$\boxed{H \equiv \mathcal{E} + PV.} \tag{5.46}$$

Show that this definition can be regarded as a Legendre transformation that converts from the energy representation of thermodynamics with $\mathcal{E}(V, S, N)$ as the fundamental potential, to an *enthalpy representation* with $H(P, S, N)$ as the fundamental potential. More specifically, show that the first law, reexpressed in terms of H, takes the form

$$\boxed{dH = VdP + TdS + \tilde{\mu}dN;} \tag{5.47}$$

and then explain why this first law dictates that $H(P, S, N)$ be taken as the fundamental potential.

(b) For a nonrelativistic system, it is conventional to remove the particle rest masses from the enthalpy just as one does from the energy, but by contrast with energy, we do not change notation for the enthalpy:

$$H_{\text{nonrelativistic}} \equiv H_{\text{relativistic}} - Nm = E + PV. \qquad (5.48)$$

What is the form of the first law (5.47) for the nonrelativistic H?

(c) There is an equilibrium statistical mechanics ensemble associated with the enthalpy representation. Show that each system of this ensemble (fluctuationally) exchanges volume and energy with a surrounding *pressure bath* but does not exchange heat or particles, so the exchanged energy is solely that associated with the exchanged volume, $d\mathcal{E} = -PdV$, and the enthalpy H does not fluctuate. Note that P is the common pressure of the bath and the system.

(d) Show that this ensemble's distribution function is $\rho = e^{-S/k_B} = \text{constant}$ for those states in phase space that have a specified number of particles N and a specified enthalpy H. Why do we not need to allow for a small range δH of H, by analogy with the small range $\mathcal{E}$ for the microcanonical ensemble (Sec. 4.5 and Ex. 4.7)?

(e) What equations of state can be read off from the enthalpy first law? What are the Maxwell relations between these equations of state?

(f) What is the Euler equation for H in terms of a sum of products of extensive and intensive variables?

(g) Show that the system's enthalpy is equal to its total (relativistic) inertial mass (multiplied by the speed of light squared); cf. Exs. 2.26 and 2.27.

(h) As another interpretation of the enthalpy, think of the system as enclosed in an impermeable box of volume V. Inject into the box a "sample" of additional material of the same sort as is already there. (It may be helpful to think of the material as a gas.) The sample is to be put into the same thermodynamic state (i.e., macrostate) as that of the box's material (i.e., it is to be given the same values of temperature T, pressure P, and chemical potential $\tilde{\mu}$). Thus, the sample's material is indistinguishable in its thermodynamic properties from the material already in the box, except that its extensive variables (denoted by Δ) are far smaller: $\Delta V/V = \Delta\mathcal{E}/\mathcal{E} = \Delta S/S \ll 1$. Perform the injection by opening up a hole in one of the box's walls, pushing aside the box's material to make a little cavity of volume ΔV equal to that of the sample, inserting the sample into the cavity, and then closing the hole in the wall. The box now has the same volume V as before, but its energy has changed. Show that the energy change (i.e., the energy required to create the sample and perform the injection) is equal to the enthalpy ΔH of the sample. Thus, *enthalpy has the physical interpretation of energy of injection at fixed volume V*. Equivalently, if a sample of material is ejected from the system, the total energy that will come out (including the work done on the sample by the system during the ejection) is the sample's enthalpy ΔH. From this viewpoint, enthalpy is the system's free energy.

5.5

5.5 Gibbs Ensemble and Representation of Thermodynamics; Phase Transitions and Chemical Reactions

Next consider systems in which the temperature T and pressure P are both being controlled by an external environment (bath) and thus are treated as independent variables in the fundamental potential. This is the situation in most laboratory experiments and geophysical situations and is common in elementary chemistry but not chemical engineering.

In this case each of the systems has a fixed number of particles N_I for the various independent species I, and it can exchange heat and volume with its surroundings. (We explicitly allow for more than one particle species, because a major application of the Gibbs representation will be to chemical reactions.) There might be a membrane separating each system from its bath—a membrane impermeable to particles but through which heat can pass, and with negligible surface tension so the system and the bath can buffet each other freely, causing fluctuations in the system's volume. For example, this is the case for a so-called "constant-pressure balloon" of the type used to lift scientific payloads into the upper atmosphere. Usually, however, there is no membrane between system and bath. Instead, gravity might hold the system together because it has higher density than the bath (e.g., a liquid in a container); or solid-state forces might hold the system together (e.g., a crystal); or we might just introduce a conceptual, imaginary boundary around the system of interest—one that comoves with some set of particles.

The equilibrium ensemble for this type of system is that of Gibbs, with distribution function

$$\rho_n = e^{G/(k_B T)} e^{-(\mathcal{E}_n + P V_n)/(k_B T)} \tag{5.49}$$

[Eq. (4.25b), to which we have added the normalization constant $e^{G/(k_B T)}$]. As for the canonical and grand canonical distributions, the quantity G in the normalization constant becomes the fundamental potential for the Gibbs representation of thermodynamics. It is called the *Gibbs potential*, and also, sometimes, the *Gibbs free energy* or *chemical free energy* (see Ex. 5.6). It is a function of the systems' fixed numbers of particles N_I and of the bath's temperature T and pressure P, which appear in the Gibbs distribution function: $G = G(N_I, T, P)$.

The Gibbs potential can be evaluated by a sum over quantum states that follows from $\sum_n \rho_n = 1$:

Gibbs potential as sum over states

$$e^{-G/(k_B T)} = \sum_n e^{-(\mathcal{E}_n + P V_n)/(k_B T)}. \tag{5.50}$$

See Ex. 5.7 for an example. This sum has proved to be less useful than the canonical and grand canonical sums, so in most statistical mechanics textbooks there is little or no discussion of the Gibbs ensemble. By contrast, the Gibbs representation of thermodynamics is extremely useful, as we shall see, so textbooks pay a lot of attention to it.

We can deduce the equations of the Gibbs representation by the same method as we used for the canonical and grand canonical representations. We begin by writing down a Legendre transformation that takes us from the energy representation to the Gibbs representation. As for the canonical and grand canonical cases, that Legendre transformation can be inferred from the equilibrium ensemble's entropy, $S = -k_B \overline{\ln \rho} = -(G - \overline{\mathcal{E}} - P\bar{V})/T$ [cf. Eq. (5.49) for ρ]. Solving for G, we get

$$G = \overline{\mathcal{E}} + P\bar{V} - TS. \tag{5.51}$$

Legendre transformation

Once we are in the thermodynamic domain (as opposed to statistical mechanics), we can abandon the distinction between expectation values of quantities and fixed values; that is, we can remove the bars and write this Legendre transformation as $G = \mathcal{E} + PV - TS$.

Differentiating this Legendre transformation and combining with the energy representation's first law (5.8), we obtain the first law in the Gibbs representation:

$$dG = VdP - SdT + \sum_I \tilde{\mu}_I dN_I. \tag{5.52}$$

first law in Gibbs representation

From this first law we read out the independent variables of the Gibbs representation, namely $\{P, T, N_I\}$ (in case we have forgotten them!), and the values of its generalized forces [equations of state; e.g., $V = (\partial G/\partial P)_{T,N_I}$]. From the equality of mixed partial derivatives, we read off the Maxwell relations. By imagining building up a large system from many tiny subsystems (all with the same, fixed, intensive variables P, T, and $\tilde{\mu}_I$) and applying the first law (5.52) to this buildup, we obtain the Euler relation

$$G = \sum_I \tilde{\mu}_I N_I. \tag{5.53}$$

Euler relation for Gibbs potential

This Euler relation will be very useful in Sec. 5.5.3, when we discuss chemical reactions.

As with previous representations of thermodynamics, to obtain the Newtonian version of all of this section's equations, we simply remove the particle rest masses from $\tilde{\mu}_I$ (which then becomes μ_I), from $\mathcal{E}$ (which then becomes E), and from G (which does not change notation).

EXERCISES

Exercise 5.6 *Problem: Gibbs Potential Interpreted as Chemical Free Energy*
In Sec. 5.4.1, we explained the experimental meaning of the free energy F for a system

in contact with a heat bath so its temperature is held constant, and in Ex. 5.5h we did the same for contact with a pressure bath. By combining these, give an experimental interpretation of the Gibbs potential G as the free energy for a system in contact with a heat and pressure bath—the "chemical free energy."

Exercise 5.7 *Problem and Practice: Ideal Gas Equation of State from Gibbs Ensemble* For a nonrelativistic, classical, ideal gas (no interactions between particles), evaluate the statistical sum (5.50) to obtain $G(P, T, N)$, and from it deduce the standard formula for the ideal-gas equation of state $P\bar{V} = Nk_B T$.

5.5.1 Out-of-Equilibrium Ensembles and Their Fundamental Thermodynamic Potentials and Minimum Principles

Despite its lack of usefulness in practical computations of the Gibbs potential G, the Gibbs ensemble plays an important conceptual role in a *minimum principle for G,* which we now derive.

Consider an ensemble of systems, each of which is immersed in an identical heat and volume bath, and assume that the ensemble begins with some arbitrary distribution function ρ_n, one that is not in equilibrium with the baths. As time passes, each system will interact with its bath and will evolve in response to that interaction. Correspondingly, the ensemble's distribution function ρ will evolve. At any moment of time the ensemble's systems will have some mean (ensemble-averaged) energy $\bar{\mathcal{E}} \equiv \sum_n \rho_n \mathcal{E}_n$ and volume $\bar{V} \equiv \sum_n \rho_n V_n$, and the ensemble will have some entropy $S = -k_B \sum_n \rho_n \ln \rho_n$. From these quantities (which are well defined even though the ensemble may be very far from statistical equilibrium), we can compute a Gibbs potential G for the ensemble. This out-of-equilibrium G is defined by the analog of the equilibrium definition (5.51),

out-of-equilibrium Gibbs potential

$$G \equiv \bar{\mathcal{E}} + P_b \bar{V} - T_b S, \tag{5.54}$$

where P_b and T_b are the pressure and temperature of the identical baths with which the ensemble's systems are interacting.[9] As the evolution proceeds, the total entropy of the baths' ensemble plus the systems' ensemble will continually increase, until equilibrium is reached. Suppose that during a short stretch of evolution the systems' mean energy changes by $\Delta\bar{\mathcal{E}}$, their mean volume changes by $\Delta \bar{V}$, and the entropy of the ensemble

---

9. Notice that, because the number N of particles in the system is fixed (as is the bath temperature T_b), the evolving Gibbs potential is proportional to

$$\frac{G}{Nk_B T_b} = \frac{\bar{E}}{Nk_B T_b} + \frac{P_b \bar{V}}{Nk_B T_b} - \frac{S}{Nk_B}.$$

This quantity is dimensionless and generally of order unity. Note that the last term is the dimensionless entropy per particle [Eq. (4.44) and associated discussion].

changes by ΔS. Then, by conservation of energy and volume, the baths' mean energy and volume must change by

$$\Delta \overline{\mathcal{E}}_b = -\Delta \overline{\mathcal{E}}, \qquad \Delta \overline{V}_b = -\Delta \overline{V}. \tag{5.55a}$$

Because the baths (by contrast with the systems) are in statistical equilibrium, we can apply to them the first law of thermodynamics for equilibrated systems:

$$\Delta \overline{\mathcal{E}}_b = -P_b \Delta \overline{V}_b + T_b \Delta S_b + \sum_I \tilde{\mu}_{Ib} \Delta N_{Ib}. \tag{5.55b}$$

Since the N_{Ib} are not changing (the systems cannot exchange particles with their baths) and since the changes of bath energy and volume are given by Eqs. (5.55a), Eq. (5.55b) tells us that the baths' entropy changes by

$$\Delta S_b = \frac{-\Delta \overline{\mathcal{E}} - P_b \Delta \overline{V}}{T_b}. \tag{5.55c}$$

Correspondingly, the sum of the baths' entropy and the systems' entropy changes by the following amount, which cannot be negative:

$$\Delta S_b + \Delta S = \frac{-\Delta \overline{\mathcal{E}} - P_b \Delta \overline{V} + T_b \Delta S}{T_b} \geq 0. \tag{5.55d}$$

Because the baths' pressure P_b and temperature T_b are not changing (the systems are so tiny compared to the baths that the energy and volume they exchange with the baths cannot have any significant effect on the baths' intensive variables), the numerator of expression (5.55d) is equal to the evolutionary change in the ensemble's out-of-equilibrium Gibbs potential (5.54):

$$\boxed{\Delta S_b + \Delta S = \frac{-\Delta G}{T_b} \geq 0.} \tag{5.56}$$

Thus, the second law of thermodynamics for an ensemble of arbitrary systems in contact with identical heat and volume baths is equivalent to the law that *the systems' out-of-equilibrium Gibbs potential can never increase.* As the evolution proceeds and the entropy of baths plus systems continually increases, the Gibbs potential G will be driven smaller and smaller, until ultimately, when statistical equilibrium with the baths is reached, G will stop at its final, minimum value.

minimum principle for Gibbs potential

The ergodic hypothesis implies that this minimum principle applies not only to an ensemble of systems but also to a single, individual system when that system is averaged over times long compared to its internal timescales τ_{int} (but times that might be short compared to the timescale for interaction with the heat and volume bath). The system's time-averaged energy $\overline{\mathcal{E}}$ and volume $\overline{V}$, and its entropy S (as computed, e.g., by examining the temporal wandering of its state on timescales $\sim \tau_{\text{int}}$), combine with the bath's temperature T_b and pressure P_b to give an out-of-equilibrium

Gibbs potential $G = \bar{\mathcal{E}} + P_b \bar{V} - T_b S$. This G evolves on times long compared to the averaging time used to define it, and that evolution must be one of continually decreasing G. Ultimately, when the system reaches equilibrium with the bath, G achieves its minimum value.

At this point we might ask about the other thermodynamic potentials. Not surprisingly, associated with each of them is an extremum principle analogous to "minimum G":

1. For the energy potential $\mathcal{E}(V, S, N)$ (Sec. 5.2), one focuses on closed systems and switches to $S(V, \mathcal{E}, N)$. The extremum principle is then the standard second law of thermodynamics: an ensemble of closed systems of fixed $\mathcal{E}$, V, and N always must evolve toward increasing entropy S; when it ultimately reaches equilibrium, the ensemble will be microcanonical and will have maximum entropy.

2. For the physical free energy (Helmholtz free energy) $F(T_b, V, N)$ (Sec. 5.4), one can derive, in a manner perfectly analogous to the Gibbs derivation, the following minimum principle. *For an ensemble of systems interacting with a heat bath, the out-of-equilibrium physical free energy $F = \bar{\mathcal{E}} - T_b S$ will always decrease, ultimately reaching a minimum when the ensemble reaches its final, equilibrium, canonical distribution.*

3. The grand potential $\Omega(V, T_b, \tilde{\mu}_b)$ (Sec. 5.3) satisfies the analogous minimum principle. *For an ensemble of systems interacting with a heat and particle bath, the out-of-equilibrium grand potential $\Omega = \bar{\mathcal{E}} - \tilde{\mu}_b \bar{N} - T_b S$ will always decrease, ultimately reaching a minimum when the ensemble reaches its final, equilibrium, grand canonical distribution.*

4. For the enthalpy $H(P_b, S, N)$ (Ex. 5.5) the analogous extremum principle is a bit more tricky (see Ex. 5.13). *For an ensemble of systems interacting with a volume bath, as for an ensemble of closed systems, the bath's entropy remains constant, so the systems' entropy S will always increase, ultimately reaching a maximum when the ensemble reaches its final equilibrium distribution.*

Table 5.2 summarizes these extremum principles. The first column lists the quantities that a system exchanges with its bath. The second column shows the out-of-equilibrium fundamental potential for the system, which depends on the bath variables and the system's out-of-equilibrium distribution function ρ (shown explicitly) and also on whatever quantities are fixed for the system (e.g., its volume V and/or number of particles N; not shown explicitly). The third column expresses the total entropy of system plus bath in terms of the bath's out-of-equilibrium fundamental potential. The fourth column expresses the second law of thermodynamics for bath plus system in terms of the fundamental potential. We shall discuss the fifth column in Sec. 5.6, when we study fluctuations away from equilibrium.

TABLE 5.2: Descriptions of out-of-equilibrium ensembles with distribution function ρ

Quantities exchanged with bath	Fundamental potential	Total entropy $S + S_b$	Second law	Fluctuational probability
None	$S(\rho)$ with $\mathcal{E}$ constant	$S + \text{const}$	$dS \geq 0$	$\propto e^{S/k_B}$
Volume and energy with $d\mathcal{E} = -P_b dV$	$S(\rho)$ with $H = \mathcal{E} + P_b V$ constant	$S + \text{const}$ (see Ex. 5.13)	$dS \geq 0$	$\propto e^{S/k_B}$
Heat	$F(T_b; \rho) = \overline{\mathcal{E}} - T_b S$	$-F/T_b + \text{const}$	$dF \leq 0$	$\propto e^{-F/(k_B T_b)}$
Heat and volume	$G(T_b, P_b; \rho) = \overline{\mathcal{E}} + P_b \overline{V} - T_b S$	$-G/T_b + \text{const}$	$dG \leq 0$	$\propto e^{-G/(k_B T_b)}$
Heat and particle	$\Omega(T_b, \tilde{\mu}_b, \rho) = \overline{\mathcal{E}} - \tilde{\mu}_b \overline{N} - T_b S$	$-\Omega/T_b + \text{const}$	$d\Omega \leq 0$	$\propto e^{-\Omega/(k_B T_b)}$

Notes: From the distribution function ρ, one computes $S = -k_B \sum_n \rho_n \ln \rho_n$, $\overline{\mathcal{E}} = \sum_n \rho_n \mathcal{E}_n$, $\overline{V} = \sum_n \rho_n V_n$, and $\overline{N}_n = \sum_n \rho_n N_n$. The systems of each ensemble are in contact with the bath shown in column one, and T_b, P_b, and $\tilde{\mu}_b$ are the bath's temperature, pressure, and chemical potential, respectively. For ensembles in statistical equilibrium, see Table 5.1. As in that table, the nonrelativistic formulas are the same as above but with the rest masses of particles removed from the chemical potentials ($\tilde{\mu} \to \mu$) and from all fundamental potentials except Ω ($\mathcal{E} \to E$, but no change of notation for H, F, and G).

5.5.2 Phase Transitions

5.5.2

phase transitions

The minimum principle for the Gibbs potential G is a powerful tool in understanding *phase transitions*. "Phase" here refers to a specific pattern into which the atoms or molecules of a substance organize themselves. The substance H_2O has three familiar phases: water vapor, liquid water, and solid ice. Over one range of pressure P and temperature T, the H_2O molecules prefer to organize themselves into the vapor phase; over another, the liquid phase; and over another, the solid ice phase. It is the Gibbs potential that governs their preferences.

To understand this role of the Gibbs potential, consider a cup of water in a refrigerator (and because the water molecules are highly nonrelativistic, adopt the nonrelativistic viewpoint with the molecules' rest masses removed from their energy E, chemical potential μ_{H_2O}, and Gibbs potential). The refrigerator's air forms a heat and volume bath for the water in the cup (the system). There is no membrane between the air and the water, but none is needed. Gravity, together with the density difference between water and air, serves to keep the water molecules in the cup and the air above the water's surface, for all relevant purposes.

Allow the water to reach thermal and pressure equilibrium with the refrigerator's air, then turn down the refrigerator's temperature slightly and wait for the water to reach equilibrium again, and then repeat the process. Suppose that you are clever enough to compute from first principles the Gibbs potential G for the H_2O at each step of the cooling, using two alternative assumptions: that the H_2O molecules organize themselves into the liquid water phase, and that they organize themselves into the solid ice phase. Your calculations will produce curves for G as a function of the common

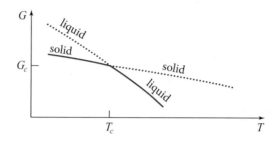

FIGURE 5.3 The Gibbs potential $G(T, P, N)$ for H_2O as a function of temperature T, with fixed P and N, near the freezing point $T = T_c = 273$ K. The solid curves correspond to the actual path traversed by the H_2O if the phase transition is allowed to proceed. The dotted curves correspond to superheated solid ice and supercooled liquid water that are unstable against the phase transition because their Gibbs potentials are higher than those of the other phase. Note that G tends to decrease with increasing temperature. This is caused by the $-TS$ term in $G = E + PV - TS$.

bath and H_2O temperature $T_b = T$ at fixed (atmospheric) pressure, with the shapes shown in Fig. 5.3. At temperatures $T > T_c = 273$ K the liquid phase has the lower Gibbs potential G, and at $T < T_c$ the solid phase has the lower G. Correspondingly, when the cup's temperature sinks slightly below 273 K, the H_2O molecules have a statistical preference for reorganizing themselves into the solid phase. The water freezes, forming ice.

It is a familiar fact that ice floats on water (i.e., ice is less dense than water), even when they are both precisely at the phase-transition temperature of 273 K. Correspondingly, when our sample of water freezes, its volume increases discontinuously by some amount ΔV; that is, when viewed as a function of the Gibbs potential G, the volume V of the statistically preferred phase is discontinuous at the phase-transition point (see Fig. 5.4a). It is also a familiar fact that when water freezes, it releases heat into its surroundings. This is why the freezing requires a moderately long time: the solidifying water can remain at or below its freezing point and continue to solidify only if the surroundings carry away the released heat, and the surroundings typically cannot carry it away quickly. It takes time to conduct heat through the ice and convect it through the water. The heat ΔQ released during the freezing (the *latent heat*) and the volume change ΔV are related to each other in a simple way (see Ex. 5.8, which focuses on the latent heat per unit mass Δq and the density change $\Delta \rho$ instead of on ΔQ and ΔV).

first-order phase transitions

Phase transitions like this one, with finite volume jumps $\Delta V \neq 0$ and finite latent heat $\Delta Q \neq 0$, are called *first-order*. The van der Waals gas (Sec. 5.3.2) provides an analytic model for another first-order phase transition: that from water vapor to liquid water; but we delay studying this model (Sec. 5.7) until we have learned about fluctuations of systems in statistical equilibrium (Sec. 5.6), which the van der Waals gas also illustrates.

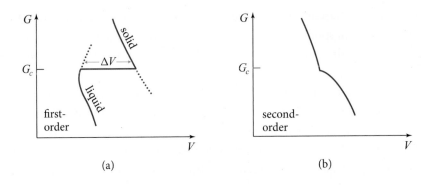

FIGURE 5.4 The changes of volume (plotted rightward) with increasing Gibbs function (plotted upward) at fixed P and N for (a) a first-order phase transition and (b) a second-order phase transition. The critical value of the Gibbs potential at which the transition occurs is G_c.

second-order phase transitions

Less familiar, but also important, are *second-order phase transitions*. In such transitions, the volumes V of the two phases are the same at the transition point, but their rates of change dV/dG are different (and this is so whether one holds P fixed as G decreases, holds T fixed, or holds some combination of P and T fixed); see Fig. 5.4b.

Crystals provide examples of both first-order and second-order phase transitions. A crystal can be characterized as a 3-dimensional repetition of a unit cell, in which ions are distributed in some fixed way. For example, Fig. 5.5a shows the unit cell for a $BaTiO_3$ (barium titanate) crystal at relatively high temperatures. This unit cell has a cubic symmetry. The full crystal can be regarded as made up of such cells stacked side by side and one on another. A first-order phase transition occurs when, with decreasing temperature, the Gibbs potential G of some other ionic arrangement,

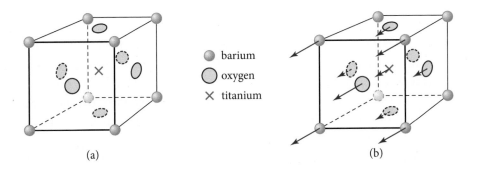

FIGURE 5.5 (a) The unit cell for a $BaTiO_3$ crystal at relatively high temperatures. (b) The displacements of the titanium and oxygen ions relative to the corners of the unit cell that occur in this crystal with falling temperature when it undergoes its second-order phase transition. The magnitudes of the displacements are proportional to the amount $T_c - T$ by which the temperature T drops below the critical temperature T_c, for small $T_c - T$.

with a distinctly different unit cell, drops below the G of the original arrangement. Then the crystal can spontaneously rearrange itself, converting from the old unit cell to the new one with some accompanying release of heat and some discontinuous change in volume.

$BaTiO_3$ does not behave in this way. Rather, as the temperature falls a bit below a critical value, the unit cell begins to elongate parallel to one of its edges (i.e., the cell's atoms get displaced as indicated in Fig. 5.5b). If the temperature is only a tiny bit below critical, they are displaced by only a tiny amount. When the temperature falls further, their displacements increase. If the temperature is raised back up above critical, the ions return to the standard, rigidly fixed positions shown in Fig. 5.5a. The result is a discontinuity, at the critical temperature, in the rate of change of volume dV/dG (Fig. 5.4b), but there is no discontinuous jump of volume and no latent heat.

This $BaTiO_3$ example illustrates a frequent feature of phase transitions: when the transition occurs (i.e., when the atoms start to move), the unit cell's cubic symmetry gets broken. The crystal switches discontinuously to a lower type of symmetry, a tetragonal one in this case. Such spontaneous symmetry breaking is a common occurrence in phase transitions not only in condensed matter physics but also in fundamental particle physics.

Bose-Einstein condensation of a bosonic atomic gas in a magnetic trap (Sec. 4.9) is another example of a phase transition. As we saw in Ex. 4.13, for Bose-Einstein condensation the specific heat of the atoms changes discontinuously (in the limit of an arbitrarily large number of atoms) at the critical temperature; this, or often a mild divergence of the specific heat, is characteristic of second-order phase transitions. Ferromagnetism also exhibits a second-order phase transition, which we explore in Secs. 5.8.3 and 5.8.4 using two powerful computational techniques: the renormalization group and Monte Carlo methods.

EXERCISES

Exercise 5.8 *Example: The Clausius-Clapeyron Equation for Two Phases in Equilibrium with Each Other*

(a) Consider H_2O in contact with a heat and volume bath with temperature T and pressure P. For certain values of T and P the H_2O will be liquid water; for others, ice; for others, water vapor—and for certain values it may be a two- or three-phase mixture of water, ice, and/or vapor. Show, using the Gibbs potential and its Euler equation, that, if two phases a and b are present and in equilibrium with each other, then their chemical potentials must be equal: $\mu_a = \mu_b$. Explain why, for any phase a, μ_a is a unique function of T and P. Explain why the condition $\mu_a = \mu_b$ for two phases to be present implies that the two-phase regions of the T-P plane are lines and the three-phase regions are points (see Fig. 5.6). The three-phase region is called the "triple point." The volume V of the two- or three-phase system will vary, depending on how much of each phase is present, since the density of each phase (at fixed T and P) is different.

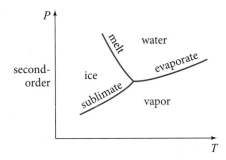

FIGURE 5.6 Phase diagram for H_2O. The temperature of the triple point (where the three phases meet) is 273.16 K and has been used to define the absolute scale of temperature.

(b) Show that the slope of the ice-water interface curve in Fig. 5.6 (the "melting curve") is given by the Clausius-Clapeyron equation

$$\frac{dP_{melt}}{dT} = \frac{\Delta q_{melt}}{T}\left(\frac{\rho_{ice}\,\rho_{water}}{\rho_{ice} - \rho_{water}}\right), \qquad (5.57a)$$

where ρ is density (mass per unit volume), and Δq_{melt} is the latent heat per unit mass for melting (or freezing)—the amount of heat required to melt a unit mass of ice, or the amount released when a unit mass of water freezes. Notice that, because ice is less dense than water, the slope of the melting curve is negative. [Hint: Compute dP/dT by differentiating $\mu_a = \mu_b$, and then use the thermodynamic properties of $G_a = \mu_a N_a$ and $G_b = \mu_b N_b$.]

(c) Suppose that a small amount of water is put into a closed container of much larger volume than the water. Initially there is vacuum above the water's surface, but as time passes some of the liquid water evaporates to establish vapor-water equilibrium. The vapor pressure will vary with temperature in accord with the *Clausius-Clapeyron equation*

$$\frac{dP_{vapor}}{dT} = \frac{\Delta q_{evaporate}}{T}\left(\frac{\rho_{water}\,\rho_{vapor}}{\rho_{water} - \rho_{vapor}}\right). \qquad (5.57b)$$

Now suppose that a foreign gas (not water vapor) is slowly injected into the container. Assume that this gas does not dissolve in the liquid water. Show that, as the pressure P_{gas} of the foreign gas gradually increases, it does not squeeze water vapor into the water, but rather it induces more water to vaporize:

$$\left(\frac{dP_{vapor}}{dP_{total}}\right)_{T\ fixed} = \frac{\rho_{vapor}}{\rho_{water}} > 0, \qquad (5.57c)$$

where $P_{total} = P_{vapor} + P_{gas}$.

5.5.3 Chemical Reactions

A second important application of the Gibbs potential is to the study of chemical reactions. Here we generalize the term "chemical reactions," to include any change in the constituent particles of the material being studied, including for example the joining of atoms to make molecules, the liberation of electrons from atoms in an ionization process, the joining of two atomic nuclei to make a third kind of nucleus, and the decay of a free neutron to produce an electron and a proton. In other words, the "chemical" of chemical reactions encompasses the reactions studied by nuclear physicists and elementary-particle physicists as well as those studied by chemists. The Gibbs representation is the appropriate one for discussing chemical reactions, because such reactions generally occur in an environment ("bath") of fixed temperature and pressure, with energy and volume being supplied and removed as needed.

As a specific example, in Earth's atmosphere, consider the breakup of two molecules of water vapor to form two hydrogen molecules and one oxygen molecule: $2H_2O \rightarrow 2H_2 + O_2$. The inverse reaction $2H_2 + O_2 \rightarrow 2H_2O$ also occurs in the atmosphere,[10] and it is conventional to write down the two reactions simultaneously in the form

$$2H_2O \leftrightarrow 2H_2 + O_2. \tag{5.58}$$

A chosen (but arbitrary) portion of the atmosphere, with idealized walls to keep all its molecules in, can be regarded as a system. (The walls are unimportant in practice, but are pedagogically useful.) The kinetic motions of this system's molecules reach and maintain statistical equilibrium, at fixed temperature T and pressure P, far more rapidly than chemical reactions can occur. Accordingly, if we view this system on timescales short compared to that $\tau_{\rm react}$ for the reactions (5.58) but long compared to the kinetic relaxation time, then we can regard the system as in *partial statistical equilibrium*, with fixed numbers of water molecules N_{H_2O}, hydrogen molecules N_{H_2}, and oxygen molecules N_{O_2}, and with a Gibbs potential whose value is given by the Euler relation (5.53):

$$G = \tilde{\mu}_{H_2O} N_{H_2O} + \tilde{\mu}_{H_2} N_{H_2} + \tilde{\mu}_{O_2} N_{O_2}. \tag{5.59}$$

(Here, even though Earth's atmosphere is highly nonrelativistic, we include rest masses in the chemical potentials and in the Gibbs potential; the reason will become evident at the end of this section.)

When one views the sample over a longer timescale, $\Delta t \sim \tau_{\rm react}$, one discovers that these molecules are not inviolate; they can change into one another via the reactions (5.58), thereby changing the value of the Gibbs potential (5.59). The changes of G are more readily computed from the Gibbs representation of the first law, $dG =$

---

10. In the real world these two reactions are made complicated by the need for free-electron intermediaries, whose availability is influenced by external factors, such as ultraviolet radiation. This, however, does not change the issues of principle discussed here.

$VdP - SdT + \sum_I \tilde{\mu}_I dN_I$, than from the Euler relation (5.59). Taking account of the constancy of P and T and the fact that the reactions entail transforming two water molecules into two hydrogen molecules and one oxygen molecule (or conversely), so that

$$dN_{H_2} = -dN_{H_2O}, \quad dN_{O_2} = -\frac{1}{2}dN_{H_2O}, \tag{5.60a}$$

the first law says

$$dG = (2\tilde{\mu}_{H_2O} - 2\tilde{\mu}_{H_2} - \tilde{\mu}_{O_2})\frac{1}{2}dN_{H_2O}. \tag{5.60b}$$

The reactions (5.58) proceed in both directions, but statistically there is a preference for one direction over the other. The preferred direction, of course, is the one that reduces the Gibbs potential (i.e., increases the entropy of the molecules and their bath). Thus, if $2\tilde{\mu}_{H_2O}$ is larger than $2\tilde{\mu}_{H_2} + \tilde{\mu}_{O_2}$, then water molecules preferentially break up to form hydrogen plus oxygen; but if $2\tilde{\mu}_{H_2O}$ is less than $2\tilde{\mu}_{H_2} + \tilde{\mu}_{O_2}$, then oxygen and hydrogen preferentially combine to form water. As the reactions proceed, the changing N_I values produce changes in the chemical potentials $\tilde{\mu}_I$. [Recall the intimate connection

$$N_I = g_s \frac{(2\pi m_I k_B T)^{3/2}}{h^3} e^{\mu_I/(k_B T)} V \tag{5.61}$$

between $\mu_I = \tilde{\mu}_I - m_I c^2$ and N_I for a gas in the nonrelativistic regime; Eq. (3.39a).] These changes in the N_I and $\tilde{\mu}_I$ values lead ultimately to a macrostate (thermodynamic state) of minimum Gibbs potential G—a state in which the reactions (5.58) can no longer reduce G. In this final state of full statistical equilibrium, the dG of expression (5.60b) must be zero. Correspondingly, the chemical potentials associated with the reactants must balance:

$$2\tilde{\mu}_{H_2O} = 2\tilde{\mu}_{H_2} + \tilde{\mu}_{O_2}. \tag{5.62}$$

The above analysis shows that the "driving force" for the chemical reactions is the combination of chemical potentials in the dG of Eq. (5.60b). Notice that this combination has coefficients in front of the $\tilde{\mu}_I$ terms that are identical to the coefficients in the reactions (5.58) themselves, and the equilibrium relation (5.62) also has the same coefficients as the reactions (5.60b). It is easy to convince oneself that this is true in general. Consider any chemical reaction. Write the reaction in the form

$$\sum_j v_j^L A_j^L \leftrightarrow \sum_j v_j^R A_j^R. \tag{5.63}$$

Here the superscripts L and R denote the "left" and "right" sides of the reaction, the A_js are the names of the species of particle or atomic nucleus or atom or molecule involved in the reaction, and the v_js are the number of such particles (or nuclei or atoms or molecules) involved. Suppose that this reaction is occurring in an environment of fixed temperature and pressure. Then to determine the direction in which the reaction preferentially goes, examine the chemical-potential sums for the two sides of the reaction,

direction of a chemical reaction governed by chemical-potential sums

$$\sum_j v_j^L \tilde{\mu}_j^L, \qquad \sum_j v_j^R \tilde{\mu}_j^R. \tag{5.64}$$

The reaction will proceed from the side with the larger chemical-potential sum to the side with the smaller; and ultimately, the reaction will bring the two sides into equality. That final equality is the state of full statistical equilibrium. Exercises 5.9 and 5.10 illustrate this behavior.

rationale for including rest masses in chemical potentials

When dealing with chemical reactions between extremely nonrelativistic molecules and atoms (e.g., water formation and destruction in Earth's atmosphere), one might wish to omit rest masses from the chemical potentials. If one does so, and if one wishes to preserve the criterion that the reaction goes in the direction of decreasing $dG = (2\mu_{H_2O} - 2\mu_{H_2} - \mu_{O_2})\frac{1}{2}dN_{H_2O}$ [Eq. (5.60b) with tildes removed], then one must choose as the "rest masses" to be subtracted values that do not include chemical binding energies; that is, the rest masses must be defined in such a way that $2m_{H_2O} = 2m_{H_2} + m_{O_2}$. This delicacy can be avoided by simply using the relativistic chemical potentials. The derivation of the Saha equation (Ex. 5.10) is an example.

EXERCISES

Exercise 5.9 **Example: Electron-Positron Equilibrium at "Low" Temperatures**
Consider hydrogen gas in statistical equilibrium at a temperature $T \ll m_e c^2/k_B \simeq 6 \times 10^9$ K. Electrons at the high-energy end of the Boltzmann energy distribution can produce electron-positron pairs by scattering off protons:

$$e^- + p \rightarrow e^- + p + e^- + e^+. \tag{5.65}$$

(There are many other ways of producing pairs, but in analyzing statistical equilibrium we get all the information we need—a relation among the chemical potentials—by considering just one way.)

(a) In statistical equilibrium, the reaction (5.65) and its inverse must proceed on average at the same rate. What does this imply about the relative magnitudes of the electron and positron chemical potentials $\tilde{\mu}_-$ and $\tilde{\mu}_+$ (with rest masses included)?

(b) Although these reactions require an e^- that is relativistic in energy, almost all the electrons and positrons will have kinetic energies of magnitude $E \equiv \mathcal{E} - m_e c^2 \sim k_B T \ll m_e c^2$, and thus they will have $\mathcal{E} \simeq m_e c^2 + \mathbf{p}^2/(2m_e)$. What are the densities in phase space $\mathcal{N}_\pm = dN_\pm/(d^3x d^3p)$ for the positrons and electrons in terms of $\mathbf{p}$, $\tilde{\mu}_\pm$, and T? Explain why for a hydrogen gas we must have $\tilde{\mu}_- > 0$ and $\tilde{\mu}_+ < 0$.

(c) Assume that the gas is very dilute, so that $\eta \ll 1$ for both electrons and positrons. Then integrate over momenta to obtain the following formula for the number

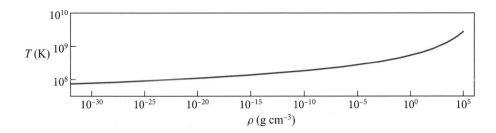

FIGURE 5.7 The temperature T_p at which electron-positron pairs form in a dilute hydrogen plasma, plotted as a function of density ρ. This is the correct upper limit (upper dashed line in Fig. 3.7) on the region where the plasma can be considered fully nonrelativistic. Above this curve, although T may be $\ll m_e c^2/k_B \simeq 6 \times 10^9$ K, a proliferation of electron-positron pairs radically changes the properties of the plasma.

densities in physical space of electrons and positrons:

$$n_\pm = \frac{2}{h^3}(2\pi m_e k_B T)^{3/2} \exp\left(\frac{\tilde{\mu}_\pm - m_e c^2}{k_B T}\right). \tag{5.66}$$

In cgs units, what does the dilute-gas assumption $\eta \ll 1$ correspond to in terms of $n_\pm$? What region of hydrogen mass density ρ and temperature T is the dilute-gas region?

(d) Let n be the number density of protons. Then by charge neutrality, $n = n_- - n_+$ will also be the number density of "ionization electrons" (i.e., of electrons that have been ionized off of hydrogen). Show that the ratio of positrons (and hence of pairs) to ionization electrons is given by

$$\frac{n_+}{n} = \frac{1}{2y[y + (1 + y^2)^{\frac{1}{2}}]}, \tag{5.67a}$$

where

$$y \equiv \frac{1}{4}n\lambda^3 e^{m_e c^2/(k_B T)}, \quad \text{and} \quad \lambda \equiv \frac{h}{\sqrt{2\pi m_e k_B T}} \tag{5.67b}$$

is the thermal deBroglie wavelength of the electrons. Figure 5.7 shows the temperature T_p at which, according to this formula, $n_+ = n$ (and $y = 0.354$), as a function of mass density $\rho \simeq m_{\text{proton}} n$. This T_p can be thought of as the temperature at which pairs form in a dilute plasma. Somewhat below T_p there are hardly any pairs; somewhat above, the pairs are profuse.

(e) Note that at low densities pairs form at temperatures $T \sim 10^8$ K $\simeq 0.02 m_e c^2/k_B$. Explain qualitatively in terms of available phase space why the formation temperature is so low.

Exercise 5.10 **_Example: Saha Equation for Ionization Equilibrium_
Consider an optically thick hydrogen gas in statistical equilibrium at temperature T. ("Optically thick" means that photons can travel only a small distance compared to

the size of the system before being absorbed, so they are confined by the hydrogen and kept in statistical equilibrium with it.) Among the reactions that are in statistical equilibrium are $H + \gamma \leftrightarrow e + p$ (ionization and recombination of H, with the H in its ground state) and $e + p \leftrightarrow e + p + \gamma$ (emission and absorption of photons by "bremsstrahlung," i.e., by the Coulomb-force-induced acceleration of electrons as they fly past protons). Let $\tilde{\mu}_\gamma, \tilde{\mu}_H, \tilde{\mu}_e$, and $\tilde{\mu}_p$ be the chemical potentials including rest masses; let m_H, m_e, and m_p be the rest masses; denote by $I \, (\equiv 13.6 \text{ eV})$ the ionization energy of H, so that $m_H c^2 = m_e c^2 + m_p c^2 - I$; denote $\mu_j \equiv \tilde{\mu}_j - m_j c^2$; and assume that $T \ll m_e c^2 / k_B \simeq 6 \times 10^9$ K and that the density is low enough that the electrons, protons, and H atoms can be regarded as nondegenerate (i.e., as distinguishable, classical particles).

(a) What relationships hold between the chemical potentials $\tilde{\mu}_\gamma, \tilde{\mu}_H, \tilde{\mu}_e$, and $\tilde{\mu}_p$?

(b) What are the number densities n_H, n_e, and n_p expressed in terms of $T, \tilde{\mu}_H, \tilde{\mu}_e$, and $\tilde{\mu}_p$—taking account of the fact that the electron and proton both have spin $\frac{1}{2}$, and including for H all possible electron and nuclear spin states?

(c) Derive the Saha equation for ionization equilibrium:

$$\frac{n_e n_p}{n_H} = \frac{(2\pi m_e k_B T)^{3/2}}{h^3} e^{-I/(k_B T)}. \tag{5.68}$$

This equation is widely used in astrophysics and elsewhere.

5.6 Fluctuations away from Statistical Equilibrium

5.6

As we saw in Chap. 4, statistical mechanics is built on a distribution function ρ that is equal to the probability of finding a chosen system in a quantum state at some chosen location in the system's phase space. For systems in statistical equilibrium, this probability is given by the microcanonical, canonical, grand canonical, Gibbs, or other distribution, depending on the nature of the system's interactions with its surroundings. Classical thermodynamics makes use of only a tiny portion of the information in this probability distribution: the mean values of a few macroscopic parameters (energy, entropy, volume, pressure, etc.). Also contained in the distribution function, but ignored by classical thermodynamics, is detailed information about fluctuations of a system away from its mean values.

As an example, consider a microcanonical ensemble of boxes, each with volume V and each containing precisely N identical, nonrelativistic, classical gas particles and containing energy (excluding rest mass) between E and $E + \delta E$, where $\delta E \ll E$. (Remember the kludge that was necessary in Ex. 4.7). Consider a quantity y that is not fixed by the set $\{E, V, N\}$. That quantity might be discrete (e.g., the total number

N_R of particles in the right half of the box). Alternatively, it might be continuous (e.g., the total energy E_R in the right half).

In the discrete case, the total number of quantum states that correspond to specific values of y is related to the entropy S by the standard microcanonical relation $N_{states}(E, V, N; y) = \exp[S(E, V, N; y)/k_B]$; and correspondingly, since all states are equally probable in the microcanonical ensemble, the probability of finding a system of the ensemble to have the specific value y is

$$p(E, V, N; y) = \frac{N_{states}(E, V, N; y)}{\sum_y N_{states}(E, V, N; y)} = \text{const} \times \exp\left[\frac{S(E, V, N; y)}{k_B}\right]. \quad (5.69a)$$

probabilities for fluctuations in closed systems

Here the entropy S is to be computed via statistical mechanics (or, when possible, via thermodynamics) not for the original ensemble of boxes in which y was allowed to vary, but for an ensemble in which y is fixed at a chosen value.

The continuous case (e.g., $y = E_R$) can be converted into the discrete case by dividing the range of y into intervals that all have the same infinitesimal width δy. Then the probability of finding y in one of these intervals is $[dp(E, V, N; y \text{ in } \delta y)/dy]\delta y = \text{const} \times \exp[S(E, V, N; y \text{ in } \delta y)]$. Dividing both sides by δy and absorbing δy on the right-hand side into the constant, we obtain

$$\frac{dp(E, V, N; y)}{dy} = \text{const} \times \exp\left[\frac{S(E, V, N; y \text{ in } \delta y)}{k_B}\right]. \quad (5.69b)$$

Obviously, if we are interested in the joint probability for a set of ys, some discrete (e.g., $y_1 = N_R$) and some continuous (e.g., $y_2 = E_R$), that probability will be given by

$$\frac{dp(E, V, N; y_1, y_2, \ldots, y_r)}{dy_q \cdots dy_r} = \text{const} \times \exp\left[\frac{S(E, V, N; y_j)}{k_B}\right], \quad (5.69c)$$

where we keep in mind (but now omit from our notation) the fact that continuous variables are to be given values y_j in some arbitrary but fixed infinitesimal range δy_j.

The probability distribution (5.69c), though exact, is not terribly instructive. To get better insight we expand S in powers of the deviations of the y_j from the values $\bar{y}_j$ that maximize the entropy (these will turn out also to be the means of the distribution). Then for small $|y_j - \bar{y}_j|$, Eq. (5.69c) becomes

$$\frac{dp(E, V, N; y_j)}{dy_q \ldots dy_r} = \text{const} \times \exp\left[\frac{1}{2k_B}\left(\frac{\partial^2 S}{\partial y_j \partial y_k}\right)(y_j - \bar{y}_j)(y_k - \bar{y}_k)\right]. \quad (5.69d)$$

Gaussian approximation to fluctuation probabilities

Here the second partial derivative of the entropy is to be evaluated at the maximum-entropy location, where $y_j = \bar{y}_j$ for all j. Expression (5.69d) is a (multidimensional) Gaussian probability distribution for which the means are obviously $\bar{y}_j$, as predicted. (That this had to be Gaussian follows from the central limit theorem, Sec. 6.3.2.)

The last entry in the first line of Table 5.2 summarizes the above equations: *for a closed system, the probability of some fluctuation away from equilibrium is proportional* *to e^{S/k_B}, where S is the total entropy for the out-of-equilibrium fluctuational macrostate* (e.g., the macrostate with N_R particles in the right half of the box).

For the specific example where $y_1 \equiv N_R =$ (number of perfect-gas particles in right half of box) and $y_2 \equiv E_R =$ (amount of energy in right half of box), we can infer $S(E, V, N; N_R, E_R)$ from the Sackur-Tetrode equation (4.42) as applied to the two halves of the box and then added:[11]

$$S(E, V, N; E_R, N_R) = k_B N_R \ln\left[\left(\frac{4\pi m}{3h^2}\right)^{3/2} e^{5/2} \frac{V}{2} \frac{E_R^{3/2}}{N_R^{5/2}}\right]$$

$$+ k_B(N - N_R) \ln\left[\left(\frac{4\pi m}{3h^2}\right)^{3/2} e^{5/2} \frac{V}{2} \frac{(E - E_R)^{3/2}}{(N - N_R)^{5/2}}\right]. \quad (5.70a)$$

It is straightforward to compute the values $\bar{E}_R$ and $\overline{N}_R$ that maximize this entropy:

$$\bar{E}_R = \frac{E}{2}, \qquad \overline{N}_R = \frac{N}{2}. \quad (5.70b)$$

Thus, in agreement with intuition, the mean values of the energy and particle number in the right half-box are equal to half of the box's total energy and particle number. It is also straightforward to compute from expression (5.70a) the second partial derivatives of the entropy with respect to E_R and N_R, evaluate them at $E_R = \bar{E}_R$ and $N_R = \overline{N}_R$, and plug them into the probability distribution (5.69d) to obtain

$$\frac{dp_{N_R}}{dE_R} = \text{const} \times \exp\left(\frac{-(N_R - N/2)^2}{2(N/4)} + \frac{-[(E_R - E/2) - (E/N)(N_R - N/2)]^2}{2(N/6)(E/N)^2}\right).$$

$$(5.70c)$$

This Gaussian distribution has the following interpretation. (i) There is a correlation between the energy E_R and the particle number N_R in the right half of the box, as one might have expected: if there is an excess of particles in the right half, then we must expect an excess of energy there as well. (ii) The quantity that is not correlated with N_R is $E_R - (E/N)N_R$, again as one might have expected. (iii) For fixed N_R, dp_{N_R}/dE_R is Gaussian with mean $\bar{E}_R = E/2 + (E/N)(N_R - N/2)$ and with rms fluctuation (standard deviation; square root of variance) $\sigma_{E_R} = (E/N)\sqrt{N/6}$. (iv) After integrating over E_R, we obtain

$$p_{N_R} = \text{const} \times \exp\left[\frac{-(N_R - N/2)^2}{2N/4}\right]. \quad (5.70d)$$

This is Gaussian with mean $\overline{N}_R = N/2$ and rms fluctuation $\sigma_{N_R} = \sqrt{N/4}$. By contrast, if the right half of the box had been in equilibrium with a bath far larger than itself, N_R would have had an rms fluctuation equal to the square root of its mean, $\sigma_{N_R} = \sqrt{N/2}$

---

11. Note that the derivation of Eq. (4.42), as specialized to the right half of the box, requires the same kind of infinitesimal range $\delta y_2 = \delta E_R$ as we used to derive our fluctuational probability equation (5.69d).

(see Ex. 5.11). The fact that the companion of the right half has only the same size as the right half, rather than being far larger, has reduced the rms fluctuation of the number of particles in the right half from $\sqrt{N/2}$ to $\sqrt{N/4}$.

Notice that the probability distributions (5.70c) and (5.70d) are exceedingly sharply peaked about their means. Their standard deviations divided by their means (i.e., the magnitude of their fractional fluctuations) are all of order $1/\sqrt{\overline{N}}$, where $\overline{N}$ is the mean number of particles in a system; and in realistic situations $\overline{N}$ is very large. (For example, $\overline{N}$ is of order 10^{26} for a cubic meter of Earth's atmosphere, and thus the fractional fluctuations of thermodynamic quantities are of order 10^{-13}.) It is this extremely sharp peaking that makes classical thermodynamics insensitive to the choice of type of equilibrium ensemble—that is, sensitive only to means and not to fluctuations about the means.

The generalization of this example to other situations should be fairly obvious; see Table 5.2. When a system is in some out-of-equilibrium macrostate, the total entropy $S + S_b$ of the system and any bath with which it may be in contact is, up to an additive constant, either the system's entropy S or the negative of its out-of-equilibrium potential divided by the bath's temperature ($-F/T_b + \text{const}$, $-G/T_b + \text{const}$, or $-\Omega/T_b + \text{const}$; see column 3 of Table 5.2). Correspondingly, the probability of a fluctuation from statistical equilibrium to this out-of-equilibrium macrostate is proportional to the exponential of this quantity in units of Boltzmann's constant (e^{-S/k_B}, $e^{-F/(k_BT_b)}$, $e^{-G/(k_BT_b)}$, or $e^{-\Omega/(k_BT_b)}$; column 5 of Table 5.2). By expanding the quantity in the exponential around the equilibrium state to second order in the fluctuations, one obtains a Gaussian probability distribution for the fluctuations, like Eq. (5.69d).

probabilities for fluctuations in systems interacting with baths

As examples, in Ex. 5.11 we study fluctuations in the number of particles in a cell that is immersed in a heat and particle bath, so the starting point is the out-of-equilibrium grand potential Ω. And in Ex. 5.12, we study temperature and volume fluctuations for a system in contact with a heat and volume bath; so the starting point is the out-of-equilibrium Gibbs function G.

EXERCISES

Exercise 5.11 *Example: Probability Distribution for the Number of Particles in a Cell*
Consider a cell with volume V, like those of Fig. 5.1, that has imaginary walls and is immersed in a bath of identical, nonrelativistic, classical perfect-gas particles with temperature T_b and chemical potential μ_b. Suppose that we make a large number of measurements of the number of particles in the cell and that from those measurements we compute the probability p_N for that cell to contain N particles.

(a) How widely spaced in time must the measurements be to guarantee that the measured probability distribution is the same as that computed, using the methods of this section, from an ensemble of cells (Fig. 5.1) at a specific moment of time?

(b) Assume that the measurements are widely enough separated for this criterion to be satisfied. Show that p_N is given by

$$p_N \propto \exp\left[\frac{-\Omega(V, T_b, \mu_b; N)}{k_B T_b}\right]$$

$$\equiv \frac{1}{N!} \int \frac{d^{3N}x \, d^{3N}p}{h^{3N}} \exp\left[\frac{-E_n + \mu_b N_n}{k_B T_b}\right] \qquad (5.71)$$

$$= \frac{1}{N!} \int \frac{d^{3N}x \, d^{3N}p}{h^{3N}} \exp\left[\frac{-\left(\sum_{i=1}^{N} \mathbf{p}_i^2/(2m)\right) + \mu_b N}{k_B T_b}\right],$$

where $\Omega(V, T_b, \mu_b; N)$ is the grand potential for the ensemble of cells, with each cell constrained to have precisely N particles in it (cf. the last entry in Table 5.2).

(c) By evaluating Eq. (5.71) exactly and then computing the normalization constant, show that the probability p_N for the cell to contain N particles is given by the *Poisson distribution*

$$p_N = e^{-\overline{N}}(\overline{N}^N/N!), \qquad (5.72a)$$

where $\overline{N}$ is the mean number of particles in a cell,

$$\overline{N} = (\sqrt{2\pi m k_B T_b}/h)^3 e^{\mu_b/(k_B T_b)} V$$

[Eq. (3.39a)].

(d) Show that for the Poisson distribution (5.72a), the expectation value is $\langle N \rangle = \overline{N}$, and the rms deviation from this is

$$\sigma_N \equiv \langle (N - \overline{N})^2 \rangle^{\frac{1}{2}} = \overline{N}^{\frac{1}{2}}. \qquad (5.72b)$$

(e) Show that for $N - \overline{N} \lesssim \sigma_N$, this Poisson distribution is exceedingly well approximated by a Gaussian with mean $\overline{N}$ and variance σ_N.

Exercise 5.12 *Example: Fluctuations of Temperature and Volume in an Ideal Gas*
Consider a gigantic container of gas made of identical particles that might or might not interact. Regard this gas as a bath, with temperature T_b and pressure P_b. Pick out at random a sample of the bath's gas containing precisely N particles, with $N \gg 1$. Measure the volume V of the sample and the temperature T inside the sample. Then pick another sample of N particles, and measure its V and T, and repeat over and over again. Thereby map out a probability distribution $dp/dT dV$ for V and T of N-particle samples inside the bath.

(a) Explain in detail why

$$\frac{dp}{dTdV} = \text{const} \times \exp\left[-\frac{1}{2k_BT_b}\left(\frac{\partial^2 G}{\partial V^2}(V-\bar{V})^2 + \frac{\partial^2 G}{\partial T^2}(T-T_b)^2\right.\right.$$
$$\left.\left. + 2\frac{\partial^2 G}{\partial T\partial V}(V-\bar{V})(T-T_b)\right)\right],$$

(5.73a)

where $G(N, T_b, P_b; T, V) = E(T, V, N) + P_bV - T_bS(T, V, N)$ is the out-of-equilibrium Gibbs function for a sample of N particles interacting with this bath (next-to-last line of Table 5.2), $\bar{V}$ is the equilibrium volume of the sample when its temperature and pressure are those of the bath, and the double derivatives in Eq. (5.73a) are evaluated at the equilibrium temperature T_b and pressure P_b.

(b) Show that the derivatives, evaluated at $T = T_b$ and $V = \bar{V}$, are given by

$$\left(\frac{\partial^2 G}{\partial T^2}\right)_{V,N} = \frac{C_V}{T_b}, \quad \left(\frac{\partial^2 G}{\partial V^2}\right)_{T,N} = \frac{1}{\kappa}, \quad \text{and} \quad \left(\frac{\partial^2 G}{\partial T\partial V}\right)_N = 0, \quad (5.73b)$$

where C_V is the gas sample's specific heat at fixed volume and κ is its compressibility at fixed temperature β, multiplied by V/P:

$$C_V \equiv \left(\frac{\partial E}{\partial T}\right)_{V,N} = T\left(\frac{\partial S}{\partial T}\right)_{V,N}, \quad \kappa \equiv \beta V/P = -\left(\frac{\partial V}{\partial P}\right)_{T,N}, \quad (5.73c)$$

both evaluated at temperature T_b and pressure P_b. [Hint: Write $G = G_{\text{eq}} + (P_b - P)V - (T_b - T)S$, where G_{eq} is the equilibrium Gibbs function for the gas samples.] Thereby conclude that

$$\frac{dp}{dTdV} = \text{const} \times \exp\left[-\frac{(V-\bar{V})^2}{2k_BT_b\kappa} - \frac{C_V(T-T_b)^2}{2k_BT_b^2}\right]. \quad (5.73d)$$

(c) This probability distribution says that the temperature and volume fluctuations are uncorrelated. Is this physically reasonable? Why?

(d) What are the rms fluctuations of the samples' temperature and volume, σ_T and σ_V? Show that σ_T scales as $1/\sqrt{N}$ and σ_V as $\sqrt{N}$, where N is the number of particles in the samples. Are these physically reasonable? Why?

Exercise 5.13 *Example and Derivation: Evolution and Fluctuations*
of a System in Contact with a Volume Bath
Exercise 5.5 explored the enthalpy representation of thermodynamics for an equilibrium ensemble of systems in contact with a volume bath. Here we extend that analysis to an ensemble out of equilibrium. We denote by P_b the bath pressure.

(a) The systems exchange volume but not heat or particles with the bath. Explain why, even though the ensemble may be far from equilibrium, any system's volume change dV must be accompanied by an energy change $d\mathcal{E} = -P_b dV$. This implies that the system's enthalpy $H = \mathcal{E} + P_b V$ is conserved. All systems in the ensemble are assumed to have the same enthalpy H and the same number of particles N.

(b) Using equilibrium considerations for the bath, show that interaction with a system cannot change the bath's entropy.

(c) Show that the ensemble will always evolve toward increasing entropy S, and that when the ensemble finally reaches statistical equilibrium with the bath, its distribution function will be that of the enthalpy ensemble (Table 5.1): $\rho = e^{-S/k_B} =$ const for all regions of phase space that have the specified particle number N and enthalpy H.

(d) Show that fluctuations away from equilibrium are described by the probability distributions (5.69a) and (5.69c), but with the system energy E replaced by the system enthalpy H and the system volume V replaced by the bath pressure P_b (cf. Table 5.2).

5.7 Van der Waals Gas: Volume Fluctuations and Gas-to-Liquid Phase Transition

The van der Waals gas studied in Sec. 5.3.2 provides a moderately realistic model for real gases such as H_2O and their condensation (phase transition) into liquids, such as water.

The equation of state for a van der Waals gas is

$$\left(P + \frac{a}{v^2} \right) (v - b) = k_B T \tag{5.74}$$

[Eq. (5.26)]. Here a and b are constants, and $v \equiv V/N$ is the specific volume (the inverse of the number density of gas particles). In Fig. 5.8a we depict this equation of state as curves (*isotherms*) of pressure P versus specific volume v at fixed temperature T. Note [as one can easily show from Eq. (5.74)] that there is a critical temperature $T_c = 8a/(27bk_B)$ such that for $T > T_c$ the isotherms are monotonic decreasing; for $T = T_c$ they have an inflection point [at $v = v_c \equiv 3b$ and $P = P_c = a/(27b^2)$]; and for $T < T_c$ they have a maximum and a minimum.

From Eq. (5.73d), derived in Ex. 5.12, we can infer that the probability dp/dv for fluctuations of the specific volume of a portion of this gas containing N particles is

probability for volume fluctuations in van der Waals gas

$$\frac{dp}{dv} = \text{const} \times \exp\left[\frac{N(\partial P/\partial v)_T}{2k_B T} (v - \bar{v})^2 \right]. \tag{5.75}$$

This probability is controlled by the slope $(\partial P/\partial v)_T$ of the isotherms. Where the slope is negative, the volume fluctuations are small; where it is positive, the fluid is unstable: its volume fluctuations grow. Therefore, for $T < T_c$, the region of an isotherm between its minimum M and its maximum X (Fig. 5.8b) is unphysical; the fluid cannot exist stably there. Evidently, at $T < T_c$ there are two phases: one with low density ($v > v_X$)

Chapter 5. Statistical Thermodynamics

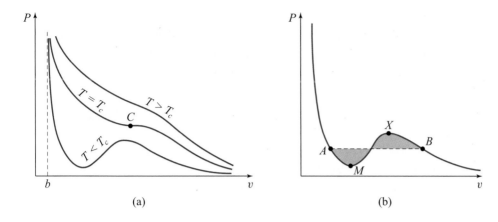

FIGURE 5.8 (a) The van der Waals equation of state $P(N, V, T)$ plotted as pressure P versus specific volume $v \equiv V/N$ at fixed temperature T, for various values of T. (b) The route of a phase transition in the van der Waals gas. The transition is a discontinuous jump from point A to point B.

is gaseous; the other with high density ($v < v_M < v_c = 3b$) is liquid. [Recall, from comment (i) at the end of Sec. 5.3.2, that $b/4$ is the volume of each of the material's particles, so in the high-density phase the particles' separations are not much larger than their diameter; this is characteristic of a fluid.]

Hold the temperature T fixed at $T < T_c$, and gradually increase the density from zero (decrease the specific volume v from infinity). At low densities, the material will be gaseous, and at high densities, it will be liquid. The phase transition from gas to liquid involves a discontinuous jump from some point B in Fig. 5.8b to another point A. The Gibbs potential controls the location of those points.

Since the two phases are in equilibrium with each other at A and B, their Gibbs potential $G = \mu N$ must be the same, which means their chemical potentials must be the same, $\mu_A = \mu_B$—as, of course, must be their temperatures $T_A = T_B$ (they lie on the same isotherm). This in turn implies their pressures must be the same, $P_A = P(\mu_A, T) = P(\mu_B, T) = P_B$. Therefore, the points A and B in Fig 5.8b are connected by a horizontal line (the dashed line in the figure). Let us use the first law of thermodynamics in the Gibbs representation to compute the change in the chemical potential μ as one moves along the isotherm from point A to point B. The first law says $dG = -SdT + VdP + \mu dN$. Focusing on a sample of the material containing N particles, and noting that along the isotherm the sample has $G = \mu N$, $dN = 0$, $dT = 0$, and $V = vN$, we obtain $d\mu = vdP$. Integrating this relation along the isotherm from A to B, we obtain

$$0 = \mu_B - \mu_A = \int_A^B d\mu = \int_A^B vdP. \tag{5.76}$$

This integral is the area of the right green region in Fig. 5.8b minus the area of the left green region. Therefore, these two areas must be equal, which tells us the location of the points A and B that identify the two phases (liquid and gaseous) when the phase transition occurs.

Consider again volume fluctuations. Where an isotherm is flat, $(\partial P/\partial v)_T = 0$, large volume fluctuations occur [Eq. (5.75)]. For $T < T_c$, the isotherm is flat at the minimum M and the maximum X, but these do not occur in Nature—unless the phase transition is somehow delayed as one compresses or expands the material. However, for $T = T_c$, the isotherm is flat at its inflection point $v = v_c$, $P = P_c$ (the material's critical point C in Fig. 5.8a); so the volume fluctuations will be very large there.

At some temperatures T and pressures P, it is possible for two phases, liquid and gas, to exist; at other T and P, only one phase exists. The dividing line in the T-P plane between these two regions is called a *catastrophe*—a term that comes from *catastrophe theory*. We explore this in Ex. 7.16, after first introducing some ideas of catastrophe theory in the context of optics.

EXERCISES

Exercise 5.14 **Example: Out-of-Equilibrium Gibbs Potential for Water; Surface Tension and Nucleation*[12]

Water and its vapor (liquid and gaseous H_2O) can be described moderately well by the van der Waals model, with the parameters $a = 1.52 \times 10^{-48}$ J m^3 and $b = 5.05 \times 10^{-29}$ m^3 determined by fitting to the measured pressure and temperature at the critical point (inflection point C in Fig. 5.8a: $P_c = a/(27b^2) = 22.09$ MPa, $T_c = 8a/(27bk_B) = 647.3$ K). [Note: 1 MPa is 10^6 Pascal; and 1 Pascal is the SI unit of pressure, 1 kg m s^{-2}.]

(a) For an out-of-equilibrium sample of N atoms of H_2O at temperature T and pressure P that has fluctuated to a specific volume v, the van-der-Waals-modeled Gibbs potential is

$$G(N, T, P; v) = N\left[-k_B T + Pv - a/v + k_B T \ln\left(\lambda_{T\,dB}{}^3/(v - b)\right)\right],$$

$$\lambda_{T\,dB} \equiv h/\sqrt{2\pi mk_B T}. \tag{5.77}$$

Verify that this Gibbs potential is minimized when v satisfies the van der Waals equation of state (5.74).

(b) Plot the chemical potential $\mu = G/N$ as a function of v at room temperature, $T = 300$ K, for various pressures in the vicinity of 1 atmosphere $= 0.1013$ MPa. Adjust the pressure so that the two phases, liquid and gaseous, are in equilibrium (i.e., so the two minima of the curve have the same height). [Answer: The required pressure is about 3.6 atmospheres, and the chemical-potential curve is shown in Fig. 5.9. If the gas is a mixture of air and H_2O rather than pure H_2O, then the required pressure will be lower.]

(c) Compare the actual densities of liquid water and gaseous H_2O with the predictions of Fig. 5.9. They agree moderately well but not very well.

---

12. Exercise adapted from Sethna (2006, Ex. 11.3 and Sec. 11.3).

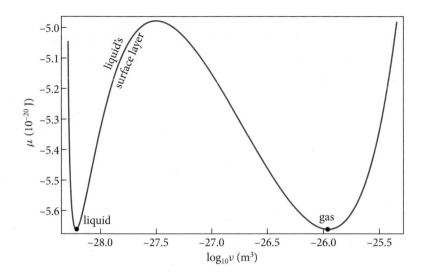

FIGURE 5.9 The out-of-equilibrium chemical potential for a van der Waals gas, with its parameters fitted to the properties of H_2O and with temperature $T = 300$ K and pressure $P = 3.6$ atmospheres at the liquid-gas interface.

(d) At the liquid's surface there is a surface layer a few molecules thick, in which the attractive force between the water molecules, $F = -6\varepsilon_o r_o^6/r^7 = -[27/(2\pi^2)]ab/r^7$ [Eqs. (5.22b) and (5.24b)], produces surface tension. This surface tension is a force per unit length, γ, that the surface molecules on one side of any line lying in the surface exert on the molecules on the other side (Box 16.4). Explain by a simple physical argument why there is an energy γA associated with this surface tension, stored in any area A of the surface. This is a free energy at fixed T, P (a Gibbs free energy) in excess of the free energy that the surface's water molecules would have if they were in the bulk liquid or the bulk gas. This excess free energy shows up in the chemical potential of Fig. 5.9, and the numbers in that figure can be used to estimate the water's surface tension γ. Show that $\gamma \sim \Delta\mu h/v$, where h and v are the thickness and specific volume, respectively, of the surface layer, and $\Delta\mu$ is the difference between the chemical potential in the surface layer and in the bulk water and gas. Estimate γ using numbers from Fig. 5.9 and compare with the measured surface tension, $\gamma \simeq 0.0723$ N/m at $T = 300$ K.

(e) In a cloud or fog, when water vapor is cooled below its equilibrium temperature with liquid water T_e, water drops try to form, that is, *nucleate*. However, there is a potential barrier against nucleation due to the surface tension of an incipient drop. If R is the drop's radius, show that the Gibbs free energy of a droplet (the sum of contributions from its surface layer and its interior) minus the free energy that the droplet's molecules will have if they remain gaseous, is

$$\Delta G = 4\pi R^2\gamma - \left(\frac{4\pi}{3}R^3\right)\frac{q\,\Delta T}{v_\ell T_e}. \tag{5.78}$$

Here q is the latent heat per molecule that is released when the vapor liquifies, v_ℓ is the liquid's specific volume, and ΔT is the amount by which the temperature has dropped below the equilibrium point T_e for the two phases. Plot this $\Delta G(R)$, and explain why (i) there is a minimum droplet radius $R_{\min}$ for nucleation to succeed, and (ii) for the droplet to form with this minimum size, thermal fluctuations must put into it some excess energy, B, above statistical equilibrium. Derive these formulas:

$$R_{\min} = \frac{2\gamma T_e v_\ell}{q \Delta T}, \qquad B = \frac{16\pi \gamma^3 T_e^2 v_\ell^2}{3q^2 \Delta T^2}. \tag{5.79}$$

Explain why the rate at which nucleation occurs must scale as $\exp[-B/(k_B T)]$, which is generally an exceedingly small number. Show that, if the nucleation occurs on a leaf or blade of grass or on the surface of a dust grain, so the drop's interface with the vapor is a hemisphere rather than a sphere, then the energy barrier B is reduced by a factor 8, and the rate of nucleation is enormously increased. That is why nucleation of water droplets almost always occurs on solid surfaces.

5.8

5.8 Magnetic Materials [T2]

The methods we have developed in this chapter can be applied to systems very different from the gases and liquids studied thus far. In this section, we focus on magnetic materials as an example, and we use this example to illustrate two powerful, modern computational techniques: the renormalization group and Monte Carlo methods.

model for magnetic material

We consider, for concreteness, the simplest type of magnetic material: one consisting of a cubic lattice of N identical atoms, each with spin 1/2 and magnetic moment m_o. The material is immersed in a uniform external magnetic field **B**, so each atom

quantum number s_i for spin orientation

(labeled by an index i) has two possible quantum states: one with its spin parallel to **B** (quantum number $s_i = +1$), the other antiparallel to **B** ($s_i = -1$). The energies of these states are $E_{s_i} = -m_o B s_i$. The atoms interact with one another's magnetic fields with a pairwise interaction energy $E_{s_i s_j}$ that we shall make more concrete in Sec. 5.8.2 below. The material's total energy, when the atoms are in the state $|n\rangle = |s_1, s_1, \ldots, s_n\rangle$, is $E_n - M_n B$, where

internal energy and magnetization

$$E_n = \sum_{i>j}^N \sum_{j=1}^N E_{s_i s_j}, \qquad M_n = m_o \sum_{j=1}^N s_j \tag{5.80a}$$

are the material's self-interaction energy (internal energy) and *magnetization*.

The atoms interact with a heat bath that has temperature T and with the external magnetic field B, which can be thought of as part of the bath.[13] When they reach sta-

---

13. Arranging the electromagnetic environment to keep B fixed is a choice similar to arranging the thermal environment to keep T fixed. Other choices are possible.

tistical equilibrium with this heat and magnetic bath, the probability for the material (all N atoms) to be in state $|n\rangle = |s_1, s_1, \ldots, s_n\rangle$ is, of course,

$$p_n = e^{G(N,B,T)/(k_B T)} e^{-(E_n - BM_n)/(k_B T)}. \tag{5.80b}$$

Here the first term is the normalization constant, which depends on the number N of atoms in the sample and the bath's B and T, and $G(N, B, T)$ is the fundamental thermodynamic potential for this system, which acts as the normalizing factor for the probability:

$$e^{-G(N,B,T)/(k_B T)} = \sum_n e^{-(E_n - BM_n)/(k_B T)}. \tag{5.80c}$$

Gibbs potential for magnetic material

We have denoted this potential by G, because it is analogous to the Gibbs potential for a gas, but with the gas's volume V_n replaced by minus the magnetization $-M_n$ and the gas bath's pressure P replaced by the material bath's magnetic field strength B. Not surprisingly, the Gibbs thermodynamic formalism for this magnetic material is essentially the same as for a gas, but with $V \to -M$ and $P \to B$.

5.8.1 Paramagnetism; The Curie Law [T2]

Paramagnetic materials have sufficiently weak self-interaction that we can set $E_n = 0$ and focus solely on the atoms' interaction with the external B field. The magnetic interaction tries to align each atom's spin with B, while thermal fluctuations try to randomize the spins. As a result, the stronger is B (at fixed temperature), the larger will be the mean magnetization $\bar{M}$. From Eq. (5.80b) for the spins' probability distribution we can compute the mean magnetization:

$$\bar{M} = \sum_n p_n M_n = e^{G/(k_B T)} \sum_n M_n e^{BM_n/(k_B T)}$$

$$= e^{G/(k_B T)} k_B T \left(\frac{\partial}{\partial B} \right)_{N,T} \sum_n e^{BM_n/(k_B T)}. \tag{5.81a}$$

The last sum is equal to $e^{-G/(k_B T)}$ [Eq. (5.80c) with $E_n = 0$], so Eq. (5.81a) becomes

$$\bar{M} = - \left(\frac{\partial G}{\partial B} \right)_{N,T}. \tag{5.81b}$$

This is obviously our material's analog of $\bar{V} = (\partial G / \partial P)_{N,T}$ for a gas [which follows from the Gibbs representation of the first law, Eq. (5.52)].

To evaluate $\bar{M}$ explicitly in terms of B, we must first compute the Gibbs function from the statistical sum (5.80c) with $E_n = 0$. Because the magnetization M_n in state $|n\rangle = |s_1, s_2, \ldots, s_N\rangle$ is the sum of contributions from individual atoms [Eq. (5.80a)], this sum can be rewritten as the product of identical contributions from each of the N atoms:

$$e^{-G/(k_B T)} = \left(e^{-Bm_o/k_B T} + e^{+Bm_o/k_B T} \right)^N = \left(2 \cosh[Bm_o/(k_B T)] \right)^N. \tag{5.81c}$$

(In the second expression, the first term is from state $s_i = -1$ and the second from $s_i = +1$.) Taking the logarithm of both sides, we obtain

$$G(B, T, N) = -Nk_B T \ln\left(2 \cosh[Bm_o/(k_B T)]\right). \tag{5.82}$$

Differentiating with respect to B [Eq. (5.81b)], we obtain for the mean magnetization

$$\bar{M} = Nm_o \tanh\left[Bm_o/(k_B T)\right]. \tag{5.83}$$

At high temperatures, $k_B T \gg Bm_o$, the magnetization increases linearly with the applied magnetic field (the atoms begin to align with $\mathbf{B}$), so the magnetic susceptibility is independent of B:

$$\chi_M \equiv \left(\frac{\partial \bar{M}}{\partial B}\right)_{T,N} \simeq Nm_o^2/(k_B T). \tag{5.84}$$

The proportionality $\chi_M \propto 1/T$ for a paramagnetic material at high temperature is called *Curie's law*. At low temperatures ($k_B T \ll Bm_o$), the atoms are essentially all aligned with $\mathbf{B}$, and the magnetization saturates at $\bar{M} = Nm_o$.

5.8.2 Ferromagnetism: The Ising Model [T2]

Turn now to a magnetic material for which the spins' interactions are strong, and there is no external B field. In such a material, at high temperatures the spin directions are random, while at low enough temperatures the interactions drive neighboring spins to
align with one another, producing a net magnetization. This is called *ferromagnetism*, because it occurs rather strongly in iron. The transition between the two regimes is sharp (i.e., it is a phase transition).

In this section, we introduce a simple model for the spins' interaction: the Ising model.[14] For simplicity, we idealize to two spatial dimensions. The corresponding 3-dimensional model is far more difficult to analyze. Studying the 2-dimensional model carefully brings out many general features of phase transitions, and the intuition it cultivates is very helpful in thinking about experiments and simulations.

In this model, the atoms are confined to a square lattice that lies in the x-y plane, and their spins can point up (along the $+z$ direction) or down. The pairwise interaction energy is nonzero only for nearest neighbor atoms:

$$E_{s_i, s_j} = -J s_i s_j \quad \text{for nearest neighbors;} \tag{5.85}$$

it vanishes for all other pairs. Here J is a positive constant (which depends on the lattice's specific volume $v = V/N$, but that will not be important for us). Note that the interaction energy $-J s_i s_j$ is negative if the spins are aligned ($s_i = s_j$) and positive if they are opposite ($s_i = -s_j$), so like spins attract and opposite spins repel. Although the Ising model does not explicitly include more distant interactions, they are present indirectly: the "knock-on" effect from one spin to the next, as we shall see, introduces

---

14. Named for Ernst Ising, who first investigated it, in 1925.

long-range organization that propagates across the lattice when the temperature is reduced below a critical value T_c, inducing a second-order phase transition. We use the dimensionless parameter

$$K \equiv J/(k_B T) \tag{5.86}$$

to characterize the phase transition. For $K \ll 1$ (i.e., $k_B T \gg J$), the spins will be almost randomly aligned, and the total interaction energy will be close to zero. When $K \gg 1$ (i.e., $k_B T \ll J$), the strong coupling will drive the spins to align over large (2-dimensional) volumes. At some critical intermediate temperature $K_c \sim 1$ [and corresponding temperature $T_c = J/(k_B K_c)$], the phase transition will occur.

critical temperature T_c and critical parameter K_c

We compute this critical K_c, and macroscopic properties of the material near it, using two modern, sophisticated computational techniques: renormalization methods in Sec. 5.8.3 and Monte Carlo methods in Sec. 5.8.4. We examine the accuracy of these methods by comparing our results with an exact solution for the 2-dimensional Ising model, derived in a celebrated paper by Lars Onsager (1944).

5.8.3 Renormalization Group Methods for the Ising Model T2

The key idea behind the renormalization group approach to the Ising model is to try to replace the full lattice by a sparser lattice that has similar thermodynamic properties, and then to iterate, making the lattice more and more sparse; see Fig. 5.10.[15]

idea behind renormalization group

We implement this procedure using the statistical sum (5.80c) for the Gibbs potential, except that here the external magnetic field B vanishes, so the bath is purely thermal and its potential is $F(N, V, T)$—the physical free energy, not G—and the statistical sum (5.80c) reads $e^{-F/(k_B T)} \equiv z = \sum_n e^{-E_n/(k_B T)}$. For our Ising model, with its nearest-neighbor interaction energies (5.85), this becomes

$$e^{-F(N,V,T)/(k_B T)} \equiv z = \sum_{\{s_1 = \pm 1, s_2 = \pm 1, \ldots\}} e^{K \Sigma^1 s_i s_j}. \tag{5.87a}$$

Here in the exponential Σ^1 means a sum over all pairs of nearest neighbor sites $\{i, j\}$. The dependence on the material's number of atoms N appears in the number of terms in the big sum; the dependence on V/N is via the parameter J, and on T is via the parameter $K = J/(k_B T)$.

The first step in the renormalization group method is to rewrite Eq. (5.87a) so that each of the open-circle spins of Fig. 5.10 (e.g., s_5) appears in only one term in the exponential. Then we explicitly sum each of those spins over ± 1, so they no longer appear in the summations:

$$z = \sum_{\{\ldots, s_4 = \pm 1, s_5 = \pm 1, s_6 = \pm 1, \ldots\}} \cdots e^{K(s_1 + s_2 + s_3 + s_4)s_5} \cdots$$

$$= \sum_{\{\ldots, s_4 = \pm 1, s_6 = \pm 1, \ldots\}} \cdots \left[e^{K(s_1 + s_2 + s_3 + s_4)} + e^{-K(s_1 + s_2 + s_3 + s_4)} \right] \cdots . \tag{5.87b}$$

---

15. This section is based in part on Maris and Kadanoff (1978) and Chandler (1987, Sec. 5.7).

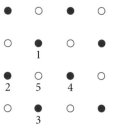

FIGURE 5.10 Partition of a square lattice into two interlaced square lattices (solid circles and open circles). In the renormalization group approach, the open-circle spins are removed from the lattice, and all their interactions are replaced by modified interactions between the remaining solid-circle spins. The new lattice is rotated by $\pi/4$ with respect to the original lattice, and the lattice spacing increases by a factor $\sqrt{2}$.

(This rewriting of z is possible because each open-circle spin interacts only with solid-circle spins.) The partition function z is now a product of terms like those in the square brackets, one for each open-circle lattice site that we have "removed." We would like to rewrite each square-bracketed term in a form involving solely nearest-neighbor interactions of the solid-circle spins, so that we can then iterate our procedure. Such a rewrite, however, is not possible; after some experimentation, one can verify that the rewrite also requires next-nearest-neighbor interactions and four-site interactions:

$$\left[e^{K(s_1+s_2+s_3+s_4)} + e^{-K(s_1+s_2+s_3+s_4)} \right]$$
$$= f(K) e^{\left[\frac{1}{2}K_1(s_1s_2+s_2s_3+s_3s_4+s_4s_1)+K_2(s_1s_3+s_2s_4)+K_3s_1s_2s_3s_4\right]}. \tag{5.87c}$$

We can determine the four functions $K_1(K)$, $K_2(K)$, $K_3(K)$, $f(K)$ by substituting each of the four possible distinct combinations of $\{s_1, s_2, s_3, s_4\}$ into Eq. (5.87c). Those four combinations, arranged in the pattern of the solid circles of Fig. 5.10, are $+\overset{+}{\underset{+}{}}+$, $-\overset{+}{\underset{+}{}}-$, $+\overset{+}{\underset{-}{}}-$, and $+\overset{+}{\underset{+}{}}-$. [Rotating the pattern or changing all signs leaves both sides of Eq. (5.87c) unchanged.] By inserting these combinations into Eq. (5.87c) and performing some algebra, we obtain

$$K_1 = \frac{1}{4}\ln\cosh(4K),$$

$$K_2 = \frac{1}{8}\ln\cosh(4K),$$

$$K_3 = \frac{1}{8}\ln\cosh(4K) - \frac{1}{2}\ln\cosh(2K), \text{ and}$$

$$f(K) = 2[\cosh(2K)]^{1/2}[\cosh(4K)]^{1/8}. \tag{5.87d}$$

By inserting expression (5.87c) and the analogous expressions for the other terms into Eq. (5.87b), we obtain the partition function for our original N-spin lattice of open and closed circles, expressed as a sum over the $(N/2)$-spin lattice of closed circles:

$$z(N, K) = [f(K)]^{N/2} \sum e^{[K_1\Sigma^1 s_i s_j + K_2\Sigma^2 s_i s_j + K_3\Sigma^3 s_i s_j s_k s_l]}. \tag{5.87e}$$

Here the symbol Σ^1 still represents a sum over all nearest neighbors but now in the $N/2$ lattice, Σ^2 is a sum over the four next nearest neighbors, and Σ^3 is a sum over spins located at the vertices of a unit cell. [The reason we defined K_1 with the 1/2 in Eq. (5.87c) was because each nearest neighbor interaction appears in two adjacent squares of the solid-circle lattice, thereby converting the 1/2 to a 1 in Eq. (5.87e).]

So far, what we have done is exact. We now make two drastic approximations that Onsager did not make in his exact treatment, but are designed to simplify the remainder of the calculation and thereby elucidate the renormalization group method. First, in evaluating the partition function (5.87e), we drop completely the quadruple interaction (i.e., we set $K_3 = 0$). This is likely to be decreasingly accurate as we lower the temperature and the spins become more aligned. Second, we assume that near the critical point, in some average sense, the degree of alignment of next nearest neighbors (of which there are as many as nearest neighbors) is "similar" to that of the nearest neighbors, so that we can set $K_2 = 0$ but increase K_1 to

$$K' = K_1 + K_2 = \frac{3}{8} \ln \cosh(4K). \tag{5.88}$$

(If we simply ignored K_2 we would not get a phase transition.) This substitution ensures that the energy of a lattice with $N/2$ *aligned* spins—and therefore N nearest neighbor and N next nearest neighbor bonds, namely, $-(K_1 + K_2)Nk_BT$—is the same as that of a lattice in which we just include the nearest neighbor bonds but strengthen the interaction from K_1 to K'. Clearly this will be unsatisfactory at high temperature, but we only need it to be true near the phase transition's critical temperature.

These approximations bring the partition function (5.87e) into the form

$$z(N, K) = [f(K)]^{N/2} z(N/2, K'), \tag{5.89a}$$

which relates the partition function (5.87a) for our original Ising lattice of N spins and interaction constant K to that of a similar lattice with $N/2$ spins and interaction constant K'.

As the next key step in the renormalization procedure, we note that because the free energy, $F = -k_BT \ln z$, is an extensive variable, $\ln z$ must increase in direct proportion to the number of spins, so that it must have the form

$$-F/(k_BT) \equiv \ln z(N, K) = Ng(K), \tag{5.89b}$$

for some function $g(K)$. By combining Eqs. (5.89a) and (5.89b), we obtain a relation for the function $g(K)$ (the free energy, aside from constants) in terms of the function $f(K)$:

$$g(K') = 2g(K) - \ln f(K), \quad \text{where } f(K) = 2[\cosh(2K)]^{1/2}[\cosh(4K)]^{1/8} \tag{5.90}$$

[cf. Eq. (5.87d)].

Equations (5.88) and (5.90) are the fundamental equations that allow us to calculate thermodynamic properties. They are called the *renormalization group equations*, because their transformations form a mathematical group, and they are a scheme

renormalization group equations

for determining how the effective coupling parameter K changes (gets renormalized) when one views the lattice on larger and larger distance scales. Renormalization group equations like these have been widely applied in elementary-particle theory, condensed-matter theory, and elsewhere. Let us examine these in detail.

The iterative map (5.88) expresses the coupling constant K' for a lattice of enlarged physical size and reduced number of particles $N/2$ in terms of K for the smaller lattice with N particles. [And the associated map (5.90) expresses the free energy when the lattice is viewed on the larger scale in terms of that for a smaller scale.] The map (5.88) has a fixed point that is obtained by setting $K' = K$ [i.e., $K_c = \frac{3}{8} \ln \cosh(4K_c)$], which implies

$$K_c = 0.507. \tag{5.91}$$

This fixed point corresponds to the critical point for the phase transition, with critical temperature T_c such that $K_c = J/(k_B T_c)$.

We can infer that this is the critical point by the following physical argument. Suppose that T is slightly larger than T_c, so K is slightly smaller than K_c. Then, when we make successive iterations, because $dK'/dK > 1$ at $K = K_c$, K decreases with each step, moving farther from K_c; the fixed point is unstable. What this means is that, when $T > T_c$, as we look on larger and larger scales, the effective coupling constant K becomes weaker and weaker, so the lattice becomes more disordered. Conversely, below the critical temperature ($T < T_c$ and $K > K_c$), the lattice becomes more ordered with increasing scale. Only when $K = K_c$ does the lattice appear to be comparably disordered on all scales. It is here that the increase of order with lengthscale changes from greater order at smaller scales (for high temperatures) to greater order at larger scales (for low temperatures).

To demonstrate more explicitly that $K = K_c$ is the location of a phase transition, we compute the lattice's specific heat in the vicinity of K_c. The first step is to compute the lattice's entropy, $S = -(\partial F/\partial T)_{V,N}$. Recalling that $K \propto 1/T$ at fixed V, N [Eq. (5.86)] and using expression (5.89b) for F, we see that

$$S = -\left(\frac{\partial F}{\partial T}\right)_{V,N} = Nk_B\left[g - K\left(\frac{dg}{dK}\right)\right]. \tag{5.92a}$$

The specific heat at constant volume is then, in turn, given by

$$C_V = T\left(\frac{\partial S}{\partial T}\right)_{V,N} = Nk_B K^2 \frac{d^2g}{dK^2}. \tag{5.92b}$$

Next we note that, because the iteration (5.88) is unstable near K_c, the inverse iteration

$$K = \frac{1}{4}\cosh^{-1}[\exp(8K'/3)] \tag{5.92c}$$

is stable. The corresponding inverse transformation for the function $g(K)$ is obtained from Eq. (5.90), with K in the function f converted to K' using Eq. (5.92c):

$$g(K) = \frac{1}{2}g(K') + \frac{1}{2}\ln\{2\exp(2K'/3)[\cosh(4K'/3)]^{1/4}\}. \tag{5.92d}$$

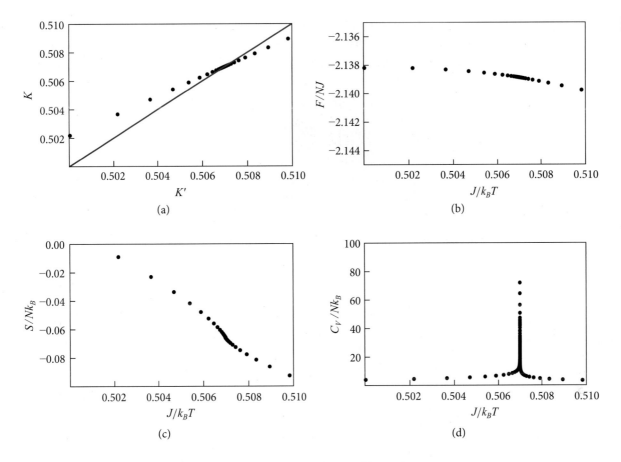

FIGURE 5.11 (a) Iteration map $K(K')$ in the vicinity of the critical point. (b) Free energy per spin. (c) Entropy per spin. (d) Specific heat per spin. Recall that $J/(k_B T) = K$.

Now, we know that at low temperature, $T \ll T_c$ and $K \gg K_c$, all the spins are aligned; correspondingly, in the statistical sum (5.87a) the two terms with all s's identical dominate, giving $z = e^{-F/(k_B T)} = e^{Ng} \simeq 2e^{2NK}$, whence $g(K) \simeq 2K$. Conversely, at high temperature, there is complete disorder, and $K \to 0$. This means that every one of the 2^N terms in the statistical sum (5.87a) is unity, giving $z = e^{Ng} = 2^N$, whence $g(K) \simeq \ln 2$. We can therefore use the iterative map, Eqs. (5.92c) and (5.92d), to approach $K = K_c$ from either side, starting with the high-temperature and low-temperature limits of $g(K)$ and evaluating thermodynamic quantities at each step. More specifically, at each step, we evaluate $g(K)$, dg/dK, and d^2g/dK^2 numerically, and from them we compute F, S, and C_V using Eqs. (5.89b), (5.92a), and (5.92b).

The iterated values of these quantities are plotted as points in Fig. 5.11. Note that the entropy S is continuous at K_c (panel c), but its derivative, the specific heat (panel d), diverges at K_c, as $K \to K_c$ from either side. This is characteristic of a second-order phase transition.

To calculate the explicit form of this divergence, suppose that $g(K)$ is a sum of an analytic (infinitely differentiable) function and a nonanalytic part. Suppose that near the critical point, the nonanalytic part behaves as $g(K) \sim |K - K_c|^{2-\alpha}$ for some "critical exponent" α. This implies that C_V diverges $\propto |K - K_c|^{-\alpha} \propto |T - T_c|^{-\alpha}$. Now, from Eq. (5.92d), we have that

$$|K' - K_c|^{2-\alpha} = 2|K - K_c|^{2-\alpha}, \tag{5.93a}$$

or equivalently,

$$\frac{dK'}{dK} = 2^{1/(2-\alpha)}. \tag{5.93b}$$

Evaluating the derivative at $K = K_c$ from Eq. (5.92c), we obtain

$$\alpha = 2 - \frac{\ln 2}{\ln(dK'/dK)_c} = 0.131, \tag{5.93c}$$

which is consistent with the numerical calculation.

exact analysis of Ising model compared to our approximate analysis

The exact Onsager (1944) analysis of the Ising model gives $K_c = 0.441$ compared to our $K_c = 0.507$, and $C_V \propto -\ln |T - T_c|$ compared to our $C_V \propto |T - T_c|^{-0.131}$. Evidently, our renormalization group approach (formulated by Maris and Kadanoff, 1978) gives a fair approximation to the correct answers but not a good one.

Our approach appears to have a serious problem in that it predicts a negative value for the entropy in the vicinity of the critical point (Fig. 5.11c). This is surely unphysical. (The entropy becomes positive farther away, on either side of the critical point.) This is an artificiality associated with our approach's ansatz [i.e., associated with our setting $K_2 = K_3 = 0$ and $K' = K_1 + K_2$ in Eq. (5.88)]. It does not seem easy to cure this in a simple renormalization-group approach.

Nonetheless, our calculations exhibit the conceptual and physical essentials of the renormalization-group approach to phase transitions.

Why did we bother to go through this cumbersome procedure when Onsager has given us an exact analytical solution to the Ising model? The answer is that it is not possible to generalize the Onsager solution to more complex and realistic problems. In particular, it has not even been possible to find an Onsager-like solution to the 3-dimensional Ising model. However, once the machinery of the renormalization group has been mastered, it can produce approximate answers, with an accuracy that can be estimated, for a variety of problems. In the following section we look at a quite different approach to the same 2-dimensional Ising problem with exactly the same motivation in mind.

EXERCISES

Exercise 5.15 *Example: One-Dimensional Ising Lattice* T2

(a) Write down the partition function for a 1-dimensional Ising lattice as a sum over terms describing all possible spin organizations.

(b) Show that by separating into even and odd numbered spins, it is possible to factor the partition function and relate $z(N, K)$ exactly to $z(N/2, K')$. Specifically, show that

$$z(N, K) = f(K)^{N/2} z(N/2, K') \qquad (5.94)$$

where $K' = \ln[\cosh(2K)]/2$, and $f(K) = 2[\cosh(2K)]^{1/2}$.

(c) Use these relations to demonstrate that the 1-dimensional Ising lattice does not exhibit a second-order phase transition.

5.8.4 Monte Carlo Methods for the Ising Model T2

In this section, we explore the phase transition of the 2-dimensional Ising model using our second general method: the *Monte Carlo* approach.[16] This method, like the renormalization group, is a powerful tool for a much larger class of problems than just phase transitions.[17]

The Monte Carlo approach is much more straightforward in principle than the renormalization group. We set up a square lattice of atoms and initialize their spins randomly.[18] We imagine that our lattice is in contact with a heat bath with a fixed temperature T (it is one member of a canonical ensemble of systems), and we drive it to approach statistical equilibrium and then wander onward through an equilibrium sequence of states $|n_1\rangle, |n_2\rangle, \ldots$ in a prescribed, ergodic manner. Our goals are to visualize typical equilibrium states (see Fig. 5.12 later in this section) and to compute thermodynamic quantities using $\bar{X} = z^{-1} \sum_n e^{-E_n/(k_B T)} X_n$, where the sum is over the sequence of states $|n_1\rangle, |n_2\rangle, \ldots$. For example, we can compute the specific heat (at constant volume) from

Monte Carlo method for Ising model

$$C_V = \frac{d\bar{E}}{dT} = \frac{\partial}{\partial T} \left(\frac{\sum_n e^{-E_n/(k_B T)} E_n}{\sum_n e^{-E_n/(k_B T)}} \right) = \frac{\overline{E^2} - \bar{E}^2}{k_B T^2}. \qquad (5.95)$$

[Note that a singularity in the specific heat at a phase transition will be associated with large fluctuations in the energy, just as it is associated with large fluctuations of temperature; Eq. (5.73d).]

In constructing our sequence of lattice states $|n_1\rangle, |n_2\rangle, \ldots$, we obviously cannot visit all 2^N states even just once, so we must sample them fairly. How can we prescribe the rule for changing the spins when going from one state in our sample to the next, so as to produce a fair sampling? There are many answers to this question; we describe

---

16. The name "Monte Carlo" is a sardonic reference to the casino whose patrons believe they will profit by exploiting random processes.
17. We shall meet it again in Sec. 28.6.1.
18. This and other random steps that follow are performed numerically and require a (pseudo) random number generator. Most programming languages supply this utility, which is mostly used uncritically, occasionally with unintended consequences. Defining and testing "randomness" is an important topic which, unfortunately, we shall not address. See, for example, Press et al. (2007).

and use one of the simplest, due to Metropolis et al. (1953). To understand this, we must appreciate that we don't need to comprehend the detailed dynamics by which a spin in a lattice actually flips. All that is required is that the rule we adopt, for going from one state to the next, should produce a sequence that is in statistical equilibrium.

Let us denote by $p_{nn'}$ the conditional probability that, if the lattice is in state $|n\rangle$, then the next step will take it to state $|n'\rangle$. For statistical equilibrium, it must be that the probability that any randomly observed step takes us out of state $|n\rangle$ is equal to the probability that it takes us into that state:

$$\rho_n \sum_{n'} p_{nn'} = \sum_{n'} \rho_{n'} p_{n'n}. \tag{5.96}$$

(Here ρ_n is the probability that the lattice was in state $|n\rangle$ just before the transition.) We know that in equilibrium, $\rho_{n'} = \rho_n \, e^{(E_n - E_{n'})/(k_B T)}$, so our conditional transition probabilities must satisfy

$$\sum_{n'} p_{nn'} = \sum_{n'} p_{n'n} e^{(E_n - E_{n'})/(k_B T)}. \tag{5.97}$$

Metropolis rule for transition probabilities

The Metropolis rule is simple:

$$\text{if } E_n > E_m, \text{ then } p_{nm} = 1;$$

$$\text{and if } E_n < E_m, \text{ then } p_{nm} = \exp[(E_n - E_m)/(k_B T)] \tag{5.98}$$

up to some normalization constant. It is easy to show that this satisfies the statistical equilibrium condition (5.97) and that it drives an initial out-of-equilibrium system toward equilibrium.

The numerical implementation of the Metropolis rule (5.98) is as follows: Start with the lattice in an initial, random state, and then choose one spin at random to make a trial flip. If the new configuration has a lower energy, we always accept the change. If it has a higher energy, we only accept the change with a probability given by $\exp[-\Delta E/(k_B T)]$, where $\Delta E > 0$ is the energy change.[19] In this way, we produce a sequence of states that will ultimately have the equilibrium distribution function, and we can perform our thermodynamic averages using this sequence in an unweighted fashion. This is a particularly convenient procedure for the Ising problem, because, by changing just one spin at a time, ΔE can only take one of five values ($-4, -2, 0, +2, +4$ in units of J), and it is possible to change from one state to the next very quickly. (It also helps to store the two acceptance probabilities $e^{-2J/(k_B T)}$ and $e^{-4J/(k_B T)}$ for making an energy-gaining transition, so as to avoid evaluating exponentials at every step.)

---

19. There is a small subtlety here. The probability of making a given transition is actually the product of the probability of making the trial flip and of accepting the trial. However, the probability of making a trial flip is the same for all the spins that we might flip ($1/N$), and these trial probabilities cancel, so it is only the ratio of the probabilities of acceptance that matters.

 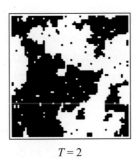 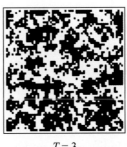

| $T = 1$ | $T = 2$ | $T = 3$ |

FIGURE 5.12 Typical equilibrium Ising lattices for temperatures $T = 1, 2, 3$ in units of J/k_B. The black regions have spins $s = +1$; the white, $s = -1$.

How big a lattice do we need, and how many states should we consider? The lattice size can be surprisingly small to get qualitatively correct results, if we adopt periodic boundary conditions. That is to say, we imagine a finite tiling of our actual finite lattice, and every time we need to know the spin at a site beyond the tiling's last column (or row), we use the corresponding spin an equal distance beyond the first column (or row). This device minimizes the effects of the boundary on the final answer. Lattices as small as 32×32 can be useful. The length of the computation depends on the required accuracy. (In practice, this is usually implemented the other way around. The time available on a computer of given speed determines the accuracy.) One thing should be clear. It is necessary that we explore a reasonable volume of state space to be able to sample it fairly and compute meaningful estimates of thermodynamic quantities. The final lattice should exhibit no vestigial patterns from the state when the computation was half complete. In practice, it is this consideration that limits the size of the lattice, and it is one drawback of the Metropolis algorithm that the step sizes are necessarily small. There is a large bag of tricks for Monte Carlo simulations that can be used for variance reduction and estimation, but we only concern ourselves here with the general method.

Returning to the Ising model, we show in Fig. 5.12 typical equilibrium states (snapshots) for three temperatures, measured in units of J/k_B. Recall that the critical temperature is $T_c = J/(k_B K_c) = J/(0.441 k_B) = 2.268 J/k_B$. Note the increasingly long-range order as the temperature is reduced below T_c.

Monte Carlo results for Ising model

We have concluded this chapter with an examination of a very simple system that can approach equilibrium according to specified rules and that can exhibit strong fluctuations. In the following chapter, we examine fluctuations more systematically.

EXERCISES

Exercise 5.16 *Practice: Direct Computation of Thermodynamic Integrals* **T2**
Estimate how long it would take a personal computer to calculate the partition function for a 32×32 Ising lattice by evaluating every possible state.

Exercise 5.17 *Example: Monte Carlo Approach to Phase Transition* T2
Write a simple computer program to compute the energy and the specific heat of a 2-dimensional Ising lattice as described in the text. Examine the accuracy of your answers by varying the size of the lattice and the number of states sampled. (You might also try to compute a formal variance estimate.)

Exercise 5.18 *Problem: Ising Lattice with an Applied Magnetic Field* T2
Modify your computer program from Ex. 5.17 to deal with the 2-dimensional Ising model augmented by an externally imposed, uniform magnetic field [Eqs. (5.80)]. Compute the magnetization and the magnetic susceptibility for wisely selected values of $m_o B/J$ and $K = J/(k_B T)$.

Bibliographic Note

Most statistical mechanics textbooks include much detail on statistical thermo-dynamics. Among those we have found useful at an elementary level are Kittel and Kroemer (1980), and at more advanced levels, Chandler (1987), Sethna (2006), Kardar (2007), Reif (2008), and Pathria and Beale (2011). Chandler (1987) and Sethna (2006) are particularly good for phase transitions. Our treatment of the renormalization group in Sec. 5.8.3 is adapted in part from Chandler, who also covers Monte Carlo methods.

6

Random Processes

These motions were such as to satisfy me, after frequently repeated observation, that they arose
neither from currents in the fluid, nor from its gradual evaporation, but belonged to the particle itself.

ROBERT BROWN (1828)

6.1 Overview

6.1

In this chapter we analyze, among others, the following issues:

- What is the time evolution of the distribution function for an ensemble of
  systems that begins out of statistical equilibrium and is brought to equilibrium through contact with a heat bath?

- How can one characterize the noise introduced into experiments or observations by noisy devices, such as resistors and amplifiers?

- What is the influence of such noise on one's ability to detect weak signals?

- What filtering strategies will improve one's ability to extract weak signals
  from strong noise?

- Frictional damping of a dynamical system generally arises from coupling to
  many other degrees of freedom (a bath) that can sap the system's energy.
  What is the connection between the fluctuating (noise) forces that the bath
  exerts on the system and its damping influence?

The mathematical foundation for analyzing such issues is the *theory of random
processes* (i.e., of functions that are random and unpredictable but have predictable
probabilities for their behavior). A portion of the theory of random processes is the
theory of stochastic differential equations (equations whose solutions are probability
distributions rather than ordinary functions). This chapter is an overview of these
topics, sprinkled throughout with applications.

In Sec. 6.2, we introduce the concept of random processes and the various probability distributions that describe them. We introduce restrictions that we shall adhere
to—the random processes that we study are stationary and ergodic—and we introduce an example that we return to time and again: a *random-walk* process, of which
Brownian motion is an example. In Sec. 6.3, we discuss two special classes of random
processes: Markov processes and Gaussian processes; we also present two important
theorems: the central limit theorem, which explains why random processes so often

- Relativity does not enter into this chapter.

- This chapter does not rely in any major way on previous chapters, but it does make occasional reference to results from Chaps. 4 and 5 about statistical equilibrium and fluctuations in and away from statistical equilibrium.

- No subsequent chapter relies in any major way on this chapter. However:
  - The concepts of spectral density and correlation function, developed in Sec. 6.4, will be used in Ex. 9.8 when treating coherence properties of radiation, in Sec. 11.9.2 when studying thermal noise in solids, in Sec. 15.4 when studying turbulence in fluids, in Sec. 23.2.1 in treating the quasilinear formalism for weak plasma turbulence, and in Sec. 28.6.1 when discussing observations of the anisotropy of the cosmic microwave background radiation.
  - The fluctuation-dissipation theorem, developed in Sec. 6.8, will be used in Sec. 11.9.2 for thermoelastic noise in solids, and in Sec. 12.5 for normal modes of an elastic body.
  - The Fokker-Planck equation, developed in Sec. 6.9, will be referred to or used in Secs. 20.4.3, 20.5.1, and 28.6.3 and Exs. 20.8 and 20.10 when discussing thermal equilibration in a plasma and thermoelectric transport coefficients, and it will be used in Sec. 23.3.3 in developing the quasilinear theory of wave-particle interactions in a plasma.

have Gaussian probability distributions, and Doob's theorem, which says that all the statistical properties of a Markov, Gaussian process are determined by just three parameters. In Sec. 6.4, we introduce two powerful mathematical tools for the analysis of random processes: the correlation function and spectral density, and prove the Wiener-Khintchine theorem, which relates them. As applications of these tools, we use them to prove Doob's theorem and to discuss optical spectra, noise in interferometric gravitational wave detectors, and fluctuations of cosmological mass density and of the distribution of galaxies in the universe. In Secs. 6.6 and 6.7, we introduce another powerful tool, the filtering of a random process, and we use these tools to develop the theory of noise and techniques for extracting weak signals from large noise. As applications we study shot noise (which is important, e.g., in measurements with laser light), frequency fluctuations of atomic clocks, and also the Brownian motion of a dust particle buffeted by air molecules and its connection to random walks. In Sec. 6.8, we develop another powerful tool, the fluctuation-dissipation theorem, which quantifies the relationship between the fluctuations and the dissipation (friction) produced by one and the same heat bath. As examples, we explore Brownian motion (once again), Johnson noise in a resistor and the voltage fluctuations it produces in electric circuits, thermal noise in high-precision optical measurements, and quantum limits on the accuracy of high-precision measurements and how to circumvent them. Finally, in Sec. 6.9 we derive and discuss the Fokker-Planck equation, which governs the evolution of Markov random processes, and we illustrate it with the random motion of an atom that is being cooled by interaction with laser beams (so-called *optical molasses*) and with thermal noise in a harmonic oscillator.

6.2 Fundamental Concepts

In this section we introduce a number of fundamental concepts about random processes.

6.2.1 Random Variables and Random Processes

RANDOM VARIABLE

A (1-dimensional) *random variable* is a (scalar) function $y(t)$, where t is usually time, for which the future evolution is not determined uniquely by any set of initial data—or at least by any set that is knowable to you and me. In other words, *random variable* is just a fancy phrase that means "unpredictable function." Throughout this chapter, we insist for simplicity that our random variables y take on a continuum of *real* values ranging over some interval, often but not always $-\infty$ to $+\infty$. The generalizations to y with complex or discrete (e.g., integer) values, and to independent variables other than time, are straightforward.

Examples of random variables are: (i) the total energy $E(t)$ in a cell of gas that is in contact with a heat bath; (ii) the temperature $T(t)$ at the corner of Main Street and Center Street in Logan, Utah; (iii) the price per share of Google stock $P(t)$; (iv) the mass-flow rate $\dot{M}(t)$ from the Amazon River into the Atlantic Ocean. One can also deal with random variables that are vector or tensor functions of time; in Track-Two portions of this chapter we do so.

RANDOM PROCESS

A (1-dimensional) *random process* (also called "stochastic process") is an ensemble $\mathcal{E}$ of real random variables $y(t)$ that, in a physics context, all represent the same kind of physical entity. For example, each $y(t)$ could be the longitude of a particular oxygen molecule undergoing a random walk in Earth's atmosphere. The individual random variables $y(t)$ in the ensemble $\mathcal{E}$ are often called *realizations* of the random process.

As an example, Fig. 6.1 shows three realizations $y(t)$ of a random process that represents the random walk of a particle in one dimension. For details, see Ex. 6.4, which shows how to generate realizations like these on a computer.

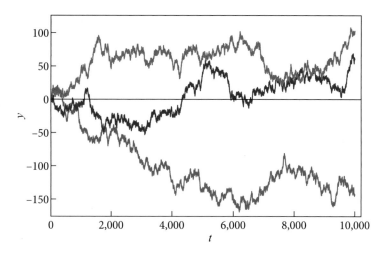

FIGURE 6.1 Three different realizations $y(t)$ of a random process that describes the location y of a particle at time t, when it is undergoing a random walk in 1 dimension (e.g., an atmospheric oxygen molecule's east-west motion).

6.2.2

6.2.2 Probability Distributions

PROBABILITY DISTRIBUTIONS FOR A RANDOM PROCESS

Since the precise time evolution of a random variable $y(t)$ is not predictable, if one wishes to make predictions, one can do so only probabilistically. The foundation for probabilistic predictions is a set of probability functions for the random process (i.e., for the ensemble $\mathcal{E}$ of its realizations).

More specifically, the most general (1-dimensional) random process is fully characterized by the set of probability distributions $p_1, p_2, p_3, \ldots$ defined as

probability distributions for a random process

$$p_n(y_n, t_n; \ldots; y_2, t_2; y_1, t_1)dy_n \ldots dy_2 dy_1. \tag{6.1}$$

Equation (6.1) tells us the probability that a realization $y(t)$, drawn at random from the process (the ensemble $\mathcal{E}$), (i) will take on a value between y_1 and $y_1 + dy_1$ at time t_1, (ii) also will take on a value between y_2 and $y_2 + dy_2$ at a later time $t_2, \ldots$, and (iii) also will take on a value between y_n and $y_n + dy_n$ at a later time t_n. (Note that the subscript n on p_n tells us how many independent values of y appear in p_n, and that earlier times are placed to the right—a practice common for physicists, particularly when dealing with propagators.) If we knew the values of all the process's probability distributions (an infinite number of p_ns!), then we would have full information about its statistical properties. Not surprisingly, it will turn out that, if the process in some sense is in statistical equilibrium, then we can compute all its probability distributions from a very small amount of information. But that comes later; first we must develop more formalism.

ENSEMBLE AVERAGES

From the probability distributions, we can compute ensemble averages (denoted by brackets). For example, the quantities

$$\langle y(t_1) \rangle \equiv \int y_1 p_1(y_1, t_1) dy_1 \quad \text{and} \quad \sigma_y^2(t_1) \equiv \left\langle [y(t_1) - \langle y(t_1) \rangle]^2 \right\rangle \qquad (6.2a)$$

ensemble average and variance

are the ensemble-averaged value of y and the variance of y at time t_1. Similarly,

$$\langle y(t_2) y(t_1) \rangle \equiv \int y_2 y_1 p_2(y_2, t_2; y_1, t_1) dy_2 dy_1 \qquad (6.2b)$$

is the average value of the product $y(t_2) y(t_1)$.

CONDITIONAL PROBABILITIES

Besides the (absolute) probability distributions p_n, we also find useful an infinite series of *conditional* probability distributions $P_2, P_3, \ldots$, defined as

$$P_n(y_n, t_n | y_{n-1}, t_{n-1}; \ldots; y_1, t_1) dy_n. \qquad (6.3)$$

conditional probability distributions

This distribution is the probability that, *if* $y(t)$ took on the values $y_1, y_2, \ldots, y_{n-1}$ at times $t_1, t_2, \ldots, t_{n-1}$, then it will take on a value between y_n and $y_n + dy_n$ at a later time t_n.

It should be obvious from the definitions of the probability distributions that

$$p_n(y_n, t_n; \ldots; y_1, t_1)$$
$$= P_n(y_n, t_n | y_{n-1}, t_{n-1}; \ldots; y_1, t_1) p_{n-1}(y_{n-1}, t_{n-1}; \ldots; y_1, t_1). \qquad (6.4)$$

Using this relation, one can compute all the conditional probability distributions P_n from the absolute distributions $p_1, p_2, \ldots$. Conversely, using this relation recursively, one can build up all the absolute probability distributions p_n from $p_1(y_1, t_1)$ and all the conditional distributions $P_2, P_3, \ldots$.

STATIONARY RANDOM PROCESSES

A random process is said to be *stationary* if and only if its probability distributions p_n depend just on time differences and not on absolute time:

stationary random process

$$p_n(y_n, t_n + \tau; \ldots; y_2, t_2 + \tau; y_1, t_1 + \tau) = p_n(y_n, t_n; \ldots; y_2, t_2; y_1, t_1). \quad (6.5)$$

If this property holds for the absolute probabilities p_n, then Eq. (6.4) guarantees it also will hold for the conditional probabilities P_n.

Nonstationary random processes arise when one is studying a system whose evolution is influenced by some sort of clock that registers absolute time, not just time differences. For example, the speeds $v(t)$ of all oxygen molecules in downtown St. Anthony, Idaho, make up random processes regulated in part by the atmospheric temperature and therefore by the rotation of Earth and its orbital motion around the Sun. The

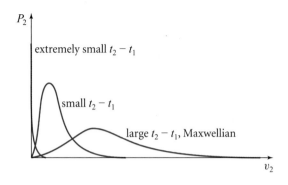

FIGURE 6.2 The probability $P_2(v_2, t_2|0, t_1)$ that a molecule with vanishing speed at time t_1 will have speed v_2 (in a unit interval dv_2) at time t_2. Although the molecular speed is a stationary random process, this probability evolves in time.

influence of these clocks makes $v(t)$ a nonstationary random process. Stationary random processes, by contrast, arise in the absence of any regulating clocks. An example is the speeds $v(t)$ of all oxygen molecules in a room kept at constant temperature.

Stationarity does not mean "no time evolution of probability distributions." For example, suppose one knows that the speed of a specific oxygen molecule vanishes at time t_1, and one is interested in the probability that the molecule will have speed v_2 at time t_2. That probability, $P_2(v_2, t_2|0, t_1)$, is sharply peaked around $v_2 = 0$ for extremely small time differences $t_2 - t_1$ and is Maxwellian for large time differences $t_2 - t_1$ (Fig. 6.2). Despite this evolution, the process is stationary (assuming constant temperature) in the sense that it does not depend on the specific time t_1 at which v happened to vanish, only on the time difference $t_2 - t_1$: $P_2(v_2, t_2|0, t_1) = P_2(v_2, t_2 - t_1|0, 0)$.

Henceforth, throughout this chapter, we restrict attention to random processes that are stationary (at least on the timescales of interest to us); and, accordingly, we use

$$p_1(y) \equiv p_1(y, t_1) \tag{6.6a}$$

for the probability, since it does not depend on the time t_1. We also denote by

$$P_2(y_2, t|y_1) \equiv P_2(y_2, t|y_1, 0) \tag{6.6b}$$

the probability that, if a (realization of a) random process begins with the value y_1, then after the lapse of time t it has the value y_2.

6.2.3

6.2.3 Ergodic Hypothesis

ergodic hypothesis

A (stationary) random process (ensemble $\mathcal{E}$ of random variables) is said to satisfy the *ergodic hypothesis* (or, for brevity, it will be called *ergodic*) if and only if it has the following property.

Let $y(t)$ be a random variable in the ensemble $\mathcal{E}$ (i.e., let $y(t)$ be any realization of the process). Construct from $y(t)$ a new ensemble $\mathcal{E}'$ whose members are

$$Y^K(t) \equiv y(t + KT), \qquad (6.7)$$

where K runs over all integers, negative and positive, and where T is some very large time interval. Then $\mathcal{E}'$ has the same probability distributions p_n as $\mathcal{E}$; that is, $p_n(Y_n, t_n; \ldots; Y_1, t_1)$ has the same functional form as $p_n(y_n, t_n; \ldots; y_1, t_1)$ for all times such that $|t_i - t_j| < T$.

This is essentially the same ergodic hypothesis as we met in Sec. 4.6.

Henceforth we restrict attention to random processes that satisfy the ergodic hypothesis (i.e., that are ergodic). This, in principle, is a severe restriction. In practice, for a physicist, it is not severe at all. In physics one's objective, when defining random variables that last forever ($-\infty < t < +\infty$) and when introducing ensembles, is usually to acquire computational techniques for dealing with a single, or a small number of, random variables $y(t)$, studied over finite lengths of time. One acquires those techniques by defining conceptual infinite-duration random variables and ensembles in such a way that they satisfy the ergodic hypothesis.

As in Sec. 4.6, because of the ergodic hypothesis, time averages defined using any realization $y(t)$ of a random process are equal to ensemble averages:

$$\bar{F} \equiv \lim_{T \to \infty} \frac{1}{T} \int_{-T/2}^{T/2} F(y(t))dt = \langle F(y) \rangle \equiv \int F(y)p_1(y)dy, \qquad (6.8)$$

for any function $F = F(y)$. In this sense, each realization of the random process is representative, when viewed over sufficiently long times, of the statistical properties of the process's entire ensemble—and conversely. Correspondingly, we can blur the distinction between the random process and specific realizations of it—and we often do so.

6.3 Markov Processes and Gaussian Processes

6.3.1 Markov Processes; Random Walk

A random process $y(t)$ is said to be *Markov* (also sometimes called "Markovian") if and only if all of its future probabilities are determined by its most recently known value:

Markov random process

$$P_n(y_n, t_n | y_{n-1}, t_{n-1}; \ldots; y_1, t_1) = P_2(y_n, t_n | y_{n-1}, t_{n-1}) \quad \text{for all } t_n \geq \ldots \geq t_2 \geq t_1. \qquad (6.9)$$

This relation guarantees that any Markov process (which, of course, we require to be stationary without saying so) is completely characterized by the probabilities

$$p_1(y) \text{ and } P_2(y_2, t | y_1) \equiv \frac{p_2(y_2, t; y_1, 0)}{p_1(y_1)}. \qquad (6.10)$$

From $p_1(y)$ and $P_2(y_2, t|y_1)$ one can reconstruct, using the Markov relation (6.9) and the general relation (6.4) between conditional and absolute probabilities, all distribution functions of the process.

Actually, for any random process that satisfies the ergodic hypothesis (which means all random processes considered in this chapter), $p_1(y)$ is determined by the conditional probability $P_2(y_2, t|y_1)$ [Ex. 6.1], so for any Markov (and ergodic) process, all the probability distributions follow from $P_2(y_2, t|y_1)$ alone!

An example of a Markov process is the x component of velocity $v_x(t)$ of a dust particle in an arbitrarily large room,[1] filled with constant-temperature air. Why? Because the molecule's equation of motion is[2] $m\,dv_x/dt = F'_x(t)$, and the force $F'_x(t)$ is due to random buffeting by other molecules that are uncorrelated (the kick now is unrelated to earlier kicks); thus, there is no way for the value of v_x in the future to be influenced by any earlier values of v_x except the most recent one.

By contrast, the position $x(t)$ of the particle is not Markov, because the probabilities of future values of x depend not just on the initial value of x, but also on the initial velocity v_x—or, equivalently, the probabilities depend on the values of x at two initial, closely spaced times. The pair $\{x(t), v_x(t)\}$ is a 2-dimensional Markov process (see Ex. 6.23).

THE SMOLUCHOWSKI EQUATION

Choose three (arbitrary) times t_1, t_2, and t_3 that are ordered, so $t_1 < t_2 < t_3$. Consider a (realization of an) arbitrary random process that begins with a known value y_1 at t_1, and ask for the probability $P_2(y_3, t_3|y_1)$ (per unit y_3) that it will be at y_3 at time t_3. Since the realization must go through some value y_2 at the intermediate time t_2 (though we don't care what that value is), it must be possible to write the probability to reach y_3 as

$$P_2(y_3, t_3|y_1, t_1) = \int P_3(y_3, t_3|y_2, t_2; y_1, t_1) P_2(y_2, t_2|y_1, t_1)dy_2,$$

where the integration is over all allowed values of y_2. This is not a terribly interesting relation. Much more interesting is its specialization to the case of a Markov process. In that case $P_3(y_3, t_3|y_2, t_2; y_1, t_1)$ can be replaced by $P_2(y_3, t_3|y_2, t_2) = P_2(y_3, t_3 - t_2|y_2, 0) \equiv P_2(y_3, t_3 - t_2|y_2)$, and the result is an integral equation involving only P_2. Because of stationarity, it is adequate to write that equation for the case $t_1 = 0$:

$$\boxed{P_2(y_3, t_3|y_1) = \int P_2(y_3, t_3 - t_2|y_2) P_2(y_2, t_2|y_1)dy_2.} \qquad (6.11)$$

Smoluchowski equation

This is the *Smoluchowski equation* (also called *Chapman-Kolmogorov equation*). It is valid for any Markov random process and for times $0 < t_2 < t_3$. We shall discover its power in our derivation of the Fokker-Planck equation in Sec. 6.9.1.

---

1. The room must be arbitrarily large so the effects of the floor, walls, and ceiling can be ignored.
2. By convention, primes are used to identify stochastic forces (i.e., forces that are random processes).

Exercise 6.1 ***Example: Limits of P_2*
Explain why, for any (stationary) random process,

$$\lim_{t \to 0} P_2(y_2, t|y_1) = \delta(y_2 - y_1).$$ (6.12a)

Use the ergodic hypothesis to argue that

$$\lim_{t \to \infty} P_2(y_2, t|y_1) = p_1(y_2).$$ (6.12b)

Thereby conclude that, for a Markov process, all the probability distributions are determined by the conditional probability $P_2(y_2, t|y_1)$. Give an algorithm for computing them.

Exercise 6.2 *Practice: Markov Processes for an Oscillator*
Consider a harmonic oscillator (e.g., a pendulum), driven by bombardment with air molecules. Explain why the oscillator's position $x(t)$ and velocity $v(t) = dx/dt$ are random processes. Is $x(t)$ Markov? Why? Is $v(t)$ Markov? Why? Is the pair $\{x(t), v(t)\}$ a 2-dimensional Markov process? Why? We study this 2-dimensional random process in Ex. 6.23.

Exercise 6.3 ***Example: Diffusion of a Particle; Random Walk*
In Ex. 3.17, we studied the diffusion of particles through an infinite 3-dimensional medium. By solving the diffusion equation, we found that, if the particles' number density at time $t = 0$ was $n_o(\mathbf{x})$, then at time t it has become

$$n(\mathbf{x}, t) = [1/(4\pi Dt)]^{3/2} \int n_o(\mathbf{x}')e^{-(\mathbf{x}-\mathbf{x}')^2/(4Dt)}d^3x',$$

where D is the diffusion coefficient [Eq. (3.73)].

(a) For any one of the diffusing particles, the location $y(t)$ in the y direction (one of three Cartesian directions) is a 1-dimensional random process. From the above $n(\mathbf{x}, t)$, infer that the conditional probability distribution for y is

$$P_2(y_2, t|y_1) = \frac{1}{\sqrt{4\pi Dt}}e^{-(y_2-y_1)^2/(4Dt)}.$$ (6.13)

(b) Verify that the conditional probability (6.13) satisfies the Smoluchowski equation (6.11). [Hint: Consider using symbol-manipulation computer software that quickly can do straightforward calculations like this.]

At first this may seem surprising, since a particle's position y is not Markov. However (as we explore explicitly in Sec. 6.7.2), the diffusion equation from which we derived this P_2 treats as negligibly small the timescale τ_r on which the velocity dy/dt thermalizes. It thereby wipes out all information about what the particle's actual velocity is, making y effectively Markov, and forcing its P_2 to satisfy the Smoluchowski equation. See Ex. 6.10, where we shall also discover that this diffusion is an example of a random walk.

6.3.2 Gaussian Processes and the Central Limit Theorem; Random Walk

GAUSSIAN PROCESSES

Gaussian random process

A random process is said to be Gaussian if and only if all of its (absolute) probability distributions are Gaussian (i.e., have the following form):

$$p_n(y_n, t_n; \ldots; y_2, t_2; y_1, t_1) = A \exp\left[-\sum_{j=1}^{n}\sum_{k=1}^{n}\alpha_{jk}(y_j - \bar{y})(y_k - \bar{y})\right], \quad \text{(6.14a)}$$

where (i) A and α_{jk} depend only on the time differences $t_2 - t_1, t_3 - t_1, \ldots, t_n - t_1$; (ii) A is a positive normalization constant; (iii) $[\alpha_{jk}]$ is a *positive-definite*, symmetric matrix (otherwise p_n would not be normalizable); and (iv) $\bar{y}$ is a constant, which one readily can show is equal to the ensemble average of y,

$$\bar{y} \equiv \langle y \rangle = \int y p_1(y)\, dy. \quad \text{(6.14b)}$$

Since the conditional probabilities are all computable as ratios of absolute probabilities [Eq. (6.4)], the conditional probabilities of a Gaussian process will be Gaussian.

Gaussian random processes are very common in physics. For example, the total number of particles $N(t)$ in a gas cell that is in statistical equilibrium with a heat bath is a Gaussian random process (Ex. 5.11d); and the primordial fluctuations that gave rise to structure in our universe appear to have been Gaussian (Sec. 28.5.3). In fact, as we saw in Sec. 5.6, macroscopic variables that characterize huge systems in statistical equilibrium always have Gaussian probability distributions. The underlying reason is that, *when a random process is driven by a large number of statistically independent, random influences, its probability distributions become Gaussian*. This general fact is a consequence of the *central limit theorem* of probability. We state and prove a simple variant of this theorem.

CENTRAL LIMIT THEOREM (A SIMPLE VERSION)

central limit theorem

Let y be a random quantity [not necessarily a random variable $y(t)$; there need not be any times involved; however, our applications will be to random variables]. Suppose that y is characterized by an arbitrary probability distribution $p(y)$ (e.g., that of Fig. 6.3a), so the probability of the quantity taking on a value between y and $y + dy$ is $p(y)dy$. Denote by $\bar{y}$ the mean value of y, and by σ_y its standard deviation (also called its rms fluctuation and the square root of its variance):

$$\bar{y} \equiv \langle y \rangle = \int y p(y) dy, \quad (\sigma_y)^2 \equiv \langle (y - \bar{y})^2 \rangle = \langle y^2 \rangle - \bar{y}^2. \quad \text{(6.15a)}$$

Randomly draw from this distribution a large number N of values $\{y_1, y_2, \ldots, y_N\}$, and average them to get a number

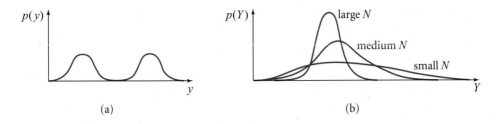

FIGURE 6.3 Example of the central limit theorem. (a) The random variable y with the probability distribution $p(y)$. (b) This variable produces, for various values of N, the variable $Y = (y_1 + \cdots + y_N)/N$ with the probability distributions $p(Y)$. In the limit of very large N, $p(Y)$ is a Gaussian.

$$Y \equiv \frac{1}{N} \sum_{i=1}^{N} y_i. \tag{6.15b}$$

Repeat this process many times, and examine the resulting probability distribution for Y. In the limit of arbitrarily large N, that distribution will be Gaussian with mean and standard deviation

$$\bar{Y} = \bar{y} \quad \text{and} \quad \sigma_Y = \frac{\sigma_y}{\sqrt{N}}, \tag{6.15c}$$

that is, it will have the form

$$p(Y) = \frac{1}{\sqrt{2\pi\sigma_Y^2}} \exp\left[-\frac{(Y - \bar{Y})^2}{2\sigma_Y^2}\right], \tag{6.15d}$$

with $\bar{Y}$ and σ_Y given by Eq. (6.15c). See Fig. 6.3b.

Proof of Central Limit Theorem. The key to proving this theorem is the Fourier transform of the probability distribution. (That Fourier transform is called the distribution's *characteristic function,* but in this chapter we do not delve into the details of characteristic functions.) Denote the Fourier transform of $p(y)$ by[3]

$$\tilde{p}_y(f) \equiv \int_{-\infty}^{+\infty} e^{i2\pi f y} p(y) dy = \sum_{n=0}^{\infty} \frac{(i2\pi f)^n}{n!} \langle y^n \rangle. \tag{6.16a}$$

The second expression follows from a power series expansion of $e^{i2\pi f y}$ in the first. Similarly, since a power series expansion analogous to Eq. (6.16a) must hold for $\tilde{p}_Y(f)$ and since $\langle Y^n \rangle$ can be computed from

$$\langle Y^n \rangle = \langle N^{-n}(y_1 + y_2 + \cdots + y_N)^n \rangle$$

$$= \int N^{-n}(y_1 + \cdots + y_N)^n p(y_1) \ldots p(y_N) dy_1 \ldots dy_N, \tag{6.16b}$$

---

3. See the beginning of Sec. 6.4.2 for the conventions we use for Fourier transforms.

it must be that

$$\tilde{p}_Y(f) = \sum_{n=0}^{\infty} \frac{(i2\pi f)^n}{n!} \langle Y^n \rangle$$

$$= \int \exp[i2\pi f N^{-1}(y_1 + \cdots + y_N)] p(y_1) \ldots p(y_N) dy_1 \ldots dy_n$$

$$= \left[\int e^{i2\pi f y/N} p(y) dy \right]^N = \left[1 + \frac{i2\pi f \bar{y}}{N} - \frac{(2\pi f)^2 \langle y^2 \rangle}{2N^2} + O\left(\frac{1}{N^3}\right) \right]^N$$

$$= \exp\left[i2\pi f \bar{y} - \frac{(2\pi f)^2 (\langle y^2 \rangle - \bar{y}^2)}{2N} + O\left(\frac{1}{N^2}\right) \right]. \tag{6.16c}$$

Here the last equality can be obtained by taking the logarithm of the preceding quantity, expanding in powers of $1/N$, and then exponentiating. By inverting the Fourier transform (6.16c) and using $(\sigma_y)^2 = \langle y^2 \rangle - \bar{y}^2$, we obtain for $p(Y)$ the Gaussian (6.15d). ∎

This proof is a good example of the power of Fourier transforms, a power that we exploit extensively in this chapter. As an important example to which we shall return later, Ex. 6.4 analyzes the simplest version of a random walk.

EXERCISES

Exercise 6.4 **Example: Random Walk with Discrete Steps of Identical Length*
This exercise is designed to make the concept of random processes more familiar and also to illustrate the central limit theorem.

A particle travels in 1 dimension, along the y axis, making a sequence of steps Δy_j (labeled by the integer j), each of which is $\Delta y_j = +1$ with probability 1/2, or $\Delta y_j = -1$ with probability 1/2.

(a) After $N \gg 1$ steps, the particle has reached location $y(N) = y(0) + \sum_{j=1}^{N} \Delta y_j$. What does the central limit theorem predict for the probability distribution of $y(N)$? What are its mean and its standard deviation?

(b) Viewed on lengthscales $\gg 1$, $y(N)$ looks like a continuous random process, so we shall rename $N \equiv t$. Using the (pseudo)random number generator from your favorite computer software language, compute a few concrete realizations of $y(t)$ for $0 < t < 10^4$ and plot them.[4] Figure 6.1 above shows one realization of this random process.

(c) Explain why this random process is Markov.

---

4. If you use Mathematica, the command RandomInteger[] generates a pseudorandom number that is 0 with probability 1/2 or 1 with probability 1/2. Therefore, the following simple script will carry out the desired computation: y = Table[0, {10000}]; For[t = 1, t < 10000, t++, y[[t + 1]] = y[[t]] + 2 RandomInteger[] - 1]; ListPlot[y, Joined -> True]. This was used to generate Fig. 6.1.

(d) Use the central limit theorem to infer that the conditional probability P_2 for this random process is

$$P_2(y_2, t|y_1) = \frac{1}{\sqrt{2\pi t}} \exp\left[-\frac{(y_2 - y_1)^2}{2t}\right]. \qquad (6.17)$$

(e) Notice that this is the same probability distribution as we encountered in the diffusion exercise (Ex. 6.3) but with $D = 1/2$. Why did this have to be the case?

(f) Using an extension of the computer program you wrote in part (b), evaluate $y(t = 10^4)$ for 1,000 realizations of this random process, each with $y(0) = 0$, then bin the results in bins of width $\delta y = 10$, and plot the number of realizations $y(10^4)$ that wind up in each bin. Repeat for 10,000 realizations. Compare your plots with the probability distribution (6.17).

6.3.3 Doob's Theorem for Gaussian-Markov Processes, and Brownian Motion

A large fraction of the random processes that one meets in physics are Gaussian, and many are Markov. Therefore, the following remarkable theorem is very important. *Any 1-dimensional random process $y(t)$ that is both Gaussian and Markov has the following form for its conditional probability distribution P_2:*

$$P_2(y_2, t|y_1) = \frac{1}{[2\pi\sigma_{y_t}^2]^{\frac{1}{2}}} \exp\left[-\frac{(y_2 - \bar{y}_t)^2}{2\sigma_{y_t}^2}\right], \qquad (6.18a)$$

where the mean $\bar{y}_t$ and variance $\sigma_{y_t}^2$ at time t are given by

$$\bar{y}_t = \bar{y} + e^{-t/\tau_r}(y_1 - \bar{y}), \quad \sigma_{y_t}^2 = (1 - e^{-2t/\tau_r})\sigma_y^2. \qquad (6.18b)$$

Here $\bar{y}$ and σ_y^2 are respectively the process's equilibrium mean and variance (the values at $t \to \infty$), and τ_r is its *relaxation time*. This result is *Doob's theorem*.[5] We shall prove it in Ex. 6.5, after we have developed some necessary tools.

Note the great power of Doob's theorem: Because $y(t)$ is Markov, all of its probability distributions are computable from this P_2 (Ex. 6.1), which in turn is determined by $\bar{y}, \sigma_y$, and τ_r. Correspondingly, all statistical properties of a Gaussian-Markov process are determined by just three parameters: its (equilibrium) mean $\bar{y}$ and variance σ_y^2,

---

5. It is so named because it was first formulated and proved by J. L. Doob (1942).

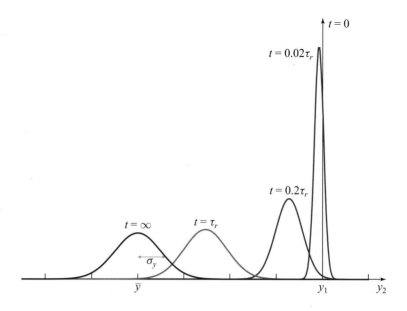

FIGURE 6.4 Evolution of the conditional probability $P_2(y_2, t|y_1)$ for a Gaussian-Markov random process [Eq. (6.18a)], as predicted by Doob's theorem. The correlation function and spectral density for this process are shown later in the chapter in Fig. 6.8.

and its relaxation time τ_r. As an example, the first absolute probability distribution is

$$p_1(y) = \lim_{t \to \infty} P_2(y, t|y_1) = \frac{1}{\sqrt{2\pi \sigma_y{}^2}} \exp\left[-\frac{(y - \bar{y})^2}{2\sigma_y{}^2}\right]. \tag{6.18c}$$

The time evolution of P_2 [Eqs. (6.18a,b)] is plotted in Fig. 6.4. At $t = 0$ it is a delta function at y_1, in accord with Eq. (6.12a). As t increases, its peak (its mean) moves toward $\bar{y}$, and it spreads out. Ultimately, at $t = \infty$, its peak asymptotes to $\bar{y}$, and its standard deviation (half-width) asymptotes to σ_y, so $P_2 \to p_1$—in accord with Eqs. (6.12b) and (6.18c).

Brownian motion

An example that we explore in Sec. 6.7.2 is a dust particle being buffeted by air molecules in a large, constant-temperature room (Brownian motion). As we discussed near the beginning of Sec. 6.3.1, any Cartesian component v of the dust particle's velocity is a Markov process. It is also Gaussian (because its evolution is influenced solely by the independent forces of collisions with a huge number of independent air molecules), so $P_2(v, t|v_1)$ is given by Doob's theorem. In equilibrium, positive and negative values of the Cartesian velocity component v are equally probable, so $\bar{v} = 0$, which means that $\frac{1}{2}m\sigma_v{}^2 = \frac{1}{2}m\overline{v^2}$, which is the equilibrium mean kinetic energy— a quantity we know to be $\frac{1}{2}k_BT$ from the equipartition theorem (Sec. 4.4.4); thus, $\bar{v} = 0$, and $\sigma_v = \sqrt{k_BT/m}$. The relaxation time τ_r is the time required for the particle to change its velocity substantially, due to collisions with air molecules; we compute it in Sec. 6.8.1 using the fluctuation-dissipation theorem; see Eq. (6.78).

6.4 Correlation Functions and Spectral Densities

6.4.1 Correlation Functions; Proof of Doob's Theorem

Let $y(t)$ be a (realization of a) random process with time average $\bar{y}$. Then the correlation function of $y(t)$ is defined by

$$C_y(\tau) \equiv \overline{[y(t) - \bar{y}][y(t + \tau) - \bar{y}]} \equiv \lim_{T \to \infty} \frac{1}{T} \int_{-T/2}^{+T/2} [y(t) - \bar{y}][y(t + \tau) - \bar{y}]dt.$$

correlation function

(6.19)

This quantity, as its name suggests, is a measure of the extent to which the values of y at times t and $t + \tau$ tend to be correlated. The quantity τ is sometimes called the *delay time*, and by convention it is taken to be positive. [One can easily see that, if one also defines $C_y(\tau)$ for negative delay times τ by Eq. (6.19), then $C_y(-\tau) = C_y(\tau)$. Thus nothing is lost by restricting attention to positive delay times.]

As an example, for a Gaussian-Markov process with P_2 given by Doob's formula (6.18a) (Fig. 6.4), we can compute $C(\tau)$ by replacing the time average in Eq. (6.19) with an ensemble average: $C_y(\tau) = \int y_2\, y_1\, p_2(y_2, \tau; y_1)\, dy_1\, dy_2$. If we use $p_2(y_2, \tau; y_1) = P_2(y_2, \tau; y_1)\, p_1(y_1)$ [Eq. (6.10)], insert P_2 and p_1 from Eqs. (6.18), and perform the integrals, we obtain

$$C_y(\tau) = \sigma_y^2 e^{-\tau/\tau_r}.$$

(6.20)

This correlation function has two properties that are quite general:

1. The following is true for all (ergodic and stationary) random processes:

   $$C_y(0) = \sigma_y^2,$$

   properties of correlation function

   (6.21a)

   as one can see by replacing time averages with ensemble averages in definition (6.19); in particular, $C_y(0) \equiv \overline{(y - \bar{y})^2} = \langle(y - \bar{y})^2\rangle$, which by definition is the variance σ_y^2 of y.

2. In addition, we have that

   $$\boxed{C_y(\tau) \text{ asymptotes to zero for } \tau \gg \tau_r,}$$

   (6.21b)

   where τ_r is the process's *relaxation time* or *correlation time* (see Fig. 6.5). This is true for all ergodic, stationary random processes, since our definition of ergodicity in Sec. 6.2.3 relies on each realization $y(t)$ losing its memory of earlier values after some sufficiently long time T. Otherwise, it would not be possible to construct the ensemble $\mathcal{E}'$ of random variables $Y^K(t)$ [Eq. (6.7)] and have them behave like independent random variables.

   relaxation time

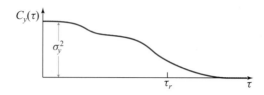

FIGURE 6.5 Properties (6.21) of correlation functions.

Exercise 6.5 *Derivation: Proof of Doob's Theorem*

Prove Doob's theorem. More specifically, for any Gaussian-Markov random process, show that $P_2(y_2, t|y_1)$ is given by Eqs. (6.18a,b).

[Hint: For ease of notation, set $y_{\text{new}} = (y_{\text{old}} - \bar{y}_{\text{old}})/\sigma_{y_{\text{old}}}$, so $\bar{y}_{\text{new}} = 0$ and $\sigma_{y_{\text{new}}} = 1$. If the theorem is true for y_{new}, then by the rescalings inherent in the definition of $P_2(y_2, t|y_1)$, it will also be true for y_{old}.]

(a) Show that the Gaussian process y_{new} has probability distributions

$$p_1(y) = \frac{1}{\sqrt{2\pi}} e^{-y^2/2}, \tag{6.22a}$$

$$p_2(y_2, t_2; y_1, t_1) = \frac{1}{\sqrt{(2\pi)^2(1 - C_{21}{}^2)}} \exp\left[-\frac{y_1{}^2 + y_2{}^2 - 2C_{21}y_1y_2}{2(1 - C_{21}{}^2)}\right]; \tag{6.22b}$$

and show that the constant C_{21} that appears here is the correlation function $C_{21} = C_y(t_2 - t_1)$.

(b) From the relationship between absolute and conditional probabilities [Eq. (6.4)], show that

$$P_2(y_2, t_2|y_1, t_1) = \frac{1}{\sqrt{2\pi(1 - C_{21}{}^2)}} \exp\left[-\frac{(y_2 - C_{21}y_1)^2}{2(1 - C_{21}{}^2)}\right]. \tag{6.22c}$$

(c) Show that for any three times $t_3 > t_2 > t_1$,

$$C_{31} = C_{32}C_{21}; \quad \text{i.e.,} \quad C_y(t_3 - t_1) = C_y(t_3 - t_2)C_y(t_2 - t_1). \tag{6.22d}$$

To show this, you could (i) use the relationship between absolute and conditional probabilities and the Markov nature of the random process to infer that

$$p_3(y_3, t_3; y_2, t_2; y_1, t_1) = P_3(y_3, t_3|y_2, t_2; y_1, t_1)p_2(y_2, t_2; y_1, t_1)$$
$$= P_2(y_3, t_3|y_2, t_2)p_2(y_2, t_2; y_1, t_1);$$

then (ii) compute the last expression explicitly, getting

$$\frac{1}{\sqrt{2\pi(1-C_{32}{}^2)}} \exp\left[-\frac{(y_3 - C_{32}y_2)^2}{2(1-C_{32}{}^2)}\right]$$

$$\times \frac{1}{\sqrt{(2\pi)^2(1-C_{21}{}^2)}} \exp\left[-\frac{(y_1{}^2 + y_2{}^2 - 2C_{21}y_1y_2)}{2(1-C_{21}{}^2)}\right];$$

(iii) then using this expression, evaluate

$$C_y(t_3 - t_1) \equiv C_{31} \equiv \langle y(t_3)y(t_1)\rangle = \int p_3(y_3, t_3; y_2, t_2; y_1, t_1) y_3 y_1 dy_3 dy_2 dy_1.$$

(6.22e)

The result should be $C_{31} = C_{32}C_{21}$.

(d) Argue that the unique solution to this equation, with the "initial condition" that $C_y(0) = \sigma_y{}^2 = 1$, is $C_y(\tau) = e^{-\tau/\tau_r}$, where τ_r is a constant (which we identify as the relaxation time). Correspondingly, $C_{21} = e^{-(t_2-t_1)/\tau_r}$.

(e) By inserting this expression into Eq. (6.22c), complete the proof for $y_{\text{new}}(t)$, and thence conclude that Doob's theorem is also true for our original, unrescaled $y_{\text{old}}(t)$.

6.4.2 Spectral Densities

There are several different normalization conventions for Fourier transforms. In this chapter, we adopt a normalization that is commonly (though not always) used in the theory of random processes and that differs from the one common in quantum theory. Specifically, instead of using the angular frequency ω, we use the ordinary frequency $f \equiv \omega/(2\pi)$. We define the Fourier transform of a function $y(t)$ and its inverse by

$$\boxed{\tilde{y}(f) \equiv \int_{-\infty}^{+\infty} y(t)e^{i2\pi ft}dt, \qquad y(t) \equiv \int_{-\infty}^{+\infty} \tilde{y}(f)e^{-i2\pi ft}df.}$$

(6.23)

Fourier transform

Notice that with this set of conventions, there are no factors of $1/(2\pi)$ or $1/\sqrt{2\pi}$ multiplying the integrals. Those factors have been absorbed into the df of Eq. (6.23), since $df = d\omega/(2\pi)$.

The integrals in Eq. (6.23) are not well defined as written because a random process $y(t)$ is generally presumed to go on forever so its Fourier transform $\tilde{y}(f)$ is divergent. One gets around this problem by crude trickery. From $y(t)$ construct, by truncation, the function

$$y_T(t) \equiv \begin{cases} y(t) & \text{if } -T/2 < t < +T/2, \\ 0 & \text{otherwise.} \end{cases}$$

(6.24a)

Then the Fourier transform $\tilde{y}_T(f)$ is finite, and by Parseval's theorem (e.g., Arfken, Weber, and Harris, 2013) it satisfies

$$\int_{-T/2}^{+T/2} [y(t)]^2 dt = \int_{-\infty}^{+\infty} [y_T(t)]^2 dt = \int_{-\infty}^{+\infty} |\tilde{y}_T(f)|^2 df = 2 \int_0^\infty |\tilde{y}_T(f)|^2 \, df.$$

(6.24b)

In the last equality we have used the fact that because $y_T(t)$ is real, $\tilde{y}_T^*(f) = \tilde{y}_T(-f)$, where * denotes complex conjugation. Consequently, the integral from $-\infty$ to 0 of $|\tilde{y}_T(f)|^2$ is the same as the integral from 0 to $+\infty$. Now, the quantities on the two sides of (6.24b) diverge in the limit as $T \to \infty$, and it is obvious from the left-hand side that they diverge linearly as T. Correspondingly, the limit

$$\lim_{T\to\infty} \frac{1}{T} \int_{-T/2}^{+T/2} [y(t)]^2 dt = \lim_{T\to\infty} \frac{2}{T} \int_0^\infty |\tilde{y}_T(f)|^2 df$$

(6.24c)

is convergent.

These considerations motivate the following definition of the *spectral density* (also sometimes called the *power spectrum*) $S_y(f)$ of the random process $y(t)$:

$$\boxed{S_y(f) \equiv \lim_{T\to\infty} \frac{2}{T} \left| \int_{-T/2}^{+T/2} [y(t) - \bar{y}] e^{i2\pi f t} dt \right|^2.}$$

(6.25)

Notice that the quantity inside the absolute value sign is just $\tilde{y}_T(f)$, but with the mean of y removed before computation of the Fourier transform. (The mean is removed to avoid an uninteresting delta function in $S_y(f)$ at zero frequency.) Correspondingly, by virtue of our motivating result (6.24c), the spectral density satisfies $\int_0^\infty S_y(f) df = \lim_{T\to\infty} \frac{1}{T} \int_{-T/2}^{+T/2} [y(t) - \bar{y}]^2 dt = \overline{(y - \bar{y})^2} = \sigma_y^2$, or

$$\boxed{\int_0^\infty S_y(f) df = C_y(0) = \sigma_y^2.}$$

(6.26)

Thus the integral of the spectral density of y over all positive frequencies is equal to the variance of y.

By convention, our spectral density is defined only for nonnegative frequencies f. This is because, were we to define it also for negative frequencies, the fact that $y(t)$ is real would imply that $S_y(f) = S_y(-f)$, so the negative frequencies contain no new information. Our insistence that f be positive goes hand in hand with the factor 2 in the $2/T$ of definition (6.25): that factor 2 folds the negative-frequency part onto the positive-frequency part. This choice of convention is called the *single-sided spectral density*. Sometimes one encounters a *double-sided spectral density*,

$$S_y^{\text{double-sided}}(f) = \frac{1}{2} S_y(|f|),$$

(6.27)

in which f is regarded as both positive and negative, and frequency integrals generally run from $-\infty$ to $+\infty$ instead of 0 to ∞ (see, e.g., Ex. 6.7).

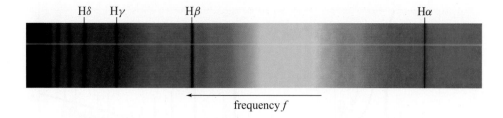

$H\delta$ $H\gamma$ $H\beta$ $H\alpha$

frequency f

FIGURE 6.6 A spectrum obtained by sending light through a diffraction grating. The intensity of the image is proportional to $d\mathcal{E}/dtdf$, which, in turn, is proportional to the spectral density $S_E(f)$ of the electric field $E(t)$ of the light that entered the diffraction grating.

Notice that the spectral density has units of y^2 per unit frequency; or, more colloquially (since frequency f is usually measured in Hertz, i.e., cycles per second), its units are y^2/Hz.

6.4.3 Physical Meaning of Spectral Density, Light Spectra, and Noise in a Gravitational Wave Detector

6.4.3

We can infer the physical meaning of the spectral density from previous experience with light spectra. Specifically, consider the scalar electric field[6] $E(t)$ of a plane-polarized light wave entering a telescope from a distant star, galaxy, or nebula. (We must multiply this $E(t)$ by the polarization vector to get the vectorial electric field.) This $E(t)$ is a superposition of emission from an enormous number of atoms, molecules, and high-energy particles in the source, so it is a Gaussian random process. It is not hard to convince oneself that $E(t)$'s spectral density $S_E(f)$ is proportional to the light power per unit frequency $d\mathcal{E}/dtdf$ (the light's power spectrum) entering the telescope. When we send the light through a diffraction grating, we get this power spectrum spread out as a function of frequency f in the form of spectral lines superposed on a continuum, as in Fig. 6.6. The amount of light power in this spectrum, in some narrow bandwidth Δf centered on some frequency f, is $(d\mathcal{E}/dtdf)\Delta f \propto S_E(f)\Delta f$ (assuming S_E is nearly constant over that band).

Another way to understand this role of the spectral density $S_E(f)$ is by examining the equation for the variance of the oscillating electric field E as an integral over frequency, $\sigma_E{}^2 = \int_0^\infty S_E(f)df$. If we filter the light so only that portion at frequency f, in a very narrow bandwidth Δf, gets through the filter, then the variance of the filtered, oscillating electric field will obviously be the portion of the integral coming from this frequency band. The rms value of the filtered electric field will be the square root of this—and similarly for any other random process $y(t)$:

$$\left(\begin{array}{c} \text{rms value of } y\text{'s oscillations} \\ \text{at frequency } f \text{ in a very narrow bandwidth } \Delta f \end{array} \right) \simeq \sqrt{S_y(f)\Delta f}. \qquad (6.28)$$

rms oscillation

---

6. In this section, and only here, E represents the electric field rather than (nonrelativistic) energy.

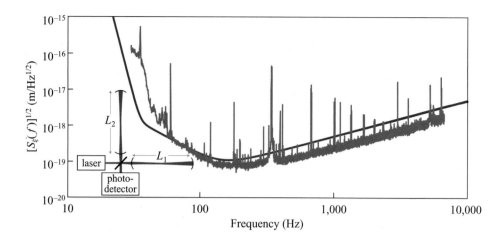

FIGURE 6.7 The square root of the spectral density of the time-varying arm-length difference $\xi(t) = L_1 - L_2$ (see inset), in the Laser Interferometer Gravitational-wave Observatory (LIGO) interferometer at Hanford, Washington, as measured on February 22, 2010. See Sec. 9.5 and Fig. 9.13. The dark blue curve is the noise that was specified as this instrument's goal. The narrow spectral lines (sharp spikes in the spectrum produced by internal resonances in the instrument) contain negligible power, and so can be ignored for our purposes. At high frequencies, $f \gtrsim 150$ Hz, the noise is due to randomness in arrival times of photons used to measure the mirror motions (photon shot noise, Sec. 6.7.4). At intermediate frequencies, 40 Hz $\lesssim f \lesssim 150$ Hz, it is primarily thermal noise (end of Sec. 6.8.2). At low frequencies, $f \lesssim 40$ Hz, it is primarily mechanical vibrations that sneak through a vibration isolation system ("seismic" noise).

(In Sec. 6.7.1, we develop a mathematical formalism to describe this type of filtering).

As a practical example, consider the output of an interferometric gravitational wave detector (to be further discussed in Secs. 9.5 and 27.6). The gravitational waves from some distant source (e.g., two colliding black holes) push two mirrors (hanging by wires) back and forth with respect to each other. Laser interferometry is used to monitor the difference $\xi(t) = L_1 - L_2$ between the two arm lengths. (Here L_1 is the separation between the mirrors in one arm of the interferometer, and L_2 is that in the other arm; see inset in Fig. 6.7.) The measured $\xi(t)$ is influenced by noise in the instrument as well as by gravitational waves. Figure 6.7 shows the square root of the spectral density of the noise-induced fluctuations in $\xi(t)$. Note that this $\sqrt{S_\xi(f)}$ has units of meters/$\sqrt{\text{Hertz}}$ (since ξ has units of meters).

The minimum of the noise power spectrum is at $f \simeq 150$ Hz. If one is searching amidst this noise for a broadband gravitational-wave signal, then one might filter the interferometer output so one's data analysis sees only a frequency band of order the frequency of interest: $\Delta f \simeq f$. Then the rms noise in this band will be $\sqrt{S_\xi(f) \times f} \simeq 10^{-19}$ m/$\sqrt{\text{Hz}} \times \sqrt{150 \text{ Hz}} \simeq 10^{-18}$ m, which is $\sim 1/1{,}000$ the diameter of a proton. If a gravitational wave with frequency ~ 150 Hz changes the mirrors' separations by much more than this miniscule amount, it should be detectable!

6.4.4 The Wiener-Khintchine Theorem; Cosmological Density Fluctuations

The Wiener-Khintchine theorem says that, for any random process $y(t)$, the correlation function $C_y(\tau)$ and the spectral density $S_y(f)$ are the cosine transforms of each other and thus contain precisely the same information:

$$C_y(\tau) = \int_0^\infty S_y(f)\cos(2\pi f\tau)df, \qquad S_y(f) = 4\int_0^\infty C_y(\tau)\cos(2\pi f\tau)d\tau.$$

(6.29)

The factor 4 results from our folding negative frequencies into positive in our definition of the spectral density.

Proof of Wiener-Khintchine Theorem. This theorem is readily proved as a consequence of Parseval's theorem: Assume, from the outset, that the mean has been subtracted from $y(t)$, so $\bar{y} = 0$. (This is not really a restriction on the proof, since C_y and S_y are insensitive to the mean of y.) Denote by $y_T(t)$ the truncated y of Eq. (6.24a) and by $\tilde{y}_T(f)$ its Fourier transform. Then the generalization of Parseval's theorem[7]

$$\int_{-\infty}^{+\infty}(gh^* + hg^*)dt = \int_{-\infty}^{+\infty}(\tilde{g}\tilde{h}^* + \tilde{h}\tilde{g}^*)df$$

(6.30a)

[with $g = y_T(t)$ and $h = y_T(t+\tau)$ both real and with $\tilde{g} = \tilde{y}_T(f), \tilde{h} = \tilde{y}_T(f)e^{-i2\pi f\tau}$], states

$$\int_{-\infty}^{+\infty} y_T(t)y_T(t+\tau)dt = \int_{-\infty}^{+\infty} \tilde{y}_T^*(f)\tilde{y}_T(f)e^{-i2\pi f\tau}\,df.$$

(6.30b)

By dividing by T, taking the limit as $T \to \infty$, and using Eqs. (6.19) and (6.25), we obtain the first equality of Eqs. (6.29). The second follows from the first by Fourier inversion. ∎

The Wiener-Khintchine theorem implies (Ex. 6.6) the following formula for the ensemble averaged self-product of the Fourier transform of the random process $y(t)$:

$$2\langle\tilde{y}(f)\tilde{y}^*(f')\rangle = S_y(f)\delta(f - f').$$

(6.31)

This equation quantifies the strength of the infinite value of $|\tilde{y}(f)|^2$, which motivated our definition (6.25) of the spectral density.

As an application of the Wiener-Khintchine theorem, we can deduce the spectral density $S_y(f)$ for any Gaussian-Markov process by performing the cosine transform of its correlation function $C_y(\tau) = \sigma_y{}^2 e^{-\tau/\tau_r}$ [Eq. (6.20)]. The result is

$$S_y(f) = \frac{(4/\tau_r)\sigma_y{}^2}{(2\pi f)^2 + (1/\tau_r)^2};$$

(6.32)

see Fig. 6.8.

---

7. This follows by subtracting Parseval's theorem for g and for h from Parseval's theorem for $g + h$.

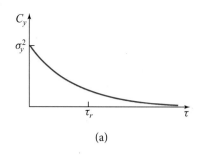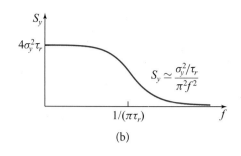

(a) (b)

FIGURE 6.8 (a) The correlation function (6.20) and (b) the spectral density (6.32) for a Gaussian-Markov process. The conditional probability $P_2(y_2, \tau | y_1)$ for this process is shown in Fig. 6.4.

As a second application, in Ex. 6.7 we explore fluctuations in the density of galaxies in the universe, caused by gravity pulling them into clusters.

EXERCISES

Exercise 6.6 *Derivation: Spectral Density as the Mean Square of Random Process's Fourier Transform*
Derive Eq. (6.31).

[Hint: Write $\langle \tilde{y}(f)\tilde{y}^*(f') \rangle = \int_{-\infty}^{+\infty} \int_{-\infty}^{+\infty} \langle y(t)y(t') \rangle e^{i2\pi f t} e^{-i2\pi f' t'} dt\,dt'$. Then set $t' = t + \tau$, and express the expectation value as $C_y(\tau)$, and use an expression for the Dirac delta function in terms of Fourier transforms.]

Exercise 6.7 **Example: Cosmological Density Fluctuations*[8]
Random processes can be stochastic functions of some variable or variables other than time. For example, it is conventional to describe fractional fluctuations in the large-scale distribution of mass in the universe, or the distribution of galaxies, using the quantity

$$\delta(\mathbf{x}) \equiv \frac{\rho(\mathbf{x}) - \langle \rho \rangle}{\langle \rho \rangle}, \quad \text{or} \quad \delta(\mathbf{x}) \equiv \frac{n(\mathbf{x}) - \langle n \rangle}{\langle n \rangle} \tag{6.33}$$

(not to be confused with the Dirac delta function). Here $\rho(\mathbf{x})$ is mass density and $n(\mathbf{x})$ is the number density of galaxies, which we assume, for didactic purposes, to have equal mass and to be distributed in the same fashion as the dark matter (Sec. 28.3.2). This $\delta(\mathbf{x})$ is a function of 3-dimensional position rather than 1-dimensional time, and $\langle \cdot \rangle$ is to be interpreted conceptually as an ensemble average and practically as a volume average (ergodic hypothesis!).

(a) Define the Fourier transform of δ over some large averaging volume V as

$$\tilde{\delta}_V(\mathbf{k}) = \int_V e^{i\mathbf{k}\cdot\mathbf{x}} \delta(\mathbf{x}) d^3 x, \tag{6.34a}$$

8. Discussed further in Sec. 28.5.4.

and define its spectral density (Sec. 28.5.4) by

$$P_\delta(\mathbf{k}) \equiv \lim_{V \to \infty} \frac{1}{V} |\tilde{\delta}_V(\mathbf{k})|^2. \tag{6.34b}$$

(Note that we here use cosmologists' "double-sided" normalization for P_δ, which is different from our normalization for a random process in time; we do not fold negative values of the Cartesian components k_j of $\mathbf{k}$ onto positive values.) Show that the two-point correlation function for cosmological density fluctuations, defined by

$$\xi_\delta(\mathbf{r}) \equiv \langle \delta(\mathbf{x})\delta(\mathbf{x} + \mathbf{r}) \rangle, \tag{6.34c}$$

is related to $P_\delta(\mathbf{k})$ by the following version of the Wiener-Khintchine theorem:

$$\xi_\delta(\mathbf{r}) = \int P_\delta(\mathbf{k}) e^{-i\mathbf{k}\cdot\mathbf{r}} \frac{d^3k}{(2\pi)^3} = \int_0^\infty P_\delta(k) \, \text{sinc}(kr) \frac{k^2 dk}{2\pi^2}, \tag{6.35a}$$

$$P_\delta(\mathbf{k}) = \int \xi_\delta(\mathbf{r}) e^{i\mathbf{k}\cdot\mathbf{r}} d^3x = \int_0^\infty \xi_\delta(r) \, \text{sinc}(kr) 4\pi r^2 dr, \tag{6.35b}$$

where $\text{sinc}\, x \equiv \sin x / x$. In deriving these expressions, use the fact that the universe is isotropic to infer that ξ_δ can depend only on the distance r between points and not on direction, and P_δ can depend only on the magnitude k of the wave number and not on its direction.

(b) Figure 6.9 shows observational data for the galaxy correlation function $\xi_\delta(r)$. These data are rather well approximated at $r < 20$ Mpc by

$$\xi_\delta(r) = (r_o/r)^\gamma, \qquad r_o \simeq 7 \text{ Mpc}, \qquad \gamma \simeq 1.8. \tag{6.36}$$

(Here 1 Mpc means 1×10^6 parsecs or about 3×10^6 light-years.) Explain why this implies that galaxies are strongly correlated (they cluster together strongly) on lengthscales $r \lesssim r_o \simeq 7$ Mpc. (Recall that the distance between our Milky Way galaxy and the nearest other large galaxy, Andromeda, is about 0.8 Mpc.) Use the Wiener-Khintchine theorem to compute the spectral density $P_\delta(k)$ and then the rms fractional density fluctuations at wave number k in bandwidth $\Delta k = k$. From your answer, infer that the density fluctuations are very large on lengthscales $\lambda = 1/k < r_o$.

(c) As a more precise measure of these density fluctuations, show that the variance of the total number $N(R)$ of galaxies inside a sphere of radius R is

$$\sigma_N^2 = \langle n \rangle^2 \int_0^\infty \frac{dk}{2\pi^2} k^2 P_\delta(k) \, W^2(kR), \tag{6.37a}$$

where

$$W(x) = \frac{3(\text{sinc}\, x - \cos x)}{x^2}. \tag{6.37b}$$

Evaluate this for the spectral density $P_\delta(r)$ that you computed in part (b). [Hint: Although it is straightforward to calculate this directly, it is faster to regard the

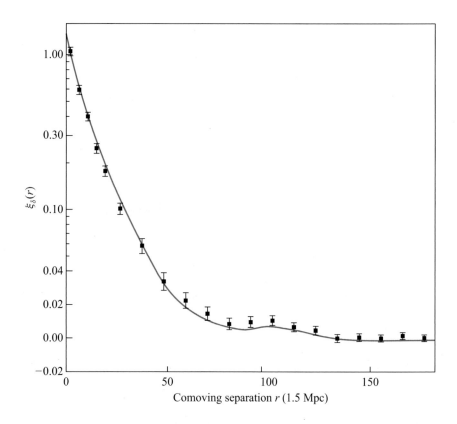

FIGURE 6.9 The galaxy correlation function $\xi_\delta(r)$ [defined in Eq. (6.34c)], as measured in the Sloan Digital Sky Survey. Notice that the vertical scale is linear for $\xi_\delta \lesssim 0.04$ and logarithmic for larger ξ_δ. Adapted from Eisenstein et al. (2005).

distribution of galaxies inside the sphere as an infinite distribution multiplied by a step function in radius and to invoke the convolution theorem.]

6.5 2-Dimensional Random Processes T2

Sometimes two (or more) random processes are closely related, and one wants to study their connections. An example is the position $x(t)$ and momentum $p(t)$ of a harmonic oscillator (Ex. 6.23), and we will encounter other examples in Secs. 28.5 and 28.6. Such pairs can be regarded as a 2-dimensional random process. In this Track-Two section, we generalize the concepts of correlation function and spectral density to such related processes.

6.5.1 Cross Correlation and Correlation Matrix T2

If $x(t)$ and $y(t)$ are two random processes, then by analogy with the correlation function $C_y(\tau)$ we define their *cross correlation* as

cross correlation
$$C_{xy}(\tau) \equiv \overline{x(t)y(t+\tau)}. \tag{6.38a}$$

When $x = y$, the cross correlation function becomes the autocorrelation function, $C_{yy}(\tau)$ interchangeable here with $C_y(\tau)$. The matrix

$$
\begin{bmatrix} C_{xx}(\tau) & C_{xy}(\tau) \\ C_{yx}(\tau) & C_{yy}(\tau) \end{bmatrix} \equiv \begin{bmatrix} C_x(\tau) & C_{xy}(\tau) \\ C_{yx}(\tau) & C_y(\tau) \end{bmatrix} \tag{6.38b}
$$

correlation matrix

can be regarded as a correlation matrix for the 2-dimensional random process $\{x(t), y(t)\}$. Notice that the elements of this matrix satisfy

$$
C_{ab}(-\tau) = C_{ba}(\tau). \tag{6.39}
$$

6.5.2 Spectral Densities and the Wiener-Khintchine Theorem T2

If $x(t)$ and $y(t)$ are two random processes, then by analogy with the spectral density $S_y(f)$ we define their *cross spectral density* as

$$
S_{xy}(f) = \lim_{T \to \infty} \frac{2}{T} \int_{-T/2}^{+T/2} [x(t) - \bar{x}] e^{-2\pi i f t} dt \int_{-T/2}^{+T/2} [y(t') - \bar{y}] e^{+2\pi i f t'} dt'. \tag{6.40a}
$$

cross spectral density

Notice that the cross spectral density of a random process with itself is equal to its spectral density, $S_{yy}(f) = S_y(f)$, and is real, but if $x(t)$ and $y(t)$ are different random processes, then $S_{xy}(f)$ is generally complex, with

$$
S_{xy}^*(f) = S_{xy}(-f) = S_{yx}(f). \tag{6.40b}
$$

This relation allows us to confine attention to positive f without any loss of information. The Hermitian matrix

$$
\begin{bmatrix} S_{xx}(f) & S_{xy}(f) \\ S_{yx}(f) & S_{yy}(f) \end{bmatrix} = \begin{bmatrix} S_x(f) & S_{xy}(f) \\ S_{yx}(f) & S_y(f) \end{bmatrix} \tag{6.40c}
$$

spectral density matrix

can be regarded as a spectral density matrix that describes how the power in the 2-dimensional random process $\{x(t), y(t)\}$ is distributed over frequency.

A generalization of the 1-dimensional Wiener-Khintchine Theorem (6.29) states that *for any two random processes $x(t)$ and $y(t)$, the cross correlation function $C_{xy}(\tau)$ and the cross spectral density $S_{xy}(f)$ are Fourier transforms of each other and thus contain precisely the same information:*

Wiener-Khintchine theorem for cross spectral density

$$
C_{xy}(\tau) = \frac{1}{2} \int_{-\infty}^{+\infty} S_{xy}(f) e^{-i 2\pi f \tau} df = \frac{1}{2} \int_0^\infty \left[S_{xy}(f) e^{-i 2\pi f \tau} + S_{yx}(f) e^{+i 2\pi f \tau} \right] df,
$$

$$
S_{xy}(f) = 2 \int_{-\infty}^{\infty} C_{xy}(\tau) e^{i 2\pi f \tau} d\tau = 2 \int_0^\infty \left[C_{xy}(f) e^{+i 2\pi f \tau} + C_{yx}(f) e^{-i 2\pi f \tau} \right] df.
$$

$$\tag{6.41}$$

The factors 1/2 and 2 in these formulas result from folding negative frequencies into positive in our definitions of the spectral density. Equations (6.41) can be proved by the same Parseval-theorem-based argument as we used for the 1-dimensional Wiener-Khintchine theorem (Sec. 6.4.4).

cross spectral density
as mean product of
random processes' Fourier
transforms

The Wiener-Khintchine theorem implies the following formula for the ensemble averaged product of the Fourier transform of the random processes $x(t)$ and $y(t)$:

$$2\langle \tilde{x}(f)\tilde{y}^*(f')\rangle = S_{xy}(f)\delta(f - f').$$

(6.42)

This can be proved by the same argument as we used in Ex. 6.6 to prove its single-process analog, $2\langle \tilde{y}(f)\tilde{y}^*(f')\rangle = S_y(f)\delta(f - f')$ [Eq. (6.31)].

EXERCISES

Exercise 6.8 *Practice: Spectral Density of the Sum of Two Random Processes* T2
Let u and v be two random processes. Show that

$$S_{u+v}(f) = S_u(f) + S_v(f) + S_{uv}(f) + S_{vu}(f) = S_u(f) + S_v(f) + 2\Re S_{uv}(f),$$

(6.43)

where $\Re$ denotes the real part of the argument.

6.6

6.6 Noise and Its Types of Spectra

Experimental physicists and engineers encounter random processes in the form of noise that is superposed on signals they are trying to measure. Examples include:

1. In radio communication, static on the radio is noise.

2. When modulated laser light is used for optical communication, random fluctuations in the arrival times of photons always contaminate the signal; the effects of such fluctuations are called "shot noise" and will be studied in Sec. 6.6.1.

3. Even the best of atomic clocks fail to tick with absolutely constant angular frequencies ω. Their frequencies fluctuate ever so slightly relative to an ideal clock, and those fluctuations can be regarded as noise.

Sometimes the signal that one studies amidst noise is actually itself some very special noise. (One person's noise is another person's signal.) An example is the light passing through an optical telescope and diffraction grating, discussed in Sec. 6.4.3. There the electric field $E(t)$ of the light from a star is a random process whose spectral density the astronomer measures as a function of frequency, studying with great interest features in the spectral lines and continuum. When the source is dim, the astronomer must try to separate its spectral density from those of noise in the photodetector and noise of other sources in the sky.

6.6.1

6.6.1 Shot Noise, Flicker Noise, and Random-Walk Noise; Cesium Atomic Clock

Physicists, astronomers, and engineers give names to certain shapes of noise spectra:

$$\boxed{S_y(f) \text{ independent of } f\text{—white noise spectrum,}}$$

(6.44a)

$$\boxed{S_y(f) \propto 1/f\text{—flicker noise spectrum,}}$$

(6.44b)

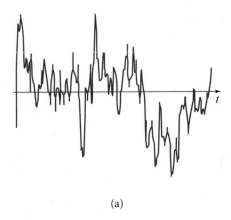
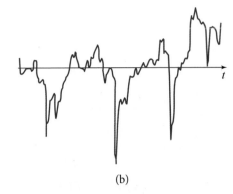

(a)   (b)

FIGURE 6.10 Examples of two random processes that have flicker noise spectra, $S_y(f) \propto 1/f$. Adapted from Press (1978).

$$S_y(f) \propto 1/f^2 \text{—random-walk spectrum.} \qquad (6.44c)$$

White noise, S_y independent of f, is called "white" because it has equal amounts of power per unit frequency S_y at all frequencies, just as white light has roughly equal powers at all light frequencies. Put differently, if $y(t)$ has a white-noise spectrum, then its rms fluctuations in fixed bandwidth Δf are independent of frequency f (i.e., $\sqrt{S_y(f)\Delta f}$ is independent of f).

Flicker noise, $S_y \propto 1/f$, gets its name from the fact that, when one looks at the time evolution $y(t)$ of a random process with a flicker-noise spectrum, one sees fluctuations ("flickering") on all timescales, and the rms amplitude of flickering is independent of the timescale one chooses. Stated more precisely, choose any timescale Δt and then choose a frequency $f \sim 3/\Delta t$, so one can fit roughly three periods of oscillation into the chosen timescale. Then the rms amplitude of the fluctuations observed will be $\sqrt{S_y(f)f/3}$, which is a constant independent of f when the spectrum is that of flicker noise, $S_y \propto 1/f$. In other words, flicker noise has the same amount of power in each octave of frequency. Figure 6.10 is an illustration: both graphs shown there depict random processes with flicker-noise spectra. (The differences between the two graphs will be explained in Sec. 6.6.2.) No matter what time interval one chooses, these processes look roughly periodic with one, two, or three oscillations in that time interval; and the amplitudes of those oscillations are independent of the chosen time interval. Flicker noise occurs widely in the real world, at low frequencies, for instance, in many electronic devices, in some atomic clocks, in geophysics (the flow rates of rivers, ocean currents, etc.), in astrophysics (the light curves of quasars, sunspot numbers, etc.); even in classical music. For an interesting discussion, see Press (1978).

Random-walk noise, $S_y \propto 1/f^2$, arises when a random process $y(t)$ undergoes a random walk. In Sec. 6.7.2, we explore an example: the time evolving position $x(t)$ of a dust particle buffeted by air molecules—the phenomenon of Brownian motion.

white noise

flicker noise

random-walk noise

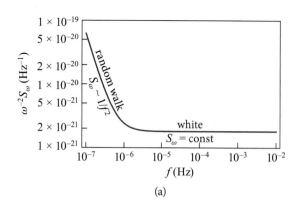
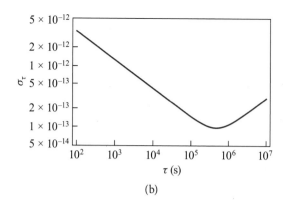

(a)

(b)

FIGURE 6.11 (a) Spectral density of the fluctuations in angular frequency ω of a typical cesium atomic clock. (b) Square root of the Allan variance for the same clock; see Ex. 6.13. Adapted from Galleani (2012). The best cesium clocks in 2016 (e.g., the U.S. primary time and frequency standard) have amplitude noise, $\sqrt{S_\omega}$ and σ_τ, 1000 times lower than this.

Notice that for a Gaussian-Markov process, the spectrum [Eq. (6.32) and Fig. 6.8b] is white at frequencies $f \ll 1/(2\pi\tau_r)$, where τ_r is the relaxation time, and it is random-walk at frequencies $f \gg 1/(2\pi\tau_r)$. This is typical: random processes encountered in the real world tend to have one type of spectrum over one large frequency interval and then switch to another type over another large interval. The angular frequency ω of ticking of a cesium atomic clock is another example.[9] It fluctuates slightly with time, $\omega = \omega(t)$, with the fluctuation spectral density shown in Fig. 6.11. At low frequencies, $f \lesssim 10^{-6}$ Hz (over long timescales $\Delta t \gtrsim 2$ weeks), ω exhibits random-walk noise. At higher frequencies, $f \gtrsim 10^{-6}$ Hz (timescales $\Delta t \lesssim 2$ weeks), it exhibits white noise—which is just the opposite of a Gaussian-Markov process (see, e.g., Galleani, 2012).

6.6.2 Information Missing from Spectral Density

In experimental studies of noise, attention focuses heavily on the spectral density $S_y(f)$ and on quantities that can be computed from it. In the special case of a Gaussian-Markov process, the spectrum $S_y(f)$ and the mean $\bar{y}$ together contain full information about all statistical properties of the random process. However, most random processes that one encounters are not Markov (though most are Gaussian). (When the spectrum deviates from the special form shown in Fig. 6.8, one can be sure the process is not Gaussian-Markov.) Correspondingly, for most processes the spectrum contains only a tiny part of the statistical information required to characterize

---

9. The U.S. national primary time and frequency standard is currently (2016) a cesium atomic clock; but it might be replaced, in a few years, by an atomic clock that oscillates at optical frequencies rather than the cesium clock's microwave frequencies. In their current (2016) experimental form, such clocks have achieved frequency stabilities and accuracies as good as a few $\times 10^{-18}$, nearly 100 times better than the current U.S. standard. Optical-frequency combs (Sec. 9.4.3) can be used to lock microwave-frequency oscillators to such optical-frequency clocks.

the process. The two random processes shown in Fig. 6.10 are good examples. They were constructed on a computer as superpositions of pulses $F(t - t_o)$ with random arrival times t_o and with identical forms

$$F(t) = 0 \text{ for } t < 0, \qquad F(t) = K/\sqrt{t} \text{ for } t > 0 \qquad (6.45)$$

(cf. Sec. 6.7.4). The two $y(t)$s look very different, because the first (Fig. 6.10a) involves frequent small pulses, while the second (Fig. 6.10b) involves less frequent, larger pulses. These differences are obvious to the eye in the time evolutions $y(t)$. However, they do not show up at all in the spectra $S_y(f)$, which are identical: both are flicker spectra (Ex. 6.15). Moreover, the differences do not show up in $p_1(y_1)$ or in $p_2(y_2, t_2; y_1, t_1)$, because the two processes are both superpositions of many independent pulses and thus are Gaussian, and for Gaussian processes p_1 and p_2 are determined fully by the mean and the correlation function, or equivalently by the mean and spectral density, which are the same for the two processes. Thus, the differences between the two processes show up only in the probabilities p_n of third order and higher, $n \geq 3$ [as defined in Eq. (6.1)].

6.7 Filtering Random Processes

6.7.1 Filters, Their Kernels, and the Filtered Spectral Density

In experimental physics and engineering, one often takes a signal $y(t)$ or a random process $y(t)$ and filters it to produce a new function $w(t)$ that is a *linear functional* of $y(t)$:

$$w(t) = \int_{-\infty}^{+\infty} K(t - t')y(t')dt'. \qquad (6.46)$$

The quantity $y(t)$ is called the filter's *input*; $K(t - t')$ is the filter's *kernel*, and $w(t)$ is its *output*. We presume throughout this chapter that the kernel depends only on the time difference $t - t'$ and not on absolute time. When this is so, the filter is said to be *stationary*; and when it is violated so $K = K(t, t')$ depends on absolute time, the filter is said to be nonstationary. Our restriction to stationary filters goes hand-in-hand with our restriction to stationary random processes, since if $y(t)$ is stationary and the filter is stationary (as we require), then the filtered process $w(t) = \int_{-\infty}^{+\infty} K(t - t')y(t')dt'$ is also stationary.

a filter and its input, output, and kernel

Some examples of kernels and their filtered outputs are

$$K(\tau) = \delta(\tau): \quad w(t) = y(t),$$
$$K(\tau) = \delta'(\tau): \quad w(t) = dy/dt,$$
$$\qquad\qquad\qquad\qquad\qquad\qquad\qquad\qquad (6.47)$$
$$K(\tau) = 0 \text{ for } \tau < 0 \text{ and } 1 \text{ for } \tau > 0: \quad w(t) = \int_{-\infty}^{t} y(t')dt'.$$

Here $\delta'(\tau)$ denotes the derivative of the Dirac δ-function.

As with any function, a knowledge of the kernel $K(\tau)$ is equivalent to a knowledge of its Fourier transform:

Fourier transform of filter's kernel

$$\tilde{K}(f) \equiv \int_{-\infty}^{+\infty} K(\tau)e^{i2\pi f\tau}d\tau. \tag{6.48}$$

This Fourier transform plays a central role in the theory of filtering (also called the theory of *linear signal processing*): the convolution theorem of Fourier transform theory states that, if $y(t)$ is a function whose Fourier transform $\tilde{y}(f)$ exists (converges), then the Fourier transform of the filter's output $w(t)$ [Eq. (6.46)] is given by

$$\tilde{w}(f) = \tilde{K}(f)\tilde{y}(f). \tag{6.49}$$

Similarly, by virtue of the definition (6.25) of spectral density in terms of Fourier transforms, if $y(t)$ is a random process with spectral density $S_y(f)$, then the filter's output $w(t)$ will be a random process with spectral density

spectral density of filter's output

$$S_w(f) = |\tilde{K}(f)|^2 S_y(f). \tag{6.50}$$

[Note that, although $\tilde{K}(f)$, like all Fourier transforms, is defined for both positive and negative frequencies, when its modulus is used in Eq. (6.50) to compute the effect of the filter on a spectral density, only positive frequencies are relevant; spectral densities are strictly positive-frequency quantities.]

easy way to compute kernel's squared Fourier transform

The quantity $|\tilde{K}(f)|^2$ that appears in the very important relation (6.50) is most easily computed not by evaluating directly the Fourier transform (6.48) and then squaring, but rather by sending the function $e^{i2\pi ft}$ through the filter (i.e., by computing the output w that results from the special input $y = e^{i2\pi ft}$), and then squaring the output: $|\tilde{K}(f)|^2 = |w|^2$. To see that this works, notice that the result of sending $y = e^{i2\pi ft}$ through the filter is

$$w = \int_{-\infty}^{+\infty} K(t - t')e^{i2\pi ft'}dt' = \tilde{K}^*(f)e^{i2\pi ft}, \tag{6.51}$$

which differs from $\tilde{K}(f)$ by complex conjugation and a change of phase, and which thus has absolute value squared of $|w|^2 = |\tilde{K}(f)|^2$.

For example, if $w(t) = d^n y/dt^n$, then when we set $y = e^{i2\pi ft}$, we get $w = d^n(e^{i2\pi ft})/dt^n = (i2\pi f)^n e^{i2\pi ft}$; and, accordingly, $|\tilde{K}(f)|^2 = |w|^2 = (2\pi f)^{2n}$, whence, for any random process $y(t)$, the quantity $w(t) = d^n y/dt^n$ will have $S_w(f) = (2\pi f)^{2n} S_y(f)$.

how differentiation and integration change spectral density

This example also shows that by differentiating a random process once, one changes its spectral density by a multiplicative factor $(2\pi f)^2$; for example, one can thereby convert random-walk noise into white noise. Similarly, by integrating a random process once in time (the inverse of differentiating), one multiplies its spectral density by $(2\pi f)^{-2}$. If instead one wants to multiply by f^{-1}, one can achieve that using the filter whose kernel is

$$K(\tau) = 0 \text{ for } \tau < 0, \qquad K(\tau) = \sqrt{\frac{2}{\tau}} \text{ for } \tau > 0; \tag{6.52a}$$

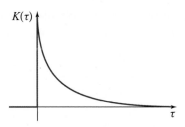

FIGURE 6.12 The kernel (6.52a), whose filter multiplies the spectral density by a factor $1/f$, thereby converts white noise into flicker noise and flicker noise into random-walk noise.

see Fig. 6.12. Specifically, it is easy to show, by sending a sinusoid through this filter, that

$$w(t) \equiv \int_{-\infty}^{t} \sqrt{\frac{2}{t - t'}} y(t') dt' \tag{6.52b}$$

has spectral density

$$S_w(f) = \frac{1}{f} S_y(f). \tag{6.52c}$$

Thus by filtering in this way, one can convert white noise into flicker noise and flicker noise into random-walk noise.

Exercise 6.9 *Derivations and Practice: Examples of Filters*

(a) Show that the kernels $K(\tau)$ in Eq. (6.47) produce the indicated outputs $w(t)$. Deduce the ratio $S_w(f)/S_y(f) = |\tilde{K}(f)|^2$ in two ways: (i) by Fourier transforming each $K(\tau)$; (ii) by setting $y = e^{i2\pi f t}$, deducing the corresponding filtered output w directly from the expression for w in terms of y, and then squaring to get $|\tilde{K}(f)|^2$.

(b) Derive Eqs. (6.52b) and (6.52c) for the kernel (6.52a).

EXERCISES

6.7.2 Brownian Motion and Random Walks

6.7.2

As an example of the uses of filtering, consider the motion of an organelle derived from pollen (henceforth "dust particle") being buffeted by thermalized air molecules—the phenomenon of Brownian motion, named for the Scottish botanist Robert Brown (1828), one of the first to observe it in careful experiments. As we discussed in Sec. 6.3.1 and in greater detail at the end of Sec. 6.3.3, any Cartesian component $v(t)$ of the particle's velocity is a Gaussian-Markov process, whose statistical properties are all determined by its equilibrium mean $\bar{v} = 0$ and standard deviation $\sigma_v = \sqrt{k_B T / m}$, and its relaxation time τ_r (which we compute in Sec. 6.8.1). Here m is the particle's

Brownian motion

mass, and T is the temperature of the air molecules that buffet it. The conditional probability distribution P_2 for v is given by Doob's theorem:

$$P_2(v_2, t|v_1) = \frac{e^{-(v_2-\bar{v}_t)^2/(2\sigma_{v_t}^2)}}{[2\pi\sigma_{v_t}^2]^{\frac{1}{2}}}, \quad \bar{v}_t = v_1 e^{-t/\tau_r},$$

$$\sigma_{v_t}^2 = (1 - e^{-2t/\tau_r})\sigma_v^2, \quad \sigma_v = \sqrt{\frac{k_B T}{m}} \tag{6.53a}$$

[Eqs. (6.18)], and its corresponding correlation function and spectral density have the standard forms (6.20) and (6.32) for a Gaussian-Markov process:

$$C_v(\tau) = \sigma_v^2 e^{-\tau/\tau_r}, \quad S_v(f) = \frac{4\sigma_v^2/\tau_r}{(2\pi f)^2 + (1/\tau_r)^2}. \tag{6.53b}$$

The Cartesian coordinate (position) of the dust particle, $x(t) = \int v\,dt$, is of special interest. Its spectral density can be deduced by applying the time-integral filter $|\tilde{K}(f)|^2 = 1/(2\pi f)^2$ to $S_v(f)$. The result, using Eq. (6.53b), is

$$S_x(f) = \frac{4\tau_r\sigma_v^2}{(2\pi f)^2[1 + (2\pi f\tau_r)^2]}. \tag{6.53c}$$

Notice that at frequencies $f \ll 1/\tau_r$ (corresponding to times long compared to the relaxation time), our result (6.53c) reduces to the random-walk spectrum $S_x = 4\sigma_v^2\tau_r/(2\pi f)^2$. From this spectrum, we can compute the rms distance $\sigma_{\Delta x}$ in the x direction that the dust particle travels in a time interval $\Delta\tau \gg \tau_r$. That $\sigma_{\Delta x}$ is the standard deviation of the random process $\Delta x(t) \equiv x(t + \Delta\tau) - x(t)$. The filter that takes $x(t)$ into $\Delta x(t)$ has

$$|\tilde{K}(f)|^2 = |e^{i2\pi f(t+\Delta\tau)} - e^{i2\pi ft}|^2 = 4\sin^2(\pi f\Delta\tau). \tag{6.54a}$$

Correspondingly, $\Delta x(t)$ has spectral density

$$S_{\Delta x}(f) = |\tilde{K}(f)|^2 S_x(f) = 4\sigma_v^2\tau_r(\Delta\tau)^2 \operatorname{sinc}^2(\pi f\Delta\tau) \tag{6.54b}$$

(where, again, $\operatorname{sinc} u \equiv \sin u/u$), so the variance of Δx (i.e., the square of the rms distance traveled) is

$$(\sigma_{\Delta x})^2 = \int_0^\infty S_{\Delta x}(f)\,df = 2(\sigma_v\tau_r)^2\frac{\Delta\tau}{\tau_r}. \tag{6.54c}$$

This equation has a simple physical interpretation. The damping time τ_r is the time required for collisions to change substantially the dust particle's momentum, so we can think of it as the duration of a single step in the particle's random walk. The particle's mean speed is roughly $\sqrt{2}\sigma_v$, so the distance traveled during each step (the particle's mean free path) is roughly $\sqrt{2}\sigma_v\tau_r$. (The $\sqrt{2}$ comes from our analysis; this physical argument could not have predicted it.) Therefore, during a time interval $\Delta\tau$ long compared to a single step τ_r, the rms distance traveled in the x direction by the random-walking dust particle is about one mean-free path $\sqrt{2}\sigma_v \tau_r$, multiplied by the square root of the mean number of steps taken, $\sqrt{\Delta\tau/\tau_r}$:

$$\sigma_{\Delta x} = \sqrt{2}\sigma_v \tau_r \sqrt{\Delta\tau/\tau_r}. \tag{6.55}$$

rms distance traveled in Brownian motion

This "*square root of the number of steps taken*" *behavior is a universal rule of thumb for random walks;* one meets it time and again in science, engineering, and mathematics. We have met it previously in our studies of diffusion (Exs. 3.17 and 6.3) and of the elementary "unit step" random walk problem that we studied using the central limit theorem in Ex. 6.4. We could have guessed Eq. (6.55) from this rule of thumb, up to an unknown multiplicative factor of order unity. Our analysis has told us that factor: $\sqrt{2}$.

EXERCISES

Exercise 6.10 *Example: Position, Viewed on Timescales* $\Delta\tau \gg \tau_r$, *as a Markov Process*

(a) Explain why, physically, when the Brownian motion of a particle (which starts at $x = 0$ at time $t = 0$) is observed only on timescales $\Delta\tau \gg \tau_r$ corresponding to frequencies $f \ll 1/\tau_r$, its position $x(t)$ must be a Gaussian-Markov process with $\bar{x} = 0$. What are the spectral density of $x(t)$ and its relaxation time in this case?

(b) Use Doob's theorem to compute the conditional probability $P_2(x_2, \tau | x_1)$. Your answer should agree with the result deduced in Ex. 6.3 from the diffusion equation, and in Ex. 6.4 from the central limit theorem for a random walk.

6.7.3 Extracting a Weak Signal from Noise: Band-Pass Filter, Wiener's Optimal Filter, Signal-to-Noise Ratio, and Allan Variance of Clock Noise

6.7.3

In experimental physics and engineering, one often meets a random process $Y(t)$ that consists of a sinusoidal signal on which is superposed noise $y(t)$:

$$Y(t) = \sqrt{2}Y_s \cos(2\pi f_o t + \delta_o) + y(t). \tag{6.56a}$$

[The factor $\sqrt{2}$ is included in Eq. (6.56a) because the time average of the square of the cosine is 1/2; correspondingly, with the factor $\sqrt{2}$ present, Y_s is the rms signal amplitude.] We assume that the frequency f_o and phase δ_o of the signal are known, and we want to determine the signal's rms amplitude Y_s. The noise $y(t)$ is an impediment to the determination of Y_s. To reduce that impediment, we can send $Y(t)$ through a *band-pass filter* centered on the signal frequency f_o (i.e., a filter with a shape like that of Fig. 6.13).

band-pass filter

For such a filter, with central frequency f_o and with bandwidth $\Delta f \ll f_o$, the bandwidth is defined by

bandwidth

$$\Delta f \equiv \frac{\int_0^\infty |\tilde{K}(f)|^2 df}{|\tilde{K}(f_o)|^2}. \tag{6.56b}$$

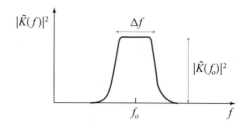

FIGURE 6.13 A band-pass filter centered on frequency f_o with bandwidth Δf.

The output $W(t)$ of such a filter, when the input is $Y(t)$, has the form

$$W(t) = |\tilde{K}(f_o)|\sqrt{2}Y_s \cos(2\pi f_o t + \delta_1) + w(t), \qquad (6.56c)$$

where the first term is the filtered signal and the second is the filtered noise. The output signal's phase δ_1 may be different from the input signal's phase δ_o, but that difference can be evaluated in advance for one's filter and can be taken into account in the measurement of Y_s; thus it is of no interest to us. Assuming (as we shall) that the input noise $y(t)$ has spectral density S_y that varies negligibly over the small bandwidth of the filter, the filtered noise $w(t)$ will have spectral density

spectral density of band-pass filter's output

$$S_w(f) = |\tilde{K}(f)|^2 S_y(f_o). \qquad (6.56d)$$

This means that $w(t)$ consists of a random superposition of sinusoids all with nearly—but not quite—the same frequency f_o; their frequency spread is Δf. Now, when one superposes two sinusoids with frequencies that differ by $\Delta f \ll f_o$, the two beat against each other, producing a modulation with period $1/\Delta f$. Correspondingly, with its random superposition of many such sinusoids, the noise $w(t)$ will have the form

$$w(t) = w_o(t) \cos[2\pi f_o t + \phi(t)], \qquad (6.56e)$$

with amplitude $w_o(t)$ and phase $\phi(t)$ that fluctuate randomly on timescales

$$\boxed{\Delta t \sim 1/\Delta f,} \qquad (6.56f)$$

but that are nearly constant on timescales $\Delta t \ll 1/\Delta f$.

band-pass filter's output in time domain   The filter's net output $W(t)$ thus consists of a precisely sinusoidal signal at frequency f_o, with known phase δ_1 and with an amplitude that we wish to determine, plus noise $w(t)$ that is also sinusoidal at frequency f_o but with amplitude and phase that wander randomly on timescales $\Delta t \sim 1/\Delta f$. The rms output signal is

$$S \equiv |\tilde{K}(f_o)|Y_s \qquad (6.56g)$$

316   Chapter 6. Random Processes

[Eq. (6.56c)], while the rms output noise is

$$N \equiv \sigma_w = \left[\int_0^\infty S_w(f) df \right]^{\frac{1}{2}}$$

$$= \sqrt{S_y(f_o)} \left[\int_0^\infty |\tilde{K}(f)|^2 df \right]^{\frac{1}{2}} = |\tilde{K}(f_o)| \sqrt{S_y(f_o) \Delta f}, \qquad \text{(6.56h)}$$

where the first integral follows from Eq. (6.26), the second from Eq. (6.56d), and the third from the definition (6.56b) of the bandwidth Δf. The ratio of the rms signal (6.56g) to the rms noise (6.56h) after filtering is

$$\boxed{\frac{S}{N} = \frac{Y_s}{\sqrt{S_y(f_o) \Delta f}}.} \qquad \text{(6.57)}$$

signal-to-noise ratio for
band-pass filter

Thus, the rms output $S + N$ of the filter is the signal amplitude to within an rms fractional error N/S given by the reciprocal of (6.57). Notice that the narrower the filter's bandwidth, the more accurate will be the measurement of the signal. In practice, of course, one does not know the signal frequency with complete precision in advance; correspondingly, one does not want to make one's filter so narrow that the signal might be lost from it.

A simple example of a band-pass filter is the following *finite Fourier transform filter*:

$$w(t) = \int_{t-\Delta t}^t \cos[2\pi f_o(t - t')] y(t') dt', \quad \text{where } \Delta t \gg 1/f_o. \qquad \text{(6.58a)}$$

In Ex. 6.11 it is shown that this is indeed a band-pass filter, and that the integration time Δt used in the Fourier transform is related to the filter's bandwidth by

$$\Delta f = 1/\Delta t. \qquad \text{(6.58b)}$$

Often the signal one seeks amidst noise is not sinusoidal but has some other known form $s(t)$. In this case, the optimal way to search for it is with a so-called *Wiener filter* (an alternative to the band-pass filter); see the very important Ex. 6.12.

EXERCISES

Exercise 6.11 *Derivation and Example: Bandwidths of a Finite-Fourier-Transform Filter and an Averaging Filter*

(a) If y is a random process with spectral density $S_y(f)$, and $w(t)$ is the output of the finite-Fourier-transform filter (6.58a), what is $S_w(f)$?

(b) Sketch the filter function $|\tilde{K}(f)|^2$ for this finite-Fourier-transform filter, and show that its bandwidth is given by Eq. (6.58b).

(c) An "averaging filter" is one that averages its input over some fixed time interval Δt:

$$w(t) \equiv \frac{1}{\Delta t} \int_{t-\Delta t}^t y(t') dt'. \qquad \text{(6.59a)}$$

What is $|\tilde{K}(f)|^2$ for this filter? Draw a sketch of this $|\tilde{K}(f)|^2$.

(d) Suppose that $y(t)$ has a spectral density that is very nearly constant at all frequencies, $f \lesssim 1/\Delta t$, and that this y is put through the averaging filter (6.59a). Show that the rms fluctuations in the averaged output $w(t)$ are

$$\sigma_w = \sqrt{S_y(0)\Delta f}, \tag{6.59b}$$

where Δf, interpretable as the bandwidth of the averaging filter, is

$$\Delta f = \frac{1}{2\Delta t}. \tag{6.59c}$$

(Recall that in our formalism we insist that f be nonnegative.) Why is there a factor $1/2$ here and not one in the equation for an averaging filter [Eq. (6.58b)]? Because here, with f restricted to positive frequencies and the filter centered on zero frequency, we see only the right half of the filter: $f \geq f_o = 0$ in Fig. 6.13.

Exercise 6.12 **Example: Wiener's Optimal Filter*
Suppose that you have a noisy receiver of weak signals (e.g., a communications receiver). You are expecting a signal $s(t)$ with finite duration and known form to come in, beginning at a predetermined time $t = 0$, but you are not sure whether it is present. If it is present, then your receiver's output will be

$$Y(t) = s(t) + y(t), \tag{6.60a}$$

where $y(t)$ is the receiver's noise, a random process with spectral density $S_y(f)$ and with zero mean, $\bar{y} = 0$. If it is absent, then $Y(t) = y(t)$. A powerful way to find out whether the signal is present is by passing $Y(t)$ through a filter with a carefully chosen kernel $K(t)$. More specifically, compute the quantity

$$W \equiv \int_{-\infty}^{+\infty} K(t)Y(t)dt. \tag{6.60b}$$

If $K(t)$ is chosen optimally, then W will be maximally sensitive to the signal $s(t)$ in the presence of the noise $y(t)$. Correspondingly, if W is large, you infer that the signal was present; if it is small, you infer that the signal was either absent or so weak as not to be detectable. This exercise derives the form of the *optimal filter*, $K(t)$ (i.e., the filter that will most effectively discern whether the signal is present). As tools in the derivation, we use the quantities S and N defined by

$$S \equiv \int_{-\infty}^{+\infty} K(t)s(t)dt, \quad N \equiv \int_{-\infty}^{+\infty} K(t)y(t)dt. \tag{6.60c}$$

Note that S is the filtered signal, N is the filtered noise, and $W = S + N$. Since $K(t)$ and $s(t)$ are precisely defined functions, S is a number. But since $y(t)$ is a random process, the value of N is not predictable, and instead is given by some probability distribution $p_1(N)$. We shall also need the Fourier transform $\tilde{K}(f)$ of the kernel $K(t)$.

(a) In the measurement being done one is not filtering a function of time to get a new function of time; rather, one is just computing a number, $W = S + N$.

Nevertheless, as an aid in deriving the optimal filter it is helpful to consider the time-dependent output of the filter that results when noise $y(t)$ is fed continuously into it:

$$N(t) \equiv \int_{-\infty}^{+\infty} K(t - t')y(t')dt'. \qquad (6.61a)$$

Show that this random process has a mean squared value

$$\overline{N^2} = \int_0^\infty |\tilde{K}(f)|^2 S_y(f)df. \qquad (6.61b)$$

Explain why this quantity is equal to the average of the number N^2 computed using (6.60c) in an ensemble of many experiments:

$$\overline{N^2} = \langle N^2 \rangle \equiv \int p_1(N)N^2 dN = \int_0^\infty |\tilde{K}(f)|^2 S_y(f)df. \qquad (6.61c)$$

(b) Show that of all choices of the kernel $K(t)$, the one that will give the largest value of

$$\frac{S}{\langle N^2 \rangle^{\frac{1}{2}}} \qquad (6.61d)$$

is Norbert Wiener's (1949) *optimal filter*, whose kernel has the Fourier transform

$$\boxed{\tilde{K}(f) = \text{const} \times \frac{\tilde{s}(f)}{S_y(f)},} \qquad (6.62a) \qquad \text{**Wiener's optimal filter**}$$

where $\tilde{s}(f)$ is the Fourier transform of the signal $s(t)$, and $S_y(f)$ is the spectral density of the noise. Note that when the noise is white, so $S_y(f)$ is independent of f, this optimal kernel is just $K(t) = \text{const} \times s(t)$ (i.e., one should simply multiply the known signal form into the receiver's output and integrate). By contrast, when the noise is not white, the optimal kernel (6.62a) is a distortion of $\text{const} \times s(t)$ in which frequency components at which the noise is large are suppressed, while frequency components at which the noise is small are enhanced.

(c) Show that when the optimal kernel (6.62a) is used, the square of the signal-to-noise ratio is

$$\boxed{\frac{S^2}{\langle N^2 \rangle} = 4 \int_0^\infty \frac{|\tilde{s}(f)|^2}{S_y(f)}df.} \qquad (6.62b) \qquad \text{**signal-to-noise ratio for Wiener's optimal filter**}$$

(d) As an example, suppose the signal consists of n cycles of some complicated waveform with frequencies spread out over the range $f_o/2$ to $2f_o$ with amplitude $\sim A$ for its entire duration. Also suppose that S_y is approximately constant (near white noise) over this frequency band. Show that $S/\langle N^2 \rangle^{1/2} \sim 2nA/\sqrt{f_o S_y(f_o)}$, so the amplitude signal to noise increases linearly with the number of cycles in the signal.

(e) Suppose (as an idealization of searches for gravitational waves in noisy LIGO data) that (i) we do not know the signal $s(t)$ in advance, but we do know that it is from a set of N distinct signals, all of which have frequency content concentrated around some f_o; (ii) we do not know when the signal will arrive, but we search for it for a long time τ_s (say, a year); and (iii) the noise superposed on the signal is Gaussian. Show that, in order to have 99% confidence that any signal found is real, it must have amplitude signal-to-noise ratio of $S/\langle N^2 \rangle^{1/2} \gtrsim [2 \ln(H/\sqrt{2 \ln H})]^{1/2}$, where $H = 100 N f_o \tau_s$. For $N \sim 10^4$, $f_o \sim 100$ Hz, $\tau_s \sim 1$ yr, this says $S/\langle N^2 \rangle^{1/2} \gtrsim 8.2$. This is so small because the Gaussian probability distribution falls off so rapidly. If the noise is non-Gaussian, then the minimum detectable signal will be larger than this, possibly much larger.

Exercise 6.13 **Example: Allan Variance of Clocks*

Highly stable clocks (e.g., cesium clocks, hydrogen maser clocks, or quartz crystal oscillators) have angular frequencies ω of ticking that tend to wander so much over very long timescales that their variances diverge. For example, a cesium clock has random-walk noise on very long timescales (low frequencies)

$$S_\omega(f) \propto 1/f^2 \quad \text{at low } f;$$ (6.63a)

and correspondingly,

$$\sigma_\omega{}^2 = \int_0^\infty S_\omega(f) df = \infty$$ (6.63b)

(cf. Fig. 6.11 and associated discussion). For this reason, clock makers have introduced a special technique for quantifying the frequency fluctuations of their clocks. Using the phase

$$\phi(t) = \int_0^t \omega(t') dt',$$ (6.64a)

they define the quantity

$$\Phi_\tau(t) = \frac{[\phi(t + 2\tau) - \phi(t + \tau)] - [\phi(t + \tau) - \phi(t)]}{\sqrt{2}\bar{\omega}\tau},$$ (6.64b)

where $\bar{\omega}$ is the mean frequency. Aside from the $\sqrt{2}$, this $\Phi_\tau(t)$ is the fractional difference of clock readings for two successive intervals of duration τ. (In practice the measurement of t is made by a clock more accurate than the one being studied; or, if a more accurate clock is not available, by a clock or ensemble of clocks of the same type as is being studied.)

(a) Show that the spectral density of $\Phi_\tau(t)$ is related to that of $\omega(t)$ by

$$S_{\Phi_\tau}(f) = \frac{2}{\bar{\omega}^2}\left[\frac{\cos 2\pi f\tau - 1}{2\pi f\tau}\right]^2 S_\omega(f)$$

$$\propto f^2 S_\omega(f) \text{ at } f \ll 1/(2\pi\tau)$$ (6.65)

$$\propto f^{-2} S_\omega(f) \text{ at } f \gg 1/(2\pi\tau).$$

Note that $S_{\Phi_\tau}(f)$ is much better behaved (more strongly convergent when integrated) than $S_\omega(f)$ is, both at low frequencies and at high.

(b) The *Allan variance* of the clock is defined as

$$\sigma_\tau{}^2 \equiv [\text{ variance of } \Phi_\tau(t)] = \int_0^\infty S_{\Phi_\tau}(f)df. \qquad (6.66)$$

Show that

$$\sigma_\tau = \left[\alpha \frac{S_\omega(1/(2\tau))}{\bar\omega^2} \frac{1}{2\tau}\right]^{\frac{1}{2}}, \qquad (6.67)$$

where α is a constant of order unity that depends on the spectral shape of $S_\omega(f)$ near $f = 1/(2\tau)$. Explain why, aside from the factor α, the right-hand side of Eq. (6.67) is the rms fractional fluctuation of ω at frequency $1/(2\tau)$ in bandwidth $1/(2\tau)$.

(c) Show that, if ω has a white-noise spectrum, then the clock stability is better for long averaging times than for short; if ω has a flicker-noise spectrum, then the clock stability is independent of averaging time; and if ω has a random-walk spectrum, then the clock stability is better for short averaging times than for long. (See Fig. 6.11.)

6.7.4 Shot Noise

A specific kind of noise that one frequently meets and frequently wants to filter is *shot noise*. A random process $y(t)$ is said to consist of shot noise if it is a random superposition of a large number of finite-duration pulses. In this chapter, we restrict attention to a simple version of shot noise in which the pulses all have the same shape, $F(\tau)$ (e.g., Fig. 6.14a), but their arrival times t_i are random

$$y(t) = \sum_i F(t - t_i), \qquad (6.68a)$$

as may be the case for individual photons arriving at a photodetector. We denote by $\mathcal{R}$ the mean rate of pulse arrivals (the mean number per second). From the definition (6.25) of spectral density, it is straightforward to see that the spectral density of y is

$$\boxed{S_y(f) = 2\mathcal{R}|\tilde F(f)|^2,} \qquad (6.68b)$$

where $\tilde F(f)$ is the Fourier transform of $F(\tau)$ (Fig. 6.14). See Ex. 6.14 for proof. If the pulses are broadband bursts without much substructure in them (as in Fig. 6.14a), then the duration τ_p of the pulse is related to the frequency $f_{\max}$ at which the spectral density starts to cut off by $f_{\max} \sim 1/\tau_p$. Since the correlation function is the cosine transform of the spectral density, the correlation's relaxation time is $\tau_r \sim 1/f_{\max} \sim \tau_p$ (Ex. 6.14).

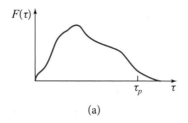
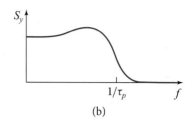

(a) (b)

FIGURE 6.14 (a) A broadband pulse that produces shot noise by arriving at random times. (b) The spectral density of the shot noise produced by that pulse.

In the common (but not universal) case that many pulses are on at once on average ($\mathcal{R}\tau_p \gg 1$), then at any moment of time $y(t)$ is the sum of many random processes. Correspondingly, the central limit theorem guarantees that y is a Gaussian random process. Over time intervals smaller than $\tau_p \sim \tau_r$ the process will not generally be Markov, because a knowledge of both $y(t_1)$ and $y(t_2)$ gives some rough indication of how many pulses happen to be on and how many new ones have turned on during the time interval between t_1 and t_2 and thus are still in their early stages at time t_3; this knowledge helps one predict $y(t_3)$ with greater confidence than if one knew only $y(t_2)$. In other words, $P_3(y_3, t_3|y_2, t_2; y_1, t_1)$ is not equal to $P_2(y_3, t_3|y_2, t_2)$, which implies non-Markov behavior.

By contrast, if many pulses are on at once, and if one takes a coarse-grained view of time, never examining time intervals as short as τ_p or shorter, then a knowledge of $y(t_1)$ is of no help in predicting $y(t_2)$. All correlations between different times are lost, so the process is Markov, and (because it is a random superposition of many independent influences) it is also Gaussian—an example of the central limit theorem at work. Thus it must have the standard Gaussian-Markov spectral density (6.32) with vanishing correlation time τ_r (i.e., it must be white). Indeed, it is: For $f \ll 1/\tau_p$, the limit of Eq. (6.68b) for S_y and the corresponding correlation function are

$$S_y(f) = 2\mathcal{R}|\tilde{F}(0)|^2, \quad C_y(\tau) = \mathcal{R}|\tilde{F}(0)|^2\delta(\tau). \tag{6.68c}$$

This formula remains true if the pulses have different shapes, so long as their Fourier transforms at zero frequency, $\tilde{F}_j(0) = \int_{-\infty}^{\infty} F_j dt$, are all the same; see Ex. 6.14b.

As an important example, consider a (nearly) monochromatic beam of light with angular frequency $\omega_o \sim 10^{15}$ s^{-1} and with power (energy per unit time) $W(t)$ that is being measured by a photodetector. The arriving light consists of discrete photons, each with its own pulse shape $W_j(t - t_j)$,[10] which lasts for a time τ_p long compared to the light's period ($\sim 3 \times 10^{-15}$ s) but short compared to the inverse frequency f^{-1} at which we measure the photon shot noise. The Fourier transform of W_j at zero

---

10. For a single photon, $W_j(t)$ is the probability per unit time for the photon's arrival, times the photon's energy $\hbar\omega_o$.

frequency is just $\tilde{W}_j(0) = \int_0^\infty W_j dt = \hbar\omega$ (the total energy carried by the photon), which is the same for all pulses; the rate of arrival of photons is $\mathcal{R} = \overline{W}/\hbar\omega_o$. Therefore the spectral density of the intensity measured by the photodetector is

$$S_W(f) = 2\overline{W}\,\hbar\omega. \qquad (6.69)$$

In the LIGO interferometer, whose noise power spectrum is shown in Fig. 6.7, this photon shot noise dominates in the frequency band $f \gtrsim 200$ Hz. (Although S_W for the laser light has white noise, when passed through the interferometer as a filter, it produces $S_x \propto f^2$.)

Exercise 6.14 *Derivation: Shot Noise*

(a) Show that for shot noise, $y(t) = \sum_i F(t - t_i)$, the spectral density $S_y(f)$ is given by Eq. (6.68b). Show that the relaxation time appearing in the correlation function is approximately the duration τ_p of $F(t)$.

(b) Suppose the shapes of $F_j(t - t_j)$ are all different instead of being identical but all last for times $\lesssim \tau_p$, and all have the same Fourier transform at zero frequency, $\tilde{F}_j(0) = \int_{-\infty}^\infty F_j dt = \tilde{F}(0)$. Show that the shot noise at frequencies $f \ll 1/\tau_p$ is still given by Eq. (6.68c).

Exercise 6.15 *Example: Shot Noise with Flicker Spectrum*

(a) Show that for shot noise with identical pulses that have the infinitely sharply peaked shape of Eq. (6.45), the power spectrum has the flicker form $S_y \propto 1/f$ for all f.

(b) Construct realizations of shot noise with flicker spectrum [Eq. (6.68a) with pulse shape (6.45)] that range from a few large pulses in the time interval observed to many small pulses, and describe the visual differences; cf. Fig. 6.10 and the discussion in Sec. 6.6.2.

6.8 Fluctuation-Dissipation Theorem

6.8

6.8.1 Elementary Version of the Fluctuation-Dissipation Theorem; Langevin Equation, Johnson Noise in a Resistor, and Relaxation Time for Brownian Motion

6.8.1

Friction is generally caused by interaction with the huge number of degrees of freedom of some sort of bath (e.g., the molecules of air against which a moving ball or dust particle pushes). Those degrees of freedom also produce fluctuating forces. In this section, we study the relationship between the friction and the fluctuating forces when the bath is thermalized at some temperature T (so it is a heat bath).

For simplicity, we restrict ourselves to a specific generalized coordinate q of the system that will interact with a bath (e.g., the x coordinate of the ball or dust particle).

We require just one special property for q: its time derivative $\dot{q} = dq/dt$ must appear in the system's lagrangian as a kinetic energy,

$$E_{\text{kinetic}} = \frac{1}{2}m\dot{q}^2, \tag{6.70}$$

and in no other way. Here m is a (generalized) mass associated with q. Then the equation of motion for q will have the simple form of Newton's second law, $m\ddot{q} = F$, where F includes contributions $\mathcal{F}$ from the system itself (e.g., a restoring force in the case of a normal mode), plus a force F_{bath} due to the heat bath (i.e., due to all the degrees of freedom in the bath). This F_{bath} is a random process whose time average is a frictional (damping) force proportional to $\dot{q}$:

$$\bar{F}_{\text{bath}} = -R\dot{q}, \quad F_{\text{bath}} \equiv \bar{F}_{\text{bath}} + F'. \tag{6.71}$$

Here R is the coefficient of friction. The fluctuating part F' of F_{bath} is responsible for driving q toward statistical equilibrium.

examples:

Three specific examples, to which we shall return below, are as follows.

dust particle

1. The system might be a dust particle with q its x coordinate and m its mass. The heat bath might be air molecules at temperature T, which buffet the dust particle, producing Brownian motion.

electric circuit

2. The system might be an L-C-R circuit (i.e., an electric circuit containing an inductance L, a capacitance C, and a resistance R) with q the total electric charge on the top plate of the capacitor. The bath in this case would be the many mechanical degrees of freedom in the resistor. For such a circuit, the "equation of motion" is

$$L\ddot{q} + C^{-1}q = F_{\text{bath}}(t) = -R\dot{q} + F', \tag{6.72}$$

so the effective mass is the inductance L; the coefficient of friction is the resistance (both denoted R); $-R\dot{q} + F'$ is the total voltage across the resistor; and F' is the fluctuating voltage produced by the resistor's internal degrees of freedom (the bath) and so might better be denoted V'.

normal mode of crystal

3. The system might be the fundamental mode of a 10-kg sapphire crystal with q its generalized coordinate (cf. Sec. 4.2.1). The heat bath might be all the other normal modes of vibration of the crystal, with which the fundamental mode interacts weakly.

LANGEVIN EQUATION

In general, the equation of motion for the generalized coordinate $q(t)$ under the joint action of (i) the bath's damping force $-R\dot{q}$, (ii) the bath's fluctuating forces F', and (iii) the system's internal force $\mathcal{F}$ will take the form [cf. Eq. (6.71)]

$$m\ddot{q} + R\dot{q} = \mathcal{F} + F'(t). \tag{6.73}$$

The internal force $\mathcal{F}$ is derived from the system's hamiltonian or lagrangian in the absence of the heat bath. For the L-C-R circuit of Eq. (6.72) that force is $\mathcal{F} = -C^{-1}q$; for the dust particle, if the particle were endowed with a charge Q and were in an external electric field with potential $\Phi(t, x, y, z)$, it would be $\mathcal{F} = -Q\partial\Phi/\partial x$; for the normal mode of a crystal, it is $\mathcal{F} = -m\omega^2 q$, where ω is the mode's eigenfrequency.

Because the equation of motion (6.73) involves a driving force $F'(t)$ that is a random process, one cannot solve it to obtain $q(t)$. Instead, one must solve it in a statistical way to obtain the evolution of q's probability distributions $p_n(q_1, t_1; \ldots; q_n, t_n)$. This and other evolution equations involving random-process driving terms are called by modern mathematicians *stochastic differential equations,* and there is an extensive body of mathematical formalism for solving them. In statistical physics the specific stochastic differential equation (6.73) is known as the *Langevin equation.*

stochastic differential equation

Langevin equation

ELEMENTARY FLUCTUATION-DISSIPATION THEOREM

Because the damping force $-R\dot{q}$ and the fluctuating force F' both arise from interaction with the same heat bath, there is an intimate connection between them. For example, the stronger the coupling to the bath, the stronger will be the coefficient of friction R and the stronger will be F'. The precise relationship between the dissipation embodied in R and the fluctuations embodied in F' is given by the following *fluctuation-dissipation theorem:* At frequencies

elementary fluctuation-dissipation theorem

$$f \ll 1/\tau_r, \tag{6.74a}$$

where τ_r is the (very short) relaxation time for the fluctuating force F', the fluctuating force has the spectral density

$$\boxed{S_{F'}(f) = 4R\left(\frac{1}{2}hf + \frac{hf}{e^{hf/(k_BT)} - 1}\right)} \quad \text{in general,} \tag{6.74b}$$

$$\boxed{S_{F'}(f) = 4Rk_BT \quad \text{in the classical domain, } k_BT \gg hf.} \tag{6.74c}$$

Here T is the temperature of the bath, and h is Planck's constant.

Notice that in the classical domain, $k_BT \gg hf$, the spectral density has a white-noise spectrum. In fact, since we are restricting attention to frequencies at which F' has no self-correlations ($f^{-1} \gg \tau_r$), F' is Markov; and since it is produced by interaction with the huge number of degrees of freedom of the bath, F' is also Gaussian. Thus, in the classical domain F' is a Gaussian-Markov, white-noise process.

classical fluctuations accompanying $-R\dot{q}$ damping are Gaussian, Markov, white-noise processes

At frequencies $f \gg k_BT/h$ (quantum domain), in Eq. (6.74b) the term $S_{F'} = 4R\frac{1}{2}hf$ is associated with vacuum fluctuations of the degrees of freedom that make up the heat bath (one-half quantum of fluctuations per mode as for any quantum mechanical simple harmonic oscillator). In addition, the second term, $S_{F'}(f) = 4Rhfe^{-hf/(k_BT)}$, associated with thermal excitations of the bath's degrees of freedom, is exponentially suppressed because at these high frequencies, the bath's modes have exponentially small probabilities of containing any quanta at all. Since in this quantum domain $S_{F'}(f)$ does not have the standard Gaussian-Markov frequency dependence (6.32), in the quantum domain F' is not a Gaussian-Markov process.

Proof of the Fluctuation-Dissipation Theorem.

proof of elementary fluctuation-dissipation theorem

In principle, we can alter the system's internal restoring force $\mathcal{F}$ without altering its interactions with the heat bath [i.e., without altering R or $S_{F'}(f)$]. For simplicity, we set $\mathcal{F}$ to zero so q becomes the coordinate of a free mass. The basic idea of our proof is to choose a frequency f_o at which to evaluate the spectral density of F', and then, in an idealized thought experiment, very weakly couple a harmonic oscillator with eigenfrequency f_o to q. Through that coupling, the oscillator is indirectly damped by the resistance R of q and is indirectly driven by R's associated fluctuating force F', which arises from a bath with temperature T. After a long time, the oscillator will reach thermal equilibrium with that bath and will then have the standard thermalized mean kinetic energy ($\bar{E} = k_B T$ in the classical regime). We shall compute that mean energy in terms of $S_{F'}(f_o)$ and thereby deduce $S_{F'}(f_o)$.

The Langevin equation (6.73) and equation of motion for the coupled free mass and harmonic oscillator are

$$m\ddot{q} + R\dot{q} = -\kappa Q + F'(t), \quad M\ddot{Q} + M\omega_o^2 Q = -\kappa q. \tag{6.75a}$$

Here M, Q, and $\omega_o = 2\pi f_o$ are the oscillator's mass, coordinate, and angular eigenfrequency, and κ is the arbitrarily small coupling constant. (The form of the coupling terms $-\kappa Q$ and $-\kappa q$ in the two equations can be deduced from the coupling's interaction hamiltonian $H_I = \kappa q Q$.) Equations (6.75a) can be regarded as a filter to produce from the fluctuating-force input $F'(t)$ a resulting motion of the oscillator, $Q(t) = \int_{-\infty}^{+\infty} K(t - t') F'(t') dt'$. The squared Fourier transform $|\tilde{K}(f)|^2$ of this filter's kernel $K(t - t')$ is readily computed by the standard method [Eq. (6.51) and associated discussion] of inserting a sinusoid $e^{-i\omega t}$ (with $\omega = 2\pi f$) into the filter [i.e., into the differential equations (6.75a)] in place of F', then solving for the sinusoidal output Q, and then setting $|\tilde{K}|^2 = |Q|^2$. The resulting $|\tilde{K}|^2$ is the ratio of the spectral densities of input and output. We carefully manipulate the resulting $|\tilde{K}|^2$ so as to bring it into the following standard resonant form:

$$S_q(f) = |\tilde{K}(f)|^2 S_{F'}(f) = \frac{|B|^2}{(\omega - \omega_o')^2 + (2M\omega_o^2 R|B|^2)^2]} S_{F'}(f). \tag{6.75b}$$

Here $B = \kappa/[2M\omega_o(m\omega_o^2 + iR\omega_o)]$ is arbitrarily small because κ is arbitrarily small; and $\omega_o'^2 = \omega_o^2 + 4mM\omega_o^4|B|^2$ is the oscillator's squared angular eigenfrequency after coupling to q, and is arbitrarily close to ω_o^2 because $|B|^2$ is arbitrarily small. In these equations we have replaced ω by ω_o everywhere except in the resonance term $(\omega - \omega_o')^2$ because $|\tilde{K}|^2$ is negligibly small everywhere except near resonance, $\omega \cong \omega_o$.

The mean energy of the oscillator, averaged over an arbitrarily long timescale, can be computed in either of two ways.

Chapter 6. Random Processes

1. Because the oscillator is a mode of some boson field and is in statistical equilibrium with a heat bath, its mean occupation number must have the standard Bose-Einstein value $\eta = 1/[e^{\hbar\omega_o/(k_B T)} - 1]$, and since each quantum carries an energy $\hbar\omega_o$, the mean energy is

$$\bar{E} = \frac{\hbar\omega_o}{e^{\hbar\omega_o/(k_B T)} - 1} + \frac{1}{2}\hbar\omega_o. \tag{6.75c}$$

Here we have included the half-quantum of energy associated with the mode's vacuum fluctuations.

2. Because on average the energy is half potential and half kinetic, and the mean potential energy is $\frac{1}{2}m\omega_o^2 \overline{Q^2}$, and because the ergodic hypothesis tells us that time averages are the same as ensemble averages, it must be that

$$\bar{E} = 2\frac{1}{2}M\omega_o^2\omega^2\langle Q^2\rangle = M\omega_o^2 \int_0^\infty S_Q(f)\, df. \tag{6.75d}$$

By inserting the spectral density (6.75b) and performing the frequency integral with the help of the narrowness of the resonance, we obtain

$$\bar{E} = \frac{S_{F'}(f_o)}{4R}. \tag{6.75e}$$

Equating this to our statistical-equilibrium expression (6.75c) for the mean energy, we see that at the frequency $f_o = \omega_o/(2\pi)$ the spectral density $S_{F'}(f_o)$ has the form (6.74b) claimed in the fluctuation-dissipation theorem. Moreover, since f_o can be chosen to be any frequency in the range (6.74a), the spectral density $S_{F'}(f)$ has the claimed form anywhere in this range. ∎

Let us discuss two examples of the elementary fluctuation-dissipation theorem (6.74):

JOHNSON NOISE IN A RESISTOR

For the L-C-R circuit of Eq. (6.72), $R\dot{q}$ is the dissipative voltage across the resistor, and $F'(t)$ is the fluctuating voltage [more normally denoted $V'(t)$] across the resistor. The fluctuating voltage is called *Johnson noise*, and the fluctuation-dissipation relationship $S_V(f) = 4Rk_B T$ (classical regime) is called *Nyquist's theorem*, because John Johnson (1928) discovered the voltage fluctuations $V'(t)$ experimentally, and Harry Nyquist (1928) derived the fluctuation-dissipation relationship for a resistor to explain them. The fluctuation-dissipation theorem as formulated here is a generalization of Nyquist's original theorem to any system with kinetic energy $\frac{1}{2}m\dot{q}^2$ associated with a generalized coordinate q and with frictional dissipation produced by a heat bath.

Nyquist's theorem for Johnson noise

BROWNIAN MOTION

In Secs. 6.3.3 and 6.7.2, we have studied the Brownian motion of a dust particle being buffeted by air molecules, but until now we omitted any attempt to deduce the motion's relaxation time τ_r. We now apply the fluctuation-dissipation theorem to deduce τ_r, using a model in which the particle is idealized as a sphere with mass m and radius a that, of course, is far larger than the air molecules.

The equation of motion for the dust particle, when we ignore the molecules' fluctuating forces, is $m dv/dt = -Rv$. Here the resistance (friction) R due to interaction with the molecules has a form that depends on whether the molecules' mean free path λ is small or large compared to the particle. From the kinetic-theory formula $\lambda = 1/(n\sigma_{mol})$, where n is the number density of molecules and σ_{mol} is their cross section to scatter off each other (roughly their cross sectional area), we can deduce that for air $\lambda \sim 0.1\,\mu$m. This is tiny compared to a dust particle's radius $a \sim 10$ to $1{,}000\,\mu$m. This means that, when interacting with the dust particle, the air molecules will behave like a fluid. As we shall learn in Chap. 15, the friction for a fluid depends on whether a quantity called the Reynolds number, $\mathrm{Re} = va/\nu$, is small or large compared to unity; here $\nu \sim 10^{-5}\,\mathrm{m^2\,s^{-1}}$ is the kinematic viscosity of air. Inserting numbers, we see that $\mathrm{Re} \sim (v/0.1\,\mathrm{m\,s^{-1}})(a/100\,\mu\mathrm{m})$. The speeds v of dust particles being buffeted by air are far smaller than $0.1\,\mathrm{m\,s^{-1}}$ as anyone who has watched them in a sunbeam knows, or as you can estimate from Eq. (6.53a). Therefore, the Reynolds number is small. From an analysis carried out in Sec. 14.3.2, we learn that in this low-Re fluid regime, the resistance (friction) on our spherical particle with radius a is [Eq. (14.34)]

$$R = 6\pi\rho\nu a, \tag{6.76}$$

where $\rho \sim 1\,\mathrm{kg\,m^{-3}}$ is the density of air. (Notice that this resistance is proportional to the sphere's radius a or circumference; if λ were $\gg a$, then R would be proportional to the sphere's cross sectional area, i.e., to a^2.)

When we turn on the molecules' fluctuating force F', the particle's equation of motion becomes $m dv/dt + Rv = F'$. Feeding $e^{i2\pi ft}$ through this equation in place of F', we get the output $v = 1/(R + i2\pi fm)$, whose modulus squared then is the ratio of S_v to $S_{F'}$. In this obviously classical regime, the fluctuation-dissipation theorem states that $S_{F'} = 4Rk_BT$. Therefore, we have

$$S_v = \frac{S_{F'}}{R^2 + (2\pi fm)^2} = \frac{4Rk_BT}{R^2 + (2\pi fm)^2} = \frac{4Rk_BT/m^2}{(2\pi f)^2 + (R/m)^2}. \tag{6.77}$$

relaxation time for Brownian motion

By comparing with the S_v that we derived from Doob's theorem, Eq. (6.53b), we can read off the particle's rms velocity (in one dimension, x or y or z), $\sigma_v = \sqrt{k_BT/m}$— which agrees with Eq. (6.53a) as it must—and we can also read off the particle's relaxation time (not to be confused with the bath's relaxation time),

$$\tau_r = m/R = m/(6\pi\rho\nu a). \tag{6.78}$$

If we had tried to derive this relaxation time by analyzing the buffeting of the particle directly, we would have had great difficulty. The fluctuation-dissipation theorem,

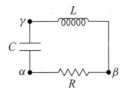

FIGURE 6.15 An L-C-R circuit. See Ex. 6.16.

Doob's theorem, and the fluid-mechanics analysis of friction on a sphere have made the task straightforward.

Exercise 6.16 *Practice: Noise in an L-C-R Circuit*

Consider an L-C-R circuit as shown in Fig. 6.15. This circuit is governed by the differential equation (6.72), where F' is the fluctuating voltage produced by the resistor's microscopic degrees of freedom (so we shall rename it V'), and $F \equiv V$ vanishes, since there is no driving voltage in the circuit. Assume that the resistor has temperature $T \gg \hbar\omega_o/k$, where $\omega_o = (LC)^{-1/2}$ is the circuit's resonant angular frequency, $\omega_o = 2\pi f_o$, and also assume that the circuit has a large quality factor (weak damping) so $R \ll 1/(\omega_o C) \simeq \omega_o L$.

(a) Initially consider the resistor R decoupled from the rest of the circuit, so current cannot flow across it. What is the spectral density $V_{\alpha\beta}$ of the voltage across this resistor?

(b) Now place the resistor into the circuit as shown in Fig. 6.15. The fluctuating voltage V' will produce a fluctuating current $I = \dot{q}$ in the circuit (where q is the charge on the capacitor). What is the spectral density of I? And what, now, is the spectral density $V_{\alpha\beta}$ across the resistor?

(c) What is the spectral density of the voltage $V_{\alpha\gamma}$ between points α and γ? and of $V_{\beta\gamma}$?

(d) The voltage $V_{\alpha\beta}$ is averaged from time $t = t_0$ to $t = t_0 + \tau$ (with $\tau \gg 1/f_o$), giving some average value U_0. The average is measured once again from t_1 to $t_1 + \tau$, giving U_1. A long sequence of such measurements gives an ensemble of numbers $\{U_0, U_1, \ldots, U_n\}$. What are the mean $\bar{U}$ and rms deviation $\Delta U \equiv \langle (U - \bar{U})^2 \rangle^{\frac{1}{2}}$ of this ensemble?

Exercise 6.17 ****Example: Detectability of a Sinusoidal Force**
that Acts on an Oscillator with Thermal Noise

To measure a very weak sinusoidal force, let the force act on a simple harmonic oscillator with eigenfrequency at or near the force's frequency, and measure the oscillator's response. Examples range in physical scale from nanomechanical oscillators ($\sim 1\,\mu$m in size) with eigenfrequency ~ 1 GHz that might play a role in future quantum information technology (e.g., Chan et al., 2011), to the fundamental mode of a ~ 10-kg sapphire crystal, to a ~ 40-kg LIGO mirror on which light pressure produces a restoring force, so its center of mass oscillates mechanically at frequency ~ 100 Hz (e.g., Abbott et al., 2009). The oscillator need not be mechanical; for example, it could be an L-C-R circuit, or a mode of an optical (Fabry-Perot) cavity.

The displacement $x(t)$ of any such oscillator is governed by the driven-harmonic-oscillator equation

$$m\left(\ddot{x} + \frac{2}{\tau_*}\dot{x} + \omega^2 x\right) = F(t) + F'(t). \tag{6.79}$$

Here m, ω, and τ_* are respectively the effective mass, angular frequency, and amplitude damping time associated with the oscillator; $F(t)$ is an external driving force; and $F'(t)$ is the fluctuating force associated with the dissipation that gives rise to τ_*. Assume that $\omega\tau_* \gg 1$ (weak damping).

(a) Weak coupling to other modes is responsible for the damping. If the other modes are thermalized at temperature T, what is the spectral density $S_{F'}(f)$ of the fluctuating force F'? What is the spectral density $S_x(f)$ of x?

(b) A very weak sinusoidal force drives the fundamental mode precisely on resonance:

$$F = \sqrt{2}F_s \cos\omega t. \tag{6.80}$$

Here F_s is the rms signal. What is the $x(t)$ produced by this signal force?

(c) A sensor with negligible noise monitors this $x(t)$ and feeds it through a narrow-band filter with central frequency $f = \omega/(2\pi)$ and bandwidth $\Delta f = 1/\hat{\tau}$ (where $\hat{\tau}$ is the averaging time used by the filter). Assume that $\hat{\tau} \gg \tau_*$. What is the rms thermal noise σ_x after filtering? Show that the strength F_s of the signal force that produces a signal $x(t) = \sqrt{2}x_s \cos(\omega t + \delta)$ with rms amplitude x_s equal to σ_x and phase δ is

$$F_s = \sqrt{\frac{8mk_BT}{\hat{\tau}\tau_*}}. \tag{6.81}$$

This is the minimum detectable force at the "one-σ level."

(d) Suppose that the force acts at a frequency ω_o that differs from the oscillator's eigenfrequency ω by an amount $|\omega - \omega_o| \lesssim 1/\tau_*$. What, then, is the minimum detectable force strength F_s? What might be the advantages and disadvantages of operating off resonance in this way, versus on resonance?

Not all generalized coordinates q have kinetic energy $\frac{1}{2}m\dot{q}^2$. An important example (Levin, 1998) arises when one measures the location of the front of a mirror by bouncing a laser beam perpendicularly off of it—a common and powerful tool in modern technology. If the mirror moves along the beam's optic axis by Δz, the distance of the bouncing light's travel changes by $2\Delta z$, and the light acquires a phase shift $(2\pi/\lambda)2\Delta z$ (with λ the light's wavelength) that can be read out via interferometry (Chap. 9). Because the front of the mirror can deform, Δz is actually the change in a spatial average of the mirror front's location $z(r, \phi; t)$, an average weighted by the number of photons that hit a given region. In other words, the (time varying) mirror position monitored by the light is

$$q(t) = \int z(r, \phi; t)\frac{e^{-(r/r_o)^2}}{\pi r_o^2}r d\phi\, dr. \tag{6.82}$$

Here (r, ϕ) are cylindrical coordinates centered on the laser beam's optic axis, and $e^{-(r/r_o)^2}$ is the Gaussian distribution of the beam's energy flux, so

$$\left[e^{-(r/r_o)^2}/(\pi r_o^2)\right]r d\phi dr$$

is the probability that a photon of laser light will hit the mirror at (r, ϕ) in the range $(dr, d\phi)$.

Because the mirror front's deformations $z(r, \phi; t)$ can be expanded in normal modes, this q is a linear superposition of the generalized coordinates $q_j(t)$ of the mirror's normal modes of oscillation and its center-of-mass displacement $q_0(t)$: $q(t) = q_0(t) + \sum_j Q_j(r, \phi)q_j(t)$, where $Q_j(r, \phi)$ is mode j's displacement eigenfunction evaluated at the mirror's face. Each of the generalized coordinates q_0 and q_j has a kinetic energy proportional to $\dot{q}_j^2$, but this q does not. Therefore, the elementary version of the fluctuation-dissipation theorem, treated in the previous section, is not valid for this q.

Fortunately, there is a remarkably powerful generalized fluctuation-dissipation theorem due to Callen and Welton (1951) that works for this q and all other generalized coordinates that are coupled to a heat bath. To formulate this theorem, we must first introduce the *complex impedance* $Z(\omega)$ for a generalized coordinate.

Let a sinusoidal external force $F = F_o e^{-i\omega t}$ act on the generalized coordinate q [so q's canonically conjugate momentum p is being driven as $(dp/dt)_{\text{drive}} = F_o e^{-i\omega t}$]. Then the velocity of the resulting sinusoidal motion will be

$$\boxed{\dot{q} \equiv \frac{dq}{dt} = -i\omega q = \frac{1}{Z(\omega)}F_o e^{-i\omega t},} \tag{6.83a}$$

where the real part of each expression is to be taken. This equation can be regarded as the definition of q's complex impedance $Z(\omega)$ (ratio of force to velocity); it is determined by the system's details. If the system were completely conservative, then the impedance would be perfectly imaginary, $Z = iI$, where I is real. For example, for a freely moving dust particle in vacuum, driven by a sinusoidal force, the momentum is $p = m\dot{q}$ (where m is the particle's mass), the equation of motion is $F_o e^{-i\omega t} = dp/dt = m(d/dt)\dot{q} = m(-i\omega)\dot{q}$, and so the impedance is $Z = -im\omega$, which is purely imaginary.

The bath prevents the system from being conservative—energy can be fed back and forth between the generalized coordinate q and the bath's many degrees of freedom. This energy coupling influences the generalized coordinate q in two important ways. First, it changes the impedance $Z(\omega)$ from purely imaginary to complex,

real and imaginary parts
of complex impedance;
resistance $R(\omega)$

$$\boxed{Z(\omega) = iI(\omega) + R(\omega),} \tag{6.83b}$$

where R is the *resistance* experienced by q. Correspondingly, when the sinusoidal force $F = F_o e^{-i\omega t}$ is applied, the resulting motions of q feed energy into the bath, dissipating power at a rate $W_{\text{diss}} = \langle \Re(F)\Re(\dot{q}) \rangle = \langle \Re(F_o e^{-i\omega t})\Re(F_o e^{-i\omega t}/Z) \rangle = \langle F_o \cos\omega t\, \Re(1/Z)\, F_o \cos\omega t \rangle$; that is,

$$\boxed{W_{\text{diss}} = \frac{1}{2}\frac{R}{|Z|^2}F_o^2.} \tag{6.84}$$

Second, the thermal motions of the bath exert a randomly fluctuating force $F'(t)$ on q, driving its generalized momentum as $(dp/dt)_{\text{drive}} = F'$.

As an example, consider the L-C-R circuit of Eq. (6.72). We can identify the generalized momentum by shutting off the bath (the resistor and its fluctuating voltage); writing down the lagrangian for the resulting L-C circuit, $\mathcal{L} = \frac{1}{2}L\dot{q}^2 - \frac{1}{2}q^2/C$; and computing $p = \partial\mathcal{L}/\partial\dot{q} = L\dot{q}$. [Equally well, we can identify p from one of Hamilton's equations for the hamiltonian $H = p^2/(2L) + q^2/(2C)$.] We evaluate the impedance $Z(\omega)$ from the equation of motion for this lagrangian with the bath's resistance restored (but not its fluctuating voltage) and with a sinusoidal voltage $V = V_o e^{-i\omega t}$ imposed:

$$L\frac{d\dot{q}}{dt} - \frac{q}{C} + R\dot{q} = \left(-i\omega L + \frac{1}{-i\omega C} + R\right)\dot{q} = V_o e^{-i\omega t}. \tag{6.85a}$$

Evidently, $V = V_o e^{-i\omega t}$ is the external force F that drives the generalized momentum $p = L\dot{q}$, and the complex impedance (ratio of force to velocity) is

$$Z(\omega) = \frac{V}{\dot{q}} = -i\omega L + \frac{1}{-i\omega C} + R. \tag{6.85b}$$

This is identical to the impedance as defined in the standard theory of electrical circuits (which is what motivates our $Z = F/\dot{q}$ definition of impedance), and as expected, the real part of this impedance is the circuit's resistance R.

Returning to our general q, the fluctuating force F' (equal to fluctuating voltage V' in the case of the circuit) and the resistance R to an external force both arise from interaction with the same heat bath. Therefore, it should not be surprising that they are connected by the generalized fluctuation-dissipation theorem:

$$S_{F'}(f) = 4R(f)\left(\frac{1}{2}hf + \frac{hf}{e^{hf/(k_B T)} - 1}\right) \quad \text{in general,} \tag{6.86a}$$

generalized fluctuation-dissipation theorem

$$S_{F'}(f) = 4R(f)k_B T \quad \text{in the classical domain, } k_B T \gg hf, \tag{6.86b}$$

which is valid at all frequencies

$$f \ll 1/\tau_r, \tag{6.87}$$

where τ_r is the (very short) relaxation time for the bath's fluctuating forces F'. Here T is the temperature of the bath, h is Planck's constant, and we have written the resistance as $R(f)$ to emphasize that it can depend on frequency $f = \omega/(2\pi)$.

A derivation of this generalized fluctuation-dissipation theorem is sketched in Ex. 6.18.

One is usually less interested in the spectral density of the bath's force F' than in that of the generalized coordinate q. The definition (6.83a) of impedance implies $-i\omega\tilde{q} = \tilde{F'}/Z(\omega)$ for Fourier transforms, whence $S_q = S_{F'}/[(2\pi f)^2|Z|^2]$. When combined with Eqs. (6.86) and (6.84), this implies

spectral density for generalized coordinate

$$S_q(f) = \frac{8W_{\text{diss}}}{(2\pi f)^2 F_o^2}\left(\frac{1}{2}hf + \frac{hf}{e^{hf/(k_B T)} - 1}\right) \quad \text{in general,} \tag{6.88a}$$

$$S_q(f) = \frac{8W_{\text{diss}}k_B T}{(2\pi f)^2 F_o^2} \quad \text{in the classical domain, } k_B T \gg hf. \tag{6.88b}$$

Therefore, to evaluate $S_q(f)$, one does not need to know the complex impedance $Z(\omega)$. Rather, one only needs the power dissipation W_{diss} that results when a sinusoidal force F_o is applied to the generalized momentum p that is conjugate to the coordinate q of interest.

The light beam bouncing off a mirror (beginning of this section) is a good example. To couple the sinusoidal force $F(t) = F_o e^{-i\omega t}$ to the mirror's generalized coordinate q, we add an interaction term $H_I = -F(t)q$ to the mirror's hamiltonian H_{mirror}. Hamilton's equation for the evolution of the momentum conjugate to q then becomes $dp/dt = -\partial[H_{\text{mirror}} - F(t)q]/\partial q = \partial H_{\text{mirror}}/\partial t + F(t)$. Thus, $F(t)$ drives p as desired. The form of the interaction term is, by Eq. (6.82) for q,

$$H_I = -F(t)q = -\int z(r,\phi)\frac{F(t)e^{-(r/r_o)^2}}{\pi r_o^2} r\,d\phi\,dr. \tag{6.89}$$

This is the mathematical description of a time varying *pressure*

$$P = F_o e^{-i\omega t} e^{-(r/r_o)^2} / (\pi r_o^2)$$

applied to the mirror face, which has coordinate location $z(r, \phi)$. Therefore, to compute the spectral density of the mirror's light-beam-averaged displacement q, at frequency $f = \omega/(2\pi)$, we can

Levin's method for computing spectral density of mirror's light-averaged displacement

1. apply to the mirror's front face a pressure with spatial shape the same as that of the light beam's energy flux (a Gaussian in our example) and with total force $F_o e^{-i\omega t}$;

2. evaluate the power dissipation W_{diss} produced by this sinusoidally oscillating pressure; and then

3. insert the ratio W_{diss}/F_o^2 into Eq. (6.88a) or Eq. (6.88b). This is called Levin's (1998) method.

In practice, in this thought experiment the power can be dissipated at many locations: in the mirror coating (which makes the mirror reflective), in the substrate on which the coating is placed (usually glass, i.e., fused silica), in the attachment of the mirror to whatever supports it (usually a wire or glass fiber), and in the supporting structure (the wire or fiber and the solid object to which it is attached). The dissipations W_{diss} at each of these locations add together, and therefore the fluctuating noises from the various dissipation locations are additive. Correspondingly, one speaks of "coating thermal noise," "substrate thermal noise," and so forth; physicists making delicate optical measurements deduce each through a careful computation of its corresponding dissipation W_{diss}.

In the LIGO interferometer, whose noise power spectrum is shown in Fig. 6.7, these thermal noises dominate in the intermediate frequency band $40\,\text{Hz} \lesssim f \lesssim 150\,\text{Hz}$.

EXERCISES

Exercise 6.18 *Derivation: Generalized Fluctuation-Dissipation Theorem* T2

By a method analogous to that used for the elementary fluctuation-dissipation theorem (Sec. 6.8.1), derive the generalized fluctuation-dissipation theorem [Eqs. (6.86)].

Hints: Consider a thought experiment in which the system's generalized coordinate q is very weakly coupled to an external oscillator that has a mass M and an angular eigenfrequency ω_o, near which we wish to derive the fluctuation-dissipation formulas (6.86). Denote by Q and P the external oscillator's generalized coordinate and momentum and by κ the arbitrarily weak coupling constant between the oscillator and q, so the hamiltonian of system plus oscillator plus fluctuating force F' acting on q is

$$H = H_{\text{system}}(q, p, \ldots) + \frac{P^2}{2M} + \frac{1}{2} M \omega_o^2 Q^2 + \kappa Q q - F'(t)\, q. \tag{6.90}$$

Here the "..." refers to the other degrees of freedom of the system, some of which might be strongly coupled to q and p (as is the case, e.g., for the laser-measured mirror discussed in the text).

T2

(a) By combining Hamilton's equations for q and its conjugate momentum p [which give Eq. (6.83a) with the appropriate driving force] with those for the external oscillator (Q, P), derive an equation that shows quantitatively how the force F', acting through q, influences the oscillator's coordinate Q:

$$\left[M(-\omega^2 + \omega_o'^2) - \frac{i\kappa^2 R}{\omega|Z|^2} \right] \tilde{Q} = \frac{\kappa}{i\omega Z} \tilde{F}'. \tag{6.91}$$

Here the tildes denote Fourier transforms; $\omega = 2\pi f$ is the angular frequency at which the Fourier transforms are evaluated; and $\omega_o'^2 = \omega_o^2 - \kappa^2 I/(\omega|Z|^2)$, with $Z = R + iI$, is the impedance of q at angular frequency ω.

(b) Show that

$$S_Q = \frac{(\kappa/\omega|Z|)^2 S_{F'}}{M^2(-\omega^2 + \omega_o'^2)^2 + \kappa^4 R^2/(\omega|Z|^2)^2}. \tag{6.92}$$

(c) Make the resonance in this equation arbitrarily sharp by choosing the coupling constant κ arbitrarily small. Then show that the mean energy in the oscillator is

$$\bar{E} = M\omega_o^2 \int_0^\infty S_Q(f)df = \frac{S_{F'}(f = \omega_o/(2\pi))}{4R}. \tag{6.93}$$

(d) By equating this to expression (6.75c) for the mean energy of any oscillator coupled to a heat bath, deduce the desired generalized fluctuation-dissipation equations (6.86).

6.9 Fokker-Planck Equation

6.9

In statistical physics, we often want to know the collective influence of many degrees of freedom (a bath) on a single (possibly vectorial) degree of freedom q. The bath might or might not be thermalized. The forces it exerts on q might have short range (as in molecular collisions buffeting an air molecule or dust particle) or long range (as in Coulomb forces from many charged particles in a plasma pushing stochastically on an electron that interests us, or gravitational forces from many stars pulling on a single star of interest). There might also be long-range, macroscopic forces that produce anisotropies and/or inhomogeneities (e.g., applied electric or magnetic fields). We might want to compute how the bath's many degrees of freedom influence, for example, the diffusion of a particle as embodied in its degree of freedom q. Or we might want to compute the statistical properties of q for a representative electron in a plasma and from them deduce the plasma's transport coefficients (diffusivity, heat conductivity, and thermal conductivity). Or we might want to know how the gravitational pulls of many stars in the vicinity of a black hole drive the collective evolution of the stars' distribution function.

The Fokker-Planck equation is a powerful tool in such situations. To apply it, we must identify a (possibly vectorial) degree of freedom q to analyze that is Markov. For the types of problems described here, this is typically the velocity (or a component of velocity) of a representative particle or star. The Fokker-Planck equation is then a differential equation for the evolution of the conditional probability distribution P_2 [Eq. (6.6b)], or other distribution function, for that Markov degree of freedom. In Sec. 6.9.1, we present the simplest, 1-dimensional example. Then in Sec. 6.9.3, we generalize to several dimensions.

crucial assumption of a Markov Process

6.9.1

6.9.1 Fokker-Planck for a 1-Dimensional Markov Process

For a 1-dimensional Markov process $y(t)$ (e.g., the x component of the velocity of a particle) being driven by a bath (not necessarily thermalized!) with many degrees of freedom, the *Fokker-Planck equation*[11] states

1-dimensional Fokker-Planck equation

$$\frac{\partial}{\partial t} P_2 = -\frac{\partial}{\partial y}[A(y)P_2] + \frac{1}{2}\frac{\partial^2}{\partial y^2}[B(y)P_2].$$

(6.94)

Here $P_2 = P_2(y, t|y_o)$ is to be regarded as a function of the variables y and t with y_o fixed; that is, Eq. (6.94) is to be solved subject to the initial condition

$$P_2(y, 0|y_o) = \delta(y - y_o).$$

(6.95)

As we shall see later, this Fokker-Planck equation is a generalized diffusion equation for the probability P_2: as time passes, the probability diffuses away from its initial location, $y = y_o$, spreading gradually out over a wide range of values of y.

In the Fokker-Planck equation (6.94) the function $A(y)$ produces a motion of the mean away from its initial location, while the function $B(y)$ produces a diffusion of the probability. If one can deduce the evolution of P_2 for very short times by some other method [e.g., in the case of a dust particle being buffeted by air molecules, by solving statistically the Langevin equation $mdv/dt + Rv = F'(t)$], then from that short-time evolution one can compute the functions $A(y)$ and $B(y)$:

$$A(y) = \lim_{\Delta t \to 0} \frac{1}{\Delta t} \int_{-\infty}^{+\infty} (y' - y)P_2(y', \Delta t|y)dy',$$

(6.96a)

$$B(y) = \lim_{\Delta t \to 0} \frac{1}{\Delta t} \int_{-\infty}^{+\infty} (y' - y)^2 P_2(y', \Delta t|y)dy'.$$

(6.96b)

11. A very important generalization of this equation is to replace the probability P_2 by a particle distribution function and the Markov process $y(t)$ by the 3-dimensional velocity or momentum of the particles. The foundations for this generalization are laid in the Track-Two Sec. 6.9.3, and an application is in Ex. 20.8.

(These equations can be deduced by reexpressing the limit as an integral of the time derivative $\partial P_2/\partial t$ and then inserting the Fokker-Planck equation and integrating by parts; Ex. 6.19.) Note that the integral (6.96a) for $A(y)$ is the mean change $\overline{\Delta y}$ in the value of y that occurs in time Δt, if at the beginning of Δt (at $t = 0$) the value of the process is precisely y; moreover (since the integral of yP_2 is just equal to y, which is a constant), $A(y)$ is also the rate of change of the mean, $d\bar{y}/dt$. Correspondingly we can write Eq. (6.96a) in the more suggestive form

$$A(y) = \lim_{\Delta t \to 0} \left(\frac{\overline{\Delta y}}{\Delta t} \right) = \left(\frac{d\bar{y}}{dt} \right)_{t=0}. \qquad (6.97a)$$

Similarly, the integral (6.96b) for $B(y)$ is the mean-squared change in y, $\overline{(\Delta y)^2}$, if at the beginning of Δt the value of the process is precisely y; and (as one can fairly easily show; Ex. 6.19) it is also the rate of change of the variance $\sigma_y^2 = \int (y' - \bar{y})^2 P_2 dy'$. Correspondingly, Eq. (6.96b) can be written as

$$B(y) = \lim_{\Delta t \to 0} \left(\frac{\overline{(\Delta y)^2}}{\Delta t} \right) = \left(\frac{d\sigma_y^2}{dt} \right)_{t=0}. \qquad (6.97b)$$

It may seem surprising that $\overline{\Delta y}$ and $\overline{(\Delta y)^2}$ can both increase linearly in time for small times [cf. the Δt in the denominators of both Eq. (6.97a) and Eq. (6.97b)], thereby both giving rise to finite functions $A(y)$ and $B(y)$. In fact, this is so: the linear evolution of $\overline{\Delta y}$ at small t corresponds to the motion of the mean (i.e., of the peak of the probability distribution), while the linear evolution of $\overline{(\Delta y)^2}$ corresponds to the diffusive broadening of the probability distribution.

DERIVATION OF THE FOKKER-PLANCK EQUATION (6.94)

Because y is Markov, it satisfies the Smoluchowski equation (6.11), which we rewrite here with a slight change of notation:

$$P_2(y, t + \tau | y_o) = \int_{-\infty}^{+\infty} P_2(y - \xi, t | y_o) P_2(y - \xi + \xi, \tau | y - \xi) d\xi. \qquad (6.98a)$$

Take τ to be small so only small ξ will contribute to the integral, and expand in a Taylor series in τ on the left-hand side of (6.98a) and in the ξ of $y - \xi$ on the right-hand side:

$$P_2(y, t | y_o) + \sum_{n=1}^{\infty} \frac{1}{n!} \left[\frac{\partial^n}{\partial t^n} P_2(y, t | y_o) \right] \tau^n$$

$$= \int_{-\infty}^{+\infty} P_2(y, t | y_o) P_2(y + \xi, \tau | y) d\xi$$

$$+ \sum_{n=1}^{\infty} \frac{1}{n!} \int_{-\infty}^{+\infty} (-\xi)^n \frac{\partial^n}{\partial y^n} [P_2(y, t | y_o) P_2(y + \xi, \tau | y)] d\xi. \qquad (6.98b)$$

In the first integral on the right-hand side the first term is independent of ξ and can be pulled out from under the integral, and the second term then integrates to one; thereby the first integral on the right reduces to $P_2(y, t|y_o)$, which cancels the first term on the left. The result is then

$$\sum_{n=1}^{\infty} \frac{1}{n!} \left[\frac{\partial^n}{\partial t^n} P_2(y, t|y_o) \right] \tau^n$$

$$= \sum_{n=1}^{\infty} \frac{(-1)^n}{n!} \frac{\partial^n}{\partial y^n} \left[P_2(y, t|y_o) \int_{-\infty}^{+\infty} \xi^n P_2(y + \xi, \tau|y) \, d\xi \right]. \qquad (6.98c)$$

Divide by τ, take the limit $\tau \to 0$, and set $\xi \equiv y' - y$ to obtain

$$\frac{\partial}{\partial t} P_2(y, t|y_o) = \sum_{n=1}^{\infty} \frac{(-1)^n}{n!} \frac{\partial^n}{\partial y^n} [M_n(y) P_2(y, t|y_o)], \qquad (6.99a)$$

where

$$M_n(y) \equiv \lim_{\Delta t \to 0} \frac{1}{\Delta t} \int_{-\infty}^{+\infty} (y' - y)^n P_2(y', \Delta t|y) \, dy' \qquad (6.99b)$$

is the nth moment of the probability distribution P_2 after time Δt. This is a form of the Fokker-Planck equation that has slightly wider validity than Eq. (6.94). Almost always, however, the only nonvanishing functions $M_n(y)$ are $M_1 \equiv A$, which describes the linear motion of the mean, and $M_2 \equiv B$, which describes the linear growth of the variance. Other moments of P_2 grow as higher powers of Δt than the first power, and correspondingly, their M_ns vanish. Thus, almost always[12] (and always, so far as we are concerned), Eq. (6.99a) reduces to the simpler version (6.94) of the Fokker-Planck equation.

TIME-INDEPENDENT FOKKER-PLANCK EQUATION

If, as we assume in this chapter, y is ergodic, then $p_1(y)$ can be deduced as the limit of $P_2(y, t|y_o)$ for arbitrarily large times t. Then (and in general) p_1 can be deduced from the time-independent Fokker-Planck equation:

time-independent Fokker-Planck equation

$$\boxed{-\frac{\partial}{\partial y}[A(y)p_1(y)] + \frac{1}{2} \frac{\partial^2}{\partial y^2}[B(y)p_1(y)] = 0.} \qquad (6.100)$$

GAUSSIAN-MARKOV PROCESS

For a Gaussian-Markov process, the mathematical form of $P_2(y_2, t|y_1)$ is known from Doob's theorem: Eqs. (6.18). In the notation of those equations, the Fokker-Planck functions A and B are

$$A(y_1) = (d\bar{y}_t/dt)_{t=0} = -(y_1 - \bar{y})/\tau_r, \quad \text{and} \quad B(y_1) = (d\sigma_{y_t}^2/dt)_{t=0} = 2\sigma_y^2/\tau_r.$$

---

12. In practice, when there are important effects not captured by A and B (e.g., in the mathematical theory of finance; Hull, 2014), they are usually handled by adding other terms to Eq. (6.94), including sometimes integrals.

Translating back to the notation of this section, we have

$$A(y) = -(y - \bar{y})/\tau_r, \qquad B(y) = 2\sigma_y^2/\tau_r. \qquad (6.101)$$

Thus, if we can compute $A(y)$ and $B(y)$ explicitly for a Gaussian-Markov process, then from them we can read off the process's relaxation time τ_r, long-time mean $\bar{y}$, and long-time variance σ_y^2. As examples, in Ex. 6.22 we revisit Brownian motion of a dust particle in air and in the next section, we analyze laser cooling of atoms. A rather different example is the evolution of a photon distribution function under Compton scattering (Sec. 28.6.3).

Exercise 6.19 *Derivation: Equations for A and B*
Derive Eqs. (6.96) for A and B from the Fokker-Planck equation (6.94), and then from Eqs. (6.96) derive Eqs. (6.97).

Exercise 6.20 *Problem: Fokker-Planck Equation as Conservation Law for Probability*
Show that the Fokker-Planck equation can be interpreted as a conservation law for probability. What is the probability flux in this conservation law? What is the interpretation of each of its two terms?

Exercise 6.21 *Example: Fokker-Planck Coefficients When There Is Direct Detailed Balance*
Consider an electron that can transition between two levels by emitting or absorbing a photon; and recall (as discussed in Ex. 3.6) that we have argued that the stimulated transitions should be microscopically reversible. This is an example of a general principle introduced by Boltzmann called *detailed balance*. In the context of classical physics, it is usually considered in the context of the time reversibility of the underlying physical equations. For example, if two molecules collide elastically and exchange energy, the time-reversed process happens with the same probability per unit time when the colliding particles are in the time-reversed initial states. However, this does not necessarily imply that the probability of a single molecule changing its velocity from $\mathbf{v}$ to $\mathbf{v}'$ is the same as that of the reverse change. (A high-energy molecule is more likely to lose energy when one averages over all collisions.)

An important simplification happens when the probability of a change in y is equal to the probability of the opposite change. An example might be a light particle colliding with a heavy particle for which the recoil can be ignored. Under this more restrictive condition, we can write that $P_2(y', \Delta t|y) = P_2(y, \Delta t|y')$. Show that the Fokker-Planck equation then simplifies (under the usual assumptions) to a standard diffusion equation:

$$\frac{\partial}{\partial t} P_2 = \frac{1}{2} \frac{\partial}{\partial y} B(y) \frac{\partial}{\partial y} P_2. \qquad (6.102)$$

Of course, there can be other contributions to the total Fokker-Planck coefficients that do not satisfy this condition, but this simplification can be very instructive. An example described in Sec. 23.3.3 is the quasilinear interaction between waves and particles in a plasma.

Exercise 6.22 *Example: Solution of Fokker-Planck Equation for Brownian Motion of a Dust Particle*

(a) Write down the explicit form of the Langevin equation for the x component of velocity $v(t)$ of a dust particle interacting with thermalized air molecules.

(b) Suppose that the dust particle has velocity v at time t. By integrating the Langevin equation, show that its velocity at time $t + \Delta t$ is $v + \Delta v$, where

$$m \Delta v + R v \Delta t + O[(\Delta t)^2] = \int_t^{t+\Delta t} F'(t') dt', \tag{6.103a}$$

with R the frictional resistance and m the particle's mass. Take an ensemble average of this and use $\overline{F'} = 0$ to conclude that the function $A(v)$ appearing in the Fokker-Planck equation (6.94) has the form

$$A(v) \equiv \lim_{\Delta t \to 0} \frac{\overline{\Delta v}}{\Delta t} = -\frac{R v}{m}. \tag{6.103b}$$

Compare this expression with the first of Eqs. (6.101) to conclude that the mean and relaxation time are $\bar{v} = 0$ and $\tau_r = m/R$, respectively, in agreement with the second of Eqs. (6.53a) in the limit $\tau \to \infty$ and with Eq. (6.78).

(c) From Eq. (6.103a) show that

$$(\Delta v)^2 = \left[-\frac{v}{\tau_r} \Delta t + O[(\Delta t)^2] + \frac{1}{m} \int_t^{t+\Delta t} F'(t') dt' \right]^2. \tag{6.103c}$$

Take an ensemble average of this expression, and use $\overline{F'(t_1) F'(t_2)} = C_{F'}(t_2 - t_1)$—together with the Wiener-Khintchine theorem—to evaluate the terms involving F' in terms of $S_{F'}$, which in turn is known from the fluctuation-dissipation theorem. Thereby obtain

$$B(v) = \lim_{\Delta t \to 0} \frac{\overline{(\Delta v)^2}}{\Delta t} = \frac{2 R k_B T}{m^2}. \tag{6.103d}$$

Combine with Eq. (6.101) and $\tau_r = m/R$ [from part (b)], to conclude that $\sigma_v^2 = k_B T/m$, in accord with the last of Eqs. (6.53a).

6.9.2 Optical Molasses: Doppler Cooling of Atoms T2

The 1997 Nobel Prize was awarded to Steven Chu, Claude Cohen-Tannoudji, and William D. Phillips for the "development of methods to cool and trap atoms with laser light" (Chu et al., 1998). In this section, we use the Fokker-Planck equation to analyze

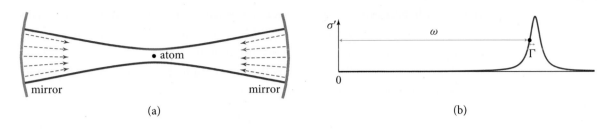

FIGURE 6.16 Doppler cooling of an atom in a Fabry-Perot cavity. (a) The cavity formed by two mirrors with laser light bouncing back and forth between them, and the sodium atom at the center. (b) The cross section $\sigma' = d\sigma/d\omega$ for the atom in its ground state to absorb a photon of laser light. The laser angular frequency ω is tuned to the off-resonance inflection point (steepest slope) of σ', indicated by the dot.

one of the most important methods they developed: *Doppler cooling*, also called *laser cooling* or *optical molasses*.

A neutral sodium atom is placed near the center (waist) of a Fabry-Perot optical cavity (Sec. 9.4.2), so it is bathed by laser light traveling in both the $+z$ and $-z$ directions (Fig. 6.16a). The atom absorbs and reemits photons and their momenta, resulting in a stochastic evolution of its z component of velocity, v. Using the Fokker-Planck equation to analyze this evolution, we shall discover that, if the light frequency and power are tuned appropriately, there is a strong resisting force ("optical molasses") on the atom as well as a randomizing force. The net effect, after the atom relaxes into equilibrium with the photon field, is a very low effective temperature ($\sim 100\ \mu$K) for the atom's motion in the z direction.

The atom has a large cross section $\sigma' \equiv d\sigma/d\omega$ to absorb a photon with angular frequency $\omega \simeq 3.20 \times 10^{15}\ \text{s}^{-1}$ (yellow light), thereby getting excited into a state with energy $\hbar\omega \simeq 2.11$ eV. The absorption cross section $\sigma'(\omega)$ has a narrow resonance (Lorentzian line shape $\propto [1 + (\omega - \omega_o)^2/\Gamma^2]^{-1}$; Fig. 6.16b) with half-width $\Gamma \simeq$ 10 MHz; and, correspondingly, the excited atom has a half-life $1/\Gamma$ to reemit a photon and return to its ground state. The laser power is adjusted to make the excitation rate $\mathcal{R}$ equal to Γ,

$$\mathcal{R} = \Gamma \simeq 10^7\ \text{s}^{-1}, \tag{6.104a}$$

thereby maximizing the rate of excitations. (At a higher power, the excitation rate will saturate at $1/\Gamma$, because the atom spends most of its time excited and waiting to reemit.)

The laser frequency is tuned to the resonance's inflection point (point of greatest slope $d\sigma'/d\omega$ on one side of the line), so that, when an atom is moving rightward with velocity v, the Doppler shift $\delta\omega/\omega = v/c$ produces a maximal fractional increase in the cross section and rate for absorbing leftward-moving photons and decrease in those for rightward-moving photons:

$$\frac{\delta\mathcal{R}}{\mathcal{R}} = \frac{\delta\sigma'}{\sigma'} = \frac{1}{\sigma'}\frac{d\sigma'}{d\omega}\left(\omega\frac{v}{c}\right) \sim \frac{\omega}{\Gamma}\frac{v}{c}. \tag{6.104b}$$

(Here and henceforth "$\sim$" means accurate to within a factor of order unity.) This results in a net resisting force $F \sim \delta R\, \hbar k$ on the atom, due to the imbalance in absorption rates for leftward and rightward photons; here $k = \omega/c = 10.7\ \mu\mathrm{m}^{-1}$ is the photons' wave number, and $\hbar k$ is the momentum absorbed from each photon. This slow-down force produces a rate of change of the atom's mean velocity

$$A = \frac{d\bar{v}}{dt} \sim -\frac{\delta R\, \hbar k}{m} \sim -\frac{\hbar k^2}{m} v. \tag{6.104c}$$

Here we have used Eqs. (6.104b) and (6.104a), and $\omega/c = k$; and we have set the slow-down rate equal to the coefficient A in the Fokker-Planck equation for v [Eq. (6.97a)].

There are two sources of randomness in the atom's velocity, both of the same magnitude: statistical randomness ($\sqrt{N}$) in the number of photons absorbed from the two directions, and randomness in the direction of reemission of photons and thence in the recoil direction. During a short time interval Δt, the mean number of absorptions and reemissions is $\sim R \Delta t$, so the rms fluctuation in the momentum transfer to the atom (along the z direction) is $\sim \hbar k \sqrt{R \Delta t}$, whence the change of the variance of the velocity is $\overline{(\Delta v)^2} \sim (\hbar k)^2 R \Delta t / m^2$. (Here $m \simeq 3.82 \times 10^{-26}$ kg is the sodium atom's mass.) Correspondingly, the B coefficient (6.103d) in the Fokker-Planck equation for v is

$$B = \frac{\overline{(\Delta v)^2}}{\Delta t} \sim \left(\frac{\hbar k}{m}\right)^2 R = \left(\frac{\hbar k}{m}\right)^2 \Gamma. \tag{6.104d}$$

From the A and B coefficients [Eqs. (6.104c) and (6.104d)] we infer, with the aid of Eqs. (6.101), the relaxation time, long-term mean, and long-term variance of the atom's velocity along the z direction, and also an effective temperature associated with the variance:[13]

$$\tau_r \sim \frac{m}{(\hbar k^2)} = 3\ \mu\mathrm{s}, \quad \bar{v} = 0, \quad \sigma_v^2 \sim \frac{\hbar \Gamma}{m} = (0.17\ \mathrm{ms}^{-1})^2,$$

$$T_{\mathrm{eff}} = \frac{m \sigma_v^2}{k_B} \sim \frac{\hbar \Gamma}{k_B} \sim 8\ \mu\mathrm{K}. \tag{6.105}$$

It is remarkable how effective this optical molasses can be!

If one wants to cool all components of velocity, one can either impose counter-propagating laser beams along all three Cartesian axes, or put the atom into a potential well (inside the Fabry Perot cavity) that deflects its direction of motion on a timescale much less than τ_r.

This optical molasses technique is widely used today in atomic physics, for example, when cooling ensembles of atoms to produce Bose-Einstein condensates

---

13. The atom's long-term, ergodically wandering velocity distribution is Gaussian rather than Maxwellian, so it is not truly thermalized. However, it has the same velocity variance as a thermal distribution with temperature $\sim \hbar \Gamma / k_B$, so we call this its "effective temperature."

(Sec. 4.9), and for cooling atoms to be used as the ticking mechanisms of atomic clocks (Fig. 6.11, footnote 9 in Sec. 6.6.1, Ex. 6.13, and associated discussions).

6.9.3 Fokker-Planck for a Multidimensional Markov Process; Thermal Noise in an Oscillator T2

Few 1-dimensional random processes are Markov, so only a few can be treated using the 1-dimensional Fokker-Planck equation. However, it is frequently the case that, if one augments additional variables into the random process, it becomes Markov. An important example is a harmonic oscillator driven by a Gaussian random force (Ex. 6.23). Neither the oscillator's position $x(t)$ nor its velocity $v(t)$ is Markov, but the pair $\{x, v\}$ is a 2-dimensional Markov process.

For such a process, and more generally for any n-dimensional Gaussian-Markov process $\{y_1(t), y_2(t), \ldots, y_n(t)\} \equiv \{\mathbf{y}(t)\}$, the conditional probability distribution $P_2(\mathbf{y}, t|\mathbf{y}_o)$ satisfies the following Fokker-Planck equation [the obvious generalization of Eq. (6.94)]:

$$\frac{\partial}{\partial t} P_2 = -\frac{\partial}{\partial y_j}[A_j(y)P_2] + \frac{1}{2}\frac{\partial^2}{\partial y_j \partial y_k}[B_{jk}(y)P_2]. \qquad (6.106a)$$

multidimensional Fokker-Planck equation

Here the functions A_j and B_{jk}, by analogy with Eqs. (6.96) and (6.97), are

$$A_j(\mathbf{y}) = \lim_{\Delta t \to 0} \frac{1}{\Delta t} \int (y'_j - y_j) P_2(\mathbf{y}', \Delta t|\mathbf{y})d^n y' = \lim_{\Delta t \to 0}\left(\frac{\overline{\Delta y_j}}{\Delta t}\right), \qquad (6.106b)$$

$$B_{jk}(y) = \lim_{\Delta t \to 0} \frac{1}{\Delta t} \int (y'_j - y_j)(y'_k - y_k) P_2(\mathbf{y}', \Delta t|\mathbf{y})d^n y' = \lim_{\Delta t \to 0}\left(\frac{\overline{\Delta y_j \Delta y_k}}{\Delta t}\right). \qquad (6.106c)$$

In Ex. 6.23 we use this Fokker-Planck equation to explore how a harmonic oscillator settles into equilibrium with a dissipative heat bath. In Ex. 20.8 we apply it to Coulomb collisions in an ionized plasma.

The multidimensional Fokker-Planck equation can be used to solve the Boltzmann transport equation (3.66) for the kinetic-theory distribution function $\mathcal{N}(\mathbf{p}, t)$ or (in the conventions of plasma physics) for the velocity distribution $f(\mathbf{v}, t)$ (Chap. 20). The reasons are (i) $\mathcal{N}(\mathbf{p}, t)$ and $f(\mathbf{v}, t)$ are the same kind of probability distribution as P_2—probabilities for a Markov momentum or velocity—with the exception that $\mathcal{N}(\mathbf{p}, t)$ and $f(\mathbf{v}, t)$ usually have different initial conditions at time $t = 0$ than P_2's delta function [in fact, P_2 can be regarded as a Green's function for $\mathcal{N}(\mathbf{p}, t)$ and $f(\mathbf{v}, t)$] and (ii) the initial conditions played no role in our derivation of the Fokker-Planck equation. In Sec. 20.4.3, we discuss the use of the Fokker-Planck equation to deduce how long-range Coulomb interactions drive the equilibration of the distribution functions $f(\mathbf{v}, t)$ for the velocities of electrons and ions in a plasma. In Sec. 23.3.3, we use the Fokker-Planck equation to study the interaction of electrons and ions with plasma waves (plasmons).

Exercise 6.23 **Example: Solution of Fokker-Planck Equation*
for Thermal Noise in an Oscillator T2

Consider a classical simple harmonic oscillator (e.g., the nanomechanical oscillator, LIGO mass on an optical spring, L-C-R circuit, or optical resonator briefly discussed in Ex. 6.17). Let the oscillator be coupled weakly to a dissipative heat bath with temperature T. The Langevin equation for the oscillator's generalized coordinate x is Eq. (6.79). The oscillator's coordinate $x(t)$ and momentum $p(t) \equiv m\dot{x}$ together form a 2-dimensional Gaussian-Markov process and thus obey the 2-dimensional Fokker-Planck equation (6.106a). As an aid to solving this Fokker-Planck equation, change variables from $\{x, p\}$ to the real and imaginary parts X_1 and X_2 of the oscillator's complex amplitude:

$$x = \Re[(X_1 + iX_2)e^{-i\omega t}] = X_1(t)\cos\omega t + X_2(t)\sin\omega t. \tag{6.107}$$

Then $\{X_1, X_2\}$ is a Gaussian-Markov process that evolves on a timescale $\sim\tau_r$.

(a) Show that X_1 and X_2 obey the Langevin equation

$$-2\omega(\dot{X}_1 + X_1/\tau_r)\sin\omega t + 2\omega(\dot{X}_2 + X_2/\tau_r)\cos\omega t = F'/m. \tag{6.108a}$$

(b) To compute the functions $A_j(\mathbf{X})$ and $B_{jk}(\mathbf{X})$ that appear in the Fokker-Planck equation (6.106a), choose the timescale Δt to be short compared to the oscillator's damping time τ_r but long compared to its period $2\pi/\omega$. By multiplying the Langevin equation successively by $\sin\omega t$ and $\cos\omega t$ and integrating from $t = 0$ to $t = \Delta t$, derive equations for the changes ΔX_1 and ΔX_2 produced during Δt by the fluctuating force $F'(t)$ and its associated dissipation. [Neglect fractional corrections of order $1/(\omega\Delta t)$ and of order $\Delta t/\tau_r$]. Your equations should be analogous to Eq. (6.103a).

(c) By the same technique as was used in Ex. 6.22, obtain from the equations derived in part (b) the following forms of the Fokker-Planck functions:

$$A_j = \frac{-X_j}{\tau_r}, \qquad B_{jk} = \frac{2k_BT}{m\omega^2\tau_r}\delta_{jk}. \tag{6.108b}$$

(d) Show that the Fokker-Planck equation, obtained by inserting functions (6.108b) into Eq. (6.106a), has the following Gaussian solution:

$$P_2(X_1, X_2, t|X_1^{(o)}, X_2^{(o)}) = \frac{1}{2\pi\sigma^2}\exp\left[-\frac{(X_1 - \bar{X}_1)^2 + (X_2 - \bar{X}_2)^2}{2\sigma^2}\right], \tag{6.109a}$$

where the means and variance of the distribution are

$$\bar{X}_j = X_j^{(o)}e^{-t/\tau_r}, \qquad \sigma^2 = \frac{k_BT}{m\omega^2}\left(1 - e^{-2t/\tau_r}\right) \simeq \begin{cases} \frac{k_BT}{m\omega^2}\frac{2t}{\tau_r} & \text{for } t \ll \tau_r \\ \frac{k_BT}{m\omega^2} & \text{for } t \gg \tau_r. \end{cases} \tag{6.109b}$$

(e) Discuss the physical meaning of the conditional probability (6.109a). Discuss its implications for the physics experiment described in Ex. 6.17, when the signal force acts for a time short compared to τ_r rather than long.

Bibliographic Note

Random processes are treated in many standard textbooks on statistical physics, typically under the rubric of fluctuations or nonequilibrium statistical mechanics (and sometimes not even using the phrase "random process"). We like Kittel (2004), Sethna (2006), Reif (2008), and Pathria and Beale (2011). A treatise on signal processing that we recommend, despite its age, is Wainstein and Zubakov (1962). There are a number of textbooks on random processes (also called "stochastic processes" in book titles), usually aimed at mathematicians, engineers, or finance folks (who use the theory of random processes to try to make lots of money, and often succeed). But we do not like any of those books as well as the relevant sections in the above statistical mechanics texts. Nevertheless, you might want to peruse Lax et al. (2006), Van Kampen (2007), and Paul and Baschnagel (2010).

REFERENCES

Abbott, B. P., R. Abbott, R. Adhikari, P. Ajith, et al. (2009). Observation of a kilogram-scale oscillator near its quantum ground state. *New Journal of Physics* **11**, 1–13.

Almheiri, A., D. Marolf, J. Polchinski, and J. Sully (2013). Black holes: Complementarity or firewalls? *Journal of High Energy Physics* **2**, 62–78.

Anderson, M. H., J. R. Ensher, M. R. Matthews, C. E. Wieman, and E. A. Cornell (1995). Observation of Bose-Einstein condensation in a dilute atomic vapor. *Science* **269**, 198–201.

Arfken, G. B., H. J. Weber, and F. E. Harris (2013). *Mathematical Methods for Physicists*. Amsterdam: Elsevier.

Bekenstein, J. (1972). Black holes and the second law. *Lettre al Nuovo Cimento* **4**, 737–740.

Bennett, C. L., M. Halpern, G. Hinshaw, and N. E. Jarosik (2003). First-year Wilkinson Microwave Anisotropy Probe (WMAP) Observations: Preliminary Maps and Basic Results. *Astrophysical Journal Supplement* **148**, 1–28.

Binney, J. J., and S. Tremaine (2003). *Galactic Dynamics*. Princeton, N.J.: Princeton University Press.

Boas, M. L. (2006). *Mathematical Methods in the Physical Sciences*. New York: Wiley.

Brown, R. (1828). XXVII. A brief account of microscopical observations made in the months of June, July and August 1827, on the particles contained in the pollen of plants; and on the general existence of active molecules in organic and inorganic bodies. *Philosophical Magazine* **4**, 161–173.

Callen, H. B., and T. A. Welton (1951). Irreversibility and generalized noise. *Physical Review* **83**, 34–40.

Carroll, S. M. (2004). *Spacetime and Geometry. An Introduction to General Relativity*. New York: Addison-Wesley.

Chan, J., T. M. Alegre, A. H. Safavi-Naeini, J. T. Hill, et al. (2011). Laser cooling of a nanomechanical oscillator into its quantum ground state. *Nature* **478**, 89–92.

Chandler, D. (1987). *Introduction to Modern Statistical Mechanics*. Oxford: Oxford University Press.

Chu, S., C. Cohen-Tannoudji, and W. D. Phillips (1998). Nobel lectures. *Reviews of Modern Physics* **70**, 685–742.

Clayton, D. D. (1968). *Principles of Stellar Evolution and Nucleosynthesis*. Chicago: University of Chicago Press.

Cohen-Tannoudji, C., B. Diu, and F. Laloë (1977). *Quantum Mechanics*. New York: Wiley.

Cornell, E. (1996). Very cold indeed: The nanokelvin physics of Bose-Einstein condensation. *Journal of Research of NIST* **101**, 419–434.

Cox, Arthur N., ed. (2000). *Allen's Astrophysical Quantities*. Cham, Switzerland: Springer.

Dalfovo, F., S. Giorgini, L. P. Pitaevskii, and S. Stringari (1999). Theory of Bose-Einstein condensation in trapped gases. *Reviews of Modern Physics* **71**, 463–512.

Davies, P. C. W. (1977). The thermodynamic theory of black holes. *Proceedings of the Royal Society A* **353**, 499–521.

Doob, J. L. (1942). The Brownian movement and stochastic equations. *Annals of Mathematics* **43**, 351–369.

Einstein, A. (1925). Quantum theory of ideal gases. *Sitzungsberichte der Preussischen Akademie der Wissenschaften* **3**, 18–25.

Eisenstein, D. J., I. Zehavi, D. W. Hogg, R. Scoccimarro, et al. (2005). Detection of the baryon acoustic peak in the large-scale correlation function of SDSS luminous red galaxies. *Astrophysical Journal* **633**, 560–574.

Ensher, J. R., D. S. Jin, M. R. Matthews, C. E. Wieman, and E. A. Cornell (1996). Bose-Einstein condensation in a dilute gas: Measurement of energy and ground-state occupation. *Physical Review Letters* **77**, 4984–4987.

Everett, A., and T. Roman (2011). *Time Travel and Warp Drives: A Scientific Guide to Shortcuts through Time and Space*. Chicago: University of Chicago Press.

Farquhar, I. E. (1964). *Ergodic Theory in Statistical Mechanics*. London: Interscience.

Feynman, R. P. (1972). *Statistical Mechanics*. New York: Benjamin.

Feynman, R. P., R. B. Leighton, and M. Sands (1964). *The Feynman Lectures on Physics*. Reading, Mass.: Addison-Wesley. Chapter 14 epigraph reprinted with permission of Caltech.

Flanders, H. (1989). *Differential Forms with Applications to the Physical Sciences*, corrected edition. Mineola, N.Y.: Courier Dover Publications.

Friedman, J., and A. Higuchi (2006). Topological censorship and chronology protection. *Annalen der Physik* **15**, 109–128.

Frolov, V. P., and I. D. Novikov (1990). Physical effects in wormholes and time machines. *Physical Review D* **42**, 1057–1065.

Frolov, V. P., and D. N. Page (1993). Proof of the generalized second law for quasistationary semiclassical black holes. *Physical Review Letters* **71**, 3902–3905.

Frolov, V. P., and A. Zelnikov (2011). *Introduction to Black Hole Physics*. Oxford: Oxford University Press.

Galleani, L. (2012). The statistics of the atomic clock noise. In L. Cohen et al. (eds.), *Classical, Semi-classical and Quantum Noise*, pp. 63–77. Cham, Switzerland: Springer Science + Business Media.

Gibbs, J. W. (1881). Letter accepting the Rumford Medal. Quoted in A. L. Mackay, *Dictionary of Scientific Quotations*. London: IOP Publishing.

Gibbs, J. W. (1902). *Elementary Principles in Statistical Mechanics*. New York: Charles Scribner's Sons.

Goldstein, H., C. Poole, and J. Safko (2002). *Classical Mechanics*. New York: Addison-Wesley.

Goodstein, D. L. (2002). *States of Matter*. Mineola, N.Y.: Courier Dover Publications.

Goodstein, D. L., and J. R. Goodstein (1996). *Feynman's Lost Lecture: The Motion of Planets around the Sun*. New York: W. W. Norton.

Grad, H. (1958). *Principles of the Kinetic Theory of Gases*. Cham, Switzerland: Springer.

Griffiths, D. J. (1999). *Introduction to Electrodynamics*. Upper Saddle River, N.J.: Prentice-Hall.

Hafele, J. C., and R. E. Keating (1972a). Around-the-world atomic clocks: Predicted relativistic time gains. *Science* **177**, 166–168.

Hafele, J. C., and R. E. Keating (1972b). Around-the-world atomic clocks: Observed relativistic time gains. *Science* **177**, 168–170.

Hartle, J. B. (2003). *Gravity: An Introduction to Einstein's General Relativity*. San Francisco: Addison-Wesley.

Hawking, S. W. (1975). Particle creation by black holes. *Communications in Mathematical Physics* **43**, 199–220.

Hawking, S. W. (1976). Black holes and thermodynamics. *Physical Review D* **13**, 191–197.

Hawking, S. W., and R. Penrose (2010). *The Nature of Space and Time*. Princeton, N.J.: Princeton University Press.

Hull, J. C. (2014). *Options, Futures and Other Derivatives*, ninth edition. Upper Saddle River, N.J.: Pearson.

Jackson, J. D. (1999). *Classical Electrodynamics*. New York: Wiley.

Johnson, J. B. (1928). Thermal agitation of electricity in conductors. *Physical Review* **32**, 97–109.

Kardar, M. (2007). *Statistical Physics of Particles*. Cambridge: Cambridge University Press.

Kay, B. S., M. J. Radzikowski, and R. M. Wald (1997). Quantum field theory on spacetimes with a compactly generated Cauchy horizon. *Communications in Mathematical Physics* **183**, 533–556.

Kim, S.-W., and K. S. Thorne (1991). Do vacuum fluctuations prevent the creation of closed timelike curves? *Physical Review D* **43**, 3929–3949.

Kittel, C. (2004). *Elementary Statistical Physics*. Mineola, N.Y.: Courier Dover Publications.

Kittel, C., and H. Kroemer (1980). *Thermal Physics*. London: Macmillan.

Kleppner, D., and R. K. Kolenkow (2013). *An Introduction to Mechanics*. Cambridge: Cambridge University Press.

Landau, L. D., and E. M. Lifshitz (1975). *The Classical Theory of Fields*, fourth English edition. Oxford: Butterworth-Heinemann.

Landau, L. D., and E. M. Lifshitz (1976). *Mechanics*. Oxford: Butterworth-Heinemann.

Landauer, R. (1961). Irreversibility and heat generation in the computing process. *IBM Journal of Research and Development* **5**, 183–191.

Landauer, R. (1991). Information is physical. *Physics Today* **44(5)**, 23–29.

Lax, M. J., W. Cai, M. Xu, and H. E. Stanley (2006). *Random Processes in Physics and Finance*. Oxford: Oxford University Press.

Levin, Y. (1998). Internal thermal noise in the LIGO test masses: A direct approach. *Physical Review D* **57**, 659–663.

Lewin, L. (1981). *Polylogarithms and Associated Functions*. New York: North Holland.

Lifshitz, E. M., and L. P. Pitaevskii (1980). *Statistical Physics, Part 1*. Oxford: Pergamon.

Lifshitz, E. M., and L. P. Pitaevskii (1981). *Physical Kinetics*. Oxford: Pergamon.

Lorentz, H. A. (1904). Electromagnetic phenomena in a system moving with any velocity smaller than that of light. *Proceedings of the Royal Netherlands Academy of Arts and Sciences (KNAW)* **6**, 809–831.

Lorentz, H. A., A. Einstein, H. Minkowski, and H. Weyl (1923). *The Principle of Relativity: A Collection of Original Memoirs on the Special and General Theory of Relativity*. Mineola, N.Y.: Courier Dover Publications.

Lynden-Bell, D. (1967). Statistical mechanics of violent relaxation in stellar systems. *Monthly Notices of the Royal Astronomical Society* **136**, 101–121.

Marion, J. B., and S. T. Thornton (1995). *Classical Dynamics of Particles and Systems*. Philadelphia: Saunders College Publishing.

Maris, H. J., and L. P. Kadanoff (1978). Teaching the renormalization group. *American Journal of Physics* **46**, 653–657.

Maxwell, J. C. (1873). Letter to William Grylls Adams (3 Dec 1873). In P. M. Harman (ed.). (1995). *The Scientific Letters and Papers of James Clerk Maxwell, Vol 2, 1862–1873*, pp. 949–950. Cambridge: Cambridge University Press.

McEliece, R. J. (2002). *The Theory of Information and Coding*. Cambridge: Cambridge University Press.

Metropolis, N., A. Rosenbluth, M. Rosenbluth, A. Teller, and E. Teller (1953). Combinatorial minimization. *Journal of Chemical Physics* **21**, 1087–1092.

Millikan, R. A. (1938). Biographical Memoir of Albert Abraham Michelson, 1852–1931. *Biographical Memoirs of the National Academy of Sciences of the United States of America* **19**, 121–146. Chapter 4 epigraph reprinted with permission of the publisher.

Minkowski, H. (1908). Space and time. Address delivered at the 80th Assembly of German Natural Scientists and Physicians, at Cologne, Germany, September 21, 1908. First German publication: *Jahresbericht der Deutschen Mathematiker-Vereinigung* **1909**, 75–88. English translation in Lorentz et al. (1923).

Misner, C. W., K. S. Thorne, and J. A. Wheeler (1973). *Gravitation*. San Francisco: Freeman.

Morris, M. S., K. S. Thorne, and U. Yurtsever (1988). Wormholes, time machines, and the weak energy condition. *Physical Review Letters* **61**, 1446–1449.

Nyquist, H. (1928). Thermal agitation of electric charge in conductors. *Physical Review* **32**, 110–113.

Ogorodnikov, K. F. (1965). *Dynamics of Stellar Systems*. Oxford: Pergamon.

Onsager, L. (1944). Crystal statistics. I. A two-dimensional model with an order-disorder transition. *Physical Review* **65**, 117–149.

Page, D. N., F. Weinhold, R. L. Moore, F. Weinhold, and R. E. Barker (1977). Thermodynamic paradoxes. *Physics Today* **30(1)**, 11.

Pais, A. (1982). *Subtle Is the Lord. . . . The Science and Life of Albert Einstein*. Oxford: Oxford University Press.

Parker, L., and D. Toms (2009). *Quantum Field Theory in Curved Spacetime: Quantized Fields and Gravity*. Cambridge: Cambridge University Press.

Pathria, R. K., and P. D. Beale (2011). *Statistical Mechanics*, third edition. Amsterdam: Elsevier.

Paul, W., and J. Baschnagel (2010). *Stochastic Processes: From Physics to Finance*. Cham, Switzerland: Springer.

Penrose, R. (1999). *The Emperor's New Mind: Concerning Computers, Minds, and the Laws of Physics*. Oxford: Oxford University Press.

Penrose, R. (2016). *Fashion, Faith and Fantasy in the New Physics of the Universe*. Princeton, N.J.: Princeton University Press.

Pierce, J. R. (2012). *An Introduction to Information Theory: Symbols, Signals and Noise*. Mineola, N.Y.: Courier Dover Publications.

Press, W. H. (1978). Flicker noises in astronomy and elsewhere. *Comments on Astrophysics* **7**, 103–119.

Press, W. H., S. A. Teukolsky, W. T. Vetterling, and B. P. Flannery (2007). *Numerical Recipes: The Art of Scientific Computing*. Cambridge: Cambridge University Press.

Raisbeck, G. (1963). *Information Theory*. Cambridge, Mass.: MIT Press.

Reichl, L. E. (2009). *A Modern Course in Statistical Physics*. London: Arnold.

Reif, F. (2008). *Fundamentals of Statistical and Thermal Physics*. Long Grove, Ill.: Waveland Press.

Schneier, B. (1997). *Applied Cryptography: Protocols, Algorithms and Source Code in C*. New York: Wiley.

Schrödinger, E. (1944). *What Is Life?* Cambridge: Cambridge University Press.

Schutz, B. (2009). *A First Course in General Relativity*. Cambridge: Cambridge University Press.

Sethna, J. P. (2006). *Statistical Mechanics: Entropy, Order Parameters, and Complexity*. Oxford: Oxford University Press.

Shannon, C. E. (1948). A mathematical theory of communication. *Bell System Technical Journal* **27**, 379–423.

Shapiro, S. L., and S. A. Teukolsky (1983). *Black Holes, White Dwarfs and Neutron Stars: The Physics of Compact Objects*. New York: Wiley.

Shkarofsky, I. P., T. W. Johnston, and M. P. Bachynski (1966). *The Particle Kinetics of Plasmas*. New York: Addison-Wesley.

Taylor, E. F., and J. A. Wheeler (1966). *Spacetime Physics*, first edition. San Francisco: Freeman.

Taylor, E. F., and J. A. Wheeler (1992). *Spacetime Physics*, second edition. San Francisco: Freeman.

ter Haar, D. (1955). Foundations of statistical mechanics. *Reviews of Modern Physics* **27**, 289–338.

Thorne, K. S. (1981). Relativistic radiative transfer—moment formalisms. *Monthly Notices of the Royal Astronomical Society* **194**, 439–473.

Thorne, K. S. (1994). *Black Holes and Time Warps: Einstein's Outrageous Legacy*. New York: W. W. Norton.

Thorne, K. S., R. H. Price, and D. A. MacDonald (1986). *Black holes: the Membrane Paradigm*. New Haven, Conn.: Yale University Press.

Tolman, R. C. (1938). *The Principles of Statistical Mechanics*. Mineola, N.Y.: Courier Dover Publications.

Tranah, D., and P. T. Landsberg (1980). Thermodynamics of non-extensive entropies II. *Collective Phenomena* **3**, 81–88.

Unruh, W. G. (1976). Notes on black hole evaporation. *Physical Review D* **14**, 870–892.

Van Kampen, N. G. (2007). *Stochastic Processes in Physics and Chemistry*. New York: North Holland.

Wainstein, L. A., and V. D. Zubakov (1962). *Extraction of Signals from Noise*. London: Prentice-Hall.

Wald, R. M. (1994). *Quantum Field Theory in Curved Spacetime and Black Hole Thermodynamics*. Chicago: University of Chicago Press.

Wald, R. M. (2001). The thermodynamics of black holes. *Living Reviews in Relativity* **4**, 6.

Weinberg, S. (2008). *Cosmology*. Oxford: Oxford University Press.

Wheeler, J. A. (2000). *Geons, Black Holes, and Quantum Foam: A Life in Physics*. New York: W. W. Norton.

Wiener, N. (1949). *The Extrapolation, Interpolation, and Smoothing of Stationary Time Series with Engineering Applications*. New York: Wiley.

Zangwill, A. (2013). *Modern Electrodynamics*. Cambridge: Cambridge University Press.

Zipf, G. K. (1935). *The Psycho-Biology of Language*. Boston: Houghton-Mifflin.

Zurek, W. H., and K. S. Thorne (1985). Statistical mechanical origin of the entropy of a rotating, charged black hole. *Physical Review Letters* **54**, 2171–2175.

NAME INDEX

Page numbers for entries in boxes are followed by "b," those for epigraphs at the beginning of a chapter by "e," those for figures by "f," and those for notes by "n."

SUBJECT INDEX

Second and third level entries are not ordered alphabetically. Instead, the most important or general entries come first, followed by less important or less general ones, with specific applications last.

Page numbers for entries in boxes are followed by "b," those for epigraphs at the beginning of a chapter by "e," those for figures by "f," for notes by "n," and for tables by "t."

baths for statistical-mechanical systems
concept of, 160
general, 172
tables summarizing, 160t, 221t, 251t
black holes
laws of black-hole mechanics and thermodynamics, 205–209
statistical mechanics of, 204–206
entropy of, 205–209
inside a box: thermal equilibrium, 206–209
quantum thermal atmosphere of, 204–205
Hawking radiation from, 204–205
accretion of gas onto, 205
blackbody (Planck) distribution and specific intensity, 113, 128, 132
Boltzmann distribution (mean occupation number), 113, 177
Boltzmann equation, collisionless, 134–135, 167, 169
derivation from Hamiltonian, 136b–137b
implies conservation of particles and 4-momentum, 135
Boltzmann transport equation, 135, 139
for photons scattered by thermalized electrons, 144–148
accuracy of solutions, 140–141
order-of-magnitude solution, 143–144
solution via Fokker-Planck equation, 343
boost, Lorentz, 64–65
Bose-Einstein condensate, 193–201
condensation process, 193, 196, 197f, 198–200
critical temperature, 196
specific heat change, 200
in cubical box, 201
Bose-Einstein ensemble
probabilistic distribution function for, 176
mean occupation number of, 112–113, 176–177
entropy of, 187
bosons, 110
bremsstrahlung, 142, 260
Brownian motion, 296, 309, 313–315. *See also* random walk
spectral density and correlation function for, 313–314
relaxation time for, 328
fluctuation-dissipation theorem applied to, 327–329

canonical ensemble, 160t, 169–172, 221t
distribution function, 171, 173
canonical transformation, 162, 164, 166
Cartesian coordinates, 16, 26, 28
central limit theorem, 292–294
examples and applications of, 261, 294–295, 322
Chapman-Kolmogorov equation. *See* Smoluchowski equation
charge density
as time component of charge-current 4-vector, 74

charge-current 4-vector
geometric definition, 78
components: charge and current density, 78
local (differential) conservation law for, 79
global (integral) conservation law for, 79, 79f
evaluation in a Lorentz frame, 81
relation to nonrelativistic conservation of charge, 81
chemical free energy (Gibbs potential), 246–249. *See also under* fundamental thermodynamic potentials; fundamental thermodynamic potentials out of statistical equilibrium
chemical potential, excluding rest mass, μ, 112, 173
chemical potential, including rest mass, $\tilde{\mu}$, 112, 172–173
chemical reactions, including nuclear and particle, 256
direction controlled by Gibbs potential (chemical-potential sum), 256–258
partial statistical equilibrium for, 256
examples
water formation from hydrogen and oxygen, 256–257
electron-positron pair formation, 258–259
emission and absorption of photons, 115–116
ionization of hydrogen: Saha equation, 259–260
nucleosynthesis in nuclear age of early universe, 192–193
chronology protection, 69
Clausius-Clapeyron equation, 254–255
clocks
ideal, 39, 39n, 49
frequency fluctuations of, 310f, 310n, 320–321
coarse graining, 183–185, 184f, 206, 210–211
communication theory, 211–217
component manipulation rules
in Euclidean space, 16–19
in spacetime with orthormal basis, 54–57
conductivity, electrical, κ_e, 139
conductivity, thermal, κ, 139
for photons scattered by thermalized electrons, 148
derivation from Boltzmann transport equation, 144–148
conservation laws
differential and integral, in Euclidean 3-space, 28
differential and integral, in spacetime, 79
contraction of tensors
formal definition, 12–13, 48
in slot-naming index notation, 19
component representation, 17, 56
coordinate independence. *See* geometric principle; principle of relativity
correlation functions
for 1-dimensional random process, 297
correlation (relaxation) time of, 297
value at zero delay is variance, 297
for 2-dimensional random process, 306–308

impedance

complex, for fluctuation-dissipation theorem, 332

index gymnastics. *See* component manipulation rules

inertial (Lorentz) coordinates, 41, 54

inertial mass density (tensorial)

definition, 87

for perfect fluid, 87

inertial reference frame. *See* Lorentz reference frame

information

definition of, 212

properties of, 216

statistical mechanics of, 211–218

per symbol in a message, 214, 215

gain defined by entropy decrease, 211–212

inner product

in Euclidean space, 10–12, 17

in spacetime, 48, 56

in quantum theory, 18b

integrals in Euclidean space

over 2-surface, 27

over 3-volume, 27

Gauss's theorem, 27

integrals in spacetime, 75–78

over 3-surface, 77, 80–81

over 4–volume, 75

Gauss's theorem, 27, 78

intensive variables, 172, 221–222

interferometer

gravitational wave. *See* laser interferometer gravitational
wave detector

interval

defined, 45

invariance of, 45–48

spacelike, timelike, and null (lightlike), 45

Ising model for ferromagnetic phase transition, 272–282

1-dimensional Ising model, 278–279

2-dimensional Ising model, 272–273

solved by Monte Carlo methods, 279–282

solved by renormalization group methods, 273–278

isothermal engine, 241

Jeans' theorem, 169

Johnson noise in a resistor, 327

Kepler's laws, 14

kernel of a filter, in theory of random processes (noise),
311–313

Lagrange multiplier, 183

Landauer's theorem in communication theory, 217–218

laplacian, 24

laser interferometer gravitational wave detector

spectral density of noise, 302

in initial LIGO detectors, 302f

latent heat, 252, 254, 255, 270

Levi-Civita tensor in Euclidean space, 24–26

product of two, 25

Levi-Civita tensor in spacetime, 71

LIGO (Laser Interferometer Gravitational-Wave
Observatory). *See also* laser interferometer
gravitational wave detector

initial LIGO detectors (interferometers)

noise in, 302f, 323, 334

signal processing for, 320, 329–330

line element, 57

Liouville equation, in statistical mechanics, 167

quantum analog of, 165b–166b

Liouville's theorem

in kinetic theory, 132–134, 133f

in statistical mechanics, 166, 168f

Lorentz contraction

of length, 66–67

of volume, 99

of rest-mass density, 81

Lorentz coordinates, 41, 54

Lorentz factor, 58

Lorentz force

in terms of electromagnetic field tensor, 53, 71

in terms of electric and magnetic fields, 6, 14, 72

geometric derivation of, 52–53

Lorentz group, 64

Lorentz reference frame, 39, 39f

slice of simultaneity (3-space) in, 58, 59f

Lorentz transformation, 63–65

boost, 64, 65f

rotation, 65

Lorenz gauge

electromagnetic, 75

magnetic materials, 270–282

paramagnetism and Curie's law, 271–272

ferromagnetism, 272–282

phase transition into, 272–273. *See also* Ising model for
ferromagnetic phase transition

magnetization

in magnetic materials, 270

Maple, 129, 132

Markov random process, 289–291

Markov, Gaussian random process

probabilities for (Doob's theorem), 295–296, 298–299

spectral density for, 303, 304f

correlation function for, 297, 304f

and fluctuation-dissipation theorem, 325
  Fokker-Planck equation for, 336–338, 343
mass conservation, 32–33, 80
mass density
  rest-mass density, 81
  as integral over distribution function, 121
mass hyperboloid, 100–101, 100f
mass-energy density, relativistic
  as component of stress-energy tensor, 83, 85
  as integral over distribution function, 126
Mathematica, 129, 132
Matlab, 129, 132
Maxwell relations, thermodynamic, 227–228, 232, 240, 247
  as equality of mixed partial derivatives of fundamental
     thermodynamic potential, 227–228
Maxwell velocity distribution for nonrelativistic thermalized
     particles, 113–114, 114f
Maxwell-Jüttner velocity distribution for relativistic
     thermalized particles, 114–115, 114f
Maxwell's equations
  in terms of electromagnetic field tensor, 73–74
  in terms of electric and magnetic fields, 74
mean free path, 140, 143–145, 146b, 149
metric tensor
  in Euclidean space
    geometric definition, 11–12
    components in orthonormal basis, 17
  in spacetime, 48
    geometric definition, 48
    components in orthonormal basis, 55
Metropolis rule in Monte Carlo computations, 280
microcanonical ensemble, 160t, 178–180, 221–228, 221t
  correlations of subensembles in, 179
  distribution function for, 179
  and energy representation of thermodynamics, 221–228
Minkowski spacetime, 1–2
modes (single-particle quantum states), 174–176
  for Bose-Einstein condensate, 194–195
momentum, relativistic, 34, 59
  relation to 4-momentum and observer, 59, 61
  of a zero-rest-mass particle, 60, 106
  Newtonian limit, 34
momentum conservation, Newtonian
  differential, 32
  integral, 32
momentum conservation, relativistic
  for particles, 60
  differential, 85
momentum density
  as component of stress-energy tensor, 83
  as integral over distribution function, 118

momentum space
  Newtonian, 98, 98f
  relativistic, 100–101, 100f
Monte Carlo methods
  origin of name, 279n
  for 2-dimensional Ising model of ferromagnetism,
     279–282
multiplicity factor for states in phase space, $\mathcal{M}$, 163
multiplicity for particle's spin states, g_s, 109

neutrinos
  chirality of, 109n
  spin-state multiplicity, 109
  in universe, evolution of
    thermodynamically isolated after decoupling, 192,
       209
neutron stars
  birth in supernovae, 111
  equation of state, 125
noise. *See also* fluctuation-dissipation theorem; spectral
     density
  as a random process, 308–313
  types of spectra (spectral densities)
    flicker noise, 308–310, 323
    random-walk noise, 308–310
    white noise, 308–310
  information missing from spectral density, 310–311
  filtering of, 311–313
  Johnson noise in a resistor, 327
  shot noise, 321–323
  thermal noise, 302f, 329–330, 334, 343–345. *See also*
     fluctuation-dissipation theorem
nuclear reactions. *See* chemical reactions, including nuclear
     and particle
nuclear reactor
  neutron diffusion in, 151–153
  xenon poisoning in, 153
nucleosynthesis, in nuclear age of early universe, 192–193
number density
  as time component of number-flux 4-vector, 79–80
  as integral over distribution function, 117, 119, 121, 126
number flux
  as spatial part of number-flux 4-vector, 79–80
  as integral over distribution function, 117
number-flux 4-vector
  geometric definition, 79–80
  as integral over distribution function, 118, 119
  components: number density and flux, 79–80
  conservation laws for, 79–80

observer in spacetime, 41

occupation number, mean
  defined, 110
  ranges, for fermions, bosons, distinguishable particles, and classical waves, 110, 111
  for cosmic X-rays, 111
  for astrophysical gravitational waves, 111
Ohm's law, 139
Olber's paradox, 138–139
optical frequency comb, 310n
optimal filtering, 318–320
orthogonal transformation, 20–21

pairs, electron-positron
  thermal equilibrium of, 258–259
  temperature-density boundary for, 259f
Parseval's theorem, 300, 303
particle conservation law
  Newtonian, 28
  relativistic, 80
particle density. *See* number density
particle kinetics
  in Euclidean space
    geometric form, 13–15
    in index notation, 19–20
  in flat spacetime
    geometric form, 49–52
    in index notation, 57–62
  in Newtonian phase space, 97–99
partition function, in statistical mechanics. *See also* fundamental thermodynamic potentials, physical-free-energy potential
  as log of physical free energy, 239
path of particle (Newtonian analog of world line), 9–10
perfect fluid (ideal fluid), 30
  Euler equation for, 33
  stress tensor for, 30–31, 32
phase mixing in statistical mechanics, 184, 184f, 210–211
phase space
  Newtonian, 98–99
  relativistic, 101–105
  in statistical mechanics, 161–163
phase transitions, 251
  governed by Gibbs potential, 251–254
  first-order, 252
    Clausius-Clapeyron equation for, 254–255
  second-order, 253
    specific heat discontinuity in, 200, 254
  triple point, 254–255, 255f
  specific examples
    water-ice, 251–252, 254–255
    water vapor–water, 255

van der Waals gas, 266–268
crystal structure change, 253–254
Bose-Einstein condensation, 196–197, 197f, 254
ferromagnetism, 272–282. *See also* Ising model for ferromagnetic phase transition
phonons
  modes for, 175, 175n
  specific heat of in an isotropic solid, 131–132
physical laws
  frameworks and arenas for, 1–3
  geometric formulation of. *See* geometric principle; principle of relativity
Planck time, 209
Poisson distribution, 264
pressure, 30
  as component of stress tensor, 30–31
  as component of stress-energy tensor, 85
  as integral over distribution function, 121, 126
primordial nucleosynthesis, 192–193
principle of relativity, 42
probability distributions, 286–288
  conditional, 287
projection tensors
  into Lorentz frame's 3-space, 61
proper time, 49

quadratic degree of freedom, 177
quantum state
  single-particle (mode), 174–175
    for Bose-Einstein condensate, 194–195
  many-particle, 175
    distribution function for, 175
quantum statistical mechanics, 165b–166b
quasars, 193, 309

radiation, equation of state for thermalized, 128–129, 132
radiative processes
  in statistical equilibrium, 115–116
  bremsstrahlung, 142, 260
  Thomson scattering, 142–144
radiative transfer, Boltzmann transport analysis of
  by two-lengthscale-expansion, 145–148
random process, 1-dimensional, 285
  stationary, 287–288
  ergodic, 288–289
  Gaussian, 292–294
  Markov, 289–290
  Gaussian, Markov. *See* Markov, Gaussian random process
random process, 3-dimensional
  cosmological density fluctuations, 304–306
random process, 2-dimensional, 306–308

statistical equilibrium for fundamental particles *(continued)*
 for identical classical particles, Boltzmann distribution, 113
statistical independence, 170
stellar dynamics. *See also under* galaxies
 statistical mechanics of galaxies and star clusters, 201–204
 evolution of cluster due to ejection of stars, 203–204
 violent relaxation of star clusters, 113n
stochastic differential equations, 325
Stokes' theorem for integrals, 27
stress tensor
 geometric definition of, 29
 components, meaning of, 30
 symmetry of, 30
 as integral over distribution function, 118
 as spatial part of relativistic stress-energy tensor, 83
 for specific entities
  electromagnetic field, 33
  perfect fluid, 30–31, 32
stress-energy tensor
 geometric definition of, 82
 components of, 82–83, 120
 symmetry of, 83–84
 as integral over distribution function, 118, 120
 and 4-momentum conservation, 84–85
 for electromagnetic field
  in terms of electromagnetic field tensor, 86
  in terms of electric and magnetic fields, 88
 for perfect fluid, 85
subensemble, 170
supernovae
 neutron stars produced in, 111
 as gravitational-wave sources, 111
system, in statistical mechanics
 defined, 157
 closed, 158–159
 semiclosed, 157–158

tangent space, 9
tangent vector, 9, 49
temperature
 definition, 168, 168n, 171
 measured by idealized thermometer, 223–224
temperature diffusion equation, 142
tensor in Euclidean space
 definition and rank, 11
 algebra of without coordinates or bases, 11–13
 expanded in basis, 16
 component representation, 17–19
tensor in quantum theory, 18b

tensor in spacetime. *See also* component manipulation rules
 definition and rank, 48
 bases for, 55
 components of, 54–57
  contravariant, covariant, and mixed components, 55
  raising and lowering indices, 55
 algebra of
  without coordinates or bases, 48, 61–62
  component representation in orthonormal basis, 54–57
tensor product, 12, 48
thermal equilibrium. *See* statistical equilibrium
thermodynamics. *See also* equations of state; fundamental thermodynamic potentials; Maxwell relations, thermodynamic
 representations of, summarized, 221t, 228
  Legendre transformation between representations, 230–232, 240, 244, 247
 energy representation, and microcanonical ensemble, 221–229
 enthalpy representation, 244–246
 grand-potential representation and grand canonical ensemble, 229–239
 physical-free-energy representation, and canonical ensemble, 239–244
 Gibbs representation and Gibbs ensemble, 246–260
 first law of, 225
  in all representations, 221t
  as mnemonic for deducing other relations, 227–228
  for black hole, 205
 second law of, 182
  underlying physics of: coarse graining and discarding correlations, 183–185, 184f, 186b–187b
  underlying quantum physics of: discarding correlations (quantum decoherence), 185, 186b–187b, 190–191
  in theory of information: when information is erased, 217–218
 of black holes, 204–209
Thomson scattering of photons by electrons, 142–144
time. *See also* clocks, ideal; simultaneity in relativity, slices of
 coordinate, of inertial frame, 39–40
 proper, 49
 imaginary, 54
time and frequency standards, 310f, 310n
time derivative
 advective (convective), 32
 with respect to proper time, 49, 52
time dilation, 66
 observations of, 70
time travel, 67–70
transformation matrices, between bases
 orthogonal, 20–21

Lorentz, 63–65, 65f
transport coefficients, 139. *See also* diffusion coefficient
triple point for phase transitions, 254–255, 255f
twins paradox, 67–70
two-lengthscale expansion, 146b
    and statistical independence in statistical mechanics, 170n
    for solving Boltzmann transport equation, 145
two-point correlation function, 305
    for galaxy clustering, 305, 306f

universe, evolution of
    formation of structure
        statistical mechanics of, 210–211
        seven ages
            nuclear age, primordial nucleosynthesis, 192–193
            atomic age, from recombination through reionization, 193
            galaxy formation, 210–211
universe, statistical mechanics of, 209–211

van der Waals gas
    equation of state for, 234
    grand potential for, 234
        derivation of, 232–238
    phase transition for, 266–268
    volume fluctuations in, 266–268
variance, 287
vector
    as arrow, 8, 40
    as derivative of a point, 9, 49
vector in Euclidean space (3–vector): components, 16
vector in quantum theory, 18b
vector in spacetime (4–vector)
    contravariant and covariant components of, 55
    raising and lowering indices of, 55

timelike, null, and spacelike, 47
velocity
    Newtonian, in Euclidean space, 9
    ordinary, in relativity, 58, 59f, 61–62. *See also* 4–velocity
violent relaxation of star distributions, 210
viscosity, shear, coefficient of, 139
    for monatomic gas, 149–150
volume in Euclidean space
    2–volume (area), 26
    vectorial surface area in 3-space, 27
    3–volume, 27
    n-volume, 24
    differential volume elements, 28
volume in phase space
    Newtonian, 98
    relativistic, 102–104
        Lorentz invariance of, 103–104, 105f
volume in spacetime, 75–77
    4–volume, 75
    vectorial 3–volume, 76–77, 77f
        positive and negative sides and senses, 76
    differential volume elements, 77

wave equations
    for electromagnetic waves. *See* electromagnetic waves
Wiener-Khintchine theorem
    for 1-dimensional random process, 303
    for 2-dimensional random process, 307–308
Wiener's optimal filter, 318–320
world line, 49, 59f
world tube, 49n, 68f, 69
wormhole, 68–69, 68f
    as time machine, 69

zero point energy, 175n

T2   Track Two; see page xvii

The study of physics (including astronomy) is one of the oldest academic enterprises. Remarkable surges in inquiry occurred in equally remarkable societies—in Greece and Egypt, in Mesopotamia, India and China—and especially in Western Europe from the late sixteenth century onward. Independent, rational inquiry flourished at the expense of ignorance, superstition, and obeisance to authority.

Physics is a constructive and progressive discipline, so these surges left behind layers of understanding derived from careful observation and experiment, organized by fundamental principles and laws that provide the foundation of the discipline today. Meanwhile the detritus of bad data and wrong ideas has washed away. The laws themselves were so general and reliable that they provided foundations for investigation far beyond the traditional frontiers of physics, and for the growth of technology.

The start of the twentieth century marked a watershed in the history of physics, when attention turned to the small and the fast. Although rightly associated with the names of Planck and Einstein, this turning point was only reached through the curiosity and industry of their many forerunners. The resulting quantum mechanics and relativity occupied physicists for much of the succeeding century and today are viewed very differently from each other. Quantum mechanics is perceived as an abrupt departure from the tacit assumptions of the past, while relativity—though no less radical conceptually—is seen as a logical continuation of the physics of Galileo, Newton, and Maxwell. There is no better illustration of this than Einstein's growing special relativity into the general theory and his famous resistance to the quantum mechanics of the 1920s, which others were developing.

This is a book about classical physics—a name intended to capture the pre-quantum scientific ideas, augmented by general relativity. Operationally, it is physics in the limit that Planck's constant $h \rightarrow 0$. Classical physics is sometimes used, pejoratively, to suggest that "classical" ideas were discarded and replaced by new principles and laws. Nothing could be further from the truth. The majority of applications of physics today are still essentially classical. This does not imply that physicists or others working in these areas are ignorant or dismissive of quantum physics. It is simply that the issues with which they are confronted are mostly addressed classically. Furthermore, classical physics has not stood still while the quantum world was being explored. In scope and in practice, it has exploded on many fronts and would now be quite unrecognizable to a Helmholtz, a Rayleigh, or a Gibbs. In this book, we have tried to emphasize these contemporary developments and applications at the expense of historical choices, and this is the reason for our seemingly oxymoronic title, *Modern Classical Physics.*

This book is ambitious in scope, but to make it bindable and portable (and so the authors could spend some time with their families), we do not develop classical mechanics, electromagnetic theory, or elementary thermodynamics. We assume the reader has already learned these topics elsewhere, perhaps as part of an undergraduate curriculum. We also assume a normal undergraduate facility with applied mathematics. This allows us to focus on those topics that are less frequently taught in undergraduate and graduate courses.

Another important exclusion is numerical methods and simulation. High-performance computing has transformed modern research and enabled investigations that were formerly hamstrung by the limitations of special functions and artificially imposed symmetries. To do justice to the range of numerical techniques that have been developed—partial differential equation solvers, finite element methods, Monte Carlo approaches, graphics, and so on—would have more than doubled the scope and size of the book. Nonetheless, because numerical evaluations are crucial for physical insight, the book includes many applications and exercises in which user-friendly numerical packages (such as Maple, Mathematica, and Matlab) can be used to produce interesting numerical results without too much effort. We hope that, via this pathway from fundamental principle to computable outcome, our book will bring readers not only physical insight but also enthusiasm for computational physics.

Classical physics as we develop it emphasizes physical phenomena on macroscopic scales: scales where the particulate natures of matter and radiation are secondary to their behavior in bulk; scales where particles' statistical—as opposed to individual—properties are important, and where matter's inherent graininess can be smoothed over.

In this book, we take a journey through spacetime and phase space; through statistical and continuum mechanics (including solids, fluids, and plasmas); and through optics and relativity, both special and general. In our journey, we seek to comprehend the fundamental laws of classical physics in their own terms, and also in relation to quantum physics. And, using carefully chosen examples, we show how the classical laws are applied to important, contemporary, twenty-first-century problems and to everyday phenomena; and we also uncover some deep relationships among the various fundamental laws and connections among the practical techniques that are used in different subfields of physics.

Geometry is a deep theme throughout this book and a very important connector. We shall see how a few geometrical considerations dictate or strongly limit the basic principles of classical physics. Geometry illuminates the character of the classical principles and also helps relate them to the corresponding principles of quantum physics. Geometrical methods can also obviate lengthy analytical calculations. Despite this, long, routine algebraic manipulations are sometimes unavoidable; in such cases, we occasionally save space by invoking modern computational symbol manipulation programs, such as Maple, Mathematica, and Matlab.

This book is the outgrowth of courses that the authors have taught at Caltech and Stanford beginning 37 years ago. Our goal was then and remains now to fill what we saw as a large hole in the traditional physics curriculum, at least in the United States:

- We believe that every masters-level or PhD physicist should be familiar with the basic concepts of all the major branches of classical physics and should have had some experience in applying them to real-world phenomena; this book is designed to facilitate this goal.

- Many physics, astronomy, and engineering graduate students in the United States and around the world use classical physics extensively in their research, and even more of them go on to careers in which classical physics is an essential component; this book is designed to expedite their efforts.

- Many professional physicists and engineers discover, in mid-career, that they need an understanding of areas of classical physics that they had not previously mastered. This book is designed to help them fill in the gaps and see the relationship to already familiar topics.

In pursuit of this goal, we seek, in this book, to *give the reader a clear understanding of the basic concepts and principles of classical physics.* We present these principles in the language of modern physics (not nineteenth-century applied mathematics), and we present them primarily for physicists—though we have tried hard to make the content interesting, useful, and accessible to a much larger community including engineers, mathematicians, chemists, biologists, and so on. As far as possible, we emphasize theory that involves general principles which extend well beyond the particular topics we use to illustrate them.

In this book, we also seek to *teach the reader how to apply the ideas of classical physics.* We do so by presenting contemporary applications from a variety of fields, such as

- fundamental physics, experimental physics, and applied physics;
- astrophysics and cosmology;
- geophysics, oceanography, and meteorology;
- biophysics and chemical physics; and

- engineering, optical science and technology, radio science and technology, and information science and technology.

Why is the range of applications so wide? Because we believe that physicists should have enough understanding of general principles to attack problems that arise in unfamiliar environments. In the modern era, a large fraction of physics students will go on to careers outside the core of fundamental physics. For such students, a broad exposure to non-core applications can be of great value. For those who wind up in the core, such an exposure is of value culturally, and also because ideas from other fields often turn out to have impact back in the core of physics. Our examples illustrate how basic concepts and problem-solving techniques are freely interchanged across disciplines.

We strongly believe that classical physics should *not* be studied in isolation from quantum mechanics and its modern applications. Our reasons are simple:

- Quantum mechanics has primacy over classical physics. Classical physics is an approximation—often excellent, sometimes poor—to quantum mechanics.

- In recent decades, many concepts and mathematical techniques developed for quantum mechanics have been imported into classical physics and there used to enlarge our classical understanding and enhance our computational capability. An example that we shall study is nonlinearly interacting plasma waves, which are best treated as quanta ("plasmons"), despite their being solutions of classical field equations.

- Ideas developed initially for classical problems are frequently adapted for application to avowedly quantum mechanical subjects; examples (not discussed in this book) are found in supersymmetric string theory and in the liquid drop model of the atomic nucleus.

Because of these intimate connections between quantum and classical physics, quantum physics appears frequently in this book.

The amount and variety of material covered in this book may seem overwhelming. If so, keep in mind the key goals of the book: to teach the fundamental concepts, which are not so extensive that they should overwhelm, and to illustrate those concepts. Our goal is not to provide a mastery of the many illustrative applications contained in the book, but rather to convey the spirit of how to apply the basic concepts of classical physics. To help students and readers who feel overwhelmed, we have labeled as "Track Two" sections that can be skipped on a first reading, or skipped entirely—but are sufficiently interesting that many readers may choose to browse or study them. Track-Two sections are labeled by the symbol T2 . To keep Track One manageable for a one-year course, the Track-One portion of each chapter is rarely longer than 40 pages (including many pages of exercises) and is often somewhat shorter. Track One is designed for a full-year course at the first-year graduate level; that is how we have mostly used it. (Many final-year undergraduates have taken our course successfully, but rarely easily.)

The book is divided into seven parts:

I. **Foundations**—which introduces our book's powerful *geometric* point of view on the laws of physics and brings readers up to speed on some concepts and mathematical tools that we shall need. Many readers will already have mastered most or all of the material in Part I and might find that they can understand most of the rest of the book without adopting our avowedly geometric viewpoint. Nevertheless, we encourage such readers to browse Part I, at least briefly, before moving on, so as to become familiar with this viewpoint. We believe the investment will be repaid. Part I is split into two chapters, Chap. 1 on Newtonian physics and Chap. 2 on special relativity. Since nearly all of Parts II–VI is Newtonian, readers may choose to skip Chap. 2 and the occasional special relativity sections of subsequent chapters, until they are ready to launch into Part VII, General Relativity. Accordingly, Chap. 2 is labeled Track Two, though it becomes Track One when readers embark on Part VII.

II. **Statistical Physics**—including kinetic theory, statistical mechanics, statistical thermodynamics, and the theory of random processes. These subjects underlie some portions of the rest of the book, especially plasma physics and fluid mechanics.

III. **Optics**—by which we mean classical waves of all sorts: light waves, radio waves, sound waves, water waves, waves in plasmas, and gravitational waves. The major concepts we develop for dealing with all these waves include geometric optics, diffraction, interference, and nonlinear wave-wave mixing.

IV. **Elasticity**—elastic deformations, both static and dynamic, of solids. Here we develop the use of tensors to describe continuum mechanics.

V. **Fluid Dynamics**—with flows ranging from the traditional ones of air and water to more modern cosmic and biological environments. We introduce vorticity, viscosity, turbulence, boundary layers, heat transport, sound waves, shock waves, magnetohydrodynamics, and more.

VI. **Plasma Physics**—including plasmas in Earth-bound laboratories and in technological (e.g., controlled-fusion) devices, Earth's ionosphere, and cosmic environments. In addition to magnetohydrodynamics (treated in Part V), we develop two-fluid and kinetic approaches, and techniques of nonlinear plasma physics.

VII. **General Relativity**—the physics of curved spacetime. Here we show how the physical laws that we have discussed in flat spacetime are modified to account for curvature. We also explain how energy and momentum generate this curvature. These ideas are developed for their principal classical applications to neutron stars, black holes, gravitational radiation, and cosmology.

It should be possible to read and teach these parts independently, provided one is prepared to use the cross-references to access some concepts, tools, and results developed in earlier parts.

Five of the seven parts (II, III, V, VI, and VII) conclude with chapters that focus on applications where there is much current research activity and, consequently, there are many opportunities for physicists.

Exercises are a major component of this book. There are five types of exercises:

1. *Practice.* Exercises that provide practice at mathematical manipulations (e.g., of tensors).

2. *Derivation.* Exercises that fill in details of arguments skipped over in the text.

3. *Example.* Exercises that lead the reader step by step through the details of some important extension or application of the material in the text.

4. *Problem.* Exercises with few, if any, hints, in which the task of figuring out how to set up the calculation and get started on it often is as difficult as doing the calculation itself.

5. *Challenge.* Especially difficult exercises whose solution may require reading other books or articles as a foundation for getting started.

We urge readers to try working many of the exercises, especially the examples, which should be regarded as continuations of the text and which contain many of the most illuminating applications. Exercises that we regard as especially important are designated by **.

A few words on units and conventions. In this book we deal with practical matters and frequently need to have a quantitative understanding of the magnitudes of various physical quantities. This requires us to adopt a particular unit system. Physicists use both Gaussian and SI units; units that lie outside both formal systems are also commonly used in many subdisciplines. Both Gaussian and SI units provide a complete and internally consistent set for all of physics, and it is an often-debated issue as to which system is more convenient or aesthetically appealing. We will not enter this debate! One's choice of units should not matter, and a mature physicist should be able to change from one system to another with little thought. However, when learning new concepts, having to figure out "where the 2π s and 4π s go" is a genuine impediment to progress. Our solution to this problem is as follows. For each physics subfield that we study, we consistently use the set of units that seem most natural or that, we judge, constitute the majority usage by researchers in that subfield. We do not pedantically convert cm to m or vice versa at every juncture; we trust that the reader can easily make whatever translation is necessary. However, where the equations are actually different—primarily in electromagnetic theory—we occasionally provide, in brackets or footnotes, the equivalent equations in the other unit system and enough information for the reader to proceed in his or her preferred scheme.

We encourage readers to consult this book's website, http://press.princeton.edu/titles/MCP.html, for information, errata, and various resources relevant to the book.

A large number of people have influenced this book and our viewpoint on the material in it. We list many of them and express our thanks in the Acknowledgments. Many misconceptions and errors have been caught and corrected. However, in a book of this size and scope, others will remain, and for these we take full responsibility. We would be delighted to learn of these from readers and will post corrections and explanations on this book's website when we judge them to be especially important and helpful.

Above all, we are grateful for the support of our wives, Carolee and Liz—and especially for their forbearance in epochs when our enterprise seemed like a mad and vain pursuit of an unreachable goal, a pursuit that we juggled with huge numbers of other obligations, while Liz and Carolee, in the midst of their own careers, gave us the love and encouragement that were crucial in keeping us going.

ACKNOWLEDGMENTS FOR *MODERN CLASSICAL PHYSICS*

This book evolved gradually from notes written in 1980–81, through improved notes, then sparse prose, and on into text that ultimately morphed into what you see today. Over these three decades and more, courses based on our evolving notes and text were taught by us and by many of our colleagues at Caltech, Stanford, and elsewhere. From those teachers and their students, and from readers who found our evolving text on the web and dove into it, we have received an extraordinary volume of feedback,[1] and also patient correction of errors and misconceptions as well as help with translating passages that were correct but impenetrable into more lucid and accessible treatments. For all this feedback and to all who gave it, we are extremely grateful. We wish that we had kept better records; the heartfelt thanks that we offer all these colleagues, students, and readers, named and unnamed, are deeply sincere.

Teachers who taught courses based on our evolving notes and text, and gave invaluable feedback, include Professors Richard Blade, Yanbei Chen, Michael Cross, Steven Frautschi, Peter Goldreich, Steve Koonin, Christian Ott, Sterl Phinney, David Politzer, John Preskill, John Schwarz, and David Stevenson at Caltech; Professors Tom Abel, Seb Doniach, Bob Wagoner, and the late Shoucheng Zhang at Stanford; and Professor Sandor Kovacs at Washington University in St. Louis.

Our teaching assistants, who gave us invaluable feedback on the text, improvements of exercises, and insights into the difficulty of the material for the students, include Jeffrey Atwell, Nate Bode, Yu Cao, Yi-Yuh Chen, Jane Dai, Alexei Dvoretsky, Fernando Echeverria, Jiyu Feng, Eanna Flanagan, Marc Goroff, Dan Grin, Arun Gupta, Alexandr Ikriannikov, Anton Kapustin, Kihong Kim, Hee-Won Lee, Geoffrey Lovelace, Miloje Makivic, Draza Markovic, Keith Matthews, Eric Morganson, Mike Morris, Chung-Yi Mou, Rob Owen, Yi Pan, Jaemo Park, Apoorva Patel, Alexander Putilin, Shuyan Qi, Soo Jong Rey, Fintan Ryan, Bonnie Shoemaker, Paul Simeon,

---

1. Specific applications that were originated by others, to the best of our memory, are acknowledged in the text.

Hidenori Sinoda, Matthew Stevenson, Wai Mo Suen, Marcus Teague, Guodang Wang, Xinkai Wu, Huan Yang, Jimmy Yee, Piljin Yi, Chen Zheng, and perhaps others of whom we have lost track!

Among the students and readers of our notes and text, who have corresponded with us, sending important suggestions and errata, are Bram Achterberg, Mustafa Amin, Richard Anantua, Alborz Bejnood, Edward Blandford, Jonathan Blandford, Dick Bond, Phil Bucksbaum, James Camparo, Conrado Cano, U Lei Chan, Vernon Chaplin, Mina Cho, Ann Marie Cody, Sandro Commandè, Kevin Fiedler, Krzysztof Findeisen, Jeff Graham, Casey Handmer, John Hannay, Ted Jacobson, Matt Kellner, Deepak Kumar, Andrew McClung, Yuki Moon, Evan O'Connor, Jeffrey Oishi, Keith Olive, Zhen Pan, Eric Peterson, Laurence Perreault Levasseur, Rob Phillips, Vahbod Pourahmad, Andreas Reisenegger, David Reis, Pavlin Savov, Janet Scheel, Yuki Takahashi, Clifford Will, Fun Lim Yee, Yajie Yuan, and Aaron Zimmerman.

For computational advice or assistance, we thank Edward Campbell, Mark Scheel, Chris Mach, and Elizabeth Wood.

Academic support staff who were crucial to our work on this book include Christine Aguilar, JoAnn Boyd, Jennifer Formicelli, and Shirley Hampton.

The editorial and production professionals at Princeton University Press (Peter Dougherty, Karen Fortgang, Ingrid Gnerlich, Eric Henney, and Arthur Werneck) and at Princeton Editorial Associates (Peter Strupp and his freelance associates Paul Anagnostopoulos, Laurel Muller, MaryEllen Oliver, Joe Snowden, and Cyd Westmoreland) have been magnificent, helping us plan and design this book, and transforming our raw prose and primitive figures into a visually appealing volume, with sustained attention to detail, courtesy, and patience as we missed deadline after deadline.

Of course, we the authors take full responsibility for all the errors of judgment, bad choices, and mistakes that remain.

Roger Blandford thanks his many supportive colleagues at Caltech, Stanford University, and the Kavli Institute for Particle Astrophysics and Cosmology. He also acknowledges the Humboldt Foundation, the Miller Institute, the National Science Foundation, and the Simons Foundation for generous support during the completion of this book. And he also thanks the Berkeley Astronomy Department; Caltech; the Institute of Astronomy, Cambridge; and the Max Planck Institute for Astrophysics, Garching, for hospitality.

Kip Thorne is grateful to Caltech—the administration, faculty, students, and staff—for the supportive environment that made possible his work on this book, work that occupied a significant portion of his academic career.